D1205476

A History of Industrial Power in the United States, 1780–1930

VOLUME ONE: WATERPOWER in the Century of the Steam Engine

A History of Industrial Power in the United States, 1780–1930

VOLUME ONE: WATERPOWER in the Century of the Steam Engine

Louis C. Hunter

Published for the Eleutherian Mills–Hagley Foundation
by the University Press of Virginia, Charlottesville

THE UNIVERSITY PRESS OF VIRGINIA

First published 1979

The illustrations from Robert L. Daugherty's *Hydraulic Turbines,* 2d ed. (New York, 1914) and his *Hydraulics*, 3d ed. (New York, 1925) are reprinted by permission of the publisher, McGraw-Hill Book Company; illustrations from Daniel W. Mead's *Water Power Engineering* (New York, 1908) and 2nd ed. (New York, 1915) and *Hydrology*, 2d ed., rev. (New York, 1950) are reprinted by permission of Mead and Hunt, Inc., Madison, Wisconsin. The photograph from Horace Kephart's *Our Southern Highlanders*, rev. ed. copyright 1913, 1922 by Macmillan Publishing Co., Inc., renewed 1941, 1950 by Laura M. Kephart, is reproduced with the permission of the publisher. The maps from Frederick H. Newell's *Irrigation in the United States*, rev. ed., 1906, are reproduced with the permission of the publisher, Thomas Y. Crowell Co. Portions of Table 18 are reproduced from Robert E. Gallman's "Commodity Output, 1839-1899," in *Trends in the American Economy in the Nineteenth Century* (Princeton, 1960) by permission of the National Bureau of Economic Research.

Library of Congress Cataloging in Publication Data

Hunter, Louis, C

A history of industrial power in the United States, 1780-1930.

Includes index.

CONTENTS: v. 1. Waterpower in the century of the steam engine.—

1. Water-power—Economic aspects—United States.—History. 2. Water rights—United States—History. 3. United States—Industries—History. I. Eleutherian Mills-Hagley Foundation, Greenville, Del. II. Title. HD1694.A5H76
338.4′0973 78-17538 ISBN 0-8139-0782-9 (v. 1)

Printed in the United States of America

To Lynwood Bryant
the imaginative and persuasive architect
of this study

Contents

Illustrations

Tables

Acknowledgments

During the many years of research, thought, and writing that have gone into this study I have become indebted to innumerable individuals for aid, encouragement, and illumination of the subject in its many facets. They have contributed in various ways by supplying information, ideas, suggestions, comments, and criticisms; providing professional expertise; calling attention to new sources of material; or suggesting alternative explanations. I would gladly reverse the conventional disclaimer and give names, credit, and responsibility to the individuals in question. Yet to single out even those of the more recent years who come most readily to mind would create a list so long as to discourage more than a simple enumeration and leave their debtor in a state of concern over probable omissions. On the other hand, it is both feasible and desirable to record the names of the scholarly institutions upon which one is almost daily dependent and which frequently become in effect collaborators in the project in hand. Yet even in such instances it seems impracticable, indeed to a degree invidious, to give the names of the more frequently consulted staff members. What follows then are the names of the libraries, historical societies, museums, and the like that have been particularly helpful and to which I am especially indebted.

Washington, D.C.

- Library of Congress, particularly the Photographic Division and the Local History Division
- Smithsonian Museum of History and Technology
- Libraries of the U.S. Geological Survey and the former Patent Office
- Library of the Industrial College of the Armed Forces
- Historic American Engineering Record, National Park Service
- National Archives and Records Service

Cambridge-Boston area

- Massachusetts Institute of Technology: the science and humanities libraries
- Libraries of Harvard University: Widener, especially the reference, union catalog, and government documents divisions; Gordon McKay, Cabot Science, Museum of Comparative Zoology, Geological Sciences, Littauer, Law School; Baker, particularly the manuscript division and the Kress Library of Business and Economics

State Library of Massachusetts
Brookline Public Library
State, Regional, and Special Libraries and Museums
American Precision Museum, Windsor, Vt.
Southern Appalachian Collection, Berea College, Berea, Ky.
Baker Library and Feldberg Library, Dartmouth College, Hanover, N.H.
Eleutherian Mills Historical Library, Wilmington, Del.
Farmers Museum of Cooperstown, N.Y.
Filson Club Library, Louisville, Ky.
Maryland Historical Society, Baltimore
Mercer Museum of the Bucks County Historical Society, Doylestown, Pa.
Merrimack Valley Textile Museum, North Andover, Mass.
New Hampshire State Library, Concord
New York State Historical Association Library, Cooperstown
Old Sturbridge Village, Sturbridge, Mass.
Thetford (Vermont) Historical Society and Museum
Vermont Historical Society, Montpelier
European Libraries and Museums
Austria: Technishes Museum für Industrie und Gewerbe, Vienna
France: Bibliothèque Nationale, Paris; Conservatorie des Arts et Métiers, Paris
Germany: Deutches Museum, Munich
Great Britain: Science Museum, London; Birmingham City Museums, Department of Science and Industry
Rumania: Muzeul Etnografic al Transilvaniei, Cluj; Muzeul Satului, Bucharest
Sweden: Skansen, Stockholm; Nordiska Museet, Stockholm

I am especially indebted through long association and repeated encouragement, stimulus, and assistance to the following scholars and friends: Lynwood Bryant, the late Arthur Harrison Cole, John J. Crnkovich, Meyer Fishbein, Eugene S. Ferguson, Edwin T. Layton, Jr., the late Frederick Merk, and the late Fritz Redlich.

In the preparation of this volume for publication I have been wholly dependent upon the staff members of the Eleutherian Mills–Hagley Foundation, the sponsoring organization, especially Barbara E. Benson and Chase Duffy for advice, suggestions, and long-continued and often arduous labors.

LOUIS C. HUNTER

Brookline, Mass.
1978

Introduction

The use of mechanical power for industrial purposes goes back some two thousand years, but the beginnings of its widespread use are associated with industrialization, which had its beginnings in Great Britain in the mid-eighteenth century. Industrialization is the long and complex process by which the manufacture of all manner of goods, long produced in homes and small workshops with hand tools and muscular energy, was progressively concentrated in mills and factories that employed machinery and mechanical power. The change transformed the modern world. The replacement of manual skills by machine operations was accompanied by remarkable advances in the efficiency of production and in the reduction of costs. The history of power in the United States since 1780 is a central feature of the nation's industrialization.

The central role of machinery and power in industrial development has long been recognized both by contemporary observers and by economic and technological historians. The monographic literature of industrial history contains a wealth of material related to the introduction and extension of the use of machinery in production and of advances in the generation and distribution of power. Yet there has been lacking a treatment of the subject that is at once comprehensive, systematic, and extensively researched. The present study undertakes to meet this need for the United States.

The industrial power with which this study is concerned is *stationary* power as employed in industrial production in mills, mines, and factories. Stationary power is to be distinguished from power in its *mobile* applications, which is more generally familiar to the public. The differences are obvious yet important. Stationary power is that generated in power plants within or adjacent to the industrial establishment in which it is used. Mobile power is that whose principal use has been in transportation, as on railways and in steam navigation during the nineteenth century and in automobiles and aircraft in the twentieth century. The development of stationary power in modern times is much less familiar or understood, and in the public mind has been overshadowed and obscured by the more highly visible and sensational mobile applications of power.

In our day, as in the past, few have had the occasion or the opportunity to visit the industrial establishments that are the habitat of stationary prime movers and the machinery which they place in motion. Occasionally in nineteenth-century factories, steam engines with their polished brightwork, decorative framing, and swiftly rotating flywheels were something of a showpiece, included in plant tours for important persons and visible at times to passersby. More commonly the engines or waterwheels driving mills and factories occupied obscure and damp sites in basements, dingy outbuildings, or grimy annexes, or, at times, were even under urban sidewalks. Moreover, the elaborate and extensive systems of shafting, pulleys, and belting through which power was distributed from engine or waterwheel in factories were more apt to confuse than to inform the visitor.

However dependent the economy has been upon mobile power in transportation for trade and travel, for securing raw materials and supplies, and for reaching markets, it was in the mines, mills, and factories of the nineteenth century where the work of industrial production was carried on. Here were produced the infinite variety and vast quantities of goods that comprised so large a part of the national output, including importantly the machinery, equipment, and fuel upon which transportation systems were dependent.

Production, or generation, receives principal attention in this history of industrial power, but two other aspects are recognized and included: transmission, or distribution, and application to industrial operations by machinery. Those uses are so infinitely diverse and the machinery employed correspondingly so varied that they can be considered only in general terms with the use of illustrative examples. The transmission of power was second in importance only to the generation of power. Transmission deals with the methods and the equipment by which power in the form of motion is conveyed from the prime mover, located typically at some central and convenient point, to the machinery employed in production. In the smallest establishments, such as common grist- and sawmills, a waterwheel was connected directly or through gearing to the machinery driven—the rotating millstones and the crank-driven up-and-down saws—and required no transmission. With the growth in the scale of industrial operations, the machines to be placed in motion increased in size, in speed, and in number. The machinery employed for the distribution of this power, known as millwork and consisting of the shafting, pulleys, and belting referred to earlier, might cost as much or more than the power plant; and its operation usually consumed from 20 to 50 percent of the power generated.

Another point emphasized throughout this study is that there is more to motive power than the prime movers that figure so prominently in industrial history. The development of waterwheels, steam engines, and their like can be understood properly only within the context of supporting facilities. This is most obvious with waterpower: waterwheels were functional only with the support of dams, raceways, and related structures. Except with the smallest and least efficient waterwheels or where there was no occasion for economy in the use of water, a millpond for the overnight accumulation and storage of streamflow—which otherwise would be wasted over the dam—was indispensable; and upstream reservoirs for storage of floodwater during wet seasons for mitigating the shortages of water during the dry months were important adjuncts of large waterpower installations. So was the accumulation of records on rainfall and streamflow, invaluable as a basis for calculations of future supply. The drainage basin of a stream was a fundamental factor in power capacity common to all waterpower sites upon the stream.

The situation with respect to steam power was much the same although in some respects not so clear-cut, since, as was usually pointed out, steam engines could be located almost anywhere and were virtually unaffected by seasonal fluctuations in the weather. Yet in the absence of a continuous supply of steam under pressure the steam engine was helpless. A steam-power plant, too, was not only dependent upon an abundant supply of water of suitable quality for making and condensing steam but also upon the arrangements for fuel supply, as well as for storage capacity at the plant against the contingencies of interrupted fuel supply. In sum, while prime movers may properly be considered as central agencies in power production, they are only a part, and in large installations only a small part, of the complete system.

Industrial power supply is most fully understood when viewed within the context of industrial production and development, which in turn must be set within the broader framework of economic history and in relation to a wide variety of technological considerations. Economic history may be described as the story of how man has secured his livelihood and met his needs for food, clothing, and shelter. More ponderously expressed, it has been defined as the history of the wealth-getting and wealth-using activities of mankind. It is subdivided into such fields as agriculture and industry, trade and transportation, manufactures and mining, money and banking, and all the rest. Closely interwoven with all those fields and factors is technology, the history of which during the past fifty years has become a major discipline in itself. Technology has to do with the tools and equipment, the machinery and materials, the skills and the expertise employed in any industrial and economic activity. An understanding of the technology of an industry or a business is essential for an understanding of the conditions of its operation, growth, and development. In important respects technology sets the limits within which each division of the enterprise functions and determines the organization of production. Power supply must conform to the requirements of machinery and processes and have a capacity adequate to all needs.

The character and content of any historical study is, of course, greatly influenced by the interests, the predilections, and the prejudices of the historian. Although the present writer thinks of himself as an economic historian, he is in no proper sense of the term an economist. He is simply an historian whose interests have fallen within the broad fields of industrial and economic development. With the background of an upbringing in one of the smaller industrial communities situated within the agricultural Middle West, his acquaintance with the manufacturing industry had its beginnings in summer jobs in factories producing agricultural machinery and equipment. Chance and circumstance directed his graduate research into heavy industry, leading to a dissertation on the early iron industry in Pittsburgh. This in turn led to research and writing on certain phases of the economic development of the Ohio River Valley, to western river transportation, and, more recently, to the study of industrial power, of which the present work is the outcome. As in his earlier writings, the objectives of this study are not those of the economist concerned with the analysis of economic forces, trends, and relationships in the light of economic theory; rather, he pursues the simpler goals of discovering and recreating a given segment of the past as experienced by those living at the time and in the light of his own understanding of the general course of American economic development. The past is viewed through the eyes, the minds, and the recorded experience of those who were participants. His concern is primarily with description within a framework of rather simple and obvious conceptions of human behavior and motivations, with principal reference to economic affairs. Analysis is largely confined to relationships within the context of a wide variety of economic institutions, activities, and influences. His goals do not include importantly the light which the study of the past may throw upon the present, although past and present are inescapably connected. In the older phrase of Leopold von Ranke, he endeavors to understand and describe the past *wie es eigentlich gewesen ist.*

While the study of industrial power is directed primarily and principally to the experience within the United States over a period of more than a century, the subject cannot be considered in isolation from the western European world. In each of the three major divisions of this study reference will frequently be made to the interacting relationships and influence in respect to industrial power between the United States and the industrially more advanced nations and regions in Europe, particularly Great Britain, France, and in varying degree certain adjoining areas on the Continent. The American colonies were a demographic and cultural extension of Europe, and with the swelling tide of European emigration to this country, especially from the British Isles, there was an accompanying flow of industrial knowledge and expertise. This was paralleled by the migration of information and ideas by way of books, periodicals, and scientific journals, and, in a limited way, despite British legislation to the contrary, of the new industrial machines and equipment. In time there was a significant counterflow by which certain American innovations in technology and industrial practice were introduced and ultimately widely adapted abroad. Technological exchange was to prove of particular importance in respect to prime movers and power transmission. All such matters will receive appropriate attention.

This *History of Industrial Power in the United States, 1780-1930,* is divided into three parts to appear in successive volumes. The first will deal with waterpower in its varied aspects—including installations ranging from the tens of thousands of common water mills developing no more than several horsepower each to the first American industrial centers, the largest of which had aggregate capacities of ten thousand horsepower and upward—and with its progressive decline and supersession by steam power during the post–Civil War decades and by hydroelectric energy from the 1890s. Consideration will be given in this first volume to the development in Great Britain and the United States of the mechanical methods of power transmission and distribution characteristic of the first Industrial Revolution. Volume two will cover in similiar fashion the introduction and development of steam power in this country during the same period, or from the general adoption of the Boulton and Watt engine at the threshold of the new century to the introduction of the steam turbine as the nineteenth century was drawing to a close. The third and final volume will cover developments between 1850 and 1930 and will be directed especially to what I have termed "the transmission revolution." In important respects the focus shifts progressively, although slowly before the 1890s, from the production of power to its transmission and distribution. Particular attention will also be given in the final volume to power problems and developments in the mining and mineral industries and to the more striking innovations accompanying the spread of electrical transmission of power and the replacement of power plants of individual industrial and commercial establishments by the electric utilities. By 1930 the electric power industry, drawing upon both steam and hydroelectric plants, had become one of the leading industries in the United States, marked by a high, and growing, degree of concentration of ownership and control. The transmission revolution in its essential character had been completed, the electrification of all segments of the economy well advanced, and the foundations laid for the lavish use of energy that was to mark the postdepression decades. The speculative extravagances of the 1920s had prepared the way for the investigations of the power industry, the regulatory controls of the New Deal years, the Tennessee Valley Authority, and the vast federal multipurpose hydroelectric projects of the Far West.

Abbreviations

DAB	*Dictionary of American Biography*
JAES	*Journal of the Association of Engineering Societies*
JBF	James Bicheno Francis
JEBH	*Journal of Economic and Business History*
JFI	*Journal of the Franklin Institute*
JWSE	*Journal of the Western Society of Engineers*
LCP	Proprietors of the Locks and Canals on the Merrimack, Baker Library, Harvard University, Cambridge, Mass.
PASA	*Proceedings of the American Statistical Association*
PASCE	*Proceedings of the American Society of Civil Engineers*
PICE	*Proceedings of the Institution of Civil Engineers*
PIME	*Proceedings of the Institution of Mechanical Engineers*
TAIEE	*Transactions of the American Institute of Electrical Engineers*
TAIME	*Transactions of the American Institute of Mining Engineers*
TASCE	*Transactions of the American Society of Civil Engineers*
TASME	*Transactions of the American Society of Mechanical Engineers*
TNS	*Transactions of the Newcomen Society*

A History of Industrial Power in the United States, 1780–1930

VOLUME ONE: WATERPOWER in the Century of the Steam Engine

1
Power in an Agricultural Economy
The Rural Water Mill

FOR TWO AND a half centuries the country water mill was the chief instrument for the generation and use of mechanical power in colonial and independent America. Serving principally the needs of pioneering and rural communities in an agricultural economy, the traditional water mill in its most common forms frequently became an entering wedge for the machine age. If, as has been remarked, the rural water mill did not sire the factory, it did in important measure prepare the way for an industrial economy by demonstrating the labor-saving benefits of machinery and mechanical power. A familiar adjunct of nearly every pioneering settlement and rural community, these rude but effective mills gave a nation of farmers and craftsmen some familiarity with the practice and problems of applying the energy of falling water to the satisfaction of human needs. The reliance of most American communities upon water mills for certain basic tasks continued well into the late nineteenth century, but from the 1860s their position was progressively undercut by the extension of the market economy and the replacement of local by factory-made products. Beginning with the first generation of settlements in Virginia and Massachusetts Bay, the common water mill became a characteristic feature of the pioneering scene across much of the continent. It continued to be an essential element in most rural communities until the self-sufficiency of the farm household gradually gave way before the growing dependence upon "store goods" attending the improvement of transportation and the advance of industrialization.

Numbers, Kinds, and Distribution

The number of such mills multiplied with the increase and spread of population, reaching totals that even before the separation from England must have ranged in the thousands. At the high tide of their usefulness in the mid-nineteenth century, country mills of the common sorts were reported by the federal census in the tens of thousands. The grand total may have approached a hundred thousand before the long, slow decline set in under the influence of an accelerating industrialization. Thousands continued to function, though hardly to flourish, during the first decades of the twentieth century. According to the federal census of manufactures of 1919, a total of 21,135 flour and grist mills existed in the United States, of which 17,708 were merchant mills producing bolted, or "refined," flour and the remainder gristmills producing in most instances unbolted meal. About half the gristmills were "small custom mills grinding the farmer's grain for a fixed toll and probably devoting most of their time to producing feed for cattle. Collectively they do not use one percent of the wheat ground in the country."[1]

Familiar as small, water-driven mills were in most parts of Europe—thousands were reported in William the Conqueror's Domesday survey, or one mill to about fifty households—their role in pioneering America was at least as important and their number in relation to population probably greater.[2] In his report on manufactures of 1791 Alexander Hamilton listed the abundance of mill seats among the advantages of the new nation for manufacturing. Thomas Jefferson in 1786 declared that no neighborhood in the United States was without its gristmill, a view which shortly found its echo in Timothy Dwight's assertion that such conveniences as water mills were almost if not literally "to be furnished in abundance in every parish in the country." For one Massachusetts county, as reported by Dwight, this was amply true; the total in 1793 was 262 for the major types—gristmills, sawmills, and fulling mills—or one mill to each 250 inhabitants.[3] This statistic lends support to a British traveler's report of about the same date that there were in New Jersey alone some 1,100 improved mill privileges; "almost innumerable mill seats" in Pennsylvania "conveniently distributed by

[1] Quoted in Charles B. Kuhlmann, *The Development of the Flour-Milling Industry in the United States* (New York, 1929), p. xiii.

[2] Margaret T. Hodgen, "Domesday Water Mills," *Antiquity* 13 (1939): 261 ff. See also H. C. Darby and I. B. Terrett, *The Domesday Geography of Midland England* (Cambridge, 1954), pp. 41 ff. Little is known of these mills' precise character. It has been assumed by reasonable but perhaps mistaken inference that most or all were water-driven. The possibility that a significant number were driven by animals, presumably oxen, need not be considered here. See also W. C. Unwin, *On the Development and Transmission of Power from Central Stations* (London, 1894), pp. 80 ff.; Victor S. Clark, *History of Manufactures in the United States, 1607–1860,* 3 vols. (New York, 1929), 1:180–81.

[3] *American State Papers, Finance*, 1:129; Thomas Jefferson to Brissot de Warville, 16 Aug. 1786, *Papers of Thomas Jefferson,* ed. Julian P. Boyd et al. (Princeton, 1950–), 10:262; Timothy Dwight, *Travels in New-England and New-York,* 4 vols. (New Haven, 1821–22), 1:379–80, and 367, citing Worcester County with a population in 1810 of 64,910.

Providence throughout the State"; a thousand others (supplied with water, alas, but eight months of the year) in Kentucky; and a grand total for the entire nation of at least 10,000 and probably nearer 20,000 developed mill seats.[4] However slim the reporter's evidential base, these estimates, judged by the census data of later years, may not have greatly exaggerated the situation.

Throughout rural America these simple but effective country mills, driven in most instances by water, nearly everwhere followed closely on the heels of settlement and persisted long after the days of pioneering had passed.[5] They rank with the plow, ax, and oxen as basic equipment of a pioneering people; they played an essential role in the subsistence phase of agricultural development; and they contributed to the transition to a market economy. Gristmills and sawmills were among the first community facilities obtained in frontier areas, in most regions taking precedence over schools, churches, and stores and coming well in advance of wagon roads. In the New York area in the majority of instances sawmills and gristmills preceded inns, schools, and stores by several years. In the Connecticut Valley gristmills preceded other types of mills, which, however, followed shortly.[6] Vested with a distinctive public interest, these mills were accorded in many colonies and states a privileged legal status, were subject usually to public regulation, and during the pioneering years were often established with community assistance in such forms as land grants, exclusive milling privileges, tax advantages, and the like. Occasionally such mills were established by public authority and at public expense.

The use of water mills was favored in most regions not only by conditions of topography and hydrology but by the absence or ineffectiveness of quasi-feudal controls, the abundance of resources awaiting exploitation, and, above all, by the perennial and often acute shortage of labor in the face of the tasks of pioneering development in a new land. Although

[4] W. Winterbotham, *An Historical, Geographical, Commercial, and Philosophical View of the American United States,* 4 vols. (London, 1795), 2:367, 404 ff., 3:178–80, 334. For my soberer estimate, see chap. 2, p. 000.

[5] Among useful general works on the subject are William B. Weeden, *Economic and Social History of New England, 1620–1789,* 2 vols. (Boston, 1891); Philip A. Bruce, *Economic History of Virginia in the Seventeenth Century,* 2 vols. (New York, 1895–96); Lewis C. Gray, *History of Agriculture in the Southern United States to 1860,* 2 vols. (Washington, D.C., 1933); Clark, *History of Manufactures,* vol. 1; R. Carlyle Buley, *The Old Northwest: Pioneer Period, 1815–1840,* 2 vols. (Indianapolis, 1950); Jacob A. Swisher, *Iowa: Land of Many Mills* (Iowa City, 1940); Edwin C. Guillet, *Early Life in Upper Canada* (Toronto, 1963).

[6] John H. French, *Gazetteer of the State of New York* (Syracuse, 1860), which dates the establishment of facilities in some 150 communities; Mary R. Pabst, *Agricultural Trends in the Connecticut Valley Region of Massachusetts, 1800–1900* (Northampton, 1940–41), p. 17.

in a few instances colonial proprietors, and later certain states, reserved the use of mill seats, the practice did not spread or persist. The Old World institution of the *banalité*, the monopoly of mills and milling rights, found no place in English America as it did in French Canada.[7] Operated commonly by the miller, himself often also a farmer, with possibly an assistant, and in many instances active only during seasons of ample stream flow, these mills served the environs on a custom basis, processing for a fee or toll the materials brought to them by all comers. A share of the material or product, a toll fixed by ordinance, statute, or custom, went to the millowner for services rendered. The amount of the toll usually ranged from one-twelfth to as much as one-fourth, according to the kind of grain in the case of gristmills, and from one-third to one-half in the case of lumber and leather.[8]

To the ubiquitous gristmills, sawmills, and tanneries and the less numerous but hardly less essential carding machines and fulling mills were added in time a variety of other small power-driven establishments, operating usually on a market rather than a community or custom-toll basis. Such were the paper, powder, and plaster mills, the linseed (flaxseed) oil and cloverseed mills, and increasingly from the early 1800s, the small textile or yarn-spinning mills.[9] In the South crop-processing requirements and the laborious nature of hand operations led to the establishment of cotton ginneries (and later presses for baling cotton), rice mills, and sugar mills. On the larger plantations such facilities, driven by animals or waterpower, became common equipment; the needs of the smaller producers came to be met by central ginneries or mills operating on a custom-toll basis or were similarly accommodated by a large plantation in the vicinity.[10]

[7] For the occasional practice of reserving mill seats when land was placed on sale, see Stevenson W. Fletcher, *Pennsylvania Agriculture and Country Life, 1640–1840,* 2 vols. (Harrisburg, 1950–55), 1:324; Guillet, *Early Life in Upper Canada,* pp. 38–39, 64, 143.

[8] See Clark, *History of Manufactures,* 1:63–64; Gray, *History of Agriculture,* 1:162; Fletcher, *Pennsylvania Agriculture,* 1:326–27; Swisher, *Iowa,* pp. 44–45; early compilations of state statutes; E. E. Calkins, *They Broke the Prairie* (New York, 1937), pp. 92 ff.; Alvin Johnson, *Pioneer's Progress: An Autobiography* (New York, 1952), pp. 88–89. There are occasional references to tolls in U.S., Secretary of the Treasury, *Documents Relative to the Manufactures in the United States,* House Ex. Doc. 308, 22d Cong., 1st sess., 2 vols. (Washington, D.C., 1833), e.g., 2:102–5.

[9] See the published federal censuses of manufactures and the unpublished census schedules, 1850 through 1870, to be found in state libraries or archives and the National Archives. State gazetteers often list the numbers of mills in each town or village and in some cases give totals by counties or for the state: see, for example, Thomas F. Gordon, *A Gazetteer of the State of Pennsylvlia* (Philadelphia, 1832). See also Table 1.

[10] The early censuses did not report separately, if at all, cotton ginneries or rice or

Standing somewhat apart from the foregoing but with some characteristics in common were the ironworks typical of the preindustrial economy. Such were the small rural blast furnaces and forges and eventually the village foundries and tilt hammers or large blacksmithies found in many of the colonies and in nearly all of the states east of the Mississippi River.[11] Initially such establishments, almost invariably driven by waterwheels, served the needs of the environs, though areas of much wider radius than those served by the more common water mills. From the blast furnaces and foundries came the hollowware—pots, kettles, pans—and other cast-iron ware for household or farm use and the mill irons and sundry equipment used by the mills themselves. The forges supplied wrought iron of different shapes and dimensions, especially the merchant bar, which, in the hands of the country or village blacksmith or occasionally the farmer himself, was shaped into the wide variety of hardware, edge tools, and nails so important in an agricultural economy.[12] These rural ironworks usually differed from the more common forms of mill industry in their requirements of much greater capital, of more elaborate and more costly equipment, and of more highly specialized and difficult skills. They were market-oriented and profit-motivated enterprises. As far as the immediate environs were concerned, they served on a barter-exchange basis, taking their pay chiefly in locally available produce.[13]

---

sugar mills. The Eighth Census, 1860, listed only 88 ginneries for the entire South (U.S., Census Office, *Manufactures of the United States in 1860* [Washington, D.C., 1865], p. 716). But see Gray, *History of Agriculture,* 1:729, 740 ff.; J. D. B. De Bow, *The Industrial Resources, Etc., of the Southern and Western States,* 3 vols. (New Orleans, 1852), 2:127–28, 403; J. Carlyle Sitterson, *Sugar Country: The Cane Sugar Industry in the South, 1753–1950* (Lexington, Ky., 1953), pp. 9 ff., 32 ff., 137 ff.; H. Weaver, *Mississippi Farmers, 1850–1860* (Nashville, 1945), p. 52; G. A. Lowry, "Ginning and Baling Cotton, from 1798 to 1898," *TASME* 19 (1897–98): 811 ff.; and Clarence H. Danhof, "Agricultural Technology to 1880," in Harold F. Williamson, ed., *The Growth of the American Economy* (New York, 1944), p. 133.

[11] U.S., Census Office, Sixth Census, 1840, *Compendium of the Enumeration of the Inhabitants and Statistics of the United States* (Washington, D.C., 1841), p. 358; Eighth Census, *Manufactures of the United States in 1860,* pp. clxxviii-clxxxiii.

[12] See Louis C. Hunter, "Influence of the Market upon Technique in the Iron Industry of Western Pennsylvania up to 1860," *JEBH* 1 (1929): 241 ff.; James D. Norris, *Frontier Iron: The Maramec Iron Works, 1826–1876* (Madison, Wis., 1964), chap. 6. On frontier blacksmithing, see Harriette S. Arnow, *Seedtime on the Cumberland* (New York, 1960), pp. 252–53; Horace Kephart, *Our Southern Highlanders* (New York, 1913). For small rural ironworks, see manuscript schedules of the federal census, 1820, for Pennsylvania, Ohio, and Kentucky, the National Archives.

[13] Louis C. Hunter, "Financial Problems of the Early Pittsburgh Iron Manufactures," *JEBH* 2 (1930):520 ff.

Although distinctive in equipment and manner of operation and differing in their relation to the environs, the various rural mills and works had much in common. Production in most instances centered in certain operations, which, though simple and repetitive, required the application of considerable force: grinding, crushing, pounding, pressing, sawing, pumping, blowing. The mills were placed in motion by mechanically generated power, save for the frequent use of oxen or horses in the smaller and simpler mills or where water or, less frequently, wind was lacking. Falling water was the common power source except in coastal regions, such as Cape Cod, Long Island, and other districts as far south as the Carolinas, where windmills were in frequent use. Before the 1850s the use of steam power in rural mill industries was with a few exceptions a rarity. The amount of power employed was typically quite small, often not more than several horsepower, save in rolling and slitting mills. A capacity of three to five horsepower was usually adequate for the requirements of gristmill, sawmill, or forge. Requiring even less power were the blowing tubs of blast furnaces making a few tons of pig iron a week, fulling stocks, carding machines, tanbark mills, and the ordinary gristmill with its tub wheel. The rotary motion of the waterwheel was conveyed by cranks, shaft-mounted cams or lifters, or gearing to provide the required up-and-down, back-and-forth, or round-and-round movements of the simple machinery of these varied establishments.

Role in Pioneering Development

The water-driven mills played an almost indispensable role in the lives of the rural and village communities they served. When the operations were performed by hand, they were not only laborious, quickly tiring, and far slower, but they turned out a product of inferior quality. In a way of life marked by a high degree of self-sufficiency covering a wide range of tasks, the removal from home and farm of some of the more laborious chores was indeed a welcome advance. The establishment of the small country water mills was commonly the first step in the slow emancipation of an agricultural people from the monotony and drudgery of rural life.

Of the contribution of rural water mills there is no want of witness. The evidence of sheer numbers and quantities in census reports is amply supported by the prominence of commonplace reference in contemporary publications, such as newspapers, gazetteers, travel accounts, and in the voluminous literature of local history and pioneering. "In all new settlements," runs a characteristic account, "mills for grinding grain and sawing logs are considered as things of the first necessity. They are part of the labor-saving machinery which civilization invented at an early period. They perform the work of many men, and do it more perfectly than it could be done by hand. Food and shelter are the first things to be provided for in a new country, and these mills are almost essential in the preparation of materials."[14] Some pioneers, declared the historian of a Kentucky community, held the gristmill "a greater necessity than a store, a courthouse or a professional physician." Writing of early Kansas, another historian remarked upon "those two great essentials to the pioneer settler—flour and building material."[15]

Sawmills, declared Albert S. Bolles in his survey of industrial development, were welcomed in America and soon came into general use throughout the colonies. "They followed the pioneer everywhere, and formed, with the grist mill, the nucleus of every settlement and neighborhood."[16] Whereas in England sawmills were long opposed and even destroyed by workers fearing loss of employment, the colonists "welcomed anything that would save labor, and so early settlement was largely directed in its course by the existence of water power."[17] "As first settlers," wrote a twentieth-century historian of early westward-bound Vermonters, "they got the pick of the land. They spotted the mill sites, the town sites, and the best stands of timber—and bought them up while they were still cheap."[18] "There was nothing more essential to the convenience and well being of the new settlements in eastern Maine," declared a local history, "than the saw and grist-mill." Of early South Carolina we

[14] Henry Bronson, *History of Waterbury, Connecticut*, 2 vols. (Waterbury, 1858), pp. 29 ff. The same point had been made more than fifty years earlier by settlers in southern Ohio seeking a federal land grant in aid of a gristmill (Alta H. Heiser, *Hamilton in the Making* [Oxford, Ohio, 1941], pp. 90 ff.).

[15] O. A. Rothert, *A History of Muhlenberg County, Kentucky* (Louisville, 1913), p. 121; L. A. Fitz, "The Development of the Milling Industry in Kansas," *Kansas Hist. Soc. Collections* 12 (1911–12):55.

[16] Albert S. Bolles, *Industrial History of the United States* (Norwich, Conn., 1878), pp. 274–75; see also a similar statement by a Canadian pioneer, Thomas Reed, in Guillet, *Early Life in Upper Canada*, p. 247. "Whatever the grain used [for bread]," declared Thomas Kettell, ". . . milling is the first necessity, and the number and capacity of the mills must always be proportioned to the numbers of the people. In a country like this, where they multiply so fast, the investments in mill property must keep pace" *(Eighty Years' Progress of the United States* [Hartford, Conn., 1868], p. 431). Kettell for many years was editor of *Hunt's Merchants' Magazine.*

[17] James E. Defebaugh, *History of the Lumber Industry of America*, 2 vols. (Chicago, 1906–7), 2:9. Perhaps the best account is Benno M. Forman's "Mill Sawing in Seventeenth Century Massachusetts," *Old-Time New England* 60 (1969–70):110–30.

[18] L. D. Stilwell, *Migration from Vermont, 1776–1890* (Montpelier, 1937), pp. 118, 141.

are told that "a mill pond and its corn mill were usually the first evidence that a section had been settled, and it was rare that the settlers were not thus provided within five years." Of North Carolina, on the other hand, an account of about 1710 noted: "The greatest failing and lack here . . . is that too few people are here and no good mills."[19] In the common experience the lack of mills was one of the most acutely felt—and most frequently mentioned—hardships of pioneering.

Bread: The Fundamental Need

A primary need everywhere was bread. For want of a gristmill to reduce grain to meal and flour, the settlers were compelled to make do with the crude, hand-operated devices known variously as samp or plumping mills, stump or chuck mortars, corn pounders, hominy blocks, graters, and the like, used where appropriate with a spring pole to lighten the labor by lifting the heavy pestle.[20] Not unlike the monotonous pounding of the foot-driven rice mill of the Asian village, the samp mills and stump mortars of the frontier "were often kept going from early morning until sundown," affording a comforting sense of neighborliness in sparsely settled communities. One authority reported that the thump of the heavy mortar could be heard a mile away and was used by neighbors for signaling.[21] Not only did the coarse meal produced by such rude devices

[19] W. D. Williamson, *History of the State of Maine* (1832), cited in Lincoln Smith, *The Power Policy of Maine* (Berkeley, Calif., 1951), pp. 25 ff.; Robert L. Meriwether, *The Expansion of South Carolina, 1729–1765* (Kingsport, Tenn., 1940), p. 173; Christoph von Graffenreid, *Account of the Founding of New Bern,* ed. V. H. Todd (Raleigh, N.C., 1920), p. 315. On local and plantation mills in the southern colonies, see Carl Bridenbaugh, *The Colonial Craftsman* (New York, 1950), pp. 18–24.

[20] Members of an affluent urban society in the twentieth century may need a reminder of the dominant role of bread in the diet of agricultural people of earlier times—and, indeed, of the peoples of underdeveloped countries and regions in the late twentieth century. See W. W. Cochrane, *The World Food Problem* (New York, 1969), pp. 58–59. On the importance of maize in the frontier diet of the southern states, see J. G. M. Ramsey, *The Annals of Tennessee to the End of the Eighteenth Century* (Charleston, S.C., 1853), pp. 718–19. According to Gray, one reason for the slow introduction of wheat and other small grains in the colonial South was that they were more difficult to prepare than corn in hand mills and mortars (*History of Agriculture,* 1:161).

[21] Gideon T. Ridlon, *Saco Valley Settlements and Their Families* (Portland, Maine, 1895), p. 191; John E. de Young, *Village Life in Modern Thailand* (Berkeley, Calif., 1955), pp. 33 ff.; Fletcher, *Pennsylvania Agriculture,* 1:399–400; see also Abby M. Hemenway, ed., *The Vermont Historical Gazetteer*, 5 vols. (Burlington, 1868–91), 4:822. In spring 1963 in the rugged hill country of southeastern Yugoslavia the writer came upon an operable (but evidently little used) spring-pole stump mortar in a farmyard.

Fig. 1. Spring-pole mortar for crushing grain (corn). This example was photographed by the author near Sarajevo in Yugoslavia. Such mills were common in frontier America.

give an unpalatable bread, but the daily labor required to supply a family's needs could hardly have been less than the two to three hours necessary with the traditional grinding stones and querns used in Mexico, Peru, and India.[22]

[22] Oscar Lewis, *Pedro Martinez: A Mexican Peasant and His Family* (New York, 1964), pp. lv–lvi; idem, *Village Life in Northern India* (New York, 1958), pp. 50–51; William W. Stein, *Hualcan: Life in the Highlands of Peru* (Ithaca, N.Y., 1961), p. 94. A report from Louisiana to the directors of the Company of the Indies urged the provision of horse mills there because "a negro must spend his day pounding [grain] in order to provide enough for two to eat" (D. Rowland and A. G. Sanders, eds., *Mississippi Provincial Archives,* vol. 2 [Jackson, 1929], p. 310). According to Denis Warner, the traditional "four tyrannies" of Chinese women before the revolution of 1949 were "the mill-stone, the grind-stone, the stove, and the oven" (*Hurricane from China* [New York, 1961], p. viii. See also William Hinton, *Fanshen: A Documentary of Revolution in a Chinese Village* (New York, 1968), p. 430. For a well-illustrated review of milling methods and equipment through the ages, see John Storck and Walter D. Teague, *Flour for Man's Bread: A History of Milling* (Minneapolis, 1952).

Indeed, there was a frontier form of the ancient quern, with the upper of two small millstones turned by hand, and especially in the southwest the Mexican type of grinding stones—the metate, or mortar, and mano, or roller—were also used. An early Pennsylvania settler recalled his family's experience: "My father erected a cabin of bark set against a large pine log, and lived in it for a year and a half. He then built a log house. In this he lived the first winter without a floor, there being no saw-mill nearer than Painted Post. For a grist mill we used a stump hollowed out by fire for a mortar, and a spring pestle. In this we pounded our samp for bread and pudding timber [*sic*] for two years. After a while several of the settlers clubbed together and purchased a pair of mill-stones two feet in diameter, which we turned by hand." This quern, or hand mill, is described in its most common form as found on the Virginia frontier in the late eighteenth century. It was, reported Samuel Kercheval, "made of two circular stones, the lowest of which was called the bed-stone, the upper one the runner. These were placed in a hoop, with a spout for discharging the meal. A staff was let into a hole in the upper surface of the runner near the outer edge, and its upper end through a hole in the board fastened to a joist above so that two persons could be employed in turning the mill at the same time."[23] This description approximates the quernlike mill still in use in eastern Kentucky about 1915 (Fig. 2).

By the time settlement moved actively into the Middle West, the Yankee-made coffee grinder often doubled as a means of producing coarse meal from grain, a marked improvement over the stump mortar at least. In Texas this general type of mill with steel or cast-iron grinding surfaces was referred to as a steel mill. G. W. Tyler tells us that in the settlement of Bell County, Texas, "almost every family brought along an 'English steel mill' for grinding corn into coarse meal. It was fastened to a post or the wall of the cabin and operated by a hand crank—a slow process." According to another Texan, the steel mill, bolted to a tree, "took the steady turning of a strong man an hour to grind a meal for a single family."[24] In his journey through

[23] W. J. McKnight, *History of Northwestern Pennsylvania* (Philadelphia, 1905), p. 617; Samuel Kercheval, *A History of the Valley of Virginia*, 3d ed. rev. (Woodstock, 1902), pp. 274–75. See also Zadock Thompson, *A Gazetteer of the State of Vermont* (Montpelier, 1824), p. 91; Arnow, *Seedtime on the Cumberland*, pp. 389 ff.; Thomas D. Clark, *A History of Kentucky* (New York, 1937), p. 238; S. P. Hildreth, *Pioneer History* (Cincinnati, 1848), pp. 360–66, 375; C. M. Walker, *History of Athens County, Ohio* (Cincinnati, 1869), pp. 127–28; and Guillet, *Early Life in Upper Canada*, pp. 194, 215, 226.

[24] G. W. Tyler, *History of Bell County* (Belton, Tex., 1966), pp. 132–33; E. F. Bates, *History and Reminiscences of Denton County* (Denton, Tex., 1918), pp. 12, 76–77, and see

Fig. 2. Hand mill, or quern, for converting corn or wheat into meal, eastern Kentucky, c. 1915. Here a hollowed-out section of a tree trunk contains the two millstones. The fixed bedstone rests beneath the runner stone. The upright wooden crank rod, grasped by the woman, fits below into a hole near the outer edge of the runner stone. The upper end is secured loosely in a hole in the framing of the wall or ceiling above, providing a convenient means for turning the runner. The ground meal issues from the inclined spout seen at the lower right. (Photo courtesy of the Appalachian Museum of Berea College, Berea, Ky.)

the southwestern Texas frontier region, Frederick Law Olmstead stopped overnight with a family whose two young boys "were immediately set to work by their father at grinding corn in the steel-mill for supper. The task seemed their usual one yet one very much too severe for their strength. Taking hold at opposite sides of the winch, they ground away outside the door for more than an hour, constantly stopping to take breath." Olmstead also observed that "the constant and severe labor of the women of the family during all our stay was the manufacture

---

also the ox mill on p. 55. See also Everett Dick, *The Sod-House Frontier, 1854–1890* (New York, 1937), p. 251; J. H. Klovstad, "The Study of Pioneer Life," *Minnesota History* 11 (1930): 70–74; French, *Gazetteer of the State of New York,* p. 489n; *Knight's American Mechanical Dictionary,* s.v. "steel mill."

of this bread," meaning tortillas, the paste for which was ground from lime-soaked corn kernels in the traditional stone metate.[25]

The most persuasive testimony of the central role of the gristmill in pioneer communities is the almost invariable references in the vast literature of reminiscence to the trouble taken by the early settlers to reach them. We meet the same tale of hardship endured on virtually every nineteenth-century frontier across the continent and persisting well into the post–Civil War decades—the tale of the settler, by foot with grain sack over shoulder or on horseback, following often little more than a tree-blazed trail, twenty, thirty, even fifty miles to the nearest mill. "A gristmill at a distance of twenty-five miles was a valuable consideration when compared with mortar and pestle." "It was not unusual . . . for farmers to go forty miles to mill and carry the grist on their shoulders." In one New York community neighbors joined in the carriage of grain and meal by canoe in a four-day round trip to the nearest mill.[26]

A traveler to Illinois in 1817 wrote, "Their bread corn must be ground thirty miles off, requiring three days to carry to the mill and bring back, a small horse-load of three bushels." Mills always followed the settlements, wrote R. C. Buley of the Old Northwest, "but for years a trip to the mill twenty or more miles away was a matter of three or four days." Writing of the mountaineers of Appalachia in 1913, Horace M. Kephart declared, "They are great walkers and carriers of burdens. Before there was a tub-mill in our settlement one of my neighbors used to go, every other week, thirteen miles to mill, carrying a two-bushel sack of corn (112 pounds) and returning with his meal on the following day. This was done without any pack-strap but simply shifting the load from one shoulder to the other, betimes." Across the border in Canada things were much the same. "Coffee-mill flour seldom made good bread," and most settlers managed to get fairly good milled flour "even if they tramped thirty miles for it." When the Canadian government erected a gristmill

[25] *Journey through Texas: A Saddle-Trip on the Southwestern Frontier*, ed. James Howard (Austin, Tex., 1962), pp. 216–17, 255. For use of the metate, see also "The Reminiscences of Mrs. Dilue Harris," *Texas Historical Association Quarterly* 4 (1900–1901):109; Gray, *History of Agriculture*, 1:161; W. A. McClintock, "Journal of a Trip through Texas and Northern Mexico in 1846–1847," *Southwestern Historical Quarterly* 34 (1931):27–28.

[26] H. C. Goodwin, *Pioneer History; or Cortland County and the Border Wars of New York* (New York, 1859), pp. 143–44; Thomas F. Gordon, *Gazetteer of the State of New York* (Philadelphia, 1836), p. 501; and French, *Gazetteer of the State of New York*, p. 201. The many town histories in Hemenway, ed., *Vermont Historical Gazetteer*, repeatedly refer to long trips to the mill—with meal and grain often backpacked.

near Kingston in 1782–83, many carried grain there over long distances, on foot and by sled in winter or by bateau or canoe when water routes were open.[27]

Mileage figures alone are not very meaningful in the motor age. Two pioneer accounts from early Pennsylvania settlements provide local color. "I had fourteen miles to go in winter to mill with an ox-team. The weather was cold and the snow deep; no roads were broken, and no bridges built across the streams. I had to wade the streams and carry the bags on my back. The ice was frozen to my coat as heavy as a bushel of corn. I worked hard all day and got only seven miles the first night, when I chained my team to a tree, and walked three miles to house myself. At the second night I reached the mill." "On the 3rd day of July," runs the second account, "I started, with my two yoke of oxen to Jersey Shore to mill, to procure flour. I crossed Pine Creek eighty times going to, and eighty times coming from mill, was gone eighteen days, broke two axle trees to my wagon, upset twice, and one wheel came off in crossing the creek. . . . the road was dreadful."[28]

Where millstreams or mill seats were wanting, horse mills were a common resort in many communities. The long hours spent at the stump mortar by the boys in a northern Illinois family in the 1820s—to supply the many travelers passing by—hastened the adoption of more effective equipment:

In the spring the question came up about going eighty miles to mill and father told the people "by the time you have anything to grind I'll have a mill to grind it"; and he set some men to cutting logs to build the mill house and Jim and I to hauling them. He went on to the prairie and picked two boulders for millstones, worked a hole into—through—one, dressed one side to a face, faced one side of the other stone. He had sent to St. Louis by Hartzell for a full set of blacksmith tools—and iron, etc., etc., and they had arrived by this

[27] Morris Birkbeck, *Notes on a Journey in America,* 4th ed. (London, 1818), p. 115; Buley, *The Old Northwest,* 1:225–27; Kephart, *Our Southern Highlanders,* p. 216; in another instance illicit distillers carried two-bushel loads of corn to a mill 20 miles distant (Kephart, *Our Southern Highlanders,* p. 140); Guillet, *Early Life in Upper Canada,* p. 217. See also William T. Utter, *The Frontier State,* vol. 2., *The History of the State of Ohio,* ed. Carl F. Wittke (Columbus, 1942), pp. 240 ff.; Dick, *Sod-House Frontier,* pp. 248, 492 ff.; Merle Curti, *The Making of an American Community* (Stanford, Calif., 1959), p. 238; Thompson, *Gazetteer of Vermont,* pp. 97, 129, 140, 194; Emory F. Skinner, *Reminiscences* (Chicago, 1908), p. 5; W. A. Goulder, *Reminiscences: Incidents in the Life of a Pioneer in Oregon and Idaho* (Boise, 1909), pp. 104–5. See also descriptions by foreign travelers in Reuben G. Thwaites, ed., *Early Western Travels,* 32 vols. (Cleveland, 1904–7), 8:270, 11:199, 230, 13:161, 14:135.

[28] McKnight, *History of Northwestern Pennsylvania,* p. 596. The first trip was made some time after 1810, the second in 1834.

Fig. 3. Horse-powered gristmill, Floyd County, Ky., 1884. (Photo courtesy of the Filson Club, Louisville.)

time. The house was built. A tramp blacksmith came along with a tramp millwright, a shaft was set up and arms morticed into it to hold the "gearing" and "sweeps" for the horses to work on like a threshing machine. The horses would work in a circle of 40 feet in diameter. . . . They whip-sawed lumber for the boulting chest and used "mull-muslin" for bolt cloth and the thing was finished and ready by fall and was a very industrious little mill for as soon as it was done with one grist of corn it began on another.[29]

Useful as horse- and ox-powered gristmills were, they were resorted to, with infrequent exceptions, only in districts or situations where small waterpowers were unavailable. Compared with all water mills, especially the early introduced and widely used tub mill, horse mills were cumbrous and costly. Gristmills were preeminently water mills.

[29] Diary of William Gallaher, manuscript in possession of Mr. and Mrs. Monroe Leigh, Washington, D.C. For the bandmill, an interesting horse mill used in Texas, see Bates, *Denton County*, pp. 333–34. The manuscript schedules of the 1850 and 1860 federal censuses contain numerous references to horse mills.

[30] See Richard B. Morris, *Government and Labor in Early America* (New York, 1946), pp. 38–39; Clark, *History of Manufactures*, 1:177; Defebaugh, *Lumber Industry of America*,

Sawmills and Shelter

In the records of frontier settlement, gristmills and sawmills were often paired, sometimes one, sometimes the other preceding according to chance and circumstance. Combination grist- and sawmills were a common feature, as the federal census schedules abundantly testify. As meal from the settler's grist was the basic article of diet, so wood was the indispensable raw material, not only for dwellings—save on the Great Plains, where sod largely took its place—but for other building uses and especially for fuel. As the gristmill was preceded by stump mortar or hand mill, so the sawmill was preceded by the common ax and broadax, and, much less frequently, by the whipsaw or pit saw. Although the two-man team of pit sawyers was found on various frontiers across the continent, the sawing of boards by hand was too laborious, too slow to be widely practiced; neither pit saws nor skill in their use was widely in evidence.[30] So it was that in the tree-covered eastern third of the continent, the log house, or shanty, enshrined in legend as the "log cabin," became the characteristic pioneer dwelling. Until the resources of a community could command a sawmill, logs in the round or flat-hewn by common ax or broadax on one or more sides and joined together by various methods of notching were the obvious and necessary method of constructing dwellings. If necessity required, the smaller and simpler forms of the log house could be erected in several days by one man skilled with the ax.[31]

2:9; Horace Greeley et al., *Great Industries of the United States* (Hartford, 1872), pp. 317–18; Goulder, *Reminiscenses,* pp. 212–14; W. J. Trimble, *Mining Advance into the Inland Empire* (Madison, Wis., 1914), p. 92. In the United States as elsewhere, pit sawing persisted in the building of wooden ships—indeed, until the end of the wooden ship (John G. B. Hutchins, *The American Maritime Industries and Public Policy, 1789–1914* [New York, 1941], p. 118). An unusual example of handsawing on a large scale was a Canadian lumber mill established near Toronto in 1832, where for twenty years "a large force of men were employed in whip-sawing until displaced by a steam sawmill" (C. P. Mulvaney et al., *History of Toronto and County of York, Ontario,* 2 vols. [Toronto, 1885], 2:159). William F. Fox, *History of Lumber Industry in the State of New York* (Washington, D.C., 1902), p. 12, gives a good description of the method. On the pit saw, see W. L. Goodman, *The History of Woodworking Tools* (London, 1964), pp. 131 ff. Pit sawing persisted in England throughout the nineteenth century. See John H. Clapham, *An Economic History of Modern Britain,* 2d ed., 3 vols. (Cambridge, 1932), 1:445; George Sturt, *The Wheelwright's Shop* (Cambridge, 1923), chap. 8; and Walter Rose, *The Village Carpenter* (Cambridge, 1937), chaps. 1–3. The writer occasionally observed this mode of sawing in the Balkan countries in 1963 and 1965.

[31] Henry C. Mercer, "The Origin of Log Houses in the United States," reprinted in *A Collection of Papers Read before the Bucks County Historical Society* 5 (1926):568; H. R.

Fig. 4. Pit sawing, Harlan County, Ky., 1886. (Photo courtesy of the Filson Club, Louisville.)

In the better-circumstanced communities the log house was a temporary structure to be replaced and abandoned to other uses as soon as conditions permitted more comfortable dwellings. These, with infrequent exceptions, were of wood-frame construction rather than masonry. In the literature of pioneering the erection of the first frame structure in the community, usually a dwelling but sometimes a barn, was an event as worthy of record as the coming of the first store or school. A Maine historian claimed that the frame house, with its greater convenience and more attractive appearance, "was an index of the progress of intelligence and refinement." "The ax produces the log hut," declared the commissioner of patents in 1850, "but not till the sawmill is introduced, do framed

---

Shurtleff, *The Log Cabin Myth* (Cambridge, Mass., 1939); H. Glassie, "The Appalachian Log Cabin," *Mountain Life and Work,* Winter 1963, pp. 9–14; Arnow, *Seedtime on the Cumberland,* pp. 256 ff.; Neil A. McNall, *An Agricultural History of the Genesee Valley, 1790–1860* (Philadelphia, 1952), p. 86; Ellen Churchill Semple, "The Anglo-Saxons of the Kentucky Mountains: A Study in Anthropo-geography," *Geographical Journal* 17 (1901): 596–97.

dwellings and villages arise; it is civilization's pioneer machine: the precursor of the carpenter, wheelwright and turner, the painter, joiner, and legions of other professions." This position was elaborated forty-five years later in an early survey of American industrial history, *Industrial Evolution of the United States* (1895), by Carroll D. Wright, a leading social scientist. Wright declared that the sawmill marked "one of the clearly defined features of industrial development. . . . When the erection of dwellings which shall last for years begins, architecture, however primitive, must be cultivated, that the dwellings may represent the taste and the intelligence of the people building them."[32]

Such was the view repeatedly expressed in histories of the pioneering years. "As time passed on and the settlers had some time to devote to the comforts of life, timber was cut and drawn to the sawmill, one, two or ten miles distant as the case might be, and sawn into lumber. Then the comforts of a board floor and other improvements were enjoyed."[33] In a new country, declared an upstate New Yorker at mid-century, a brook and a sawmill were invaluable aids, providing in the first instance boards to replace the dirt floors of the log dwelling. But "the ambition of all settlers was to displace the log by the framed house, as soon as possible; but to do this, carpenters and house-joiners and sawmills and nails were necessary and these were not so easily obtained in new settlements."[34] The experience of the older communities in the seaboard states was repeated again and again as settlement advanced westward across the continent.[35] The passing of the crude frontier

[32] Lyndon Oak, *History of Garland, Maine* (Dover, 1912), p. 36; U.S., Patent Office, *Annual Report of the Commissioner of Patents for 1850,* pt. 1 (Washington, D.C., 1851), p. 387; Carroll D. Wright, *Industrial Evolution of the United States* (Meadville, Pa., 1895), p. 75. Wright dedicated this work, published under the rubric "Chatauqua Reading Circle Literature," to Francis A. Walker.

[33] Richard W. Musgrove, *History of the Town of Bristol, Grafton County, New Hampshire,* 2 vols. (Bristol, 1904), 1:95–96. See also Dwight, *Travels,* 2:469.

[34] Henry C. Wright, *Human Life* (1849), quoted in Louis C. Jones, ed., *Growing Up in the Cooper Country* (Syracuse, N.Y., 1965), pp. 94–95. See also Henry B. Plumb, *History of Hanover Township and Wyoming Valley . . . Pennsylvania* (Wilkes-Barre, 1885), pp. 203, 256–58, 310, 329; William D. Williamson, *The History of the State of Maine,* 2 vols. (Hallowell, 1839), 2:703.

[35] See, for example, McNall, *Genesee Valley,* pp. 85 ff.; Buley, *The Old Northwest,* 1:230–32; Calkins, *They Broke the Prairie,* pp. 92 ff.; Fletcher, *Pennsylvania Agriculture,* 1:376–77; Samuel R. Brown, *The Western Gazetteer* (New York, 1817), p. 113; W. H. Smith, *History of the State of Indiana,* 2 vols. (Indianapolis, 1903), 1:372; Swisher, *Iowa,* chap. 6; Ray A. Billington, "How the Frontier Shaped the American Character," *American Heritage* 9 (April 1958): 8–9; Curti, *The Making of an American Community,* p. 239; Tyler, *History of Bell County,* pp. 132–33; and R. N. Richardson, *Frontier of Northwest Texas, 1846–1876* (Glendale, Calif., 1963), p. 290.

Fig. 5. Sawmill at Sperryville, Va. The saw was set in a heavy wooden frame and placed in an up-and-down motion by a crank on the end of the main waterwheel shaft. The log was clamped to a movable frame or carriage (*right*) that moved slowly toward the saw, powered by the main waterwheel. For an earlier sawmill see Figure 27, chapter 2. (Photos by John O. Brostrup, 1936, HABS Collection, Library of Congress.)

log dwelling in an older state was documented by the New York State census of 1855. Log dwellings then comprised only one in fifteen of all dwellings, their average value of $50 put to shame by the value of nearly $800 for the predominant type of frame-built houses. Indeed, at this date, with nearly 5,000 sawmills in the state—approximately one to every 700 people—there was little excuse save poverty for the survival of the log dwelling. In contrast, the single or double log cabins of one story remained characteristic of the southern Appalachians in eastern Kentucky in 1900, a region described by Ellen Semple as marked by isolation, poverty, and general backwardness. It was only in some of the "marginal counties of the mountain region and in the sawmill districts, [that] one sees a few two-story frame houses."[36]

[36] French, *Gazetteer of the State of New York*, pp. 108–10, 150; also *Eighty Years' Progress*

With the advance of settlement westward into the prairie and plains country, the timber cover thinned and over vast regions all but disappeared. Here to some degree the heavy burdens of pioneering shifted from the men to the women. The massive task of land clearing central to farm building in the eastern states, often prolonged for years, was on the treeless prairie reduced to the heavy double-teaming, once-over task of breaking and turning nature's thick sod. Except for cottonwoods lining the stream banks and the scraggly pines and hardwoods found in scattered patches of rough country, timber was lacking and sawmills were rare. For want of timber the settlers turned to a by-product of land breaking, and the sod house gave its name to a period and region of settlement. The housewife bore the main burden of accommodating family living to a dwelling made chiefly of mud. A traveler in northwestern Nebraska in 1889 stopped briefly at such a home; the husband was away making money for the family: "The home was exceptionally neat, but it had no floor. . . . they had been among the earliest settlers in that part of the country—had lived here for six years, each year hoping to have enough money to buy lumber for a floor, but something had always happened to make them put off the expenditure. Now that her husband was working she really felt that the lumber could be had before winter closed in. With a touch of honest unassumed yearning she said she had always wanted a floor."[37]

The sawmill was introduced to the colonies by English, Dutch, and Swedes alike. It attained a popularity and use in America not reached in England even during the course of the nineteenth century—and with good reason since the relative scarcity of timber and abundance of labor in England were dramatically reversed in America.[38] The sawmill

of the United States, pp. 353 ff.; Semple, "The Anglo-Saxons of the Kentucky Mountains," p. 596.

[37] Seth K. Humphrey, *Following the Prairie Frontier* (Minneapolis, 1931), pp. 159–60, and chap. 12.

[38] For sawmilling in seventeenth-century Massachusetts and Maine, see the excellent articles by Benno M. Forman and Richard M. Candee in *Old-Time New England* 60 (1970):110–49. The most comprehensive treatment of sawmilling is Defebaugh, *Lumber Industry of America,* vol. 2; vol. 1 deals partly with Canada. See also Clark, *History of Manufactures,* 1:175–77; Bolles, *Industrial History,* pp. 274–75, 498–501; J. Leander Bishop, A *History of American Manufactures from 1608 to 1860,* 2 vols. (Philadelphia, 1861–64), 1:104; Morris, *Government and Labor,* pp. 38–40; Weeden, *Economic and Social History,* 1:168 ff.; Bruce, *Economic History of Virginia,* 2:487–91. A good contemporary description of a Pennsylvania sawmill of the 1790s is Winterbotham, *Historical . . . View of the American U.S.,* 3:425. Hugh Jones, *The Present State of Virginia,* ed. Richard L. Morton (Chapel Hill, N.C., 1956), pp. 142–43, contrasts the great benefits of saw-

Figs. 6 and 7. In less than a generation, pioneer Americans replaced their hand-hewn log cabins with houses built of sawed lumber, as in this classic example of the Ephraim Swain Finches on the Nebraska frontier. (Photos courtesy of the Solomon D. Butcher Collection, Nebraska State Historical Society, Lincoln.)

appeared almost as early as the gristmill and, from all the evidence, multiplied even more rapidly. It provides not only the means for improved dwellings and furnishings but also the first cash crop for communities with access to coastal or inland water transportation.[39] Although the condition of the early trails and roads hardly permitted timber to be hauled to sawmills more than a few miles at best—ten miles is the maximum that has come to my attention—the economy and superior quality of the milled product over the hand-hewn were hardly less than with the grinding of bread grains. Because of the difficulties of land carriage of timber and lumber, the occasion for sawmills to meet local needs was frequently more pressing than for gristmills. Settlers could travel considerable distances to mill by horse or on foot with grain and meal, but "the transportation of lumber through a wilderness without roads, across streams without bridges and through swamps with uncertain depths of mud, involved hardships."[40] In number sawmills usually outstripped gristmills at a fairly early date, for wood was not only the basic material of frame dwellings but the most important of all fabricating materials for domestic, agricultural, and industrial purposes throughout the nineteenth century. As late as 1880 sawmills were the most numerous and widespread of all industrial establishments, accounting for one-tenth of the total number.[41]

Fulling and Carding Mills

Much less common than gristmills and sawmills, but for home manufactures hardly less useful, were the fulling mills and carding and cloth-

---

mills in Virginia about 1717 with the "prohibition" [*sic*] of their use in some parts of Britain. In his 1825 travels through England, Zachariah Allen noted the comparative infrequency of sawmills and the opposition of sawyers to their introduction (*The Practical Tourist,* 2 vols. [Providence, 1832], 1:159).

[39] See R. G. Wood, *A History of Lumbering in Maine, 1820–1861* (Orono, 1935), and Bernard Bailyn, *New England Merchants in the Seventeenth Century* (Cambridge, Mass., 1955), pp. 50–51, 70.

[40] Oak, *History of Garland, Maine,* p. 36. See also Hemenway, ed., *The Vermont Gazetteer,* 3:111, 138–39; Henry O. Severance, *The Story of a Village Community* (New York, 1931), p. 59.

[41] French, *Gazetteer of the State of New York,* giving dates of erection, indicates that in New York the sawmill usually preceded the gristmill by several years. Bishop attributed this priority in western New York to the pains taken by the authorities to encourage and assist settlers in the erection of sawmills (*History of American Manufactures,* 1:107). See also Swisher, *Iowa,* p. 65. U.S., Census Office, Tenth Census, 1880, *Compendium of the Tenth Census,* 2 pts. (Washington, D.C., 1883), pt. 2, pp. 932–33.

dressing establishments. Fulling was the traditional finishing process in making woolen cloth. In a prolonged operation combining pounding with washing, fulling freed the rough-woven cloth from the natural grease in the fibers and the oil used in carding and spinning wool. The pounding action of heavy wooden stocks or beaters in soapy water had the even more important effect of compacting the cloth, increasing its strength and durability, a process accompanied by a reduction in dimensions. With the ensuing "dressing" operations, fulling added to the softness and appearance of woolen cloth, which led to its use in clothing for Sunday and holiday use.[42] The ancient method of fulling by foot-treading was occasionally practiced in America, for want of other means, as illustrated by the following example from the early Pennsylvania frontier:

> The required quantity of flannel was laid upon the bare floor, and a quantity of soap and water thrown over it; then a number of men seated upon stools would take hold of a rope tied in a circle and begin to kick the flannel with their bare feet. When it was supposed to be fulled sufficiently, the men were released from their task, which was a tiresome one, yet a mirth provoking one, too, for, if it were possible, one or so must come from his seat, to be landed in the midst of the heap of flannel and soap, much to the merriment of the more fortunate ones.[43]

Only the better grades of cloth were fulled. In the early Middle Ages the laborious process of foot-treading or belaboring cloth with clubs gave way to the water mill. The turning waterwheel shaft, studded with projecting knobs or cams, alternately raised and let fall heavy

[42] The classic account of fulling mills in Europe, with primary reference to Britain, is E. M. Carus-Wilson, "An Industrial Revolution of the Thirteenth Century," *Economic History Review* 11(1941): 39–60. For fulling processes and mechanisms, see *Appleton's Dictionary of Machines, Mechanics, Engine-Work, and Engineering; Knight's American Mechanical Dictionary,* s.v. "Fulling"; Oliver Evans, *The Young Millwright and Miller's Guide,* various editions; *Eighty Years' Progress of the United States,* pp. 301 ff. See also manuscript schedules, federal censuses of manufactures, 1820 on, National Archives; 1850–1880, state libraries and archives; J. E. A. Smith, *History of Pittsfield, Massachusetts, 1800–1876* (Springfield, 1876), pp. 36 ff., 165 ff. On the superiority of fulled woolen cloth (especially when woven from machine-carded wool), see W. H. Parlin, *Historical Reminiscences of East Winthrop, Maine* (Waterville, 1902), pp. 6–7; and William B. Lapham, *History of Bethel . . . Maine, 1768–1890* (Augusta, 1891), pp. 383 ff. Of the total yardage of household-made cloth reported by the New York State census of 1822—some 10 million yards—linen, cotton, and other "thin" cloths accounted for 5.7 million yards; fulled woolen cloth, about 2.0 million yards; and flannel and other woolen cloths not fulled, 2.5 million yards (Horatio G. Spafford, *A Gazetteer of the State of New York* [Albany, 1824], p. 591).

[43] McKnight, *History of Northwestern Pennsylvania,* p. 481.

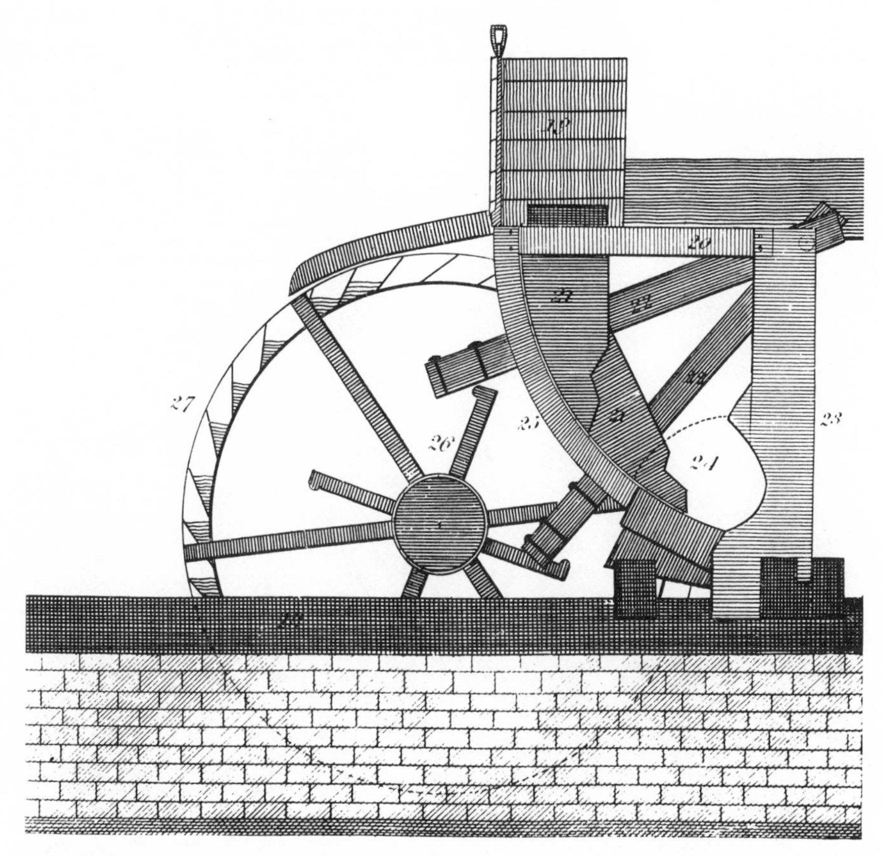

Fig. 8. Side view of a fulling mill. "The cloth is put in a loose heap in the stock *24;* the water being drawn on the wheel [*27*], the tappet-arms [*26*] lift the mallets [*21*], alternately, which strike the under part of the heap of cloth, and the upper part is continually falling over, and thereby turning and changing its position under the mallets, which are shaped in the figure to produce this effect." (Oliver Evans, *The Young Millwright and Miller's Guide* [Philadelphia, 1850].)

wooden stocks, which may still be seen (and heard) in some of the less developed parts of Europe.[44] Here was a major technical innovation, fully comparable in the relief afforded the domestic worker to the mechanization of spinning and weaving not effected until several centuries later.

Hardly less laborious was the preliminary process, long performed by hand with comblike cards, of untangling and straightening the wool fibers in preparation for spinning the yarn. So very slow and laborious

[44] See Louis C. Hunter, "The Living Past in the Appalachias of Europe: Water-Mills in Southern Europe," *Technology and Culture* 8 (1967):455–56 and figs. 4, 5, and 6.

Fig. 9. A water-powered fulling mill similar to that in Figure 8 but more crudely constructed. Photographed by the author in 1963 near Ioannina, Epirus, Greece.

was hand carding, according to North, that one pound of carded wool ready for spinning was a day's labor. Machine carding not only resulted in a great saving in this tedious labor, but the firm and uniform character of the carded "rolls" greatly facilitated the hand-spinning operation that followed.[45]

The carding machine was introduced in this country during the 1790s and during the succeeding decades spread rapidly, often as an adjunct of a fulling mill.[46] About 1810 a contemporary observer reported that "in the northern and middle states, carding machines, worked by water, are everywhere established and are rapidly being extended southward and westward . . . and as many fulling mills are erected as are required for finishing all the cloth woven in private families."[47] After 1800 carding mills were erected in almost every township in the Hudson-Mohawk region of New York, and by 1806 a Cincinnati newspaper carried an advertisement with a detailed description of a carding mill.[48] One authority held that the carding machine rather than the fulling mill was the real forerunner of the woolen mill, engaged in spinning the yarn and eventually in weaving the cloth as well. "The fulling mill and the carding mill," declared North, "were as essential to the economic complement of every considerable village locality as was the blacksmith's shop, the saw-mill and the gristmill." Fulling and carding in water-powered mills, a later census report noted, were chiefly a frontier phenomenon, "the pioneers of an advancing civilization."[49]

Tanneries

The tanning of leather was another operation early removed from the farmstead for processing with the aid of water- or horsepower. Made from the hides of domestic livestock, leather was an article of many applications

[45] S. N. D. North, "The New England Wool Manufacture," in William T. Davis, ed., *The New England States,* 4 vols. (Boston, 1897), 1:201–2. North's account of fulling and carding is particularly good. See also *Knight's American Mechanical Dictionary,* 1:467–71, for drawings of cards and carding machines.

[46] See Clark, *History of Manufactures,* 1:191, 423, 561. For emphasis on the role of John Schofield, see Smith, *History of Pittsfield,* pp. 164–68.

[47] Thomas P. Keystone, *Mirror for Americans: A Likeness of the Eastern Seaboard,* ed. Ralph H. Brown (Boston, 1935), p. 78.

[48] David M. Ellis, *Landlords and Farmers in the Hudson-Mohawk Region, 1790–1850* (Ithaca, N.Y., 1946), p. 111; Utter, *Frontier State,* p. 248.

[49] North, "The New England Wool Manufacture," pp. 201–2; U.S., Census Office, Eleventh Census, 1890, vol. 6, *Manufacturing Industries,* pt. 3, *Selected Industries* (Washington, D.C., 1895), pp. 13 ff.

Fig. 10. Carding machines of this type, either hand- or water-powered, were used for custom carding much the way local gristmills were used for grinding corn and wheat. (Abraham Rees, ed., *The Cyclopaedia,* 1st Am. ed., plates, vol. 4.)

and great utility around the farm and home. An early settler in northwestern Pennsylvania, 1811, recalled that

> the want of leather, after our first shoes were worn out, was severely felt. Neither tanner nor shoemaker lived in the country. But "necessity is the mother of invention." I made me a trough out of a big pine-tree, into which I put the hides of any cattle that dies among us. I used ashes for tanning them, instead of lime, and bear's grease for oil. The thickest served for sole leather, and the thinner ones, dressed with a drawing knife, for upper leather; and thus I made shoes for myself and neighbors.[50]

Leather production by tanning was a long, laborious, and disagreeable process, requiring bulky equipment and no little skill. Although for want of an alternative tanning was often conducted on the farm, the opportunity of escaping this chore, including the tedious grinding of the tanbark in a bark mill, was usually welcomed. The oak or chestnut bark commonly used as a tanning agent was reduced to a coarse powder in a mill consisting either of two millstones in the manner of gristmills or of a heavy stone wheel (often a discarded millstone) turning on its edge in a circular trough and producing when horse power was

[50] McKnight, *History of Northwestern Pennsylvania,* pp. 596–97.

Fig. 11. A Mexican bark mill. (*Harper's Magazine,* Oct. 1884. Photo courtesy of Charles W. Hughes, Thetford Historical Society, Thetford, Vt.)

used some half cord of ground bark daily. Cast-iron mills came into use in the early 1800s.[51] Tanneries offered a product superior to home-tanned leather, usually in exchange for a half (sometimes two-thirds) by weight of the raw hides to be processed.

Ranking next to gristmills and sawmills in number, tanneries long employed power only in the mills for preparing the tanbark. Of the tanneries reported in the manuscript schedules of the federal census of 1850 for five states, the number driven by waterpower, steam power, and horsepower, with value of product when available reported in thousands of dollars, was, respectively, as follows: Vermont, 93, 3, and 30 ($493, $36, and $59); New Hampshire, 96, 10, and 28; Massachusetts, 90, 41,

[51] Hemenway, ed., *Vermont Historical Gazetteer,* 3:150, 4:81; advertisement of the M'Clurg Foundry, Pittsburgh, in *Pittsburgh Commonwealth,* 2 Aug. 1809; Bishop, *History of American Manufactures,* 1:453–54; Utter, *Frontier State,* p. 237. Brief references to home tanning are in *History of Union County, Kentucky* (Evansville, Ind., 1886), p. 30; W. F. Rogers, "Life in East Tennessee near the End of the Eighteenth Century," *East Tennessee Historical Society Publications* 1 (1929):40–41.

and 40 ($887, $1,354, and $254); Maryland, 19, 17, and 67 ($573, $184, and $62); Kentucky, 1, 10, and 96.[52]

Public Support of Water Mills

To the testimony of contemporary witnesses and the proof of sheer numbers as evidence of the key role of water mills must be added the widespread practice of public encouragement and aid. Mill ventures in new and thinly settled districts were hardly an attractive investment to those seeking opportunities for gain.[53] It became common practice in much of New England and not unusual practice elsewhere for colonial and local authorities, such as towns or townships, to offer inducements to hasten the establishment of mill facilities. These facilities were desirable not only in themselves but were increasingly recognized as a means of attracting settlers and advancing community development and stimulating land values. Such inducements ranged from the granting of mill seats and adjoining land, the monopoly of milling rights for a stated period, and the exemption from taxation to outright money donations and exemption from military service.[54] Although rarely so common-

[52] See the description of tanning in *Andrew Ure's Dictionary of Arts, Manufactures, and Mines,* 2 vols. (New York, 1854), 2:54 ff., and articles on tanning, tanning apparatus, and tanning materials in *Knight's American Mechanical Dictionary*, 3:2489–94. Brief accounts are found in Clark, *History of Manufactures,* 1:167–68; and F. A. Michaux, *Travels to the Westward of the Alleghany Mountains* (London, 1803), reprinted in Thwaites, *Early Western Travels,* 3:[105]–306. For the importance of tanneries to rural communities, see John H. Moore, "The Textile Industry of the Old South as Described in a Letter by J. M. Wesson in 1858," *JEH* 13 (1956): 205. The data on power employed by tanneries in 1850 have been compiled from the manuscript schedules of the federal census, located in most instances in the libraries of the states cited. Microfilm copies of these schedules for manufacturing, 1850–80, are for many states available in the National Archives.

[53] Bishop, *History of American Manufactures,* 1:104, 126–27, stresses this point, as does Defebaugh, *Lumber Industry of America,* 2:271.

[54] Abundant evidence of community aid in the erection of gristmills and sawmills exists in general works, the monographic literature, and the vast body of town and county history. Among the more useful works are Roy H. Akagi, *The Town Proprietors of the New England Colonies* (Philadelphia, 1924), esp. pp. 88–92; Oscar and Mary Handlin, *Commonwealth: A Study of the Role of Government in the American Economy, 1774–1861* (New York, 1947), pp. 75–78, 110; A. C. Ford, *Colonial Precedents of Our National Land System* (Madison, Wis., 1910), p. 110; Kuhlmann, *Flour Milling Industry,* pp. 27 ff. For New England see also such local histories as Hemenway, ed., *Vermont Historical Gazetteer;* Williamson, *History of the State of Maine,* vol. 2; Ridlon, *Saco Valley Settlements;* Agnes Hannay, *A Chronicle of Industry on the Mill River* (Northampton, Mass., 1935–36), pp. 12–16.

place elsewhere as in New England, the course of such aid by public or private agency can be traced across the country.

Where the colonizing agencies did not anticipate it, they were soon made aware of the critical role of grain milling, and in less degree saw-milling, in the successful planting of settlements in America. In New England some early provision was made for the introduction of both gristmills and sawmills by the Courts of Assistance in London. Owing evidently to English inexperience, Dutch workers familiar with the art were sent over to build sawmills.[55] Very early the Virginia Company instructed the colony's governor to build water mills and blockhouses in every plantation. The injunction to erect mills led to a request that millwrights be sent, and mill irons and millstones were included among the essential supplies needed. Windmills and horse mills were introduced as well to meet the needs for meal. When the building of mills lagged behind settlement, the Virginia legislature in 1667 extended encouragement to potential owners in the form of assistance in securing mill seats.[56] Two years later Maryland also passed laws to promote the erection of water mills by facilitating the acquisition of millsites. Such aid chiefly took the form of statutory authority to the prospective millowner to secure condemnation of land belonging to others for the purpose of dam building and to permit the resulting flooding of upstream riparian lands. Occasionally aid was given to other types of mills, but gristmills were the most common beneficiaries.[57]

[55] Orra L. Stone, *History of Massachusetts Industries,* 4 vols. (Boston, 1930), 1:4–5; see also Weeden, *Economic and Social History,* 1:102 ff., 168–69, 198 ff., 306–10. Defebaugh cites numerous references to early aid to the establishment of mills by colonizers and colonial authorities (*Lumber Industry of America,* 2:21–28, 271 ff.). According to another tradition, Captain John Mason brought eight Danes to New England in the seventeenth century "to build mills, to saw timber and to make potash" (Ridlon, *Saco Valley Settlements,* p. 191).

[56] Bruce, *Economic History of Virginia,* 2:243, 487–89. For numerous references to early Virginia laws for encouragement and regulation of gristmills and other water mills, see indexes to William W. Hening, *The Statutes at Large; Being a Collection of All the Laws of Virginia . . . ,* 13 vols. (Richmond, 1809–23).

[57] Kuhlmann, *Flour-Milling Industry,* p. 28; Ford, *Colonial Precedents,* p. 109. Similar provisions are found in the digests of statutes for Alabama by John G. Aikin, for Kentucky by C. S. Morehead and Mason Brown, and for Tennessee by R. L. Caruthers and A. O. P. Nicholson. See also M. S. Heath, *Constructive Liberalism: Role of the State in Economic Development in Georgia to 1860* (Cambridge, Mass., 1954), p. 22; M. T. Thomson, "The Gristmill in Georgia," *Georgia Review* 7 (1953): 6–7; C. M. Brevard, *History of Florida* (Deland, 1924), p. 268; A. O. Tibbals, *A History of Pulaski County, Kentucky* (Bagdad, Ky., 1952), p. 69; and E. C. Barker, ed., "Minutes of the Ayuntamiento of San Felipe de Austin, 1828–1832," *Southwestern Historical Quarterly,* 23 (1919–

In some parts of New England, however, especially Maine and New Hampshire, sawmilling received much encouragement as a means of stimulating the lumber trade to the coastal towns and cities southward and the export trade to Britain and elsewhere.

The establishment of gristmills was one of the first concerns of the authorities directing the affairs of the pioneer Swedish settlements along the Delaware River in the 1640s. The proprietors of the Carolina colony of New Bern defended the expense of a gristmill and sawmill on the ground that "we were forced to it, for the people could not grind their corn." In Mississippi Territory, for want of suitable streams attention was directed to horse mills; the home authorities in France were urged to send mills suitable for such use, together with millstones.[58]

In New England to an extent not approached in most colonies to the south, assistance in the provision of mills passed from the provincial to the local, that is, the town authorities, once the colonies were securely established. The local histories of New England, especially town histories, contain innumerable references to the action of town proprietors in extending aid in the building of mills. Many examples are found in the five-volume *Vermont Historical Gazetteer*, issued serially as a magazine from 1867 to 1882. Frequently the reports of such action are emphasized by such statements as "the want of mills a serious evil"; "the deficiency of mill . . . inconsistent with the existence of civilized life"; here "the power of water was first employed to perform the labor and drudgery of civilization." In one instance the aid granted for the erection of a mill was supplemented by according the grantee the privilege of naming the settlement, which he did.

As an early historian of Norwich, Connecticut, explained:

Every enterprise which had any tendency to promote the public convenience was patronized by a grant of land. Hugh Amos, who first established a regular ferry over the Shetucket river, received one hundred acres by way of encourage-

---

20). A Maryland law of 1748 sought to encourage the production of flour for export by making grants of land to mill builders (Bishop, *History of American Manufactures*, 1:136–37).

[58] J. Thomas Scharf and Thompson Westcott, *History of Philadelphia*, 3 vols. (Philadelphia, 1884), 1:38, 130 ff., 150 ff.; George Smith, *History of Delaware County, Pennsylvania* (Philadelphia, 1862), pp. 38, 54–55; and Amandus Johnson, *The Instruction for Johan Printz, Governor of New Sweden* (Philadelphia, 1930), p. 34; Graffenreid, *Account of the Founding of New Bern*, pp. 61–62, 288, 314–15. Rowland and Sanders, eds., *Mississippi: Provincial Archives*, 2:310, 344, 465, 546, 558, 584, 591 ff., 620–23, 627.

ment. John Elderkin was repeatedly remunerated in this way for keeping the town mill. A blacksmith was induced to settle among them by a similar reward. A miller, a blacksmith, and a ferryman were important personages for the infant settlement. Saw mills met with the same liberal patronage.

In the last instance the inducement was 200 acres of land plus a mill privilege and a monopoly of sawmilling on the stream, but inaction brought a reversion of the privilege to the town. Similar efforts to obtain a fulling mill were unavailing, and the townsfolk lacked this important facility until one was eventually erected at public expense.[59]

It was not unusual for the town proprietors to have difficulty in finding someone to build the desired mill or mills or for the recipient of a grant to fail to build or complete the mill. In certain cases several years passed between the first action and the eventual securing of the mill.[60] In the case of Waterbury, Connecticut, the time elapsing between the initial offer of town aid and the actual completion of the mill was nearly a quarter century—from 1679 to 1703.[61] The statement of the settlers of Andover, New Hampshire, in November 1767 complains to the proprietors of their backwardness in promoting the settlement of the town and failure to live up to promises: "They clear us no rodes build us no bridges." The promised sawmill had not been completed, but "we suffer more abundantly for want of a grist mill we have been forced to go twenty mils to mill this year with a teem and men."[62]

Occasionally members of the community contributed labor, as in digging the millrace or building the milldam.[63] "The mill site," declared

[59] Frances M. Caulkins, *History of Norwich, Connecticut, 1660–1845* (Norwich, 1845), pp. 59–60.

[60] See, for example, William Little, *The History of Weare, New Hampshire, 1735–1888* (Lowell, Mass., 1888), pp. 72 ff., 87–89.

[61] Henry Bronson, *History of Waterbury, Connecticut,* 1:80. See also William A. Wallace, *History of Canaan, New Hampshire* (Concord, 1910), pp. 23 ff., for an account of repeated efforts to secure construction of a much desired mill.

[62] John R. Eastman, *History of the Town of Andover, New Hampshire, 1751–1906* (Concord, 1910), pp. 9–10. See also Ford, *Colonial Precedents,* p. 111; Bishop, *History of American Manufactures,* 1:104; Guillet, *Early Life in Upper Canada,* p. 247; Ray A. Billington, *Westward Expansion* (New York, 1960), pp. 218–19; D. Doggett, "Water-powered Mills of Flat Rock River," *Indiana Magazine of History* 32 (1936): 319–59; Barker, ed., "Minutes of the Ayuntamiento."

[63] Forms of group or cooperative ownership of gristmills have not been uncommon in parts of eastern Europe and doubtless elsewhere abroad (correspondence and conversations with Professor Albert Struna of Ljubljana, Yugoslavia, author of *Vodni Pogoni na Slovenskem* [Water Power in Slovenia] [Ljubljana, 1955] and Dr. Valeriu Butura, Cluj, Romania). Such practice seems to have been unusual in this country.

the Handlins, writing of colonial Massachusetts, "has always been a subject of communal concern"; to secure saw- and gristmills, "towns and proprietors had granted away the privilege of exploiting the water power, often adding other inducements as well." As early as the 1630s in what was to become Maine such proprietors as Thomas Eyre, John Mason, and Ferdinando Gorges actively promoted the establishment of sawmills as a means of advancing settlement and securing an export commodity.[64] Of New England generally, R. H. Akagi noted that "the grant by townships by the general court in different colonies always included the benefit of streams and thus arose a curious custom of 'water-rights' which the proprietors invariably claimed . . . [and] freely granted to themselves or to others with the specific purpose of building mills. . . . Together with the water right, the proprietors, in order to encourage the building of mills, always granted a necessary tract of land and often a gratuity of money."[65]

In the case of Massachusetts especially, the practice of making town grants of mill privileges and land to encourage the erection of mills was a common one. We have a fairly detailed record of the town-supported gristmill at Brookfield, sixty miles west of Boston, showing the method of contracting for major components of the mill with a breakdown of costs,

---

The single example to come to my attention was the agreement in 1768 of some twenty inhabitants of Easthampton, Long Island, to erect a horse mill to grind meal for their mutual benefit (*Records, Town of East-Hampton, Long Island* [Sag Harbor, N.Y., 1887], p. 423). Although the gristmill in one Maine community may not have been owned in common, for several years no miller was in attendance and each family tended to the grinding of its own meal (Lapham, *History of Bethel . . . Maine,* pp. 64, 311). See also Hugh D. McLellan, *History of Gorham, Me.* (Portland, Maine, 1903), pp. 256–57. For voluntary contributions of labor and materials, see W. H. Parlin, *Historical Reminiscences of East Winthrop, Maine,* p. 9; E. W. Barber, "Beginnings in Eaton County," *Historical Collections of the Pioneer and Historical Society of Michigan* 29 (1901): 363; *History of Warren County, Ohio* (Chicago, 1881), p. 613; and N. N. Hill, comp., *History of Coshocton County, Ohio* (Newark, 1881), p. 547. In the case of Tecumseh, Michigan, the settlers in 1826 agreed to contribute $200 toward the cost of a gristmill in course of erection (J. J. Hogaboam, *The Bean Creek Valley* [Hudson, Mich., 1876], p. 14).

[64] Handlin and Handlin, *Commonwealth,* pp. 12–16. Richard M. Candee, "Merchant and Millwright: The Water-powered Sawmills of the Piscataqua," *Old-Time New England* 60 (1969–70): 131 ff. Candee's article and the companion one by Benno M. Forman, "Mill Sawing in Seventeenth Century Massachusetts," pp. 110 ff., are basic documents in the founding of the lumber industry in the United States, presenting a rich body of data extending to lengthy appendixes.

[65] Akagi, *Town Proprietors of the New England Colonies,* pp. 88 ff.; see also Josiah G. Holland, *History of Western Massachusetts,* 2 vols. (Springfield, 1855), 2:18, 42–43, 217–18, 315, 329, 337, 350.

1669–70. The figures on ironwork, millwork, and man-days of employment in construction indicate that this one-run gristmill was of substantial construction and considerable outlay and represented a major effort for a small community. Millstones, ironwork, and mechanical assembly were contracted for at Northampton and Springfield, some thirty to forty miles to the west. Their cost, in rounded figures, was £21, £18, and £23, respectively, bringing the total outlay for machinery to £62, or about three-fourths of the completed mill cost of £81. Local materials and skills were relied upon for the less demanding tasks of erecting the mill building and constructing the dam and raceways; these totaled a few shillings less than £20, at a labor cost per man-day of 2 shillings fourpence. Clearly a mill representing some 600 man-days of labor was no light undertaking and in this case required not only public assistance but contracting outside for the more demanding skills and costly materials.[66]

More important than town grants in the long run were the "mill acts" adopted in one form or another, with Massachusetts taking the lead in New England with its mill act of 1714. This act granted to owners and projectors of water mills the very important (and valuable) right to erect dams, regardless of the wishes of upstream riparian owners objecting to the flooding of and injury to their lands. Such owners had to be content with payment of damages by the millowners, as assessed by a local jury. The mill acts or their legal equivalent reflected a consensus, among legislators at least, that the application of waterpower in mills was not only profitable to the owners but "vastly beneficial to the public."[67]

[66] Louis E. Roy, *Quaboag Plantation, Alias Brookfield: A Seventeenth Century Massachusetts Town* (West Brookfield, 1965), pp. 125–33. Since charges for the mill were given in the account books as town debts, Roy concludes that it was built at the town's expense. However, a grant of seven to eight acres was made to one John Pynchon for the mill. In 1670 Pynchon bought out the three "principal investors," whose combined interests came to about £55. The value of the pound in seventeenth-century Massachusetts is suggested by the description of the salary of Dedham's first minister, £60–80, as "a handsome sum in a society in which the total estate of 500 pounds meant wealth." In 1670 an estate valued at £80 was required to qualify for the franchise, yet by 1678 only a fourth of the male taxpayers in Dedham could meet this minimum (Kenneth A. Lockridge, *A New England Town, the First Hundred Years: Dedham, Massachusetts, 1636–1736* [New York, 1970], pp. 32, 48). Comparative figures are not very helpful owing to variations in capacity and construction. A sawmill built at Salem, Massachusetts, 1689, according to an itemized account of expenses, cost £53, less several shillings (Forman, "Mill Sawing," p. 113).

[67] Joseph K. Angell, *A Treatise on the Common Law, in Relation to Water-Courses,* 5th ed. rev. (Boston, 1854), p. 93.

The public character of water mills was thus widely recognized. As early as 1839 a term that a century later was to have general currency was applied to them. In his *History of the State of Maine* (1839), W. D. Williamson declared that "mills were uniformly considered as being of public utility, and their owners the objects of particular favor."[68] In many states it was necessary to obtain the approval of public authority, usually the county court, for the erection of milldams; but except where navigation was involved this procedure was evidently pro forma. It was more than offset by the readiness of the state—see, for example, the statutes of Virginia, Alabama, Kentucky, and Tennessee—to extend its powers of condemnation on behalf of a petitioning millowner to secure control for dam-building purposes of the opposite bank of a millstream owned by another and reluctant party. And as with the mill act of Massachusetts, the millowner within certain limits was given the right to flood upstream riparian lands regardless of the wishes of their owners. In both instances the injured parties had to be paid damages, but these were to be assessed by a local jury and not left to individual determination or the outcome of litigation. In most states, too, the gristmill took on a public character whenever it undertook, as long was the general practice, to grind for toll. Typically it became subject to regulation in such matters as the amount of tolls charged and in the use of proper weights and measures.

Community aids to the erection of mills continued into the nineteenth century and followed settlement into the trans-Appalachian West. Among the colonizing and land companies providing for or encouraging the erection of mills were the Connecticut Land Company in its Western Reserve lands and the Ohio Company in its Marietta settlement. A frontier historian reported that the New England Emigrant Aid Company was active in bringing both grist- and sawmills to the settlers in Kansas. He cites a Kansan to the effect that the arrival of an Emigrant Aid mill at Manhattan from Lawrence, drawn by twenty oxen, "was a greater event to the citizens in 1859 than the arrival of the Union Pacific Railway eight years later."[69] Yet the growing awareness of the

[68] Williamson, *History of the State of Maine,* 2:73. An account of Kansas during the 1850s and 1860s described the first gristmills "as a necessity to society rather than as a manufacturing enterprise for profit" (L. A. Fitz, "The Development of the Milling Industry in Kansas," pp. 53–54). On the later extension of the principles of the mill acts to other manufactures in Rhode Island, see Peter J. Coleman, *The Transformation of Rhode Island, 1790–1860* (Providence, 1963), p. 77.

[69] Dick, *Sod-House Frontier,* p. 492; William G. Rose, *Cleveland: The Making of a City* (Cleveland, 1950), pp. 23–24, 29. Also Billington, *Westward Expansion,* p. 218.

value of water privileges or mill seats and the eagerness to obtain and exploit them increasingly made such subventions unnecessary.

Numerous attempts to persuade the federal government to recognize the important role of water mills in land settlement by providing assistance in their erection found little response in Congress. For example, some sixty inhabitants of Hamilton County, Ohio, in 1800 unsuccessfully petitioned Congress for the grant of a half section of land in aid of a gristmill, since the lack of one was their greatest inconvenience. Congress did intervene in May 1800 to the extent of granting preemption rights to owners of gristmills and sawmills, but despite petitionary pressure refused in later legislation to continue this practice. An adverse congressional committee report in 1824 declared that the application of the preemption principle to improvers of mill seats on the public domain would cause "these valuable appendages [to be] universally taken up and occupied to the great injury of the government . . . and with no advantage to land purchasers except the favored individuals."[70]

With the widespread awareness of the effect of land settlement in stimulating land values, mill seats came to be regarded more as opportunities for speculative gain than as occasions for public subsidy. As an early historian of Maine remarked, grants of community aid to secure mills gave way in time to commercial goals; by 1785 lands in eastern Maine were in great demand: "there was a passion for obtaining settlers' lots, mill-sites and water privileges."[71] A few years later a traveler from France noted the practice of Pennsylvania land speculators of deceiving land buyers "with the big words, 'mill-seats,' 'timber,' et cetera."[72] With migration up the Connecticut Valley into Vermont, about 1800, timber lands and waterpower sites were a special attraction to millwrights; and later when the restless Vermonters headed westward to New York State and beyond, "they spotted the mill-sites, the town-sites and the best stands of timber and bought them up . . .

[70] Heiser, *Hamilton in the Making,* pp. 91–93; *Statutes at Large, The United States,* vol. 2, sec. 16, p. 78; *American State Papers, Public Lands,* 2:876, 3:720. One writer states that preemption rights were granted in Indiana Territory to men establishing grist- or sawmills (Doggett, "Water-powered Mills of Flat Rock River," p. 322). An exception was made in respect to the Indians, the government undertaking to erect mills for use of tribes concerned in specific negotiations (*American State Papers, Indian Affairs,* pt. 2, pp. 133, 170, 679). See also Thwaites, *Early Western Travels,* 16:277–78, 27:137–40, 148; and Angie Debo, *The Rise and Fall of the Choctaw Republic* (Norman, Okla., 1934), pp. 10–11, 112.

[71] Williamson, *History of the State of Maine* (1832), cited in Smith, *Power Policy of Maine,* pp. 25 ff.

[72] Rayner W. Kelsey, ed., *Casenove's Journal, 1794* (Haverford, Pa., 1922), p. 66.

[while] still cheap." By the mid-1830s, declared George S. Gibb, New England speculators "had come to place preposterous values on every creek and stream which could conceivably turn a water wheel."[73] In these same years a climax of sorts was reached in the rush of settlement into southern Michigan. Townsite developers were quick to recognize and occupy sites in which the advantages of mill seats were combined with those of proximity to good agricultural lands, timber, and lines of communication. By erecting or encouraging the building of gristmills and sawmills, they attracted agricultural settlement, encouraged trade, and provided opportunities for manufacturing on a modest scale.[74] Mill seats of the traditional sort were shortly eclipsed in the older states by speculative operations in waterpowers with a potential for large industrial ventures, with eastern capital, as usual, playing a leading role in their location and development.[75]

Growth in Numbers

To the testimony of community action at the time and of later reminiscence must be added the impressive evidence of numbers as reported in state and federal enumerations. For 1820–25 there are quite detailed reports for the four eastern states, as shown in Table 1. These figures, in view of the imperfections of census taking, are probably

[73] Stilwell, *Migration from Vermont,* pp. 118, 141; George S. Gibb, *The Saco-Lowell Shops: Textile Machinery Building in New England* (Cambridge, Mass., 1950), pp. 20–21, 110–11. A generation later the same process was repeated in Oregon (A. L. Throckmorton, *Oregon Argonauts* [Portland, 1961], pp. 113 ff.). See also R. P. Sweiringa, *Pioneers and Profits: Speculation on the Iowa Frontier* (Ames, 1968).

[74] See George N. Fuller, *Economic and Social Beginnings of Michigan* (Lansing, 1916), and Hogaboam, *The Bean Creek Valley,* pp. 11 ff., 33 ff., 47 ff., 67ff.

[75] On Norwich, Connecticut, see U.S., Census Office, Tenth Census, 1880, vols. 16–17, *Reports on the Water-Power of the United States* (Washington, D.C., 1885, 1887), pt. 1, pp. 202, 186; on the Rapids of the Saint Louis River (Lake Superior), ibid., pt. 2, pp. 89 ff.; on Great Falls of the Potomac River, Alexander Mackay, *The Western World; or, Travels in the United States in 1846–47,* 2 vols. (Philadelphia, 1849), 1:241; on Rochester, N.Y., Blake McKelvey, *Rochester, the Water-Power City, 1812–1854* (Cambridge, Mass., 1945), pp. 17 ff.; on Niagara Falls, Edward D. Adams, *Niagara Power: History of the Niagara Falls Power Company, 1886–1918,* 2 vols. (Niagara Falls, 1927), 1: chap. 2; in the South, Coleman, *The Transformation of Rhode Island,* p. 297. For the Rhode Island and adjoining Massachusetts textile region, see also J. Herbert Burgy, *The New England Cotton Textile Industry* (Baltimore, 1932), pp. 15–16; and *Proceedings of the General Convention of Agriculturalists and Manufacturers . . .* (Harrisburg, Pa., 1827), pp. 59–60.

Table 1. Water mills in four eastern states, 1820–25

	Maine	Vermont	New Hampshire	New York
Gristmills	524	373	697	2,140
Sawmills	746	786	964	4,321
Fulling mills	149	252	262[a]	993
Carding machines	210	216	251	1,235
Tanneries	248	275	330[b]	1,000[c]
Totals	1,877	1,902	2,504[d]	9,689
Population	298,335	235,981	244,161	1,372,812
Inhabitants per mill	160	124	98	142

SOURCES: William D. Williamson, *The History of the State of Maine* 2 vols. (Hallowell, 1839), 2:702–3; Zadock Thompson, *A Gazetteer of the State of Vermont* (Montpelier, 1824), p. 305 ff., see also p. 14; John Farmer and Jacob B. Moore, *A Gazetteer of the State of New Hampshire* (Concord, 1823), pp. 49–64; and Horatio G. Spafford, *A Gazetteer of the State of New York* (Albany, 1824), p. 591.

[a] Here listed as "clothing mills."

[b] 193 bark mills are listed separately.

[c] Estimated.

[d] Totals given elsewhere vary slightly.

an understatement. In sheer numbers the leading classes of water mills, lumping country and commercial mills together, led all other kinds of manufactories of the day, shops of craftsmen excluded, by a wide margin. At one end of the industrial spectrum they merged into such other rural-sited but business-oriented enterprises as ironworks, saltworks, paper mills, and textile mills. At the other end they were closely identified with the household manufactures in whose needs they had their origin. The first comprehensive federal census enumeration of water mills in 1840 reported some 66,000 of the common sorts of mills, distributed regionally as shown in Table 2.

The importance of these tens of thousands of water mills is not to be belittled by the circumstances of their small size and capacity, averaging no more than several horsepower and operated typically by no more than one or two men. Their prime merit indeed was in their small size and capacity, a condition of their wide distribution and their location in the midst of the communities served. In the aggregate these mills bulked large in the industrial facilities of the preindustrial age. Adding the

Table 2. Common water mills in the United States, 1840

	Population	Grist-mills	Saw-mills	Fulling mills	Tan-neries	Ratio of Mills to population[a] All mills	Ratio of Mills to population[a] Grist & saw
New England states[b]	2,234,822	2,525	5,469	951	1,486	1:215	1:280
Middle Atlantic states	5,118,076	5,399	12,896	1,327	2,733	1:230	1:280
Southern states	5,578,885	10,612	6,568	52	2,031	1:290	1:325
Western states	4,131,370	5,125	6,717	255	1,979	1:295	1:350
Totals	17,063,153	23,661	31,650	2,585	8,229	1:245	1:310

SOURCE: U.S., Census Office, Sixth Census, 1840, *Compendium of the Enumeration of the Inhabitants and Statistics of the United States* (Washington, D.C., 1841), pp. 360–64.
NOTE: In addition to the gristmills there were 4,364 flouring mills. No figures on carding mills were given; many if not all may be included with fulling mills. It is possible that these numbers are understated, owing to remoteness and isolation and to the exclusion of the smaller tub mills, serving only one family or a few neighbors in the manner persisting in the southern Appalachians into the twentieth century.
[a] Rounded to nearest 5.
[b] New England includes Vermont, New Hampshire, Maine, Massachusetts, Connecticut, and Rhode Island; Middle Atlantic includes New York, New Jersey, Pennsylvania, Delaware, Maryland, and District of Columbia; Southern includes Virginia, North and South Carolina, Georgia, Alabama, Mississippi, Louisiana, Tennessee, Arkansas, and Florida Territory; Western includes Kentucky, Ohio, Indiana, Illinois, Missouri, Michigan, Wisconsin and Iowa Territories.

several thousand "flouring mills" to the number of the commoner sorts tabulated above gives a total number of some 71,000 mills employing some 87,000 hands, subject to the errors of enumeration, which were probably on the side of omission rather than exaggeration. On the level of employment—the only feasible basis of comparison with the census data available—water mills in the aggregate stood somewhat higher than cotton textiles, the leading branch of manufactures, which had undergone industrialization by 1840. The 1,369 cotton textile mills of all kinds as reported by the 1840 census employed some 72,000 hands. Next was the iron industry with an employment in all branches of over 30,000, followed by woolen manufactures (excluding fulling mills) with 16,000 hands, and then by machinery, hardware, cutlery, and firearms with a combined total of some 20,000 hands.[76]

The several kinds of water mills reviewed here, engaged chiefly in processing farm and domestic products, virtually exhausted the feasible applications of mechanization in the multifarious operations of subsistence farming. In such major activities as clearing the land, cultivating and harvesting crops, erecting and maintaining buildings, fences, and home furniture (rough lumber apart), and providing the large and indispensable supply of domestic fuel, water mills could play no part.[77] However eager to secure the services of such mills, however welcome the relief from the burdensome labor and coarse quality of the handmade products, the contributions of these mills to the farm economies of the communities served were supplemental and quite modest. An evident delight in the harnessing of a natural force for the relief of weary muscles and in the automatic operations of these ingenious mills tended, perhaps, both to exaggerate their importance and, viewed in retrospect, to obscure the vastly more important prime movers that played an indispensable role in nineteenth-century American farming. Such were the draft animals, chiefly oxen and horses, without whose aid farming advance would have had to take the improbable labor-intensive course followed, to cite an extreme case, in such countries as China and Japan.[78]

Horses alone, in the federal census of 1840, topped the 4 million mark and together with oxen of probably equal number were in the service of some 3,700,000 persons employed in agriculture.[79] Oxen, according to Clarence Danhof, "were the typical work animals of pre-commercial farming." Easily raised, they reached useful maturity early, were hardy and easily trained, and when no longer workable, were sold or slaughtered for food with little if any loss. They were especially suitable to the needs

[76] *Compendium of the Enumeration of the Sixth Census,* 360–64. According to the census of 1860 the number of employees for the flour and meal and lumber industries combined was, rounded, 96,000; for the cotton and woolen goods industries combined, 156,000. Value of products for the two groups of industries was $342 million and $168 million, respectively, out of a total, all manufactured products, of $1,886,000,000 (Eighth Census, *Manufactures of the United States in 1860* pp. 733–42).

[77] On the cost in man-hours to provide fuel, see R. V. Reynolds and A. H. Pierson, *Fuel Wood Used in the United States, 1630–1930,* Department of Agriculture Circular no. 641 (Washington, D.C., 1942). See also Lura Beam, *A Maine Hamlet* (New York, 1957), pp. 71–72, and reports of wood fuel consumption in the 1820 federal census schedules for Maine, the National Archives.

[78] See, e.g., the report of an American agronomist, Franklin H. King, *Farmers of Forty Centuries; or Permanent Agriculture in China, Korea, and Japan* (Madison, Wis., 1911).

[79] *Compendium of the Enumeration of the Sixth Census,* p. 359. Oxen are not separated from the nearly 15 million "neat cattle" reported for agriculture.

of a pioneering economy, "pulling heavy, inefficient plows and drawing heavy loads over poor roads," and provided low-cost animal power well adapted to the conditions and requirements of subsistence farming under frontier conditions. Their slow pace favored use when fields were small, land clearance a major task, and transport requirements in marketing produce minimal. Commercial farming was accompanied by the replacement of oxen by horses, whose higher cost, greater requirements in food and care, and unacceptability as food when redundant were more than offset by their quicker pace in plowing and transport.[80]

The roles of water mills and draft animals in the farm economy reflected widely disparate needs and radically different methods of application that hardly bear direct comparison. Horse and oxen could be, and frequently were, harnessed to grain and other mills, as will be discussed in a later volume. Yet, within the scope of prevailing technologies, by no stretch of the imagination could water mills plow fields. In sheer number of units and quantities of energy developed and applied, the gap between these two motive powers was immense and noteworthy. At this point it is sufficient to note that the predominantly farm population of the country obtained assistance from not only tens of thousands of water-powered mills but from work animals rising in number over the years 1850-1900 from five to over fifteen million. During this half century, when rural population a little more than doubled, the horsepower capacity of steam engines and waterwheels in manufacturing and mining combined rose from one million to ten million.[81] Further relief had to come from the processes of economic growth and the extension of the agencies of trade, transportation, and manufacturing functioning within a market economy. The applications of mechanization and power would become increasingly remote from the agricultural communities served. These in turn could escape the limitations of a subsistence livelihood only to the extent they could obtain outlets for their produce.

[80] Clarence H. Danhof, *Change in Agriculture, the Northern United States, 1820–1870* (Cambridge, Mass, 1969), pp. 141–44.

[81] Carroll R. Daugherty, "The Development of Horsepower Equipment in the United States," in *Power Capacity and Production in the United States,* U.S. Geological Survey Water-Supply Paper no. 579 (Washington, D.C., 1928), pp. 45–47. See also Sam H. Schurr and Bruce C. Netschert, *Energy in the American Economy, 1850–1975* (Baltimore, 1960), pp. 54–55; J. Frederic Dewhurst, *America's Needs and Resources: A New Survey* (New York, 1955), pp. 1116 ff.

Mills and the Market Economy

The water mills that long played a central role not only on the frontiers of settlement but in most farming communities often became the entering wedge of a slowly emerging market economy. Early millowners were welcomed in the first instance and frequently became persons of prominence in the community. Yet they often obtained only partial support from milling and, like most others with by-occupations in the community, lived more or less by the land, no doubt bartering much of the mill tolls for other needed commodities or services. Sooner or later the point was reached where, transport conditions permitting, market outlets were sought for the mounting surplus of a growing community. It was a short and natural step for the millowner to supplement custom operations by the purchase of raw materials to process and market on his own account. With the addition of bolting equipment, grist, or "country," mills became "merchant" mills, producing flour to meet market demands while continuing to convert grain into meal for the custom trade, or "country work"—meal being the unbolted product direct from grinding between the millstones of the country mill and flour the bolted and refined product of the commercial mill. Sawmill operators added saws, increased power and capacity, and, as opportunity offered, added planing and shingle machines, engaged in logging their own woodland, or secured timber by purchase. Fulling mills often added carding machines or the finishing operations known as cloth-dressing. In time they might expand into other textile operations—spinning or weaving—beginning with the carded wool rolls or the "country" cloth received as toll for their services. In such manner growth and development proceeded in the New World much as in the Old, though in the postcolonial years probably at a somewhat faster pace.[82] Eventually the more specialized types of mills and manufactories arose under favorable conditions. Such were the mills engaged in the production of paper, powder, oil and textiles, iron forges, blast furnaces, foundries, saltworks, sugar refineries, and the like. These stood quite apart from the basic grist-, saw- and fulling mills, requiring

[82] This process may be followed in innumerable town and county histories giving attention to economic development; see, e.g., E. C. Smith and P. M. Smith, *A History of the Town of Middlefield, Massachusetts* (Menasha, Wis., 1924), and John B. Armstrong, *Factory under the Elms: A History of Harrisville, New Hampshire, 1774–1969* (Cambridge, Mass., 1969), chaps. 2, 6. See also references in Handlin and Handlin, *Commonwealth,* pp. 131, 195–96, to the role of local shopkeepers, merchants, and other small capitalists in this process. See also Percy W. Bidwell, "Rural Economy in New England at the Beginning of the Nineteenth Century," *Transactions of the Connecticut Academy of Arts and Sciences* 20 (1916):241–399.

more specialized equipment and skills, serving wider territories, and possessing little more in common than a willingness in areas where cash was scarce to take country produce in return for their wares—which in turn were often paid out in lieu of wages—for want of an adequate outlet by outright sale.

In the advance toward commercialization gristmills and sawmills before 1850 followed different courses in the regions of their earliest and most extensive development along the north Atlantic seaboard. Almost from the beginning of colonial settlement sawmilling was established on a commercial basis. In extensive regions in the Northeast commercial mills without doubt predominated in numbers and output over the small custom mills serving adjacent farm communities. Gristmilling, on the other hand, made little headway in commercialization before the middle of the eighteenth century. The development of "flouring mills," or flour milling, as this branch of the industry came to be known, awaited development of a surplus of grain and its economical transport to seaboard mills and markets. The demands of urban and overseas markets called for the refined product of flour, and this in turn required the additional milling processes of cleaning the grain, bolting the meal to remove objectionable portions, and cooling and packing the flour product. For generations to come the great majority of the farming and rural population were content with the whole meal made from their own grain, and the miller was content with his custom toll.

The situation of milling in the production of lumber was in important respects just the reverse. The relative ease with which the generally available timber could be converted into rude log shelters by the settlers themselves delayed for a time any pressing concern with securing lumber and mills for its production. At the same time, the vast forest cover extending to the water's edge along coastline and estuaries provided a seemingly inexhaustible crop ready for the harvest. Nature not only supplied the timber, but in northern latitudes the snow cover permitted ready sledging of logs to nearby streams. There they awaited spring freshets to carry them to the mills below—mills driven by the same water that would carry lumber-laden rafts or boats to the seacoast, where the lumber would be shipped to domestic and foreign markets. In regions like colonial New Hampshire and Maine sawmilling was often in the van of settlement. As the nearby timber along the shores and lower stream reaches was exhausted, the sawmilling industry moved up the rivers, taking out the best and most accessible timber, in time advancing from the main stream up tributary streams, occupying the most advantageous millsites. Sawmills multiplied in number and increased in size and capacity as

Fig. 12. Sawmill at Jefferson Mills, N.H., 1892. (Van Name Collection, Library of Congress.)

market conditions warranted. The pattern established in northern New England was, with variations in detail, applied in other parts of the country, if not in advance, on the heels of settlement. Small mills were everywhere built to meet local demand, frequently several within the same town to offset the difficulties and cost of overland hauling.[83]

[83] The foregoing discussion is based largely on two sources: the massive documentation of the increase and spread of sawmilling operations, state by state and chronologically, by Defebaugh (editor of the *American Lumberman,* a trade periodical), *Lumber Industry of America,* vol. 2. This study covers the Middle Atlantic and New England states from Pennsylvania and New Jersey to Maine, New Hampshire, and Vermont. The other source is articles in *Old-Time New England* (1969–70) by two young scholars: Forman, "Mill Sawing in Seventeenth Century Massachusetts," pp. 110–30, and Candee, "Merchant and Millwright: The Water-powered Sawmills

Water Mills in Western Europe

In its long-continued reliance on water mills this country was in most respects simply following the pattern of the western European world and adopting the essential features of the traditional technology. In the absence of census counts the overall number of such mills in Britain is a matter for conjecture. Rural England continued to rely largely on small mills for meal and flour well toward the end of the nineteenth century. The inadequacy of the supply of water mills brought into widespread use thousands of windmills, by one estimate some 5,000 as late as 1820. Sawmills made far slower progress than in the United States.[84] In France physiographic conditions were much more favorable to waterpower than in Britain, as was also the scarcity of cheap fuel in many parts of the country. At mid-century water mills *(moulins à eaux)* were much more numerous in France in proportion to area than they were in the United States, although in relation to population French mills were far fewer. In 1847 France counted some 37,000 water mills of all kinds and capacities together with some 8,700 windmills.[85] Flour and grist mills comprised about 80 percent of the water mills and more than 90 percent of the windmills.[86] The pattern of milling was not very different in Germany. Of the 54,000 gristmills in the German Reich in 1875, water mills accounted for over 60 percent, windmills for over 30 percent, and steam mills for most of the remainder.[87] Thus as the overseas extension of Europe to the west, the American colonies and nation shared a community of ex-

of the Piscataqua," pp. 131–49. The articles give particular attention to technology and sources of capital and identify and list sawmills in Massachusetts and western Maine.

[84] See the discussion of windmills in a companion volume in preparation. On British resistance to the sawmill and for the modes of use of whipsaw and pit saw, see S. W. Worssam, *On Mechanical Saws* (London, 1868), pp. 4–5; Johann Beckmann, *A History of Inventions, Discoveries, and Origins,* 4th ed. rev., 2 vols. (London, 1846), 1:220–30; and sources cited in notes 30 and 38 above.

[85] Le Ministère de l'agriculture et du commerce, *Statistique de France,* pt. 4, *Industrie,* 4 vols. (Paris, 1847), 1:357, 2:294–95, 3:450–51, 4:274–75. Fifty years later the number had not greatly changed. Of more than 46,000 manufacturing establishments in 1899 with an aggregate of nearly 500,000 horsepower, grist-, saw-, and oil mills accounted for more than 38,000 establishments and some 280,000 horsepower (France, Direction du travail, *Repartition des forces motrices à vapeur et hydrauliques en 1899,* 2 vols. [Paris, 1901], 2:vii–xx).

[86] *Annuaire officiel de la meunerie française . . . 1954* (Paris, 1955), pp. 653–54. See also Hunter, "The Living Past in the Appalachias of Europe," pp. 446–47.

[87] Ludwig Hollander, *Die Lage der deutschen Mühlenindustrie unter dem Einfluss der Handels Politik, 1879–1897* (Stuttgart, 1898), pp. 4–5.

perience with and benefits from waterpower, the leading source of mechanical power on both sides of the Atlantic as late as 1850.

The American experience had much in common with that of the Old World in that in both instances water mills were identified with the conditions and needs of a subsistence agriculture. But there were important differences, economic and institutional. The role of water mills in this country, as we have seen, reflected in part the special hardships of life on the frontiers of settlement, where to the struggle for a livelihood under exceptionally difficult conditions was added the heavy burden of farm building with the immense initial task of clearing timber from the land. There were offsetting advantages of which the most important and fundamental were the vast abundance of land and the corresponding scarcity of labor. It was necessary to offer strong inducements to attract the settlers essential to develop resources and to increase land values, the twin goals of colonizers and land operators. This reversal of the Old World conditions of land scarcity and labor abundance effectively discouraged the extension to the English colonies of the manorial system with its subservient tenantry and the numerous rights and privileges of the lord, including what the French termed the *banalités*.[88] These included a monopoly of such basic community facilities as water mills, bake ovens, wine presses, and the like, of which the grain mill, in the words of Marc Bloch, was "probably the most ancient and certainly the most widespread." Not only did the lord have a monopoly of all water rights, mill seats, and the erection and operation of mills, but the tenants were required to make use, with proper payment, of the mill and were forbidden the use of hand mills. So deep did peasant resentment of the milling monopoly become that it served as a major grievance against the manorial system. The possession and use of hand mills, that abomination of the American frontier community and badge of backwardness, served as a principal means of evading the use of the lord's mill. Its use became a continuing source of controversy and conflict in parts of France, Germany, and Britain. Confiscated repeatedly, the hand mills were on occasion recaptured in minor uprisings and put into defiant use.[89]

In colonial North America the manorial system found a secure place in only two regions: the extensive domains of French Canada and the lower Hudson Valley of New York. The transplantation of the seignorial

[88] The discussion which follows is based particularly on Marc Bloch, *Land and Work in Medieval Europe: Selected Papers,* trans. J. E. Anderson (New York, 1969), pp. 136–68. The introductory chapters in W. B. Munro, *The Seigniorial System in Canada* (New York, 1907), provide an excellent account with special reference to France.

[89] Bloch, *Land and Work*, pp. 153–59.

system to Canada, as W. B. Munro noted in his careful study, purged a declining institution of its most odious features and gave it a new lease on life. In the short run even the hated *banalité* of milling rights demonstrated a certain usefulness. Under French Canada's formidable conditions of climate and physiography and slow increase of settlement, the provision and maintenance of mills would have imposed at least as heavy a burden on the settlers as that faced on the English frontiers. The high cost of importing essential parts and of building and operating mills often proved more of a burden than a benefit to the seigniors, and one consequently often evaded. The seigniors frequently sought to compensate for high costs and lack of capital by exacting illegally high tolls, cheating on measures, and cutting the quality of meal, forcing the provincial government to intervene on behalf of the settlers, moderating toll charges, regulating the quality of the product, and in the absence of seignorial mills allowing the settlers to build their own.[90]

Initial plans for exploiting the boundless lands of English America by huge grants to leading landed aristocrats in England held out the possibility of developments not unlike those of New France, as witness the elaborate paper plans drawn up for Carolina. But the financial and managerial costs of quasi-feudal settlements promised to be far too burdensome to win continued support. The grantees found the simpler business of land disposal on modest terms to all comers a more congenial mode of operation.[91] Only in the case of the lower Hudson Valley of New York, following the example set by the Dutch regime in the patroonship of Rensselaerswyck, were a number of great manorial estates established with milling and other rights reserved for the lord. The burdens of insecure tenure and many obligations led to continued unrest and repeated movements of protest—and an unsavory reputation that hardly invited imitation elsewhere.[92]

[90] Munro, *Seigniorial System*, chap. 6.

[91] See the discussion of land grants and colonial settlement policies in Curtis P. Nettels, *The Roots of American Civilization* (New York, 1939), pp. 125–40, 525–30. William Penn reserved for himself, for a time, exclusive rights to erect grist- and sawmills in the colony of Pennsylvania, organizing a corportion for this purpose. "In 1698 Penn abandoned his exclusive rights in milling. Thereafter grist mills increased rapidly" (Fletcher, *Pennsylvania Agriculture*, 1:324).

[92] These matters are considered in some detail in Irving Mark, *Agrarian Conflicts in Colonial New York* (New York, 1940), and Ellis, *Landlords and Farmers*.

The Decline of Water Milling

With the accelerating advance of industrialization from the 1840s and the progressive though gradual penetration of rural life by the market economy, water mills steadily declined in usefulness and importance. With the farm population more than doubling from 1840 to 1880, the number of gristmills and sawmills in this country decreased by about 10 percent, from some 55,000 to 50,000.[93] Except for the relatively small proportion of these mills driven by draft animals, chiefly in the first years of settlement or in districts with few mill seats, or by wind in relatively restricted areas, the country mills were driven by the energy of falling water. The vast majority, as we have seen, were located on small streams where power of low capacity was developed without great difficulty or large outlay. Such millstreams were subject to rather marked seasonal fluctuations in flow, but in farming communities accustomed to accommodating their activities to seasonal changes these fluctuations were for the most part a matter of inconvenience and discomfort rather than appreciable loss. When streamflow diminished or even ceased, there were the alternatives of resorting again to the hominy block or hand mill, rigging up a horse mill, or taking one's grist to the nearest working mill.[94]

By all the evidence the greater number of the common country gristmills and sawmills, equipped with a single run of millstones or a single up-and-down sash or gate saw, developed no more than several horsepower; the smaller ones no more than one to two horsepower. In the latter class more commonly than not fell not only the tub mill but the fulling mill, the carding machine, and the bark mill of the tanyard. For the most part custom mills taking their payment in a small fraction of the raw or finished product as toll, they represented investments usually measured, as evidenced by census schedules between 1800 and 1860, in hundreds of dollars, more commonly below than above a thousand dollars.[95] The

[93] U.S., Census Office, Tenth Census, 1880, vol. 2, *Manufactures* (Washington, D.C., 1883), pp. viii–ix; 24,338 gristmills and 25,708 sawmills were reported. See also the perceptive characterization of the historic role of the small water mill in American development in H. H. Bennett, "Utilization of Small Water Powers," *Transactions, Third World Power Conference, Washington, 1936*, 10 vols. (Washington, D.C., 1937), 7:473–74.

[94] One settler "had a little hand-mill which, unlike the little water-mills along the streams did not fail in dry weather, and hence became quite popular among his neighbors when the water was low" (Hill, *History of Coshocton County, Ohio,* p. 586). See also H. S. Knapp, *A History of the Pioneer and Modern Times of Ashland County* [*Ohio*] (Philadelphia, 1863), p. 436.

[95] Manuscript schedules, Fourth Census, 1820.

Fig. 13. Water mill survivals. *Above,* Piney Branch Mill near Fairfax, Va. (Photo by John O. Brostrup.); *below,* the Hollingsworth Mill near Herrick, Ala. (Photo by Alex Bush.) Both mills were photographed in 1936 for the Historic American Buildings Survey and were in working condition at that time. (HABS Collection, Library of Congress.)

large numbers and wide distribution of these water mills contrast sharply with the number and distribution of stationary steam powers. The first systematic enumeration of steam engines in this country was the Treasury Department survey of 1838. According to this survey and allowing for certain omissions, not over 2,000 stationary engines were then in use in the United States. This figure gives a ratio of one engine to every 8,000 to 9,000 persons in the nation. Employed in manufacturing and related activities, the engines had an average rating of 20 horsepower, five to ten times that of the average water mill.[96]

Although by 1840 the foundations of an active industrialization of the economy had been laid, nine-tenths of the American people still lived in rural communities. Most of them had probably never seen a steam engine in operation, unless they lived in areas near rivers navigable by steamboats.[97] In contrast, the number of persons to whom at this time the common water mill was an unfamiliar sight must have been quite small. Mechanical power, in short, to most Americans was waterpower as represented by the small mills that in the greater part of the United States were found in nearly every township and in most counties were numbered in the scores.

Contributions of the Water Mill

In sum, the common water mills, community oriented in a premarket, subsistence economy, contributed to the development of the American economy in three principal ways. First, they assisted directly and materially both in advancing the frontier and in maintaining if not raising the living standards of agricultural America during more than two hundred years of its subsistence phase. In the English colonies water mills were freed from the postfeudal restraints on milling associated with concentration of land ownership in much of Europe. Their use in America was favored not only by topography and climate and by the heavy labor demands of farm building added to subsistence requirements, but also

[96] See below, chap. 10. This estimate is based on a generous allowance of 2,000 engines (only 1,860 stationary engines were reported, based on actual enumeration supplemented by estimates for districts not visited) and a population estimated at 16,300,000 for 1838 (U.S., Secretary of the Treasury, *Report on the Steam-Engines in the United States,* House Ex. Doc. 21, 25th Cong., 3d sess. [Washington, D.C., 1839], pp. 377 ff.).

[97] The total number of steam engines of all kinds, enumerated and estimated, 1838, was 3,010, of which steamboats accounted for 800, locomotives 350, and "standing engines," 1,860 (ibid.).

by the desire of the state to promote land settlement and the desire of private capitalists to promote land sales and development. Actively present and influential, too, was the unwillingness of the pioneering settlers to abandon on beckoning frontiers accustomed levels of quality in food, shelter, and clothing, save as a temporary necessity.

Second, as a result of many generations of living and making a living in older regions and on frontiers alike, in the course of which water mills became a commonplace feature, mechanical power was a familiar and almost indispensable element in the life of nearly every community. Accepted by all and woven into the pattern of everyday living, the leading water mills—grist, saw, carding, and fulling—were early established as integral parts of a subsistence agriculture. They both provided direct relief from some of the heavy tasks of the domestic economy and preserved something of the amenities of living found in more settled societies. The advantages of water mills were emphasized in a fashion by the very limitations attending their use: the deprivations during the first years of settlement and the seasonal interruptions in availability characteristic of many millstreams and regions. Finally, habituation to use and, with many, familiarity with the construction and management of water mills served to prepare an agricultural people for eventual transition to an emerging industrial economy in which the use of machinery and power in mills, mines, and factories became a central feature. As George S. Gibb has suggested, not only were gristmills, sawmills, and fulling mills integral parts of the American community from quite early times, but in them "American mechanics learned how to harness natural forces for moving machinery too cumbersome for manual effort." By the end of the eighteenth century, he concluded, "America appears to have been as advanced as any other country in the basic applications of water power to mechanical movement, and was on the threshold of undisputed supremacy in the field of practical hydraulics."[98]

[98] George S. Gibb, "The Pre-Industrial Revolution in America," *Bulletin of the Business History Society,* 20 (1946):109.

2
The Traditional Technology

THE AGE OF MECHANICAL power in the United States had its origins, as we have seen, in the simple rural water mills introduced from Europe in the early seventeenth century. Faced with creating a viable economy in the remote, raw, and often hostile wilderness, the colonists had not only to secure the requirements for sheer physical survival but at the same time to accumulate capital savings in the form of farms, homes, and community facilities. They sought and secured a measure of relief by harnessing the natural forces of water and wind or the muscular energy of draft animals to drive the simple mill machinery by which grain was ground, lumber sawed, wool carded, cloth fulled, and hides tanned.

The technology of water mills in colonial America was wholly European in origin and character. The prime movers, the supporting hydraulic facilities, and some features even of the mill machinery went back to ancient times and the cultures of the eastern Mediterranean basin. Although the earliest extant examples of references to waterwheels date from the period of classical Greece and Rome, it was not until the fifteenth and subsequent centuries that water mills became a commonplace feature of the economies of the Western world.[1] For centuries their use

[1] There is no comprehensive treatment of waterpower in colonial or post-Independence America, but much material exists in local and regional histories and in the monographic literature of industrial history. See especially James E. Defebaugh, *History of the Lumber Industry of America,* 2 vols. (Chicago, 1906–7), vol. 2; Charles B. Kuhlmann, *The Development of the Flour-Milling Industry in the United States* (New York, 1929); J. Leander Bishop, *A History of American Manufactures from 1608 to 1860,* 3d ed. rev., 3 vols. (Philadelphia, 1868); Victor S. Clark, *History of Manufactures in the United States, 1607–1860*, 3 vols. (New York, 1929); and regional histories such as those of William B. Weeden for New England and Philip A. Bruce for Virginia. Among the most important histories of water mills in the Western world are Richard Bennett and John Elton, *History of Corn Milling,* 4 vols. (London, 1898–1904); L. A. Moritz, *Grain-Mills and Flour in Classical Antiquity* (Oxford, 1958); Lynn White, Jr., *Medieval Technology and Social Change* (Oxford, 1962); chapters on power by R. J. Forbes in Charles Singer et al., *A History of Technology,* 5 vols. (Oxford, 1954–58); E. C. Curwen, "The Problems of Early Water-Mills," *Antiquity* 18–19 (1944–45): 130–46; Bertrand Gille, "Le Moulin à eau: une revolution technique medievale," *Techniques et civilization,* vol. 3 (1954); and Marc Bloch, "The Advent and Triumph of the Water-mill," in

was confined largely, often entirely, to grinding bread grains into meal and, especially in the eastern Mediterranean basin, to raising water for irrigation by means of the Persian wheel. From the early Middle Ages there was a gradual extension of the use of waterwheels in a widening range of industrial applications in manufacturing and mining. Change proceeded at an extraordinarily slow pace. The differences between the waterwheels depicted by Agricola in his treatise on mining, *De Re Metallica* (1556), to go back no farther, and those shown by Oliver Evans in *The Young Mill-wright and Miller's Guide* (1795) were inconsiderable. Until the latter half of the nineteenth century mechanical power in the Western world was preeminently waterpower.[2]

Inescapably, the technology of colonial America in this as in most other respects was simply an extension of the technology prevalent in the Old World, reflecting in its details usages characteristic of the countries of emigrant origin. In many instances the transfer of water-milling practices was directly promoted by the sending of craftsmen or mill materials by the colonizing agency. Yet the chief vehicle of transfer over the long run was undoubtedly the colonists themselves. In their cultural baggage they brought a fund of knowledge, skills, and experience related to Old World water-milling design, construction, and use. Not until the closing years of the eighteenth century was the innovative ability of American mechanics revealed in this branch of technology.

It is important to recognize at the outset that the history of industrial motive power is much more than the history of the prime movers, which have tended to monopolize the attention not only of the practical men concerned in their use and development but of the historians who have traced their evolution. Prime movers, whether men and animals or waterwheels and steam engines, are but one important element in what may be termed a system of power. A power system begins with the sources of energy to be harnessed, which must be recognized, understood, and used to the best advantage under the prevailing economic and technological conditions. It ends with the delivery of the power to the machine

Land and Work in Medieval Europe: Selected Papers, trans. J. E. Anderson (New York, 1969).

[2] "It is only of late years," declared William Fairbairn, "that in this country [Britain] the steam engine has nearly superseded the use of air and water as a prime mover. Until recently steam has been auxiliary to water, it is now the principal source of power, and waterfalls are of comparatively small value, except in certain districts" (*Treatise on Mills and Millwork,* 2 pts. [London, 1861–63], 1:67. For an extended discussion of this matter, see Louis C. Hunter, "Waterpower in the Century of the Steam Engine," in Brooke Hindle, ed., *America's Wooden Age: Aspects of Early Technology* (Tarrytown, N.Y., 1975).

or machines to be driven, directly or by means of transmission equipment—line shafts and belting—for distribution to machinery at the points of use within mill or factory.

As described in another place, waterpower has its origin in the hydrologic cycle, in the rainfall and runoff within the drainage basin that supplies a given stream and the power installations of the mills upon it. With the vast majority of common water mills the controlling conditions of meteorology, topography, and geology were those found within the few square miles of drainage area of the typical millstream. Yet the volume and fluctuations of streamflow, which with the height of fall were the determinants of the power capacity of a millsite, often varied widely among streams in other respects similar, as engineering experience was in time to demonstrate.

The performance of a power system is judged not only by the efficiency and capacity of the waterwheel but by the delivery of the power in such manner as to meet the operational requirements of the machinery driven. With the growth and development of industry and the evolution of mechanical processes, operational requirements change, as do considerations bearing upon the supply of water to the wheels and the efficiency and regularity of their performance. In sum, the prime movers, which historically have tended to be viewed in isolation, must be seen in the context of the power system of which they are but one component, and this system must be seen in relation to the prevailing economy and technology.

The development of a specific direct-drive waterpower, whether for a mill or a factory, began with the selection of a mill seat or water privilege, that is, a stream site where an abrupt descent in the streambed, a succession of lesser falls, or an extended rapids offered a concentration of fall favorable to power development. The greater the height of the fall, generally speaking, the less was the volume and weight of water required to obtain a given amount of power, thus reducing the size and expense of the hydraulic facilities from millpond to waterwheel and tailrace. Such sites are much more numerous and more sharply defined in hilly terrain and the upper portions of a river system than in the lower valleys with their gentler slopes and typically sedimentary character. The typical mill installation consisted of a waterwheel erected beside or beneath the mill structure; a dam at some suitable point upstream to divert more or less of the flow into the headrace, to increase the amount of fall (usually also to create a pond for storage of at least some of the nighttime streamflow); a canal called a millrace, or headrace, to carry the water to the mill with minimal loss of fall; a penstock, or sluice, with gate, to convey the water to the wheel; and a tailrace to carry the water, discharged from the wheel with energy largely spent, back to the stream below the mill. There were such further appurtenances as gates where needed to regulate water flow, wasteways for the contingencies of flood and high water, and trash racks to keep floating debris and ice from clogging penstock and waterwheel.[3]

Dams, Millponds, and Races

Next to the waterwheel and appurtenances, the dam was the most prominent feature of a water mill and the least dispensable. It was a barrier constructed of locally available materials, such as stone, timber, brush, and soil, in various arrangements erected across the stream with its ends well anchored in the banks on either side. Although often a low structure intended simply to divert water into the headrace, its principal use in most installations was to increase the amount of fall provided by nature and thereby the power potential of the site: weight of water times height of fall in a given time being the measure of capacity as usually reckoned (550 foot-pounds per second equal one horsepower). At the same time that the dam raised the stream level, it backed the water upstream—it might be for hundreds of yards or scores of miles, depending upon the height of the structure and the character of the terrain—and created a storage reservoir known as a millpond. Here was accumulated the water that outside working hours would otherwise flow over the dam and run to waste. Thus the power capacity of a stream might be, and often was, doubled or more in daily use; and if the conditions of upstream terrain were favorable—for example, high banks and no low-lying lands that might be flooded with consequent expense in damages to be paid to riparian owners—a very large pond might be obtained that would for a time mitigate water shortages during periods of drought.

In its simplest form the dam was little more than a rude obstruction of tree trunks, branches, and stones, the openings chinked with gravel, sand,

[3] No aspect of waterpower development has received less attention in the history of technology and is less satisfactorily handled than the means by which the fall in the streambed that comprises a mill seat is best exploited. See Louis C. Hunter, "The Living Past in the Appalachias of Europe: Water-Mills in Southern Europe," *Technology and Culture* 8 (1967): 446–66. Two general engineering studies helpful in understanding waterpower development in the earlier years are Joseph P. Frizell, *Water-Power: An Outline of the Development and Application of the Energy of Flowing Water* (New York, 1900), chaps. 1–3, and Daniel W. Mead, *Water Power Engineering: The Theory, Investigation, and Development of Water Power* (New York, 1908), chaps. 4, 5, 23, 24.

clay, or loam. Unlike Old World practice, especially as in Britain, masonry dams were long almost unknown and never common, confined in use chiefly to the large power installations of the industrial age. Timber was the most common material. In the form of tree trunks and limbs, rudely hewn, or in planks and boards from a nearby mill, it was used in a variety of arrangements, in the more substantial structures often as cribwork filled with stones and rock (see Figs. 14-16).[4] Time and experience revealed what kinds of structures stood up best under the conditions of normal use and the stress of seasonal high waters and occasional abnormal floods. Care was necessary to obtain secure abutments at the stream banks, and, by means of "tumbling aprons" of timber, plank, or rocks, to prevent the water passing over the dam from undermining its foundations. Even the most securely built dams could not be completely protected against sporadic heavy floods that carried all before them, at times the mill no less than the dam.[5]

In the smallest country mills the supporting facilities were often of the rudest sort; a rough pile of stones and timber, filled in with brush and stones, was sufficient to raise and divert the water. Rude ditches, cleaned periodically of earth and debris, served as races to bring and carry away the water. With tub mills, dams were hardly necessary; the vigorous use of a shovel for ten minutes would be sufficient to divert water from the stream into the wooden trough serving as a penstock. There were, of course, various arrangements for using the energy of streamflow directly, thereby eliminating the wheel, as in hydraulic mining, to say nothing of exploiting the stream current in floating vessels and rafts downstream. Two of the oldest devices for developing waterpower all but dispensed with dam, pond, and races and, save for a limited shore

[4] U.S., Census Office, Tenth Census, 1880, vols. 16–17, *Reports on the Water-Power of the United States,* (Washington, 1885, 1887), contains much information on the design and construction of dams in use and under construction at the time. Numerous drawings illustrate chiefly installations serving large factories and even industrial communities.

[5] For the construction and use of dams before 1850, see Oliver Evans, *The Young Mill-wright and Miller's Guide,* 1795 and later editions, Arts. 61–65, 85–86; John Nicholson, *The Operative Mechanic, and British Machinist,* 2d Am. from 3d London ed., 2 vols. (Philadelphia, 1831), 1:104 ff.; and Zachariah Allen, *The Science of Mechanics* (Providence, 1829), pp. 197–206. A widely used handbook of a later date was that of the hydraulic turbine builders James Leffel & Co., *The Construction of Mill Dams* (Springfield, Ohio, 1881). On the use of brush dams, see Tenth Census, *Water-Power,* pt. 2, pp. 304, 314–15, 324 ff.; Jacob A. Swisher, *Iowa: Land of Many Mills* (Iowa City, 1940), pp. 46–49; T. U. Taylor, *Water Powers of Texas,* U.S. Geological Survey Water-Supply and Irrigation Paper no. 105 (Washington, D.C., 1904), pp. 94 ff., and passim.

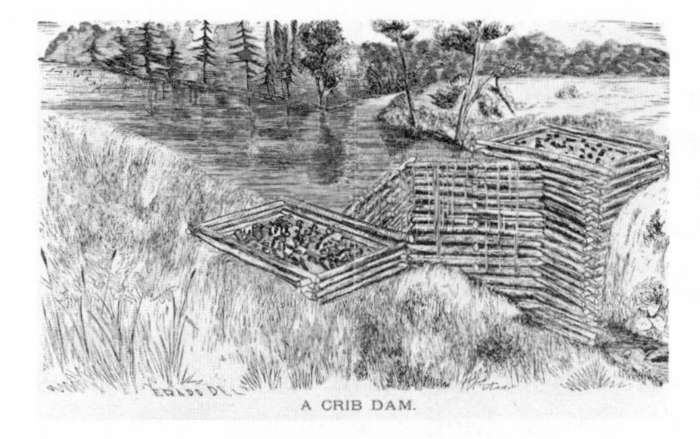

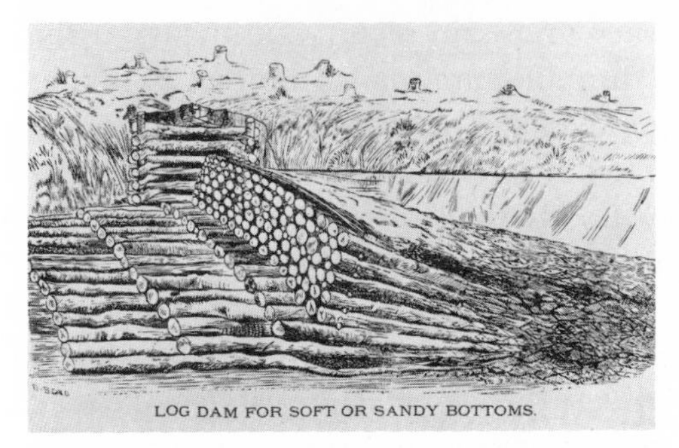

Figs. 14–16. Typical forms of milldams. (James Leffel & Co., *The Construction of Mill Dams* [Springfield, Ohio, 1874].) Of some thirty types of dams described in this work, all but a few were built chiefly of timber, supplemented by stone, gravel, and earth.

area, with the water privilege or mill seat itself. The floating, or boat, mill and the Persian wheel, survivals of which can occasionally still be seen in some parts of Europe as well as throughout the Middle East and in China, were arrangements for supporting an undershot paddle wheel at some point convenient to the shore so that the submerged floats took their motion from the stream current. Clusters of boat mills producing meal and flour were to be seen as late as the 1890s at many of the larger points on major European rivers. The often large, vertical, undershot wheel known variously as the noria, the Persian, or the Egyptian wheel goes back to the first century B.C. or earlier. With pots, gourds, or bamboo tubes attached around the circumference, alternately filling at the bottom and emptying near the top of the wheel's revolution, the noria provided a simple and cheap means of elevating water. Boat mills were poorly adapted to American needs but found occasional use; the noria, or current wheel, in its American adaptation was widely used in river-bed placer mining in California and, somewhat later, for irrigation in semiarid regions of the West and Southwest.[6]

[6] See Forbes, "Production of Flour and Bread in Roman Times," in Singer et al., *A History of Technology,* 2:103 ff. For a description of early nineteenth-century floating mills in Europe, see letter of General Tallmadge, *JFI* 20 (1835):278–79. A good drawing appears in Jacob Leupold, *Theatrum Machinarum Molarum, oder Schau-Platz der Mühlen-Bau-Kunst* (Dresden, 1767), pl. 24, figs. 1–5. On floating mills in the United

Fig. 17. *Top*, noria, southern China. Placed in motion by the stream striking the basket-work wheel floats, the noria lifts water in hollow sections of bamboo and empties it into crude troughs that carry it off to the fields. (Harry A. Franck, *Rovings through Southern China* [New York, 1925].) *Bottom,* a thirty-foot current waterwheel used for irrigation on a branch of Hat Creek, Sioux County, Neb. (Erwin H. Barbour, *Wells and Windmills in Nebraska* [Washington, D.C., 1899].)

In many situations, particularly with falls too low to provide the desired power, it was not practicable to increase the fall by means of the dam alone, and the millrace was assigned a supporting or even the principal role. The dam was placed some distance above the falls at a suitable site, and the desired fall obtained by carrying the race along the valley side on a contour course with just enough gradient to carry the required volume of water. Races a mile long were not unheard of, being resorted to so as to avoid the flooding of bordering farmland and consequent heavy damage payments. By sacrificing some of the fall thus obtained, the mill could be located well above the reach of most floods.[7] "In the western states," it was reported in 1905, "it is common to obtain water power by diverting a portion of the stream flow along in a ditch with but a small slope, until the fall in the main stream has given a sufficient difference in level to provide for the necessary power."[8] To avoid power waste, the slope of the canal bed had to be the minimum necessary to move the required volume of water. The term *millrace* was a misnomer. Opinions varied concerning the slope required for an effective volume of flow. Oliver Evans believed one inch per 100 feet was adequate in a long race, but he cited a French authority as finding one inch per 500 feet sufficient.[9] A rude ditch adequate to the needs of a common mill might

---

States, see Reuben G. Thwaites, ed., *Early Western Travels,* 32 vols. (Cleveland, 1904-7), 4:135, 8:269–70, 9:110–11, 10:225. On use of the current wheel for irrigation and riverbed mining, see *Current Wheels: Their Use in Lifting Water for Irrigation,* U.S. Office of Experiment Stations Bulletin no. 146 (Washington, D.C., 1904), and Erwin H. Barbour, *Wells and Windmills in Nebraska,* U.S. Geological Survey Water-Supply and Irrigation Paper no. 29 (Washington, D.C., 1899), pp. 67–68. See also discussion on mining to 1860 in a companion volume in preparation. Staunton in *Knight's American Mechanical Dictionary* (1877), s.v. "Noria," reported a current wheel, or noria, with a diameter of thirty feet. It was rimmed with twenty buckets, each with a capacity of six-tenths of a gallon. When running at four revolutions per minute, it raised to a height of some twenty-five feet 48 gallons of water a minute and over 300 tons of water a day. See also Thomas Ewbank, *A Descriptive and Historical Account of Hydraulic and Other Machines,* 4th ed. (New York, 1850), pp. 112–13; *Encyclopaedia Britannica,* 11th ed., s.v. "Hydromechanics."

[7] For examples, see Tenth Census, *Water-Power,* pt. 1, pp. 63, 249, 262, 375–79. Examples in the Miami Valley of Ohio and along the Iowa River appear ibid., pt. 2, pp. 379, 480.

[8] "Hydraulic Power in Great Britain," *Engineering Magazine* 28 (1904–5): 820.

[9] Evans, *Miller's Guide,* Art. 65, p. 128. Evans's co-author, Thomas Ellicott, allowed "one inch to a rod [16.5 feet = 198 inches] for fall in the races; but if they be very wide and long, less will do (ibid., p. 330)." Unless otherwise indicated, references to the *Miller's Guide* are to the 13th edition (1850).

cost but little, and the use of the land was often gratis in a community welcoming a convenient grist- or sawmill.

Still other expedients might be adopted, terrain permitting, for increasing the fall in a district without suitable sites, such as carrying the headrace across the narrow neck of an oxbow formed by a stream's meander or across a V-shaped peninsula formed by a tributary joining a larger stream.[10] An extreme instance of such an expedient was reported in Breathitt County, Kentucky, where the North Fork of the Kentucky River makes "a detour of over five miles and returns to within sixty feet of the point of departure. A tunnel through the rock partition has furnished water power, there being a difference in elevation of eight or nine feet in the stream bed." Another example of technical ingenuity—and evidence of millwrighting experience and competence—was the means by which the town of Dedham, Massachusetts, near Boston, obtained an admirable and much needed waterpower, at the expense of some ensuing litigation, in the absence of good mill seats on either of the streams passing through the town, the Charles and the Neponset. In 1637 a three-mile-long, man-made canal, known as Mother Brook, was built at town expense between points on the two streams having a difference in elevation of some sixty feet. Thus was provided at intermediate locations the waterpower capacity on which the town's industrial growth was based.[11]

The dam, millpond, and races of the water mill, to adopt a useful if imperfect analogy, functioned somewhat in the same fashion as the boiler-furnace component of the steam-power plant.[12] In their different ways they developed and stored fluids, water or steam, under pressure; this pressure was in turn converted by waterwheel or steam engine into torque, that is, turning power on the wheel shaft. In the common water mill the supporting facilities of dams, races, gates, and the like were

[10] Examples are found in Tenth Census, *Water-Power,* pt. 1, pp. 280–81, 306–7, 373, 498, 697; pt. 2, pp. 194–95, 202, 234, 351–52, 386, 457. See also Richard W. Musgrove, *History of the Town of Bristol, Grafton County, New Hampshire,* 2 vols. (Bristol, 1904), 1:134 and facing page.

[11] Mary Verhoeff, *The Kentucky River Navigation* (Louisville, 1917), p. 8. For the Dedham enterprise, see Erastus Worthington, *The History of Dedham, from the Beginning of Its Settlement in September, 1635, to May, 1827* (Boston, 1827), pp. 12–16, 37–38; Herman Mann, *Historical Annals of Dedham from Its Settlement in 1635, to 1847* (Dedham, Mass., 1847), pp. 11–40, 103 ff. By 1827 Mother Brook was supporting a number of mills and much machinery at five dams.

[12] In *Reflections on the Motive Power of Fire,* ed. E. Mendoza (New York, 1960), pp. 15 ff., Sadi Carnot drew an analogy between waterpower and steam power that is also discussed in D. S. L. Cardwell, "Power Technologies and the Advance of Science, 1700–1825," *Technology and Culture* 6 (1965): 188 ff.

usually modest affairs, accounting for no great capital outlay and expense for maintenance and repair. Even in Oliver Evans's *The Young Mill-Wright and Miller's Guide*, the millwright's bible, these matters were dismissed in several pages. As hydraulic power increased in scale and sophistication to meet the needs of large, market-serving mills and factories and, in time, industrial communities such as Lowell, Massachusetts, the facilities required to supply the huge breast wheels of many large mills came to absorb much the greater part of the capital investment and engineering skills represented by the entire power installation.

Hydraulic Prime Movers: The Waterwheels

The basic components of a waterpower to which these various structures and facilities were contributory were the fall in the streambed and the volume of streamflow. Whether abrupt or gradual, concentrated or extended, the fall was the constant in the power equation. Upon the volume of flow and its regularity—the product of rainfall and runoff and the extent of the drainage or gathering basin—depended the gross capacity and reliability of the power at a given site. Since fall and flow were accepted as determined by nature and beyond control and since the supporting facilities, such as dams and races, on most millstreams were readily manageable by the millwright, attention in the traditional technology centered upon the mechanism by which the stream was put to work—the waterwheel. The decisions as to wheel type, size, proportions, and rotating speeds—rpm—in relation to a specific mill seat and power use largely determined the effectiveness and economy of the water mill. If calculations—or, more likely, estimates—as to the volume of flow or the power required proved grossly inaccurate, the success of the enterprise would be jeopardized. As prime movers, waterwheels had two important advantages distinguishing them from steam engines and giving to their design a basic simplicity favorable alike to construction and use. These were their rotary motion and the absence of an enclosing case, the latter a complication eventually introduced in reaction wheels and hydraulic turbines. The basic wheel form of the water motor freed designers, inventors, and builders, first, from having to supply such mechanical arrangements as the valve mechanisms required in the reciprocating steam or hydraulic engine to reverse the direction of the piston's motion twice in each revolution of the wheel shaft; second, from having to use a crank to convert the reciprocating action of the piston assembly into the rotary motion of the shaft.[13] Nor did most waterwheels have the problems of balancing the weight of reciprocating parts, a factor that hampered the development of the high-speed steam engine, or of smoothing out variations in the power of the piston at different points in the cycle of each stroke, which made the use of a flywheel customary. Moreover, the waterwheel itself provided a certain flywheel action through its weight and rotary motion, although the latter was slow in wheels of any size. In the *Miller's Guide,* Article 158, Ellicott recommended that the fast-running "flutter wheels" of the sawmill "should be very heavy, that they may act as a fly, or balance, to regulate the motion, and work more powerfully." Of hardly less importance in waterwheels of the traditional kind was freedom from the steam engine's requirement of numerous parts of cast and wrought iron, finished to precise dimensions and presenting difficult problems of shaping, packing, and lubrication.

The design of the waterwheel was simplicity itself: a circular structure of varying diameter and breadth around the circumference of which at regular intervals were arranged floats or buckets for intercepting the falling water and capturing a portion of the energy produced by the fall of water in a given volume from a higher to a lower level. For the height of fall and volume of flow available at the mill seat, the capacity of a waterwheel depended upon the type and dimensions of the wheel, its efficiency in the use of water, and its design, construction, and installation. Rotating speeds, measured in revolutions per minute, varied inversely with the diameter of the wheel. The quick motion desired in small gristmills with horizontal wheels and in the undershot "flutter" wheel driving the up-and-down saw directly through connecting rod and crank in the sawmill was obtained commonly by wheels ranging roughly between three and five feet in diameter. The undershot, breast, and overshot wheels on which mills and manufactories chiefly relied were given diameters appropriate to the medium range of falls characteristic of the eastern United States. Such wheels, with diameters usually ranging from eight to thirty feet, revolved much slower, typically between twenty and five rpm, with speeds decreasing as diameters increased. Waterwheels thus moved more slowly than steam engines of comparable power before the 1850s and relied largely on gearing and on cam takeoffs on wheel shafts to obtain the operating speeds required in most manufac-

[13] Not all water motors were rotary in action. The so-called water-pressure engines introduced chiefly for mining and other pumping uses employed a piston and cylinder with a reciprocating action, as did the hydraulic cranes and elevators of a later period. See Jean F. d'Aubuisson, *A Treatise on Hydraulics for the Use of Engineers,* trans. J. Bennett (Boston, 1852), pp. 452–60; William J. M. Rankine, *A Manual of the Steam Engine and Other Prime Movers,* 16th ed. (London, 1906), chap. 4, pt. 2.

turing operations. In short, the general design of the traditional water motor was simplicity itself, a single moving part, a structure having the form of one of the oldest of mechanical devices, the wheel, built with the simplest of tools and skills from one of nature's most abundant and readily shaped materials. Building such a wheel presented little difficulty to a craftsman skilled in woodworking. But much skill and experience were required to select the wheel type and to determine the dimensions best adapted to serve the needs of a specific purpose at a designated millsite. To assist in this task was a primary purpose of millwright handbooks.

In operation a stream of water was directed from above the wheel against the floats, or paddles, or into the buckets, or troughs, arranged around the wheel's circumference. From the force of impact or the weight of the falling water, the wheel was kept in motion as a continuous succession of floats or buckets came into the path of the falling water. The two basic types of waterwheels on which industry above the small country water mill largely depended were known from the manner of construction as float and bucket wheels and from the manner of their operation as impact and gravity wheels. They made their appearance early in history and with but minor changes in the manner and materials of construction and regional variations in form underwent no major modifications during the centuries of their principal use.[14] The wheel had to be large enough to accommodate the necessary volume of water, and, other things equal, this was obtained typically by increasing the breadth of the wheel and the size of its floats or buckets. Only with the advance of industrialism and applied science in the nineteenth century were new wheel types devised to meet the operational needs and the business and engineering criteria of the new age.

The traditional types of waterwheels in much of the Old World and the New were adapted to use with what came later to be termed low to

[14] Most handbooks and treatises on mechanics, hydraulics, and millwrighting give some attention to the several types of waterwheels. In addition to Evans, *Miller's Guide,* Frederick Overman, *Mechanics for the Millwright, Machinist, Engineer, Civil Engineer, Architect & Student* (Philadelphia, 1858), and Allen, *Science of Mechanics,* see Joseph Glynn, *Rudimentary Treatise on the Power of Water,* 5th ed. (London, 1875); d'Aubuisson, *Treatise on Hydraulics;* Rankine, *Manual;* and George R. Bodmer, *Hydraulic Motors; Turbines and Pressure Engines* (London, 1889). The general histories of technology are not usually very informative beyond chronological identification. Two later, reminiscent articles by hydraulic engineers are: Joseph P. Frizell, "The Old-Time Water-Wheels of America," *TASCE* 28 (1893): 237–49, with drawings; and Samuel Webber, "Ancient and Modern Water-Wheels," *Engineering Magazine* 1 (1891):139 ff.

medium heads, that is, falls between some five to twenty-five feet. Though falls of from, say, thirty to fifty feet were occasionally put to use, they occurred infrequently except in the less accessible mountain regions. Thomas Ellicott, collaborator with Oliver Evans in *The Young Mill-wright and Miller's Guide,* headed Part 5, his "practical instructions for building mills," as "suitable to all falls of from three to thirty-six feet." This last figure was the greatest fall covered by any of the tables in the *Guide* and the range of 10.5 to 36.4 feet was applicable to overshot wheels. For the far more widely used undershot wheels the range given was one foot to twenty-five feet; for breast wheels, six to fifteen feet, and for tub wheels, eight to twenty feet. A communication to the *Journal of the Franklin Institute* in 1835 believed it to be reasonably certain "that more than nine-tenths of the water powers in the United States is confined to situations which do not afford an aggregate of ten feet, head and fall."[15] A late nineteenth-century source contrasted the waterpowers of Europe and the United States: "In the mountainous districts of Europe, as a rule the falls are high and the quantities of water small, while, on the other hand, in America the falls are moderate or low, and the quantities of water very large, so that the two may be taken as illustrating extreme phases of the occurrence of water power."[16]

The difficulties and cost of exploiting high-head waterpowers with conventional impact and bucket wheels long discouraged their development. Such heads were first exploited in the United States in the metal-mining regions of the Far West, where the tangential waterwheels of the Pelton type (see below, chap. 8) were developed to utilize heads far too high for satisfactory use with old-style wheels. Except for what came to be termed reaction wheels and the turbine wheels to be considered later,

[15] Evans, *Miller's Guide,* pp. 164, 171, 177, 182; Daniel Livermore, "Remarks upon the Employment of Pressure Engines, as a Substitute for Water Wheels," *JFI* 15 (1835):166. See also Allen, *Science of Mechanics,* p. 207. In millwrighting usage, "head and fall" was the vertical distance from the water level in the forebay, terminus of the headrace, to the tailrace at its normal level. "Head" was the distance from the penstock level to the point on the wheel where the water struck; "fall" was the remaining distance to the tailrace. By the late nineteenth century it was stated that "for all ordinary purposes, the terms 'head,' 'fall,' 'head and fall,' may be taken as synonymous" (Tenth Census, *Water-Power,* pt. 1, p. 223).

[16] Bodmer, *Hydraulic Motors,* p. 2. In the American edition of the widely used handbook Nicholson's *Operative Mechanic,* 1:116–19, the millwright tables cover falls from one to twenty feet. On one of the important smaller river systems in New England, the Blackstone in Massachusetts, eighty-six water privileges were reported, of which two-thirds had falls of less than twenty feet (Massachusetts, State Board of Health, *Seventh Annual Report* [Boston, 1876], pp. 74 ff.).

the traditional wheels discussed here ruled the field in large plants and small and in most branches of industry to the 1850s.[17]

The most widely used of the impact wheels was the undershot, but the simplest of the type was the quick-running horizontal wheel of small diameter, which had a long and important career in gristmilling throughout much of the Western world and in Asia. In the United States this type was known, from the enclosing wooden hoop frequently used to confine the water to the wheel, as the "tub wheel." Among bucket or gravity wheels, the overshot wheel was for centuries evidently the only one in wide use. The variant forms known as "breast," "pitchback," and "back-shut" wheels appear to have first obtained importance late in the eighteenth century. They proved especially well adapted to the operating requirements of the larger mills and factories, having large capacity and high efficiency but requiring the use of gearing to obtain the necessary operating speeds for most industrial purposes. Before the introduction of the hydraulic turbine, breast wheels were mainly used by large industry, but overshot wheels were much used by smaller establishments under appropriate conditions.[18]

[17] Zebulon Parker, co-developer of the Parker reaction wheel, devised a plan for developing the occasional high-head waterpowers in the East, using, among other devices, a special wheel and belt transmission operating at high velocities. He stated that his plan had been adopted at "a considerable number of mills . . . running from two to five years" (*Scientific American* 6 [24 May 1851]:288). Parker's article was part of a long series of short articles published by the *Scientific American* in 1850 and 1851 under the headings "Hydrostatics" and "Hydraulics."

[18] In 1849, before hydraulic turbines gained acceptance in Great Britain, William Fairbairn, an engineer noted for the design and installation of large bucket wheels in British textile establishments, concluded that little or no improvement had been made in the principle on which waterwheels were constructed since John Smeaton's experiments of the 1750s, and that most were improvements in details, especially in bucket form, made feasible by the shift from wood to iron in construction. Fairbairn cited *Mechaniques et inventions approuvées par l'Academie royale des sciences*, first published in 1785, to the effect that before 1700 neither overshot nor breast wheels were much in use in Europe, indeed if at all known (Fairbairn, "On Water-Wheels with Ventilated Buckets," *PICE* 8 (1849):45–64). So far as Britain is concerned, the evidence is to the contrary. See the account of the overshot wheel at the Saugus ironworks in the 1640s, below. Leslie Syson in *British Water-Mills* (London, 1965), 52, quotes descriptions of overshot and breast wheels in Justice Fitzherbert, *Booke of Surveying* (1558). A fuller treatment of waterwheels appears in Fairbairn's *Treatise on Mills and Millwork*, pt. 1, pp. 112 ff. On overshot and breast wheels, see *Scientific American* 6 (29 Mar. and 5, 12, 19 Apr. 1851). Of the works cited above, d'Aubuisson's *Treatise on Hydraulics* provides the most detailed contemporary analysis of the functioning of overshot and breast wheels; a reviewer of the American edition declared that "we can hardly conceive of a more valuable addition to our books upon practical science" (*JFI* 55 [1853]: 215–16). Much the best

With the important exception of the tub wheel, impact and gravity wheels alike turned in a vertical plane on horizontal shafts, a fact of considerable operating importance. They were very similar in construction, too, being made, as was the tub wheel, almost entirely of wood. Radial arms were mortised into or through the often massive shaft hubs and were braced and bound at their extremities to form sturdy, even massive wheels of varying diameter and breadth according to their capacity and use. The two types differed chiefly in the succession of floats (paddles, blades) of the one type and the buckets, or troughs, of the other, mounted at intervals around the wheel's circumference. Buckets required more material and greater skill in construction—read higher cost—than the flat boards commonly used as floats. The burden of water carried by the buckets of the gravity wheel as they descended demanded heavier construction. Apart from the introduction of certain components of iron, changes in the design and construction of the old-style waterwheels were relatively minor.

Undershot wheels, though ranking low in efficiency, a matter of little concern where the supply of water was abundant, were the simplest and most versatile of the old-style waterwheels. They took their name from the circumstance that the falling water directed to the paddles in the bottom quadrant of the wheel on the upstream side passed under the wheel (see Fig. 18). The strike-and-splash-off action of the water with this wheel was very wasteful of energy; the useful effect of the falling water's energy probably ranged between 15 and 30 percent, depending upon the degree of care and skill in the wheel's construction and use. With a volume of flow sufficient to offset its low efficiency in water use and given a wheel breadth adequate to handling such volume, the undershot wheel could be used to advantage over a rather wide range of fall, as we have seen. Depending upon the desired rotating speed, there was a fairly wide range of choice in wheel diameter. The undershot wheel was the one best suited for use with falls of six to eight feet or less, including situations where the wheel was moved by the force of the stream current alone, as in floating mills. Since the natural occurrence of waterfalls was, in general, in inverse ratio to the height of fall—low falls being far more common than high ones except in mountainous regions—the opportunities for the use of undershot wheels were far more numerous than for wheels moved by the water's weight.

By present evidence the overshot wheel was the earliest and long re-

---

discussion of waterwheels about 1850 in the United States is the article in *Appleton's Dictionary of Machines, Mechanics, Engine-Work, and Engineering,* 2:786 ff.

Head and fall—3 ft.
Wheel diameter—18 ft.
Wheel breadth—2 ft. for each ft. diameter of millstone

Head and fall—7 ft.
Wheel diameter—18 ft.
Wheel breadth—equal to diameter of millstone

Head and fall—15 ft.
Wheel diameter—12 ft.
Wheel breadth—½ ft. for each ft. diameter of millstone

Fig. 18. Undershot waterwheels. Dimensions prescribed by Thomas Ellicott, millwright in Oliver Evans, *Young Mill-wright and Miller's Guide* (Philadelphia, 1807).

mained the predominant type of bucket wheel. The water was carried by means of a penstock, or sluice, over the top of this wheel, entering the buckets on the *downstream* side and thus reversing the direction of rotation of the undershot wheel (see Fig. 19). After a short free fall from the penstock, required for filling the buckets properly, the water acted upon the bucket wheel by force of gravity. Apart from the impact action on the wheel following the initial free fall and the wasting of water from the buckets as they approached the bottom of descent, the action of bucket wheels was quite efficient. Under test conditions, carefully designed and well-built wheels sometimes displayed efficiencies in water use of 80 percent and occasionally higher. Under ordinary working conditions in common water mills and in general everyday industrial use, the range of efficiency was perhaps 50 to 70 percent. With medium to high falls of some fifteen to thirty-five feet, the overshot wheel with a diameter closely approaching the height of fall customarily operated at an advantage that long gave it the preference in treatises on waterpower. Competent millwrights and millowners favored it as well, especially where the volume of flow was otherwise insufficient to meet power needs. However, with falls of eight feet or less, the energy loss in the initial free fall, early wasting of water from the buckets, and the retarding effect of running against the tailwater current during floods were unduly high in proportion to the gain from gravity action in the limited distance of fall.[19]

The superiority of overshot to undershot wheels in efficiency of water use in the ratio of two to one, as demonstrated by John Smeaton in his classic and much publicized experiments, reported in 1759, became decisive where shortages in water supply were concerned.[20] In time the overshot wheel gave way to a modification termed the breast wheel, which possessed marked operating superiority under similar conditions. Construction of the two wheels was virtually the same, except that the position of the buckets and the motion of the wheels were reversed, with water admitted to the breast wheel on the upstream side (Fig. 20). This arrangement brought three important operating advantages lacking in the overshot wheel, with relatively minor reduction in efficiency. First,

[19] Overman, *Mechanics for the Millwright*, pp. 200 ff. Fairbairn gave "ordinary overshot water wheels" an efficiency of "about 60 percent"; "improved breast wheels" of his own construction, 75 percent (*Treatise on Mills and Millwork*, pt. 1, p. 123). Reporting his own experiments on a larger (37.3 foot diameter by 3.5 foot breadth) overshot wheel in Brittany in 1805, d'Aubuisson found an average efficiency in water use of virtually 75 percent (*Treatise of Hydraulics*, Arts. 364, 394).

[20] Evans, *Miller's Guide*, Arts. 67, 68, 69, includes lengthy excerpts from Smeaton's report. See also P. N. Wilson, "The Water-Wheels of John Smeaton, *TNS* 30 (1955–57): 25–48.

Low overshot wheel
Diameter—12 ft.
Breadth—same as diameter of millstone

Very high overshot wheel
Diameter—30 ft.
Breadth—3½ in. for each ft. diameter of millstone

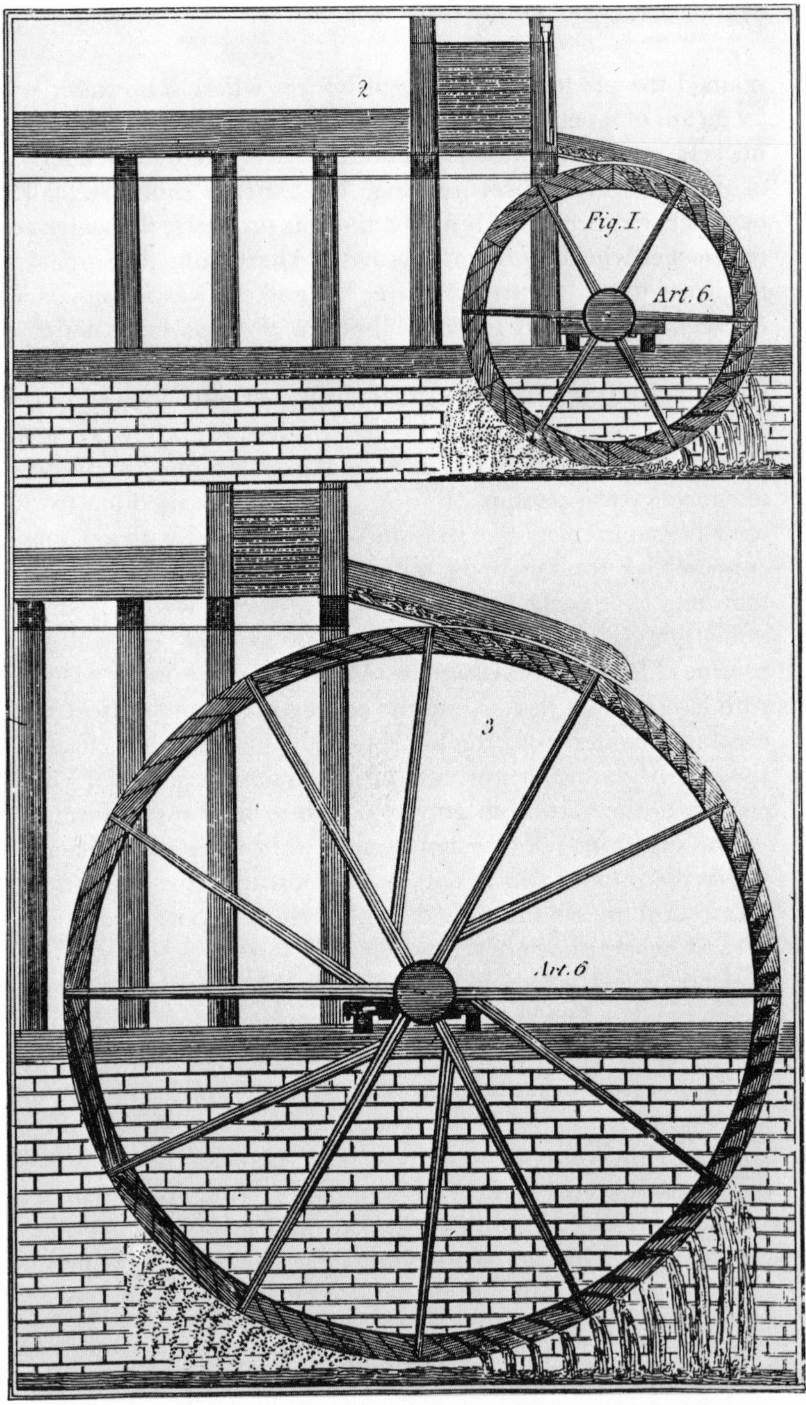

Fig. 19. Overshot waterwheels. Dimensions prescribed by Thomas Ellicott, millwright, in Oliver Evans, *Young Mill-wright and Miller's Guide* (Philadelphia, 1807).

w breast wheel
ead and fall—8 ft.
'heel diameter—18 ft.
eadth—9 in. for each ft.
diameter of millstone

iddling breast wheel
ead and fall—12 ft.
'heel diameter—18 ft.
eadth—8 in. for each ft.
diameter of millstone

gh breast wheel
ead and fall—13 ft.
heel diameter—16 ft.
eadth—7 in. for each ft.
diameter of millstone

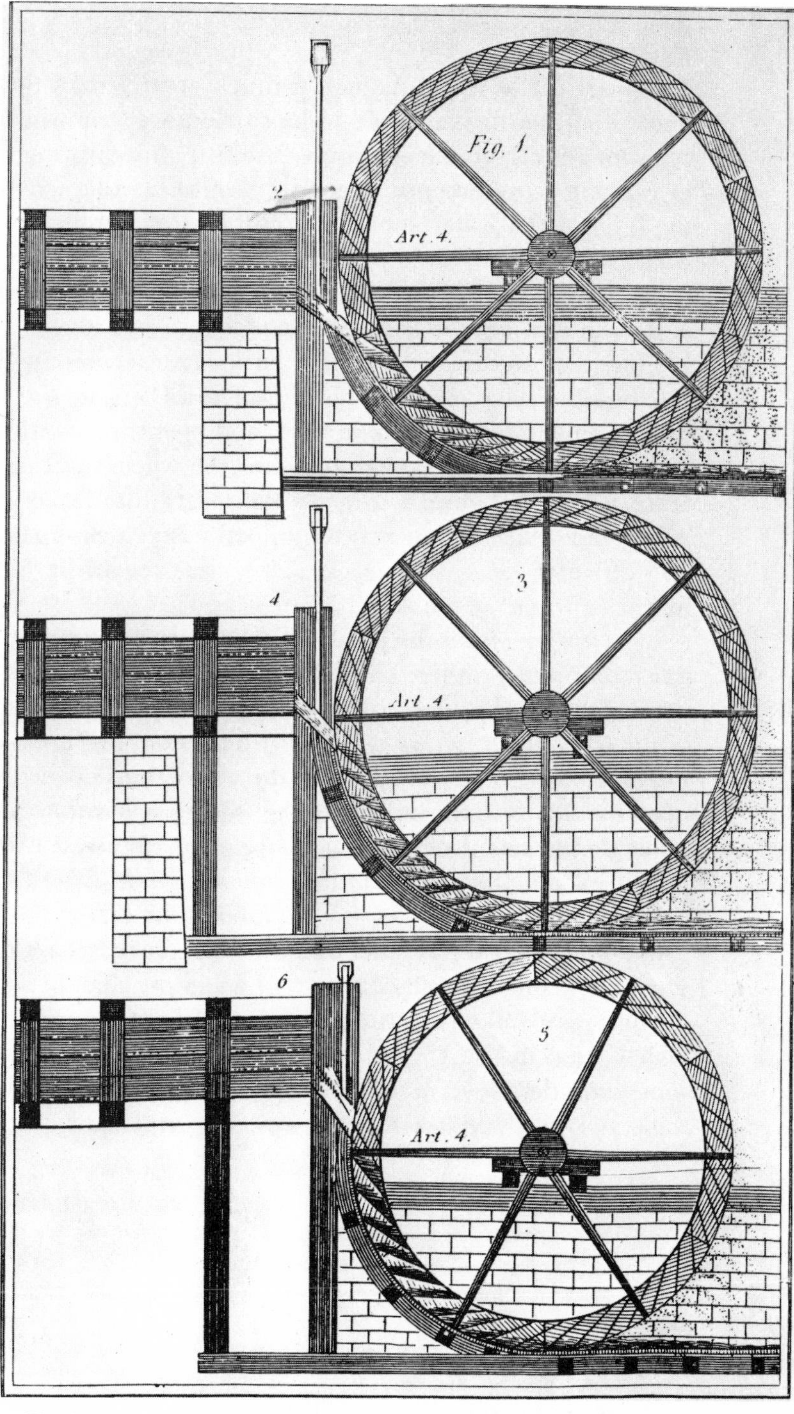

Fig. 20. Breast wheels. Dimensions prescribed by Thomas Ellicott, millwright, in Oliver Evans, *Young Mill-wright and Miller's Guide* (Philadelphia, 1807).

it allowed use of wheel diameters much greater than the height of fall, since the water did not need to be carried over the wheel's top, as with overshot wheels, giving operating flexibility in adjusting revolving speed to working requirements. Second, it could be adapted to variations in the level of the water supply in millpond and millrace resulting from seasonal and short-term variations in streamflow. Third, the rotation of the lower part of the breast wheel *with* instead of *against* the current of flow in the tailrace enabled the wheel to run longer and with much less loss of power in the so-called backwater accompanying floods than was possible with the overshot wheel revolving in the opposite direction.[21] Depending on circumstances and operating considerations, water was admitted at a level slightly below the wheel shaft up to a short distance below the wheel's crest. In the larger installations of the factory age, breast wheels came to be equipped with gates—openings for admission of water—at several levels to be used according to the prevailing height of water in the supply source.

In its original form the breast wheel, like the undershot, was built with flat radial floats rather than with the troughlike buckets later adopted; yet it obtained the effect of buckets by having the lower quarter or third of the wheel turn within a close-fitting apron, or "breast," which, with minor leakage, held the water on the wheel (and pressed with its weight upon the floats) to the bottom of the wheel's revolution, an improvement upon the overshot wheel, which began to lose water from its buckets before reaching the low point in its revolution. In Britain the close-fitting breast was customarily built of masonry; in this country it was more commonly of wood. Whether because of greater care and cost in construction or less concern with efficiency of water use than in Britain, the close-fitting breast in combination with radial floats seems not to have been widely used in the United States, although it is often referred to and sometimes described in treatises on waterpower and millwrighting. The mills at Lowell, Massachusetts, were an important exception.[22]

[21] For a somewhat different statement of the advantages of the breast wheel over the overshot, see Fairbairn, "On Water-Wheels with Ventilated Buckets," pp. 45 ff.; also his *Treatise on Mills and Millwork*, pt. 1, pp. 112 ff. Oliver Evans early called attention to "the necessity of air-holes to let air into the buckets, that the water may have liberty to escape freely" (*Miller's Guide*, Art. 72). See also d'Aubuisson, *Treatise on Hydraulics*, pp. 372–75.

[22] James B. Francis, *Lowell Hydraulic Experiments, 2d ed.* (New York, 1868), pp. 1–2. On the British preference for substantial construction with the best masonry work, see Nicholson, *Operative Mechanic*, pp. 104 ff. See also Rankine, *Manual*, pp. 160–61. For American views and practice, see Allen, *Science of Mechanics*, pp. 208–9; Jacob Bigelow, *Elements of Technology* (Boston, 1829), pp. 257 ff.; Henry Pallett. *The Miller's, Millwright's,*

The breast wheel appears not to have played an important role either in Britain or the United States until after 1800. Its significant use was identified with the emerging industrial age rather than with water milling in its traditional phase. The pattern for its use in cotton textiles in this country was set at Lowell, where the breast wheel of massive dimensions was securely established by 1830 if not earlier. From Lowell it spread to most large-scale manufactories depending upon waterpower. The breast wheels in general use at the Lowell mills were of the type known as pitchback wheels (Fig. 21). In this form of wheel the water was admitted at a point fairly close to the crest, somewhat in the manner of the overshot wheel, but on the upstream side. The pitchback wheel (in Britain known as "back-shut") had the advantage over the overshot wheel of not being encumbered "with a useless load of water at its top, where it does nothing but add to the weight upon the necks or pivots of the wheel-shaft, and to the consequent loss of power by the increased friction upon them."[23]

A Wheel in a Class by Itself

Throughout the long history of water milling before the introduction of the hydraulic turbine in France during the 1830s, there was only one wheel of importance which in variant forms and under different names turned in a horizontal plane on a vertical shaft.[24] In America as in the Old World this wheel played an important role, though it was often ignored or taken for granted. In the United States the wheel gave its name to the small tub, or grist, mill. Historically the great merit of this small waterwheel, ranging in design from the rude to the elegant, was its simplicity and low cost of construction and its admirable adapta-

---

and Engineer's Guide (Philadelphia, 1866), pp. 222–23; Overman, *Mechanics for the Millwright*, p. 205; and Webber, "Ancient and Modern Water-Wheels," pp. 139 ff. Evans discusses the pros and cons of breast wheels, *Miller's Guide*, Art. 72 (and pl. 4, fig. 31); Ellicott describes the breast wheel, indicating a certain preference for the overshot wheel under similar conditions (ibid., Arts. 122–124).

[23] Evans, *Miller's Guide*, p. 377 ff., quoting G. Mainwaring of England. Jacob Perkins is credited with calling attention first in this country to the pitchback wheel by erecting a large one in a nail factory at Newburyport, Massachusetts, in 1796 (Greville and Dorothy Bathe, *Jacob Perkins: His Inventions, His Times, and His Contemporaries* [Philadelphia, 1943], p.18). See also Fairbairn, *Treatise on Mills and Millwork*, pt. 1, pp. 122–23.

[24] See Bennett and Elton, Moritz, White, and other works on the history of water mills cited in note 1 above (pp. 51–52). See also Paul N. Wilson, *Watermills with Horizontal Wheels* (Kendal, England, 1960).

Pitchback wheel
Head and fall—19 ft.
Wheel diameter—18 ft.
Wheel breadth—6 in. for each ft. diameter of millstone

Middling overshot wheel
Wheel diameter—18 ft.
Wheel breadth—6 in. for each ft. diameter of millstone

Fig. 21. Pitchback and overshot waterwheels. Dimensions prescribed Thomas Ellicott, millwright, in Oliver Evans, *Young Mill-wright and Mil Guide* (Philadelphia, 1807).

tion to what for many centuries was by far the most important application of waterpower in the Western world—the reduction of bread grains to meal between millstones. The small mill with this simple motor, until recent years actively at work in parts of southern Appalachia and other less developed regions of the Western world, was merely a mechanized form of the ancient quern (see Appendix 2). The wooden rod by which the quern's upper stone was turned by hand was replaced by the vertical shaft rising from the wheel below. No cog or gear wheels were required, as in mills with vertical wheels, to change the direction of motion from horizontal to vertical shafts and to obtain the high rotating speeds required for effective grain milling. With its simple one-piece-in-motion construction and direct connection with the upper millstone, the tub wheel supplied the simplest—and perhaps the oldest—of water mills with the simplest of motors.

This mill and its small waterwheel represented mechanical power in its most "democratic," that is to say, egalitarian, form. Under appropriate geographic and economic conditions they placed power for the performance of an important household task at the disposal of country folk living on and directly from the land. These simple mills served not only small farm communities but frequently individual farm families. As late as the 1960s they were found in considerable numbers in the more isolated hill and mountain regions of southern and eastern Europe.[25] The multitude of small streams afforded numerous sites where the slope and volume of flow were seasonally quite adequate to local needs. The mechanical equipment and the structures of such mills are neither large nor complicated. They are made today as in the past largely of wood. For craftsmen skilled in their building, tub mills evidently required no more than one or two weeks of labor, depending on size, and an outlay of perhaps no more than several dollars to obtain the iron or ironwork required—and a similar amount for millstones if bought rather than made.

In its most common form the tub wheel resembled a sturdy wagon wheel, with or without a rim, the spokes replaced by arms with floats or buckets in a variety of forms. The spokes were wedged tightly into mortises sunk into the circumference of the large hub forming the lower end of the wheel shaft, which rested and turned upon

[25] Wilson, *Watermills with Horizontal Wheels*. See also Daniel W. Gade, "Grist Milling with the Horizonal Waterwheel in the Central Andes," and Felix F. Strauss, " 'Mills without Wheels' in the Sixteenth-Century Alps," *Technology and Culture*, 12 (1971):13 ff. and 23 ff.; Jorge Dias, "Moulins Portugais," *Revista de Etnografia,* no. 6, Museu de Etnografia e Historia; Hunter, "The Living Past in the Appalachias of Europe," p. 446 ff.

Fig. 22. A tub mill located on an Appalachian mountain stream. (Horace Kephart, *Our Southern Highlanders* [New York, 1913].)

a step (one-point) bearing attached to the bridgetree, the horizontal hinged at one end and free to move at the other (see Fig. 23). By means of a hand screw or by a simple wedging device, the entire wheelshaft and millstone assembly could be raised or lowered slightly to regulate the fineness of the meal.

In the simplest design, characteristic of the small gristmills of Scandinavia and outlying groups of islands whence presumably this wheel reached colonial America, the wheel floats, or blades, were simply short, flat boards, mortised into the hub obliquely or even parallel to the shaft.[26] In the more graceful forms, as found throughout much of southern and eastern Europe, the floats were and still are skillfully shaped in concave bowl or spoon forms, varying with the custom of the region. The floats varied in number usually from six or eight to more than twenty; overall wheel diameter ranged usually between three and six feet, although occasionally tub wheels reached eight or ten feet. They were used on falls as low as four to six feet and higher than thirty feet. Rotating speeds were adapted to the size of the millstones and ranged usually between some 60 rpm for the largest and 120 rpm for the smaller millstones. The smaller tub mills, as used on a single farm or by a small number of families, developed no more than a fractional horsepower; the larger ones, as in custom-toll mills, might develop 2 or 3 horsepower. The maximum capacity of well-built wheels probably did not often exceed 4 or 5 horsepower. The stream of water was carried to the mill often on troughs supported by trestles and then directed through an inclined open trough or closed trunk so as to strike the floats or buckets tangentially at one side of the wheel. Seasonal operations owing to wide variations of flow on many small streams were accepted as a matter of course; if mills on streams with more abundant and regular flow were not within acceptable distance, hand mills of the quern type or even stump mortars were put to use. During low periods owners of horse mills did a good business.

Owing to its small size and capacity, the tub wheel has received little attention in the literature of American water mills, although it occupies a certain prominence in studies of the preindustrial phase of technology. Its neglect probably reflects in some small part its inconspicuous

[26] The Van Name Collection of Photographs and Scrapbooks, Library of Congress, contains numerous examples of tub wheels of various forms. The Webber and Frizell articles on old-style waterwheels cited above (p. 62, n. 14) mention the tub wheel; Frizell in some detail with drawings to illustrate his discussion. For an excellent description of a type of tub wheel still in use in Appalachia, with photographs and drawings, see Eliot Wigginton, ed., *Foxfire 2* (Garden City, N.Y., 1973), pp. 142–63.

Fig. 23. Tub mill c. 1900, Madison County, N.C. Now in the Mercer Museum, Doylestown, Pa., this mill, complete save for the enclosing house and supply-trough penstock, was one of six mills found in the district. Within the mill structure it occupied a space some four to five feet square and seven to eight feet high, overall. The millhouse stood directly on a swift, narrow mountain stream six to eight inches deep. Without the aid of a dam the water was carried directly from the stream by a rough, open wooden trough on a level some fifteen feet to a small forebay. From here it was directed against one side of the wheel by a closed wooden trunk at an angle of about 45° with a fall of six feet. (Photo courtesy of the Mercer Museum, Bucks County Historial Society, Doylestown, Pa.)

position beneath the mill; the mill wheel of song and story was invariably a large overshot wheel, turning slowly beside the mill with picturesque dignity. Even in grain milling, by far its most important application, the unsuitability of so small a wheel for the demands of commercial operations in flour milling excluded it from more than the briefest of mention in discussions of power supply. An efficiency in water use of more than 15 percent was rarely accorded it by American writers, on the basis of what actual tests it is difficult to determine. James B. Francis made the conservative estimate of 10 percent efficiency. Tests by French engineers, reported by d'Aubuisson, of a form of tub wheel extensively used in southern France showed a useful effect of more than 30 percent.[27] Census enumerators in this country often ignored the tub mill, evidently owing to its predominantly noncommercial character. From its introduction in this country, as largely elsewhere, this in many respects admirable machine was identified with the self-sufficiency of a subsistence economy in newly settled or underdeveloped regions. On many frontiers it followed closely on the heels of the first settlers during the post-Revolutionary period. In regions east of the Mississippi it is often mentioned in state and regional histories published before the Civil War. The tub mill was to persist longest in the stagnant backwaters of settlement in inhospitable hill country. Brief attention to the closing phase in its use will be given in a later chapter.

The most convincing evidence of the onetime importance of the tub wheel is its inclusion in the first (1795) and all later editions of Evans's *Young Mill-wright and Miller's Guide*. In chapter 4, Article 71, entitled "Of Tub Mills," Evans gave it nearly as much space and attention as the undershot wheel with its greater utility and much wider use. He stressed the "exceeding simplicity and cheapness" of the tub mill: "If well-fixed, it will not get out of order in a long time." For milling grain, tub mills were "preferable in all seats which have a surplus of water, and above 8 feet fall." In sawmilling its most familiar application was that pictured in the *Miller's Guide* in the drawing of the up-and-down sawmill (Fig. 27). Here it served as a convenient means of powering the log-carriage return at the end of the cut—known as the "go-back."

Despite the difficulties of gearing a tub wheel to provide

[27] Francis, cited by Maine, Hydrographic Survey, *The Water-Power of Maine, by Walter Wells* (Augusta, 1869), p. 154; d'Aubuisson, *Treatise on Hydraulics*, p. 410. In another form of tub wheel the same engineers found an average efficiency of 12.5 percent (ibid., p. 416). Poncelet observed in 1825 that tub mills realized no more than one-fifteenth of the power expended when used for grinding grain, or about 7 percent efficiency (Arthur Morin, "Experiments on Water-Wheels, Having a Vertical Axis, Called Turbines," trans. Ellwood Morris, *JFI* 36 (1843):234–35).

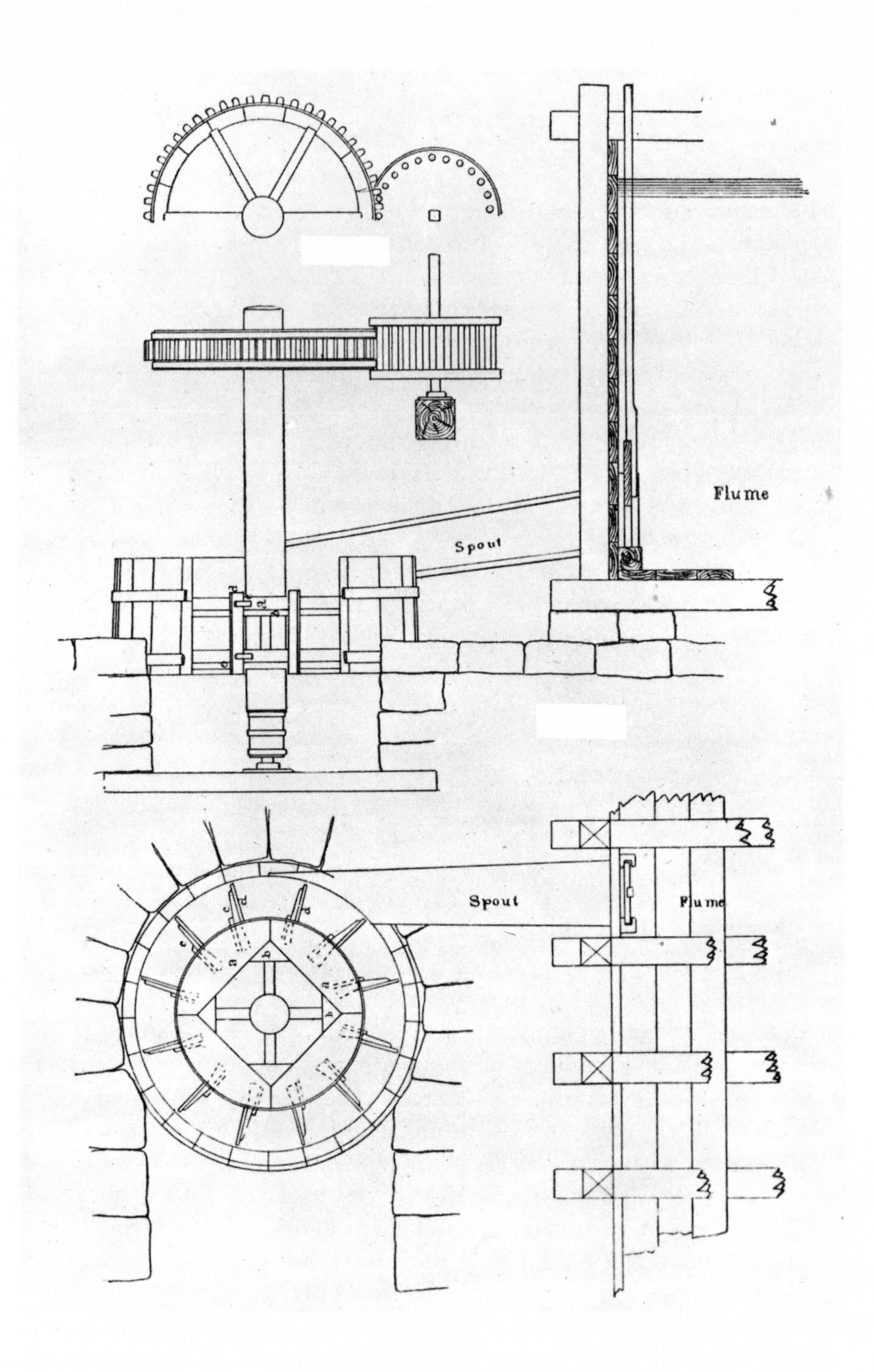

Flume
Spout
Spout
Flume

Figs. 24-26. Tub wheels showing power takeoff by cogwheel (*left*) or belt-pulley (*right*). (*Left page,* Joseph P. Frizell, "The Old-Time Water-Wheels of America," TASCE 28 [1893],) (*right page* Van Name Collection, Library of Congress.)

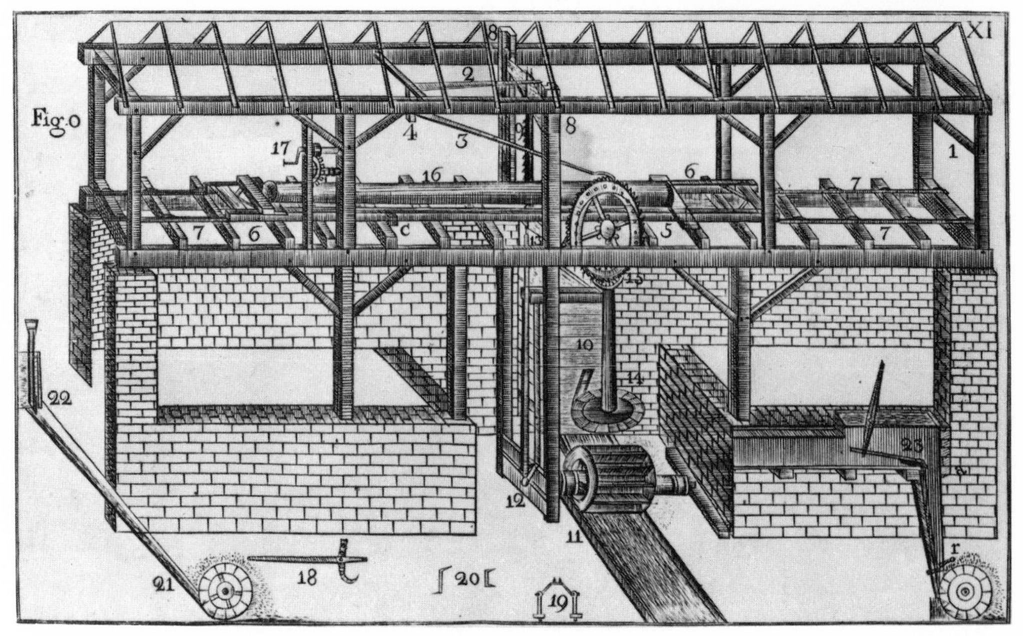

Fig. 27. The classic American up-and-down sawmill. (Oliver Evans, *Young Mill-wright and Miller's Guide* [Philadelphia, 1807].)

0–1 Frame of mill; uncovered; 52 ft. long and 12 ft. wide

2 Lever communicating power from saw-gate through feeder-pole (3) and rag-wheel (5) to move log carriage and log against saw a fraction of inch each stroke thru a pawl

5 Rag-wheel: Cogged on one side to mesh with cog-wheel on upper end tub-wheel shaft and with rounds meshing with cogs under log carriage; strong pins on opposite side rag-wheel for treading carriage back by foot when power drive is out of order

6 Log carriage: 32 ft. (29 ft.) long by 4 ft. wide; under-side cogged to mesh with rag-wheel in power drive; log held in place by means of dogs (19–20)

7 Ways, wooden, on which carriage runs, extending entire length of mill

8 Fender posts: 12 ft. long, 12 in. square; grooved for movement of saw-gate

9 Saw blade: 6 ft. long by 7–8 in. wide (new) held in tension within saw-gate

10 Forebay of water, projecting through foundation wall of mill

11 Flutter wheel: Primary mill motor driving saw-gate thru crank and pitman

12–13 Crank and pitman—poorly represented; between fender posts below saw

14 Tub wheel: "A very light wheel, 4 feet in diameter"; thru cog-wheel on upper end of its shaft reverses rag-wheel and log carriage when its gate is hand-opened

15 Cog-wheel on upper end tub-wheel shaft; meshes with cogs under log carriage

16 Log on carriage, partly sawed through

17 Crank and hand-windlass to enable one man to draw heavy logs on to the mill

18–20 Cant hook and dogs—for rolling logs and fixing in place

21–23 Different type chutes for directing water against flutter-wheel, depending on circumstances: open and inclined (21–22); vertical (23)

"The sluice drawn from the penstock 10, puts the wheel 11 in motion—the crank 12 moves the saw-gate, and saw 9, up and down, and as they rise they lift up the lever 2, which pushes forward the hand-pole 3, which moves the rag-wheel 5, which gears in the cogs of the carriage 4, and draws forward the log 16 to meet the saw, as much as is proper to cut at a stroke. When it is within 3 inches of being through the log, the cleet C, on the side of the carriage, arrives at a trigger and lets it fly, and the sluice gate shuts down; the miller instantly draws water on the wheel 14, which runs the log gently back, & c." The tub wheel (14) used for powering the log-carriage return "is a very light wheel, 4 feet in diameter, and put in motion by a motion of the foot or hand, at once throwing it in gear with the rag-wheel 5, lifting off the hand and clicks from the ratchet, and hoisting a little gate to let water on the wheel. The moment the saw stops, the carriage 6 with the log begins to move gently back again."

the reciprocating movement required in the up-and-down sawmill, this feat was early and effectively accomplished, at least in the important lumber industry of Maine. On the Saco River at Saco Falls, for example, owing to the rugged and rocky character of the terrain, the mills were "located where the power could be cheaply applied to the simple '*tub*' wheels operating under low heads of six to ten feet, the whole fall of 40 feet being divided by dams into three falls, respectively 8, 16 and 16 feet high." In 1918 in his paper on the Norse mills in America, Frederick H. Shelton reported that in Maine the tub wheels often found in use had in the past been extensively employed both inland and along the coast in sawmilling. They were, he noted, "usually used to drive saw mills by a bevel gear from the top of the vertical shaft." He described the tub wheel, no longer in use, of a woodworking factory south of Bangor as follows:

This wheel is about four and one-half feet in diameter, worked in a two-inch plank tub case about seven feet in diameter, and has six plank blades, two inches thick by about twenty inches deep and twenty-four inches long. These blades are bolted on to six faces of the heavy shaft that is at that point made hexagonal, instead of being mortised into the shaft; which hexagonal bolting arrangement makes a much easier and strong construction. Each blade moreover, carries a small apron of wood, bolted to it, on the lower edge of the near side face to hold the water a little or prevent its passing through too fast. The shaft is fourteen inches diameter and about seven or eight feet to the bevel gear above. The wheel is carried on a sole tree or sole-hurst that has no adjustment, as there is no grain grinding variation. . . . The inlet water comes into the wheel through the usual chute, at an angle of about thirty degrees. I believe this to be a typical wheel of the Maine district of the past fifty to seventy-five years or so. . . . It is a Norse wheel in a case forming a "tub" wheel.[28]

Country mills apart, the most impressive example that has come to my attention of the tub wheel's use in manufacturing was at the Springfield Armory of the federal Ordnance Department. Here in 1830 some 250 hands were engaged in the production of 12,000 stands of small arms annually with powered machinery performing a variety of forging, boring,

[28] Frederick H. Shelton, "Norse Mills of Colonial Times in Pennsylvania," *Collections of Papers Read before the Bucks County Historical Society*, 5 (1926):180–81; Maine, Hydrographic Survey, *Water-Power of Maine*, p. 59. Evans's biographers suggest that his attention was drawn to improvements in milling by "the inadequate and antiquated methods employed in grist mills which he saw everywhere around him [on the Eastern Shore of Maryland]. These small stone or log built mills, with shingle roofs, stood on almost every little stream" (Greville and Dorothy Bathe, *Oliver Evans: A Chronicle of Early American Engineering* [Philadelphia, 1935], p. 10).

grinding, milling, and polishing operations. A total of twenty-seven wheels was employed in the several "water-shops," so-termed from their locations along the millstream, which passed through the grounds. Thirteen of them were tub wheels. With an effective head of 3.1 feet the tub wheels operated with an estimated efficiency of 10 percent and developed a calculated average of 0.5 horsepower. Six overshot wheels under an effective head of 3.6 feet and an estimated efficiency of 20 percent developed an average of 2.0 horsepower.[29] One of the limiting conditions in the use of wheels of this character in manufactories of any size must have been the problem of conveying the water to and from the wheels at the point of use.

Apart from the problem of water supply and removal when used to drive individual machines as above, the horizontal wheel of this type had other shortcomings narrowly limiting its usefulness. These were its small capacity and the difficulties in power takeoff where the use or distribution of power required a change in direction from the vertical to the horizontal. Small capacity was a consequence both of low efficiency and of the difficulty of giving a wooden wheel of this basic design, resting and turning on a single-point bearing, the dimensions and strength necessary to develop at best more than several horsepower. As noted earlier, wheels above six feet in diameter were unusual; ten feet is the largest wheel that has come to my attention. Even with the most careful workmanship and skillful joinery it was difficult in wood construction to obtain the tightness and durability of joints and the structural rigidity and overall strength to meet the stresses attending the development of any considerable power. Where wheels were used in operations requiring power delivery on a horizontal shaft, capacity was further reduced by friction losses in the rude wooden gearing long relied upon. The use of belt pulleys mounted on the upper end of the vertical shaft, as shown in Figure 23, of tub wheels was quite feasible, as also was spur gearing. Bevel gearing was also feasible but was

[29] See "Report of a Survey of Sites for a National Armory on the Western Waters," *American State Papers, Military Affairs*, 4:491 ff. W. R. Bagnall has described the largest tub wheels to come to my attention as ten feet in diameter, eighteen inches deep, "and wholly of wood"; they were employed in driving a small early-nineteenth-century cotton-spinning mill (568 spindles) in Rhode Island (*The Textile Industries of the United States* [Cambridge, Mass., 1893], p. 445). In 1832 Thomas Blanchard advised the use of a ten-foot tub wheel, supplied by "spouts" directing the water to opposite sides of the wheel—a practice for increasing power to which Evans had early called attention in his *Miller's Guide*, Art. 71, p. 169. The tub wheel was to drive a circular sawmill of his design (T. Blanchard to E. B. Usher, 19 July 1832, Usher Papers, MS 201, U85, Baker Library, Harvard University, Cambridge, Mass.).

not only power wasting but noisy, troublesome, and subject to breakage.

Some engineers and historians were later to assign to the tub wheel a worthy place in the genealogy of the hydraulic turbine. In France, the land of the turbine's origin, it was indeed the starting point. Turbine and tub wheel alike were horizontal wheels with vertical shafts; of small diameter, both were quick-running in contrast to the slow motion of bucket wheels of the breast and overshot types. Yet in hydraulic terms there was little in common between the strike-and-splash-off action of the tub wheel and the pressure-reaction character of the turbine. Turbines operated to best advantage submerged; tub wheels could operate only above the tailwater. In efficiency the two wheels were at opposite poles. By 1850 the tub wheel was the element of the traditional technology most completely identified with the past, whereas the turbine had already demonstrated those characteristics and qualities that insured for waterpower a major role for the indefinite future.

Wheels with Horizontal Shafts

Among the traditional waterwheels those with horizontal shafts were largely free from the limitations of the tub wheel. The shafts of undershot, overshot, and breast wheels with two-point support could be given bearings of sufficient size and strength at each end of the shaft, with appropriately sturdy foundations, to provide stable support for as large and powerful wheels as circumstances might require. The breast wheels in general use at the largest New England textile mills before 1850 were as large as a small house and weighed twenty tons; they developed some 200 horsepower. Operationally the great merit of the vertical wheel was simplicity of power takeoff for the most common industrial uses. Three methods were available for taking power from a horizontal shaft without resort to gearing: cranks, wiper cams, and belt pulleys. Where rotary motion was required, pulleys and belting were to prove the simplest method. In the numerous instances where a reciprocating action was required there was a choice between the wiper cam and the crank. The cam (tappet, wiper) was the simplest and most versatile of power takeoff devices. Short projecting pieces or knobs of hard wood of appropriate form and dimensions were mortised securely into an extension of the heavy wooden wheel shaft at such points in the circumference and length as were necessary for the operation at hand, seen in Figures 8 and 106. As the waterwheel revolved, each cam or wiper in turn struck a matching projection on the helve of the forge hammer, the mallets of the fulling mill, or the stamps of the ore-crushing battery in a lift-and-let-fall or strike-and-rebound action. Here was a very simple and effective means of communicating force.

The most important use of the crank in water milling was in the up-and-down sawmill. During the many generations of its use, the heavy crank, usually of cast iron and weighing well over a hundred pounds, was the most critical component of the sawmill next to the saw itself. Attached to the shaft end of a small undershot wheel—known as the "flutter wheel" from the rapidity of its action and the flying water spray—the crank communicated the up-and-down motion to the saw frame through a wooden connecting rod, known appropriately from the human sawyer it replaced as the "pitman." Figure 27, the drawing of the oldstyle sawmill, remained a fixture in Evans's *Miller's Guide* from its first appearance in 1795; unfortunately the crank and pitman are barely discernible at the left end of the flutter wheel. The crank served similarly to place in motion the blast bellows of the iron forge and blast furnace.[30] Vertical wheels were the mainstay of flouring mills and the larger gristmills. The wooden gear wheels employed cogs and rounds and bore such long-obsolete names as big and little cogwheels, face wheels, trundles, wallowers, and spur wheels. For all their bulk, weight, friction, and maintenance problems they provided the necessary means for changing the direction of motion from the horizontal wheel shaft to the vertical millstone spindles and for obtaining the high rotating speeds required in grain milling.

Wheel capacity and rotating speeds were the central operating characteristics whatever the wheel type and the industrial operation. Within the limits of a mill seat's potential and a wheel's efficiency in using water, capacity could be varied simply by altering the wheel's breadth and the dimensions of the floats or buckets. As noted earlier, a wheel's rotating speed varied inversely with its diameter: the smaller the diameter, the faster it ran. The massive breast wheels of the Lowell

[30] The various power take-off devices are illustrated in such older works as Agricola, *De Re Metallica* (1556), pp. 175–99 in the Hoover edition; Bernard F. Belidor, *Architecture Hydraulique*, 2 pts. in 4 vols. (Paris, 1782–90): vol. 1; and similar works by Leupold, Ramelli, Strada, and others. Several examples are given in Maurice Daumas, ed., *A History of Technology and Invention*, 2 vols. (New York, 1969), 2:181, 208–9, 250–51. For nineteenth-century examples, see technical dictionaries and encyclopedias such as Appleton's and Knight's. In the literature of the history of technology and industry I have come upon no reference to the efficiency of the cam method of transmitting power between wheel shaft and the driven machine.

Fig. 28. Wooden gearing, 19th-century gristmill. To drive a millstone at speeds in excess of 100 rpm with a waterwheel turning at less than 10 rpm, step-up wooden gearing was used. In the photograph the small 23-in. lantern gear powered the upper millstone, not visible but attached to the spindle marked 4985. The crown gear (64-in. diameter) at the right conveyed power from a secondary shaft and represented a second acceleration in speed from the slowly turning waterwheel. These gears were originally in the Phillips Mill, Solebury, Pa. They are now in the collections of the Mercer Museum of the Bucks County Historical Society. (Photo by Charles A. Foote. Courtesy of the Mercer Museum, Bucks County Historical Society, Doylestown, Pa.)

mills, with diameters usually of about fifteen feet, ran with a customary operating speed of six rpm; and the great sixty-two-foot overshot wheel of the Burden Iron Works in Troy, New York, moved at the sedate pace of only two rpm.[31] Until wheels of the reaction and turbine types were introduced in the 1820s and 1830s, the only quick-running waterwheels were the tub wheel of the small gristmill and the flutter wheel of the sawmill (see Table 3).

The concept of useful effect, or efficiency, obtained a certain currency in engineering circles from the widely reported results of John Smeaton's experiments with waterwheels (see above, p. 67, n. 19). Although the determination of efficiency was probably outside the competence of most millwrights, Evans's summation in the *Miller's Guide* of Smeaton's findings—that impact wheels of the undershot type were only one-half as powerful as those moved by gravity—was doubtless confirmed in a general way by the millwright's own experience, as was Evans's emphasis on the low useful effect of the tub wheel.

There was little apparent concern with uniformity of motion in the traditional technology of milling. In such rude mill operations as pounding, blowing, grinding, sawing, and raising water—evidently even in spinning yarn in the early textile mills—there was little need for such uniformity, or advantage to be gained by it. In manufactories of the slowly emerging industrial age the desirability of waterwheel governors only gradually obtained recognition.[32] No instance of their use in small water mills of the traditional sort has come to my attention. Even as late as 1865 some of the waterwheels of the large textile mills of the Lawrence Manufacturing Company in Lowell, Massachusetts, had no governors.[33] Changes in wheel speed owing to

[31] Ithamar Beard, "Practical Observations of the Power Expended in Driving the Machinery of a Cotton Manufactory at Lowell," *JFI* 11 (1833):8; "Estimate of Power Required to Drive Lawrence Mill No. 5," JBF to Lawrence Manufacturing Co., 2 Feb. 1859, LCP, vol. A-181, no. 89. For the Burden wheel at Troy, see Appendix 9.

[32] See Allen, *Science of Mechanics*, p. 211; Nicholson, *Operative Mechanic*, pp. 35 ff., 45 ff. On the related flywheel action of waterwheels, see Evans, *Miller's Guide*, Arts. 22, 29, 70, 158, and appendix, pp. 373 ff. See also "On Equalizing the Motion of Machinery," *American Journal of Improvements in the Useful Arts* 1 (1828): 189–92. Allen's reference probably relates to the use of governors in cotton-spinning mills in the Rhode Island area.

[33] Memorandum of JBF for H. V. Butler and Company, 6 July 1865, LCP, D-3, 654. *Appleton's Dictionary of Machines*, s.v. "Water-Wheels" reported: "In the higher class of wheels, and especially where great steadiness of motion is required, the shuttle [varying admission of water to the wheel] is worked by a self-acting apparatus." In the article the accompanying drawing of a waterwheel governor was taken from

Table 3. Dimensional and performance data for waterwheels

Type of waterwheel	Diameter	Height of fall (in ft.)	Consumption of water (c.f.s.)	Revolutions per minute wheel	millstone	Efficiency (percent)
Flour mill						
Undershot	12–18 ft.	8–20	28.0–11.0[a]	24–26	88–106[b]	15–30
Overshot	9–30 ft.	11–36	11.4– 3.4[c]	14–8	103–99	50–70
Breast	15 ft.[d]	6–15	30.0–10.0[c]	14–19	101–98	40–60
Gristmill						
Tub wheel						
4 ft. stone	26–42 in.	8–20	17.0– 7.0	122	122	10–15
5 ft. stone	33–56 in.	8–20	–	98	98	10–15
6 ft. stone	40–66 in.	8–20	41.0–16.0	81	81	10–15
7 ft. stone	47–78 in.	8–20	–	70	70	10–15
Sawmill						
Flutter wheel (width 66–12)	32–57 in.	6–30	–	120	–[b]	15–30

SOURCE: Oliver Evans, *The Young Mill-wright and Miller's Guide* (1850), pp. 161 ff., 292-340ff.
NOTE: Tables for waterwheels used in flour mills do not give the width of wheel, this dimension being left to the discretion of the millwright but proportioned to the amount of water to be handled with a given height of fall.
[a] With a 15-foot wheel, making 17–27 rpm and driving a 5-foot stone 100–98 rpm.
[b] 120 strokes per minute.
[c] With 5-foot stones.
[d] Only diameter given.

variations in load, or in the head from changing levels in the millpond, could readily be made by hand adjustment of the gate opening in the penstock by means of a rod provided for that purpose.

---

the large waterwheel installation at Greenock, Scotland. Francis, *Lowell Hydraulic Experiments* (Boston, 1855), pl. I, gives a clear drawing of the governor of the Tremont turbine (1851). The earliest reference to the use of a waterwheel governor in the United States that has come to my attention is found in "Report on National Armory on the Western Waters," by an ordnance board, January 18, 1825: "In cases where a variety of operations are to be performed by one wheel, it is calculated to have governors (so-called) to regulate the motion. These machines are calculated to raise the gate, and give additional power by increasing the column of water on the wheel when the motion is too slow, and depress the gate and diminish the column when the motion of the wheel is too rapid. By this means a regular and nearly uniform motion is preserved in the operation of all the wheels and machinery (*American State Papers, Military Affairs,* 2:755).

A Technology of Wood

From its very beginnings water milling rested upon a technology of wood. It remains largely such in the remote regions of its survival today. This technology was indeed a fundamental condition of its practical utility and a cause of its eventual decline. Except for the stonework used in foundations, wheel pits, and the like in this country as in Europe, most mills and their machinery continued to the end of the eighteenth century to be built almost entirely of wood. This fact was owing, of course, to the scarcity, cost, and difficulty of working metal, especially iron, and the widespread availability of various woods and the ease and skill with which they could be shaped.[34] To the ancient crafts of carpenter and joiner were in time added those of shipwright, wheelwright, and eventually that predecessor of the mechanical engineer, the millwright. All these artisans were skilled in their several ways in the arts of cutting, shaping, and joining wood. In the construction of wheels and gearing the art of joinery was especially important, using with great skill and often impressive results the various forms of mortise and tenon, scarfing, and the like. Appropriately, the wooden pins were termed treenails (trenails, trunnels) and wedges.

Against wood's advantages as a structural and mechanical material must be placed its deficiencies, especially in the moving parts of machinery and as compared with iron: a lack of strength; a readiness to wear, decay, warp, and shrink, and, under conditions of intermittent use and alternate wetting and drying, to loosen at the joints. The large dimensions necessary to offset the low strength of even the best woods—ranging from one-fourth to one-tenth the strength of wrought iron, according to the character of the stress—required the use of shafts and gearing that were heavy, cumbersome, and power wasting in their friction. The creaking and rumbling of moving parts that in time became loose and ill fitting announced the presence of a water mill for hundreds of yards about on a calm day. Wheel shafts two feet in diameter were common, and in the larger mills the shafts reached three feet and more. An overshot wheel of eighteen-foot diameter, with main gearing to drive a two-run flour and grist mill, probably developing at most

[34] Use of wood as a fabricating material for many purposes in the United States was frequently remarked upon by foreign travelers. Useful statements are also found in Nelson L. Derby, *Report on Architecture and Materials of Construction, Vienna International Exhibition, 1873* (Washington, D.C., 1875); and Charles R. Dodge, *A Descriptive Catalogue of Manufactures from Native Woods . . . at the World's Industrial and Cotton Exhibition at New Orleans, La.* (Washington, D.C., 1886).

15 horsepower, weighed when dry about eight tons. That meant a weight of about one-half ton per horsepower, not including the weight of the water absorbed by the wood and carried in the buckets when the wheel was in operation.[35]

The power loss by friction in wheel and gearing may well have totaled a fourth and more of that generated. With the double-gearing usual in two-run mills—consisting in millwright's terminology of master cogwheel, countercogwheels, wallowers, and trundles—wear and tear was heavy. In busy mills the almost constant attention required to detect worn and broken cogs was a source of trouble and expense. The superintendent of the Springfield Armory reported in 1825 that wooden waterwheels would not last more than eight years without considerable repairs: "With constant expense they may be kept along with frequent interruptions two or three years longer." George Woodbury, writing from experience in rehabilitating and operating an old-style mill, reported that "coopering the wheel" was perhaps the biggest maintenance item. He cited mill books showing that annual repairs required the complete replacement of every part of the wheel in five years.[36]

Iron was used sparingly in the mill wheel, gearing, and throughout the water mill. The most advanced American practice as revealed in the 1795 edition of Evans's *Miller's Guide* confined its use principally to the journal inserts—gudgeons—on which the waterwheel and main cogwheel shafts rested and turned, together with the reinforcing bands shrunk on the shaft ends, which the use of gudgeon inserts virtually made necessary. Iron was no less essential in certain parts of the mill machinery—saws and cranks in sawmills; the ironwork of millstone

[35] My reckoning, based on the bills of materials in Evans, *Miller's Guide*, Art. 155, p. 333, and weight of American timbers tabulated in *Coal Miners' Pocketbook*, 11th ed. (New York, 1916), p. 283. A British authority gave the following estimates of weight for wooden overshot wheels: 12.5 ft. diameter by 7 ft. breadth, 5.7 tons; 16 ft. diameter by 9 ft. breadth, 10.5 tons; 24 ft. diameter by 12 ft. breadth, 23.7 tons (Robertson Buchanan, *Essay on the Shafts of Mills* [London, 1814], p. 33). One important advantage of iron over wooden waterwheels was their lesser weight in relation to size and capacity.

[36] George Woodbury, *John Goffe's Mill* (New York, 1948), pp. 98–99; Col. Roswell Lee to Col. G. Bomford, Washington, D.C., 19 July 1825, Springfield Armory Records, National Archives. See also *Scientific American* 13 (16 Jan. 1858):151; C. R. Weidner, "Test of a Steel Overshot Water Wheel," *Engineering News* 2 (Jan. 1913); Allen, *Science of Mechanics*, pp. 206–7. English experience was similar; see the account of mill-gear repairs in Walter Rose, *The Village Carpenter* (Cambridge, 1937), pp. 104 ff. By contrast, during thirty years of service, the great iron breast wheels—measuring 50 feet in diameter by 12 feet in breadth—installed by Fairbairn at the Catrine Textile Mills in Britain, 1825–27, "required little or no repairs, and . . . remain nearly as perfect as when erected" (Fairbairn, *Treatise on Mills and Millwork*, pt. 1, p. 126).

assemblies, such as spindles, rynds, steps, damsels, and bray irons; the hammerheads and anvils for the trip-hammers of forges. The cost and frequent difficulty in the less settled areas of obtaining the "hardware" of mills were long important factors in delaying the erection of mills and possibly in limiting their number. In the case of the two-run mill described in the *Miller's Guide*, the aggregate weight of ironwork, including 6 gudgeons and 22 reinforcing bands, was between 600 and 700 pounds, the eighteen-foot overshot waterwheel alone accounting for more than half of the total (see Appendix 3).[37] In view of the evidently greater size and capacity of this mill compared with those of a gristmill erected at Quaboag Plantation in central Massachusetts in 1669–70, for which a record of the ironwork exists, there had been little if any change in mill construction for well over a century. In the Quaboag mill, with an eight-foot-diameter wheel and but one run of stones, the aggregate weight of the ironwork—including hoops and gudgeons for the wheel shaft, bolts and keys, 20 small wedges, 250 spikes, and a single bar of steel weighing 162 pounds—reached 499 pounds. By the mid-1820s, however, major advances had been made in large merchant flour mills centering in the replacement of wood by iron in much of the millwork, including both shafting and gear wheels.[38]

So far as most ordinary water mills were concerned, reliance largely on wood construction in waterwheels and millwork continued to the Civil War years, as clearly revealed by the repeated reissue of Evans's *Miller's Guide* through the 1850s, evidence of the gradually extending use of iron being confined to brief news items and discussions introduced in the appendixes.[39] The principal exceptions were in the gradual replacement of the old-style wooden cog-and-stave gear wheels by cast-iron spur and bevel gear wheels. The latter were used without machining

[37] My calculation (see n. 35 above, p. 89). On the increasing use of cast iron in mill gearing and millstone accessories, see *JFI* 3 (1827):109–10. By this account there was a growing practice of using a mixture of ground emery and oil for "wearing in" mill spindles in the steps on which they rested and turned.

[38] Louis D. Roy, *Quaboag Plantation, Alias Brookfield: A Seventeenth Century Massachusetts Town* (West Brookfield, 1965), pp. 126 ff.; also Evans, *Miller's Guide*, pp. 365–69 and pl. 27.

[39] On use of iron, see W. Parkin, Sept. 1825, in Evans, *Miller's Guide*, pp. 370 ff.; also Robertson Buchanan, *Practical Essays on Mill Work and Other Machinery*, 3d ed. rev. (London, 1841), pp. 381–91. See also Allen, *Science of Mechanics*, pp. 206–7; Denison Olmsted, *Memoir of Eli Whitney, Esq.* (New Haven, 1846), pp. 68–69; Henry Burden to Col. Roswell Lee, 10 Nov. 1825, Springfield Armory Records. On the increasing use of iron, see later references in Overman, *Mechanics for the Millwright*, pp. 200 ff., 304 ff., 794 ff.; *Appleton's Dictionary of Machines*, s.v. "Water-Wheels."

directly as they came from small local foundries, for which in time they became an important product. For the waterwheels themselves the transition from the age of wood came not with the extension of the use of iron in wheel construction but in the gradual replacement from the 1820s of the old-style wheels by reaction wheels and from the 1840s by the new turbine wheels. As late as 1870 the author of a successor handbook to Evans's *Guide* remarked of sawmills in frontier districts where transport conditions were poor: "Almost the whole mill machinery must be improvised of wood on the spot. Timber, water power and space being of little value, while *iron* is a precious metal. . . . Mills are sometimes built almost entirely of wood, without bands or bolts, or even nails, with scarcely anything metallic except the saw."[40]

The Millwright's Craft

Millwrighting as the art of designing and building water mills was perhaps the most demanding craft of its time. Its practice by the millwright required familiarity not only with the building crafts of the carpenter, the joiner, and the mason but with blacksmithing, the wheelwright's trade, and the emerging craft of the machinist. Since the running of lines and the fixing of levels with fair accuracy were indispensable for the proper location and layout of mill and hydraulic facilities, the millwright had to be something of a surveyor as well. He was the key technician of the preindustrial and early industrial age, the craftsman predecessor of the civil and mechanical engineer. Many of the leading engineers of the early Industrial Revolution in Britain—the Smeatons, Rennies, Fairbairns and Bruntons—began their careers in the planning and execution of the enlarged water mills comprising the early factories. As described by William Fairbairn, who had himself traveled this route with distinction (and whose nostalgia in recollection must be allowed for), the millwright was almost the sole representative of the mechanical arts in his day. He was respected "as the authority in all applications of wind and water . . . an itinerant engineer and mechanic of high reputation."[41]

[40] David Craik, *The Practical American Millwright and Miller* (Philadelphia, 1870), pp. 100 ff.

[41] Fairbairn, *Treatise on Mills and Millwork*, pt. 1, pp. ix–xi. The word *millwright* dates back to 1481 in the *Oxford English Dictionary* and to 1640 in the *Dictionary of American English.* For British experience, see Cyril T. G. Boucher, *John Rennie, 1761–1821*

His craft went back at least to the late Middle Ages. Typically unschooled, though often apprenticed when opportunity offered, the millwright was largely self-taught, guided by the traditional lore of his vocation and the lessons of his own experience. According to Fairbairn, "He could handle the axe, hammer, and plane with equal skill and precision; he could turn, bore or forge with the ease and despatch of one brought up to these trades; and he could set out and cut in the furrows of a millstone with an accuracy equal or superior to that of the miller himself. . . . Generally he was a fair mathematician, knew something of geometry, levelling and mensuration. . . . He could calculate the velocities, strength and power of machines; he could draw in plan and section, and could construct buildings, conduits and water courses."[42] John Rennie (1761–1821), for instance, began an unusually broad training for his career at age twelve with a two-year apprenticeship under Scottish millwright Andrew Meikle. Following the apprenticeship, he attended high school and the University of Edinburgh, interspersing the practice of millwrighting with his education. His training was climaxed by a kind of journeyman's tour of major engineering works in Great Britain.[43]

Like other Old World craftsmen the millwright often found his way to America, in some instances under the auspices of colonizing agencies. Dard Hunter cites two advertisements from the *New York Gazette*, 15 August and 31 October 1768: "Thomas Shaw and Nathaniel Sedgfields, lately arrived from England, take this method of acquainting the public, that they are capable of building most sorts of mills, as grist-mills, paper and oil-mills." "Domenicus Andler, wheelwright, from Germany, acquaints the public that he can make most sorts of mills,

---

(Manchester, 1963), pp. 3–5; Joseph Glynn, *Rudimentary Treatise on the Construction of Cranes and Machinery* (London, 1859), p. v. Obituary notices in *PICE* and *PIME* frequently refer to the millwright beginnings of members; e.g., "John Chisholm (1777–1856)" (*PICE* 16 (1856–57):121); "John U. Raistrick (1780–1856)" (ibid., p. 129); "James M. Rendel (1799–1856)" (ibid., p. 133); and "Walter Hunter (1772–1853)," described as "almost the last in the race of clever English millwrights" (ibid., 12:161–63). On millwrighting in America, see William R. Bagnall, *Sketches of Manufacturing Establishments in New York City, and of Textile Establishments in the Eastern States*, ed. Victor S. Clark, 4 vols. (Washington, D.C., 1908), 4:2445; idem, *Textile Industries of the U.S.*, pp. 396–97. Robert Allison, "The Old and the New," *TASME* 16 (1894–95):750–51. On the itinerant aspect, see John Fritz, *The Autobiography of John Fritz* (New York, 1912), pp. 13, 24–25; Fred H. Colvin, *Sixty Years with Men and Machines: An Autobiography* (New York, 1947), p. 47.

[42] Fairbairn, *Treatise on Mills and Millwork*, pt. 1, pp. ix–xi.

[43] Boucher, *Rennie*, pp. 3–5.

such as grist, oil, fulling, paper and saw-mills; also forges and furnaces, to work either by water, wind or horses. He has practiced for many years in Germany, and has the honour to be employed by most of the princes."[44]

In a growing country and especially on the frontiers of settlement, even itinerant millwrights were for many years in short supply. Doubtless in many instances millwrighting must have been taken up as an occasional practice, the by-occupation of a carpenter, a blacksmith, or even a farmer with a knack for figuring and using tools. Such men probably began with the repair and advanced to the copying of existing mills, sharing in the inducements held out so often to the builders of first mills in new settlements. Of one master millwright in Maine, Ephraim Sands, it was reported of his skill in framing mill timbers with the great broadax: "So correct was his eye, and so accurate his stroke, that he refused to have his timber 'lined.' " That is, he refused to have his timber marked with a chalk or other line as a guide. "At every blow he carried his axe through the slab from the top to the bottom, and thus hewed in a day more than two ordinary axemen."[45]

For more than two centuries the building of water mills in this country proceeded without formal guidance. It is possible that occasional copies of such eighteenth-century treatises as those of Jacob Leupold (1725), Bernard de Belidor (1737), Antoine Deparcieux (1754), and Johann Beyer (1767), as well as John Smeaton's report of 1759, "The Natural Powers of Wind and Water to Turn Mills," found their way to these shores, although improbably into the calloused hands of a millwright. Well before the beginnings of colonization the basic types of water mills with their prime movers were established facts in the greater part of Europe and the British Isles. Except for the extensive use of windmills in some regions, chiefly in grain milling and pumping, falling water was the only source of mechanical power until the early 1700s. The theory and applications of hydraulics, with increasing reference to waterpower, were a subject of rising interest and study throughout the seventeenth and eighteenth centuries. A long succession of scientists and

[44] Dard Hunter, *Papermaking in Pioneer America* (Philadelphia, 1952), pp. 60–61; Defebaugh, *Lumber Industry of America*, 1:271–72, 306 ff.

[45] Gideon T. Ridlon, *Saco Valley Settlements, and Their Families* (Portland, Maine, 1895), p. 201. Another Maine historian remarked on "the immense number of millwrights, axe-men, sawyers and other laborers" employed in Maine lumber camps over two centuries (William D. Williamson, *The History of the State of Maine*, 2 vols. [Hallowell, 1839], 2:700). On millwrights in a Vermont county, see Hamilton Child, ed., *Gazetteer of Orange County, Vermont, 1762–1888* (Syracuse, N.Y., 1888), pp. 298, 301, 413.

engineers, from Jean Charles de Borda, John Smeaton, M. Fabre, and Leonhard Euler to Claude Burdin, Jean Victor Poncelet, Arthur Morin, and Benoît Fourneyron, aided at times by scientific societies and governments was active in the field.[46] Theoretical inquiries and practical experiments were directed to the improvement of waterwheels with particular reference to the forms, proportions, and operating conditions yielding the greatest useful effect from the water employed. These investigations led not only to a growing understanding of this branch of hydraulic science and to improvements in the traditional practice of water milling but led in the early nineteenth century to such important developments as Poncelet's curved-float undershot wheel, a rapidly proliferating class of so-called reaction wheels, and, most important of all, the turbine of Fourneyron, which was to open a new era in the history of industrial power.[47]

The Millwright's Companion

The first appearance in Philadelphia in 1795 of *The Young Mill-wright and Miller's Guide*, written by Oliver Evans with the collaboration of Thomas Ellicott, was a landmark in the history of American water milling. Both authors were to some extent conversant with the European literature of hydraulics, but the *Miller's Guide* reflected chiefly the practical experience of the two men. Evans was a leading engineer and inventor of his time and was also the author of the equally indispensable *Abortion of the Young Steam Engineer's Guide*, published in 1805. Ellicott, member of a family prominent in the flour-milling industry of the Middle Atlantic states, was a millwright of distinction and of wide and long experience.[48] The *Miller's Guide*, whatever its

[46] Nicholson listed fifty-two treatises on millwork distributed by date of publication as follows: the 1600s, five; 1700–49, ten; 1750–75, sixteen; 1776–99, fifteen; 1800–1826, six (*Operative Mechanic*, pp. 120–21). See also Buchanan, *Practical Essays*, pp. xxii ff.

[47] For a nineteenth-century treatment of hydraulics, see d'Aubuisson's *Treatise on Hydraulics*, described by the author as a "treatise on experimental and applied hydraulics and not on theoretical hydraulics." An extended discussion limited to motors is in *Appleton's Dictionary of Machines*, 2:786–841. A comprehensive and well-illustrated discussion of hydraulics is that by W. C. Unwin in *Encyclopaedia Britannica*, 11th ed., vol. 14. In the same volume is a brief article tracing the development of hydromechanics.

[48] See Kuhlmann, *Flour-Milling Industry*, pp. 27 ff. For the circumstances of the collaboration between the two men, see Bathe and Bathe, *Oliver Evans*, pp. 45 ff. See also Ellicott's introduction to pt. 5 of Evans, *Miller's Guide*. There is overlapping and du-

deficiencies, freed millowners and millwrights from complete dependence on the traditional lore and practice and such misconceptions and errors as these might sanction. It provided millwrights with a rational basis for the planning and design of water mills, together with much practical information for guidance in the actual construction and equipping of such mills. Although the reduction of oral tradition to the printed page might be expected to mark the beginning of its end, the failure significantly to revise the *Miller's Guide* in the sixteen editions appearing before 1860 must have served in some degree to perpetuate tradition at a new level.[49] Attention was directed almost entirely to grain milling. The principal emphasis was upon the type, size, and proportions of the waterwheel best adapted to the conditions found at a given mill seat. With its tables and explanations, its theorems and theories, its examples, calculations, and rules of thumb, its bills of material and numerous drawings, the Evans-Ellicott handbook ruled the water-milling field virtually unchallenged for more than half a century. It may be presumed to reflect primarily the conditions, needs, and practices of the leading grain-raising and flour-milling regions of the Middle Atlantic seaboard.

Of this substantial volume of some 400 pages (1850 edition), about one-fifth was devoted to "The Principles of Mechanics and Hydraulics," by Evans. Although enlivened to some extent by examples drawn from Evans's experience, much of it was too abstract and abstruse to be practically useful. One suspects that the greater part of this collection of axioms, laws, and principles—of falling bodies, of circular motion and central forces, of the motion of projectiles, of the different kinds of levers, on friction, on hydrostatic paradox, the seventh law of spouting fluids demonstrated—served more to bemuse than to enlighten most millwrights and millowners. It was commendable chiefly for the recognition given to certain aspects of the theory and principles underlying milling operations. Two briefer sections by Evans followed: "Description of the Author's Improvements" and "On the Manufacturing of Grain into Flour." These described Evans's patented flour-milling

---

plication in the discussion and tables of the collaborators, which at times may have been a source of confusion to the reader.

[49] Editions of Evans's *Miller's Guide* are listed and briefly annotated by the Bathes in *Oliver Evans,* p. 344. Changes by editors and publishers in successive editions appear to have been relatively minor. A critical collation of the editions remains to be done, together with an analysis of the successor handbooks that drew from Evans in greater or less degree. A greater need is for an engineering analysis and review of both the theoretical and practical sections of the *Miller's Guide.*

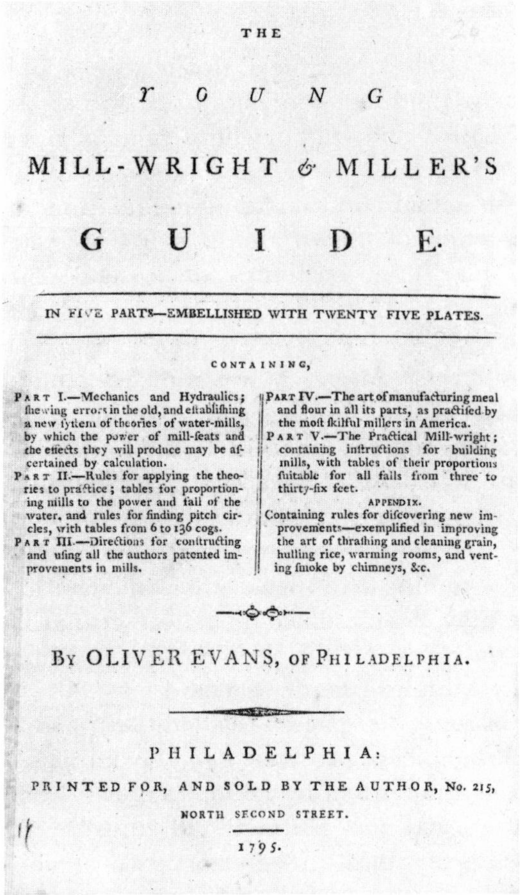

THE

YOUNG

MILL-WRIGHT & MILLER'S

GUIDE.

IN FIVE PARTS—EMBELLISHED WITH TWENTY FIVE PLATES.

CONTAINING,

PART I.—Mechanics and Hydraulics; ſhewing errors in the old, and eſtabliſhing a new ſyſtem of theories of water-mills, by which the power of mill-ſeats and the effects they will produce may be aſcertained by calculation.

PART II.—Rules for applying the theories to practice; tables for proportioning mills to the power and fall of the water, and rules for finding pitch circles, with tables from 6 to 136 cogs.

PART III.—Directions for contructing and uſing all the authors patented improvements in mills.

PART IV.—The art of manufacturing meal and flour in all its parts, as practiſed by the moſt ſkilful millers in America.

PART V.—The Practical Mill-wright; containing inſtructions for building mills, with tables of their proportions ſuitable for all falls from three to thirty-ſix feet.

APPENDIX.

Containing rules for diſcovering new improvements—exemplified in improving the art of thraſhing and cleaning grain, hulling rice, warming rooms, and venting ſmoke by chimneys, &c.

BY OLIVER EVANS, OF PHILADELPHIA.

PHILADELPHIA:

PRINTED FOR, AND SOLD BY THE AUTHOR, No. 215, NORTH SECOND STREET.

1795.

Fig. 29. Title page of the first edition of Oliver Evans's *Miller's Guide*. (Eleutherian Mills Historical Library.)

methods and equipment, the celebrated pioneering example of automatic, continuous-process manufacturing, and the concrete, everyday operations of flour milling.

For most millowners, manufacturers, and millwrights the main substance of the *Miller's Guide* centered in Part 2, by Evans, "Of the Different Kinds of Mills," and Part 5, "Ellicott's Plans for Building Mills." These two parts comprise some 150 pages—including twenty-five full-page plates with over a hundred drawings, the more important ones drawn to scale—filled with concrete, practical information and specific directions for planning and building water mills. "Of the Different Kinds of Mills" describes the several types of waterwheels and the conditions under which each could be used to advantage. The special value for millwrights of this and the somewhat overlapping and duplicating section by Ellicott lay in the "millwright tables," with explanations prepared for the several wheel types, as shown in Tables 4 and 5;

in the directions for laying out and building the different kinds of wheels and the gearing necessary for their use in gristmills and flour mills; and, for those seeking understanding as well as guidance, in such

Table 4. The millwright's table for tub mills

Total deſcent of the water, which is in this table made to ſuit the diameter of the wheel and head above it.	Diameter of the wheel.	Head above the wheel, allowing for the friction of the aperture, ſo as to give the water velocity 3 for 2 of the wheel.	Number of revolutions of the wheel per minute.	Double gear, 5 feet ſtones. No. of cogs in maſter cog-wheel.	Rounds in the wallower.	Cogs in the counter cog-wheel.	Rounds in the trundle.	Revolutions of the ſtone per minute.	Singlegear, 6 feet ſtones. Cogs in the cog-wheel.	Rounds in the trundle.	Revolutions of the ſtone per minute.	Cubic feet of water required per ſecond, for 5 feet ſtones.	Area of a ſection of the canal, allowing the velocity of the water in it to be 1 foot per ſecond.	Diameter of the pitch circle of the great cog-wheels for ſingle gear, pitch 4 1-4 inches.
feet.	ft.	feet.										cu.ft.	ſup. ft.	feet. inches.
10,51	9	1,51	14,3	54	21	44	16	102,9	60	11	78,	11.46	11,46	6:9 0-4 12-22
11,74	10	1,74	13,	54	21	48	18	98,	60	10	78,	10,3	10,3	
12,94	11	1,94	12,6	60	21	48	18	96,	66	11	75,6	9,34	9,34	7:5 1-4
14,2	12	2,2	12,	66	23	48	17	97,	66	10	79,2	8,53	8,53	
15,47	13	2,47	11,54	66	21	48	17	99,3	84	12	80,7	7,92	7,92	9:5 1-2
16,74	14	2,74	11,17	72	23	48	17	98,7	96	14	76,6	7,2	7,2	10:9 3-4 6-22
17,99	15	2,99	10,78	78	23	48	18	98,3	96	13	81,9	6,77	6,77	
19,28	16	3,28	10,4	78	23	48	17	99,5	120	16	76,	6,4	6,4	13:6 1-4 2-22
20,5	17	3,5	10,1	78	21	48	18	96,6	120	15	80,8	6,	6,	
21,8	18	3,8	9,8	84	24	48	17	97,	128	16	78,4	5,56	5,56	14:5 0-4 8-22
23,03	19	4,03	9,54	84	23	48	17	98,3	128	15	81,4	5,32	5,32	
24,34	20	4,34	9,3	88	23	48	17	100,	128	15	79,3	50,4	5,04	
25,54	21	4,54	9,1	88	23	48	17	98,3	128	15	77,6	4,81	4,81	
26,86	22	4,86	8,9	96	24	48	17	100,5	128	14	81,4	4,57	4,57	
27,99	23	4,99	8,7	96	25	54	18	100,2				4,34	4,34	
29,27	24	5,27	8,5	[illegible]	25	54	17	103,				4,19	4,19	
30,45	25	5,45	8,3	9[illegible]	25	54	17	101,				4,	4,	
31,57	26	5,57	8,19	96	25	54	17	99,[illegible]				3,82	3,82	
32,77	27	5,77	8,03	104	25	54	18	100,2				3,7	3,7	
33,96	28	5,96	7,93	104	25	54	18	99,				3,6	3,6	
35,15	29	6,15	7,75	112	26	54	18	100,1				3,4	3,4	
36,4	30	6,4	7,63	112	26	54	18	98,[illegible]				3,36	3,36	
1	2	3	4	5	6	7	8	9	10	11	12	13	14	15

SOURCE: Evans, *Miller's Guide* (1807), p. 158.

of the general discussion of mechanics and hydraulics as might be related to their own experience.

Evans gave his principal attention to the central problem of the millwright: the determination of the best type and design of wheel to use with a given fall and volume of streamflow for a water mill of

Table 5. The millwright's table for overshot mills

Head of water above the point of impact or top of the wheel.	Velocity of the water per ſecond.	Velocity of the wheel, counted at the centre of the buckets, and being ,66 velocity of the water.	Diameter of the wheel, to the centre of the buckets, for a ſtone 4 feet diameter, 122 revolutions in a minute.	Ditto for a 5 feet ſtone, to revolve 98 times in a minute.	Ditto for a 6 feet ſtone, to revolve 81 times in a minute.	Ditto for a 7 feet ſtone, to revolve 70 times in a minute.	Cubic feet of water per ſecond, required to drive the 4 feet ſtones.	Sum of the areas of the apertures of the gate for a 4 feet ſtone.	Cubic feet of water required, per ſecond, for a 6 feet ſtone.	Sum of the areas of the apertures for a 6 feet ſtone.	Area of a ſection of the canal ſufficient to bring the water to 4 feet ſtones, with a velocity of 1,5 feet per ſecond.	Ditto for 6 feet ſtones.
ft.	feet.	feet.	feet.	feet.	feet.	feet.	cub. ft.	ſu.ft.	cub. ft.	ſu.ft.	ſup. ft.	ſup. ft.
8	22,8	15,04	2,17	2,73	3,3	3,9	17,34	,76	40,9	1,79	11,56	27,3
9	24,3	16,03	2,5	3,12	3,68	4,37	15,41	,64	36,35	1,5	10,3	24,23
10	25,54	16,85	2,63	3,28	3,97	4,59	13,87	,54	32,72	1,28	9,25	21,7
11	26,73	17,64	2,75	3,44	4,15	4,8	12,61	,47	29,74	1,11	8,4	19,83
12	28,	18,48	2,9	3,6	4,34	4,9	11,56	,41	27,26	,97	7,7	18,17
13	29,16	19,24	3,01	3,74	4,53	5,24	10,67	,36	25,17	,86	7,1	16,8
14	30,2	19,93	3,12	3,9	4,7	5,43	9,9	,33	23,36	,77	6,6	15,56
15	31,34	20,68	3,24	4,03	4,87	5,67	9,24	,29	21,93	,7	6,16	14,62
16	32,4	21,38	3,34	4,12	5,01	5,83	8,67	,27	20,45	,6	5,71	13,6
17	33,32	21,99	3,43	4,25	5,18	5,95	8,16	,24	19,24	,57	5,44	12,15
18	34,34	22,66	3,54	4,41	5,32	6,18	7,7	,22	18,18	,52	5,13	12,12
19	35,18	23,21	3,63	4,52	5,47	6,33	7,3	,2	17,	,48	4,9	11,33
20	36,2	23,89	3,71	4,62	5,49	6,47	6,93	,19	16,36	,45	4,62	10,9
1	2	3	4	5	6	7	8	9	10	11	12	13

SOURCE: Evans, *Miller's Guide* (1807), p. 168.

the intended kind and capacity. The specific and detailed directions for laying out and building each type of waterwheel were supplied by Ellicott. These gave the precise number and dimensions of the scantling required in each wheel, and instructions for laying out the several wheel components and for cutting and fitting them preliminary to assembly and installation. Bills of material were itemized, covering not only waterwheel and cogwheels but the supporting mill structure for both the scantling and ironwork required. The text is clarified and amplified by the large number of drawings, gathered at the end of the volume, prepared chiefly by Ellicott. The smaller ones if not always clear to the unpracticed eye were at least suggestive. Nearly all were keyed into the text, where specific dimensions were often given.

The supporting facilities of dam, races, and related equipment were given short shrift in the *Miller's Guide*, possibly because raising barriers across streams and digging ditches seemed not to present complex problems or to require elaborate calculations. Quite possibly, too, the traditional lore was in most situations reasonably adequate to need. The problems of the nature of motion in rivers and canals and of measuring the flow of water through orifices and over weirs, which came to occupy so large a place in hydraulic engineering, found only a faint and confusing reflection confined to the theoretical part of the *Miller's Guide*. A few scattered paragraphs sufficed for suggestions respecting the dimensions and construction of the headrace and the gradient requisite for securing an adequate supply of water to the wheel with minimum loss of fall. Certain precautions were to be observed in the building of the dam to insure tightness and, it was hoped, to avoid its loss during the first heavy freshet. Evans was unimpressed by the competence of the average millwright in this as in many other matters, finding him wanting in the capacity for systematic and careful observation and only too prone to prejudice. Yet the best advice he could offer those lacking experience in the improvement of mill seats was to seek the counsel of as many millers and millwrights of experience as possible, resolving conflicts of views in common discussion. By such precautions, he stated, "those who have the least experience in the milling business, frequently build the best and most complete mills."[50]

If the *Miller's Guide* fell somewhat short in its contribution, if any, to hydraulic science as well as in supplying practical aid in building

[50] Evans, *Miller's Guide,* Art. 118.

water mills, it provided a much needed and long useful handbook for the guidance of millwrights. The authors' approach was a modest one. While seeking to share their knowledge with others, each remarked upon the limitations of his own experience and of the body of existing knowledge. Of somewhat different backgrounds, interests, and experience, Evans and Ellicott still shared a common desire to help others and to extend the boundaries of their own understanding. As Evans remarked in his preface, "They who have been versed in science and literature have not had practice and experience in the arts; and they who have had practice and experimental knowledge, have not had time to acquire science and theory, those necessary qualifications for completing the system." Except for John Smeaton, Evans had found "no authors who had joined practice and experience with theory."[51] In advancing the mastery of his craft, Ellicott had both sought out improvements by others and "often searched the book stores in expectation of finding books that might instruct me," but always met with disappointment.[52]

Yet it is important to recognize that neither of these men was greatly concerned with water mills of the common sorts. The small output and frequently intermittent operation of most one-run gristmills producing meal for a community on a custom-toll basis offered no market for the elaborate and costly milling equipment so zealously promoted by Evans. The Bathes have pointed out that by Evans's own account his agents for selling rights to the use of his flour-milling improvements carried copies of the *Guide* "to show the millers and millwrights," presenting them with "a great proportion of the edition" of 2,000 copies "by orders from me, for the purpose of getting my improvements introduced."[53] We are told by another good authority that Evans badly needed income to support his wide-ranging interests in other fields, especially steam power, and that his primary interest in publishing the *Miller's Guide* was to promote the sale of his milling improvements.[54] A testimonial to the merits of Evans's improvements appeared in the first edition of the *Miller's Guide*, signed by four members of the Ellicott family, not including Thomas, who made his own favorable comment in his prefatory remarks. This

[51] Ibid., Preface, 1795 edition. The prefatory remarks were abbreviated or entirely omitted in some of the later editions.

[52] Ibid., pt. 5, 1795 edition.

[53] Bathe and Bathe, *Oliver Evans,* p. 47.

[54] Statement of Thomas P. Jones, friend and admirer of Evans and editor of more than one later edition of the *Miller's Guide,* 1850 ed., p. iv.

testimonial declared that the Evans system of milling machinery was worthy of the attention "of any persons concerned in merchant *or even extensive country mills*" (italics supplied).[55]

Nor did Ellicott display interest in the piddling operations for whose conduct the common run of millwrights was sufficiently capable. With pride of leadership and a long record of achievement, he began Part 5 of the original edition of the *Miller's Guide* with "a short history of the rise and progress of merchant mills," which quickly reached its climax in the description of the great Ellicott-designed Occoquan mill in Virginia, with its three waterwheels and six runs of millstones.[56] Ellicott not only ignored the tub mill in the directions for building wheel and mill but also chose for detailed description and explanation an impressive four-story masonry structure, drawn to scale and equipped with two runs of stones and driven by an eighteen-foot overshot wheel with eight-foot buckets. Having many times the capacity and cost of ordinary gristmills, it was described by Ellicott as one of the most complete of "the old fashioned grist-mills, that may do merchant-work in the small way."[57]

The value of the Evans-Ellicott handbook as a mirror of American water-milling practice is limited by the authors' preoccupation with grain milling, especially flour milling as conducted in merchant mills of the Middle Atlantic seaboard. Yet in 1795 flour and grist mills together comprised at a generous estimate not more than one-third of the total number of water mills in the nation, amounting in the aggregate to some 6,000 to 8,000 of all types and sizes.[58] The other two-thirds included not only the sawmills, which outnumbered grain mills by a substantial margin, but such mills as fulling, carding, paper, bark, linseed oil, gunpowder, cotton textile or yarn, and ironworks—blast furnaces, bloomeries, forges, and rolling and slitting mills. Virtually the only space allotted these varied water mills in the *Miller's Guide* was the twelve pages of Articles 158 and 159 of chapter 22 of Ellicott's section. They include very useful dimensional data and

[55] Bathe and Bathe, *Oliver Evans,* p. 24. Evans's hopes for a profitable business in England were dashed by the report of his agent that the "footy little windmills" which did a large proportion of the grain milling in Britain at the time were quite unsuited to the installation of the Evans machinery (ibid., pp. 50–51). The same was true of common rural gristmills in this country.

[56] A portion of this statement was kept in later editions, together with a fold-in drawing (pl. 22) of the Occoquan River mill.

[57] Evans, *Miller's Guide,* Arts. 152–55, pls. 17–21.

[58] This is a rough projection backward based on the incomplete census reports of 1810 and 1820.

drawings devoted to fulling and rice mills. The sawing of wood and fulling of cloth receive only one-fourth the space given by Ellicott to flour milling alone.[59]

The theoretical and much of the applied material in the *Miller's Guide* was generally applicable to, or in the hands of an experienced and skilled millwright could be adapted to, other industrial operations. What the millwright and millowner in other branches of water milling needed most acutely and did not get was guidance not only in the choice of wheel type for a given industrial operation but in the determination of the power, wheel size, rotating speed, and manner of power takeoff. However convenient for the historian of technology a more comprehensive treatment of water milling would have been, the area to which Evans and Ellicott addressed themselves was not only the field of their greatest experience and competence but also the industry that as late as 1860 was exceeded only by cotton goods and lumber in annual net value of product.[60]

The publication of *The Young Mill-wright and Miller's Guide* in its numerous but meagerly revised editions was nonetheless a major contribution to the practice and understanding of water milling during the preindustrial years. For all the interest and experience of the authors in what during the 1780s and 1790s was, in scale of operations and degree of mechanization, the nation's leading industry, they were to make slight contribution to the wider and changing role of waterpower in the years ahead. They and their editorial successors failed to keep the *Guide* abreast of developments in the theory, practice, and agencies of hydraulic power; the latest editions differ only marginally from the first. The failure to keep up with changing times, conditions, and needs is suggested by a comparison of the 1850 and later editions with the unsigned article "water-wheels" in the standard manual of the period, *Appleton's Dictionary of Machines, Mechanics, Enginework, and Engineering*, first published in two volumes in 1850 and 1851. The two treatments of hydromechanics fall within different worlds. The 1850 edition of the

[59] Evans, *Miller's Guide*, Arts. 158–59 and pls. 23, 24. A two-page description of a mill for cleaning and hulling rice appeared in the 1795 edition as an appendix and was later incorporated into the text and supplemented by a drawing.

[60] U.S., Census Office, Eighth Census, *Manufactures of the United States in 1860* (Washington, D.C., 1865), pp. 733 ff. In total (gross) value of product, flour milling ("flour and meal") not only led the entire field but was greater than the cotton and woolen goods industries and the basic iron industries combined. Net value here is taken as (total) value of products less cost of raw materials.

Miller's Guide, including the revisions first added in 1826 by Thomas P. Jones, professor of mechanics at the Franklin Institute and editor of its *Journal*, offered as its chief concession to the march of industrial innovation the inclusion of extended excerpts from Robertson Buchanan's *Practical Essays on Millwork*, dealing with the use of cast iron in the wheels and other millwork. *Appleton's Dictionary* wryly noted: "Water power is an old if not an antiquarian subject, on which the light of modern improvement has been but feebly reflected since the days of Smeaton. With a few exceptions, it has been abandoned to those who recognise in it no principle, and no scope for improvement; and whose practice is not more opposed to improvement than it is empirical and opposite to all true principle." The statement could apply to the *Guide* as well as to the subject at hand. In reply to a reader's inquiry the *Scientific American*, 1 November 1856, characterized the *Guide* as "out of date to a very great extent."[61]

In defense of the editors and publishers of the *Guide* it must be said that the function of this classic work was clearly not to advance the water-milling art nor even to keep pace with such improvements as were made during the half century that followed its first appearance. Rather it served much in the manner intended by its authors, as a convenient and practical handbook for the assistance of the men who built, maintained, and operated the large and mounting aggregate of water mills serving the nation. In the half century from 1790 to 1840 both the total national population and the rural population alone increased more than fourfold. During this same period the number of water mills increased from an estimated total of some 7,500 in 1790 to more than 55,000 in 1840, counting at the last date grist- and sawmills alone.[62]

[61] *Appleton's Dictionary of Machines*, s.v. "Water-Wheels," 2:786–87. The excerpts from Buchanan appear in Evans, *Miller's Guide*, pp. 381–92. The author of a successor to the Evans *Guide* remarked that the improvement in machinery and mills "renders Mr. Evans's work comparatively useless, so far as the mechanical construction of the present age relates to mill-building" (William C. Hughes, *The American Miller, and Millwright's Assistant* [Philadelphia, 1851], p. 99). On the meager contribution of Americans to the applied science of hydraulics—in the industrial applications of which American mechanics and engineers had excelled in many ways—see the perceptive review of Bennett's translation of d'Aubuisson, *Treatise on Hydraulics* in *JFI* 55 (1853): 215–16.

[62] See article on the 1840 census report, "Resources of the United States," *Hunt's Merchants' Magazine* 7 (1842): 436–39. The original statistics appear in U.S., Census Office, Sixth Census, 1840, *Compendium of the Enumeration of the Inhabitants and Statistics of the United States* (Washington, D.C., 1841), pp. 358–64.

Although the *Miller's Guide* had its chief usefulness in the hands of millwrights, it doubtless had its place on the shelves of the rapidly growing professions of engineers, civil and mechanical, on whom the advance of hydraulic engineering in the service of industry chiefly depended. In certain respects the *Miller's Guide* must have hastened the decline of the traditional technology it reflected. It did this in the first instance simply by committing to the printed page what had hitherto been communicated only by example and oral tradition, undergoing in the process comparison, criticism, and selection. More than this, with its emphasis throughout upon measurement, calculation, and the quantities involved, the *Miller's Guide* reveals the approach of Evans and Ellicott to the planning and design of water mills to have been in many respects an essentially engineering one. In their preoccupation with medium to large-scale flour milling and their emphasis upon continuity of operations and economy in water use, they may be said to have given the first strong impulse in this country to the shift from water milling to waterpower.

Milling Practice in the Colonial Period

The *Miller's Guide* throws virtually no direct light on the course of American milling practice during the nearly two centuries of colonial experience. Colonial records contain a wealth of information on the introduction, increase, and spread of water mills but throw little light on the manner of their construction and equipment. Who troubles to document the commonplace? Of the types of waterwheels used, the manner of their construction, the linkage with the driven machinery, and the means used to accumulate and direct the flow of water to the mill, the record, with infrequent exceptions, is silent. We do not know what variations upon Old World forms and arrangements may have been introduced, whence and by whom, what modifications in the details of design, construction, and practice may have been promoted by colonial conditions and needs.

Two exceptions to the generally meager evidence are of particular importance because of their early date, in the 1640s. The first was in Massachusetts, near Boston; the second in a Swedish settlement in eastern Pennsylvania. The Hammersmith Ironworks, of Saugus, Massachusetts, was for its day a large and ambitious but ultimately unsuccessful corporate enterprise with colonial and English investors. The integrated works comprised a blast furnace, a bloomery forge, and a rolling and slitting mill, the last for making nailrods. Excavation discovered a sufficient portion of one waterwheel to permit its reconstruction: an overshot wheel, nearly sixteen feet in diameter and two feet wide, used for driving the blast bellows of the furnace. The main structure of the wheel was almost wholly of wood, carefully built with treenails and wedges of wood for fastening. The Pennsylvania mill, erected in 1643, was one of a number of gristmills built to serve the early Swedish settlement in the region and was said to be the first of the kind in Pennsylvania. If two historians interested in this mill are correct, this mill was a tub mill and may well mark the coming of this important type to the United States. In his *History of Delaware County, Pennsylvania* (Philadelphia, 1862), George Smith reported: "From the holes in the rocks [at the millsite] the mill must have occupied a position partly over the stream [Cobb's Creek] and was doubtless driven by a tub-wheel which required little gearing." More than half a century later a historian of Bucks County, Pennsylvania, interested in the development of the tub wheel, believed it "a moral certainty that the mill that the Swedes built here was a Norse mill." Since a form of the tub wheel had long been in general use in Sweden, it was natural that the Swedes should have brought it with them for the mills erected by them.[63]

Overall, the indirect evidence of physiographic and economic conditions, the fragments of surviving records, the inferences based on postcolonial experience on the frontier, added to the record and rationale of water milling at the end of the eighteenth century provided by the Evans *Guide*, seem reasonably conclusive. Beyond much doubt the vast majority of water mills in colonial America were driven by the simpler, more readily made impact wheels of the undershot and tub-wheel types in preference to gravity wheels. The latter's more efficient performance must have been offset by the greater care and cost in construction and limited range of usefulness. Mill seats with the minimal fall required

[63] Edward N. Hartley, *Ironworks on the Saugus* (Norman, Okla., 1957), esp. chap. 9; George Smith, *History of Delaware County, Pennsylvania* (Philadelphia, 1862), p. 38 (but see also pp. 54–55, 94–95, 191, 203 citing archival materials); Shelton, "Norse Mills," pp. 175–85, esp. pp. 181–82. Hartley's account is very thorough. On early Swedish settlements, see J. Thomas Scharf and Thompson Westcott, *History of Philadelphia*, 3 vols. (Philadelphia, 1884), 1:130 ff. Shelton's paper reviews the widespread distribution of the Norse, or tub, wheel in Europe; see below, Appendix 2, for further interest of the Bucks County Historical Society in tub wheels. Stevenson W. Fletcher, *Pennsylvania Agriculture and Country Life*, 1640–1840, 2 vols. (Harrisburg, 1950–55), 1:323, adopts the Smith interpretation of the wheel used on the first Swedish mill.

for the advantageous use of bucket wheels were far less numerous than the lesser falls to which the simpler impact wheels were suited.

The evidence of technology is largely inferential and it is also with little doubt of secondary importance to the economic considerations that in most instances must have been the controlling ones. These were the poverty of the settlers, the thin and widely scattered population on the frontiers of settlement, and the great difficulty and cost of supplying such mill components as could not be made on site with local materials and skills.

The evidence suggests a two-phase process in the transfer of water mills to the advancing frontier. The first phase of water milling, marked by large dependence on assistance from colonizing or public authority and imports of the more costly components, had by the late eighteenth century given way to a second and less demanding phase. This was illustrated in New England by the advance of settlement up the Connecticut Valley into Vermont and New Hampshire and along the coast and inland into Maine, gathering momentum from the 1760s and accelerating after independence from Britain. The initiative was taken increasingly by private land associations, the "town proprietors," and even by individual land operators who erected or gave aid to mills as a means of attracting settlers, thereby stimulating the demand for and value of their holdings. Generations of pioneering experience had brought a widespread familiarity with the techniques of land clearing, farm building, self-sufficiency, and water milling, and widely shared skills in the use of such edge tools as ax, auger, wedge, and froe. The opportunities for linking the practice of craft skills with land ownership and farming attracted craftsmen to the frontiers of settlement, especially carpenters, blacksmiths, shoemakers, and the like, including millwrights.[64]

[64] On woodworking tools in pioneering and farm life, see Albert S. Bolles, *Industrial History of the United States* (Norwich, Conn., 1878), pp. 270 ff.; Harriette S. Arnow, *Flowering of the Cumberland* (New York, 1963), pp. 254 ff., 263–65, 278–80; Floyd B. Haworth, *The Economic Development of the Woodworking Industry in Iowa* (Iowa City, 1933), pp. 12–13; Isaac L. Bell, "The Iron Manufacture of the United States, and a Comparison of It with That of Great Britain," in Great Britain, Executive Commission, Philadelphia Exhibition, *Reports on the Philadelphia Centennial Exhibition of 1876*, 3 vols. (London, 1877), 1:128–30, 135–36. For the respective roles of wood and iron in a frontier economy, see Henry Blackman Plumb, *History of Hanover Township and Wyoming Valley in Luzerne County, Pennsylvania* (Wilkes-Barre, 1885), pp. 186 ff., 209 ff. On the union of manufactures and farming see Percy Bidwell, "Rural Economy in New England at the Beginning of the Nineteenth Century," *Transactions of the Connecticut Academy of Arts and Sciences* 20 (1916): 241–399. See also U.S., Secretary of the

Not only were mill seats by this time readily identified, but general recognition of their value led to early occupation and the prospect of development. Local resources for such essentials as millstones and mill irons (blast furnaces, forges, smithies) were explored and exploited. Millstones cut from loose boulders, granite, hard sandstone, or other local stone were usually much inferior to the imported French or English ones, but in some fashion at least they ground the grain, were much cheaper, and were at hand. In the northeastern United States, at least, rock ledges were not less plentiful than mill seats, and a boulder-scattered pasture was a characteristic feature of every hill farm. With patience and labor, with hammer and improvised chisel, small millstones of a sort could be fashioned.

In their concern to obtain gristmills with minimal delay and expense, settlers in frontier areas thus defied conventional practice in respect to both the material and size of millstones. Millstones as reported in Article 63 of Evans's *Miller's Guide* (1850 edition) ranged from 3.5 to 7.0 feet in diameter. Assuming an average thickness of ten inches for the two stones, the resulting weight would vary from a half ton or so to two or three tons per millstone, presenting a problem of handling and carriage as well as cost.[65] It is a reasonable inference that the millstones of the frontier mills were as small and rudely made as the wheels that drove them—we read of one stone that could be held in the lap for dressing—and that they weighed no more than 150 to 300 pounds. References to homemade millstones usually describe them as split and roughly shaped from granite ledges or boulders, contrasting sharply with imported stones of French buhrstone, a siliceous rock that is virtually the only millstone material recognized by such standard nineteenth-century works as *Appleton's Dictionary of Machines, Appleton's Cyclopaedia of Applied Mechanics,* and *Knight's American Mechanical Dictionary.* This buhrstone, reported by *Appleton's Dictionary* to be "found in abundance

---

Treasury, *Documents Relative to the Manufactures in the United States,* House Ex. Doc. 308, 22d Cong., 1st Sess., 2 vols. (Washington, D.C., 1833), 1:625, also pp. 528–29, 746–47, 756; Albert Gallatin, Report to the House of Representatives on "American Manufactures," 19 Apr. 1810, *American State Papers, Finance,* 2:430; "Book of Mechanics and Merchants in Hall County, Georgia," 24 Nov. 1820, Returns of the Fourth Census.

[65] My calculation, based on cubic content of millstone and stone weight as given in *Coal Miners' Pocketbook,* 11th ed., p. 276. A description of a gristmill, 1868, gave the average weight of the millstones of 4 to 6 feet in diameter as 14 cwt, or nearly 1,600 pounds each. Operating at 120 rpm, a single pair of stones required four horsepower to keep in motion while grinding (*Eighty Years' Progress of the United States* [Hartford, Conn., 1868], p. 432).

only in the mineral basin of Paris and a few adjoining districts," was cut into small parallelepipeds at the quarries for export and assembly into finished stones, bound together with plaster of paris and iron hoops. About 1850 good French buhr millstones brought £48 in England for stones 6.5 feet in diameter. *Appleton's Cyclopaedia* recognized Georgia also as a source of buhrstones.[66] Two native sources for early American millstones in the 1760s were Mount Tom in western Massachusetts and Mount Esopus in New York.[67]

The situation is less clear with respect to mill irons than to millstones. Most of the smaller mill irons could be made at the common blacksmithies, which by the early 1800s followed the advance of settlement with no very wide lag. But such larger parts as sawmill crank and the heavier wheel gudgeons and millstone spindles could be forged effectively only at the larger smithies equipped with water-powered trip-hammers. The saw blade itself evidently presented less of a problem, although the sources of supply are not clear. Beyond the competence of all but the large smithies because of its length and bladelike character ("6 feet long, 7 or 8 inches wide, when new," says the *Miller's Guide*), its light weight of perhaps twelve to fifteen pounds made carriage from eastern supply centers relatively easy. Cast-iron cranks and gudgeons were usually among the products of rural blast furnaces making cast products as well as pig iron. Bloomery forges for making wrought iron direct from the ore in some instances produced heavy mill irons. Accounts of pioneering in the early decades of the nineteenth century repeat the familiar tales of laborious trips with hand-pulled sled or packhorse—even backpack—carrying mill cranks weighing 100 to 150 pounds. Since they were virtually immune to destruction by fire, flood, or wear, these cranks often served a succession of mills.[68]

[66] *Appleton's Dictionary of Machines,* 2:385. References to millstones are found in a number of the foreign travel works reprinted in Thwaites, *Early Western Travels,* 3:203, 4:308, 9:240, 13:243, 19:77, 29:321. On use of local or homemade stones, see M. H. Humphrey et al., "Extracts from the History of Three Rivers," *Hist. Collections of Michigan Pioneer and Hist. Society* 38 (1912): 420–21, 431 ff; Abby M. Hemenway, ed., *The Vermont Historical Gazetteer,* 5 vols. (Burlington, 1868–91), 3:138–39, 590; H. Hollister, *Contributions to the History of the Lackawanna Valley* (New York, 1857), pp. 97–98; Horatio G. Spafford, *A Gazetteer of the State of New York* (Albany, 1824), p. 598; Ebenezer Emmons, *Geological Report of the Midland Counties of North Carolina* (New York, 1856), p. 267.

[67] Clark, *History of Manufactures,* 1:178; see also Kuhlmann, *Flour-Milling Industry,* p. 14.

[68] See Ridlon, *Saco Valley Settlements,* pp. 193–94; Hemenway, ed., *The Vermont Historical Gazetteer,* 1:317, 357, 360, 3:435; Hollister, *Contributions to the History of the*

Just as the log house, whose erection in a few days by two or three cooperating neighbors, had obtained general acceptance on the frontier as adequate for the first years, so the provision of rude, makeshift gristmills and sawmills of low output preceded by several seasons, if not many years, the erection of more substantial and effective mills. The latter often, in time perhaps typically, awaited the opportunity of combining custom and market operations—in Ellicott's phrase in the case of grain milling, "grist-mills, that may do merchant-work in the small way." This was the phase no doubt when the simple tub mill with its gear-free wheel became a familiar feature of the landscape, especially on the frontiers. These mills, as remarked earlier, persisted longest in the more isolated and less favored of the sparsely settled regions.

What I have termed, though in no formal or necessary sense, the second phase of water milling finds abundant illustration in the prolific literature of pioneering reminiscence. There are many references in these accounts to the slight, even ramshackle character of the early mills, slow in operation, low in output, and often running only intermittently as circumstances required or as the volume of streamflow permitted.[69]

Lackawanna Valley, pp. 91–96, 100–1; Lyndon Oak, *History of Garland, Maine* (Dover, 1912), pp. 37 ff.; C. A. Miner, "The Early Grist Mills of the Wyoming Valley, Pennsylvania," *Proceedings and Collections of the Wyoming Hist. and Geolog. Soc.* 5 (1900): 113, 117, 134–35; and Orsamus Turner, *History of the Pioneer Settlement of Phelps and Gorham's Purchase* . . . (Rochester, N.Y., 1870), pp. 449, 528–29, 593. See also Defebaugh, *Lumber Industry of America*, 2:196–99; Clark, *History of Manufactures*, 1:175–76; Evans, *Miller's Guide*, Arts. 84, 102, 156, 158–59, and pl. 24. On the work of the blacksmith, see Edward Hazen, *The Panorama of Professions and Trades* (Philadelphia, 1836), pp. 203 ff.; Fletcher, *Pennsylvania Agriculture*, 1:455–56; Phil R. Jack, "A Blacksmith's Ledgers, 1861–1883," *Western Pa. Hist. Mag.* 45 (1962): 139–45; Aldren A. Watson, *The Village Blacksmith* (New York, 1968); H. R. Bradley-Smith, *Blacksmiths' and Farriers' Tools at Shelburne Museum: A History of Their Development from Forge to Factory* (Shelburne, Vt., 1966). The last two sources are especially valuable and are well illustrated.

[69] There is abundant evidence of the small scale of early millstreams and the rude character of the frontier water mills. The Genesee River long remained "unimproved" by waterpowers "because every creek in the vicinity afforded sufficient power for the wants of the people" (John H. French, *Gazetteer of the State of New York* [Syracuse, 1860], p. 404n.). For a similar statement on tributaries of the Scioto River, Ohio, see *Tenth Census, Water-Power*, pt. 2, pp. 476–77. On use of interior streams in South Carolina, see Robert Mills, *Statistics of South Carolina* (Charleston, 1826), pp. 352, 615, 632. See also H. G. Schmidt, *Rural Hunterdon* (New Brunswick, N.J., 1946), pp. 211–12; Turner, *Pioneer Settlement of Phelps and Gorham's Purchase*, pp. 529, 573; Thomas K. Cartmell, *Shenandoah Valley Pioneers and Their Descendants* (Winchester, Va., 1909), pp. 69–70; Hemenway, ed., *The Vermont Historical Gazetteer*, 3:289, 617, 751, also 76–77, 138–39; William B. Lapham, *Centennial History of Norway, Oxford County, Maine, 1786–*

Occasionally such mills were jointly owned and used by the families served. Under conditions of sparse settlement and often only seasonally passable roads, it was not uncommon to find a growing community served by two or more slight mills erected in different parts of the settlement for convenience in use.[70] In time such mills usually gave way to the more substantial structures required in the effective conversion for market disposal of local surpluses of grain, lumber, and cloth, an event welcomed, if with less enthusiasm than the acquisition of the first rude mills.

Thus by inference from indirect evidence a reasonably strong case can be made for the simple impact wheels as the motors long in general use in the common water mills. From similar evidence a case can be made for the growing use of gravity wheels of the overshot type from about 1750 in flour milling and the heavier branches of the iron industry, especially rolling and slitting mills. The rapid expansion of the flour and breadstuff export trade of the period was supported by the increase in number and capacity of large merchant flour mills, equipped with two to four and more runs of stones of large capacity and with mechanized bolting and handling equipment. Such mills were most numerous on the larger streams having outlet on tidewaters between Philadelphia and Richmond. They enjoyed convenient access to the grain-raising country of the interior and the shipping facilities of the tidal rivers and seaports.[71] Large-scale production led to rationalization of operations, which in the matter of power supply called for the selection of mill seats with a capacity appropriate to the desired output, and, where water supply fell short of such require-

---

1886 (Portland, 1886), pp. 53, 59, 62; Ridlon, *Saco Valley Settlements,* pp. 203–4; Defebaugh, *Lumber Industry of America*, 2:16, citing Dr. Douglas about 1750; Hugh D. McLellan, *History of Gorham, Me.* (Portland, 1903), pp. 256–57; Miner, "Early Grist Mills of the Wyoming Valley," pp. 116, 126.

[70] Many instances may be found in town histories, e.g., in Hemenway, ed., *The Vermont Historical Gazetteer;* also Lapham, *Centennial History of Norway, Maine;* Ridlon, *Saco Valley Settlements.*

[71] The most useful source on the development of commercial flour milling is Kuhlmann, *Flour-Milling Industry;* for the spread of merchant milling in the eighteenth and early nineteenth centuries, see esp. pp. 23–26, 38–42, 96–101. See also Clark, *History of Manufactures,* 1:178 ff.; Bishop, *History of American Manufactures,* 1:588. For contemporary references see W. Winterbotham, *An Historical, Geographical, Commercial, and Philosophical View of the American United States,* 4 vols. (London, 1795), 2:466 ff.; Rayner W. Kelsey, ed., *Casenove's Journal, 1794* (Haverford, Pa., 1922), pp. 26 ff., 40 ff., 50 ff., 64 ff., 75 ff.; Thomas Cooper, *Some Information respecting America,* 2d ed. (London, 1795), pp. 88–97, 110–11, 134 ff.

ments, to the adoption of wheels using water to the greatest advantage. The marked superiority of overshot to undershot wheels in efficiency could in many instances transform a grave power shortage into a surplus and provide the means for expanding operations. Only a handful of specific references to colonial overshot wheels has come to my attention, all but one dating from after 1750. A state legislative report, on the extensive New Jersey ironworks of Peter Hasenclever, established in the 1760s, listed among the celebrated ironmaster's "important innovations" his exclusive use of overshot wheels.[72] There is reason for believing that the larger merchant mills, which played a leading role in flour milling from the mid-eighteenth century, were usually driven by overshot wheels. In a controversy over flour inspection in the 1820s between country millers and the city millers of Baltimore, the city millers declared that through their use of overshot wheels "they got more power and could utilize their power supply to best advantage in the dry seasons."[73] From the special emphasis given to the overshot wheel by Ellicott in the 1795 edition of the *Miller's Guide*, it is a fair presumption that among the larger merchant mills at least they had become the prevailing type of motor.

From Water Milling to Waterpower

The age of water milling was followed by the age of waterpower. The first half of the nineteenth century was a period of transition as the isolation and local self-sufficiency of the pioneering way of life gradually gave way to the interdependence of a market economy. What George R. Taylor has termed the transportation revolution, resting on the extension of inland steam navigation, canal transport, and then the

[72] John B. Pearse, *A Concise History of the Iron Manufacture of the American Colonies up to the Revolution* (Philadelphia, 1876), p. 66. The legislative committee was appointed to investigate the bankruptcy of Hasenclever in 1770. Scattered mentions of overshot wheels in mid-eighteenth-century ironworks are found in Herbert C. Keith and Charles R. Harte, *The Early Iron Industry of Connecticut* (New Haven, 1935); see fold-in chart at back of book. Remains of a wooden wheel at the mid-seventeenth-century ironworks at Saugus, Massachusetts, are believed to be those of an overshot wheel; the works were built by English artisans sent for the purpose. The dimensions and kinds of waterwheels used in other operations at the Saugus works "are not definitely known, although it is clear, both from general archeological evidence and from their known and assumed functions that all were quite large, that one was an undershot, the others overshot or pitchback" (Hartley, *Ironworks on the Saugus,* pp. 182–84 and photo f. p. 113).

[73] Kuhlmann, *Flour-Milling Industry,* pp. 93 ff., also pp. 18–19, 45 ff.

railroads, provided the technological basis for sweeping economic change.[74] The work of rural water mills in processing local materials on a custom-toll or barter-exchange basis was at first supplemented and then gradually displaced by the commercial conversion of raw materials for the general market. Although water mills and the industrial factories succeeding them relied upon much the same hydraulic facilities and equipment for their power supply, the two establishments represent different stages in the industrial use of waterpower. Water mills in this country as in the Old World were a preindustrial institution devoted chiefly to meeting the needs of a noncommercial subsistence economy and, as a result, occupied an increasingly marginal position in the market economy from which industrialization gradually emerged. Considerations of monetary costs and accounting and of the rationalization of the techniques of production were almost wholly wanting in the economy of water milling, functioning as it did in the service of isolated communities, often on an intermittent or seasonal basis. Such considerations as maximum output, continuity of operations, and efficiency of equipment had little relevance to need.[75]

In one other respect, important in both the technology and economy of production, the water mill differed from the factory: in the role of power transmission. The typical factory with its varied operations and numerous machines required a system of power distribution—the millwork of shafting, pulleys, and belting—to convey power from the prime mover to the points of use throughout the plant. The ordinary water mill, as we have seen, lacked a system of power transmission as such. Engaged typically in a single and simple repetitive operation with the waterwheel closely joined to the driven machinery, the water mill in its entirety was a machine with, to use a later phrase, a built-in motor. The only use of millwork was in the employment of gear wheels where necessary to change the direction of motion or the speed of rotation.

The large merchant flour mills that came into prominence in the late

[74] George R. Taylor, *The Transportation Revolution, 1815–1860* (New York, 1951).

[75] A prominent hydraulic engineer whose early training was obtained in New England stressed the small capacity of the water mills of the colonial and early industrial periods, seldom of more than 10 to 60 horsepower at most. "On a series of rapids on a small river, or at an outlet of a lake, the sawmill, the grist mill and the carding and fulling mill, each had its own dam across the stream, and together they formed the nucleus of a small industrial community, developed in the midst of agriculture, with power units planned with extreme individuality" (John R. Freeman, "General Review of Current Water Power Practice in America," in *Transactions of the First World Power Conference, 1924,* 5 vols. [London, (1925)], 2:375, 382–83).

eighteenth century marked a kind of halfway point in the transition from water mills of the traditional kind to the factories of the new industry. Large in scale, market-oriented and profit-motivated, such mills had little more in common with the country gristmill than the material processed and the grinding operation. In the merchant mill the extension of mechanization from grinding to the cleaning, cooling, bolting, and packing operations and to handling operations throughout added very substantially to power requirements. Power was required not only to perform the several operations themselves but to meet friction losses in transmission and to accommodate the enlargement of capacity resulting from all these improvements (see below, chapter 9). Thus the merchant mill anticipated in some respects the coming of the factory age as represented by the new textile mills and ironworks introduced from Britain during the 1790s and succeeding decades.

As the new types of industrial establishments increased in scale and in sophistication of management, their power requirements and facilities gradually lost any significant resemblance to those of the water mills with which their predecessors, the fulling mills and carding machines and the small rural forges and blast furnaces, were identified. Although the basic elements remained much the same—dams, millponds, races, wheels—the scale, complexity, and refinement of detail in design and operation found in such major hydropower installations as those of the New England textile centers bore slight resemblance to the water mills in which they had their origin. The transition was reflected only in part by the shift to the more efficient types of waterwheels. In the traditional technology of water milling, as I suggested earlier, the waterwheel and its gearing were the center of attention, almost to the exclusion of the supporting hydraulic facilities, which generally received but perfunctory treatment. Yet in the large hydropower installations of the industrial age ahead, such as that at Lowell, Massachusetts, the waterwheels might almost be described as a minor but indispensable feature of the total power system. The wheel installations proper were dwarfed by the complex of hydraulic works consisting of great masonry dams and protective walls, networks of distributing canals equipped with water gates and wasteways, and, eventually, systems of upstream storage reservoirs.

3
The Geographical Basis and Institutional Framework

WITHOUT DEPRECIATING THE role of other natural resources, such as minerals, waterpower to the mid-nineteenth century may properly be ranked after land, timber, and waterways as fourth in importance in the economic life and development of the United States. Its importance derived from the great abundance, wide distribution, and range in capacities of millsites, the relative ease and low cost of their development and use, and their adaptability to the requirements of an economy evolving from the subsistence level of the advancing frontier to the mature industrialism of the north Atlantic seaboard. This power source of great simplicity and long familiarity was based upon streamflow, one of the most common and widespread of natural phenomena, renewed perennially by rainfall and snowmelt.[1]

For all its merits, waterpower was subject to limitations in use and development that in time placed it at a growing disadvantage with steam power. These limitations were both physical and institutional. The physical limitations of topography, geology, and meteorology restricted waterpower to use at sites adjacent to falls and rapids; it was transportable by

[1] This chapter is based on a variety of engineering, technical, and legal publications, the last named including treatises on the law of watercourses and a considerable body of case law, as cited below. Among the extensive sources on the physical bases of water, a particularly useful work is the federal survey of waterpower made in connection with the Tenth Census, 1880, and published as vols. 16–17, *Reports on the Water-Power of the United States,* (Washington, D.C., 1885, 1887). See Appendix 4. Two general works on hydraulic engineering written by men active in that field have been especially useful: Joseph P. Frizell, *Water-Power: An Outline of the Development and Application of the Energy of Flowing Water* (New York, 1900), and Daniel W. Mead, *Water Power Engineering: The Theory, Investigation, and Development of Water Powers* (New York, 1908). Mead's work has valuable chapter bibliographies, including books and the extensive literature of the engineering journals. The Water-Supply Papers, among other publications of the U.S. Geological Survey, are important sources, as are occasional publications of state agencies. Perhaps the earliest of the latter is Maine, Hydrographic Survey, *The Water-Power of Maine, by Walter Wells* (Augusta, 1869). A comprehensive, concise, and amply documented review of hydrology in relation to hydraulic engineering appears in Robert W. Abbett, ed., *American Civil Engineering Practice,* 3 vols. (New York, 1956–57), 2:sec. 13, pp. 1–75, with many charts, graphs, and tables, including time series.

mechanical means only for quite short distances. Within the potential of a given stream site the power capacity was fixed in amount; unlike steam power, it could not be increased by adding more engines and boilers and firing more fuel. The final deficiency, and eventually the major one for large and progressive industry, was the unreliability of waterpower, subject as it was to often irregular and unpredictable seasonal variations in supply, dependent as streamflow is upon the amount of precipitation and runoff within each drainage basin. The burden of these geographic limitations, as will be seen, was approximately in inverse ratio to the level of economic and industrial development. It mounted progressively with the gradual but slowly accelerating shift from a subsistence to a market economy and with the increasingly rapid advance of industrialization from the 1850s in the United States.

The institutional limitations of waterpower were of a subtler and more elusive yet less compelling and more manageable character. Rooted in custom and law, they were subject to discussion, debate, and modification—and a vast amount of litigation. They were a product in the first instance of the English common-law doctrine that ownership of a water privilege or mileat went with the land bordering the two sides of a stream—riparian rights, from the Latin *ripa*, "bank" or "shore." Such ownership was qualified: it did not extend to the stream itself but conveyed rights of usage only.[2] Usage in turn was subject to the rights and claims of other user interests, which if not numerous were often formidable and in varying degrees mutually incompatible. Prominent among the competing and at times combative claimants upon stream use—and abuse—were the owners of riparian farmlands above and below the water privilege; commercial and navigation interests dependent often upon quite small streams for seasonal access to markets by flatboat or raft; lumber interests similarly dependent during freshets upon uninterrupted streamflow for floating timber to market or mills; and, hardly less implacable, upstream communities dependent on the seasonal migration of fish to spawning grounds for a welcome and inexpensive change of diet.

The resolution of these varied and more or less conflicting interests, arising in many if not in most instances from the dams and millponds required to develop a water privilege as well as from internecine conflicts among millowners on the same stream over the manner of its use, gave rise to a special body of law of water rights going back to Anglo-Saxon and earlier times. Americans made their own contributions to this body of law, both by judicial interpretation and by legislation. The extraor-

[2] See under "Water Rights," *Encyclopaedia Britannica,* 11th ed.

dinary amount of litigation incident to the use of streamflow for power, especially in the more fully developed districts, generated a corresponding amount of acrimony, uncertainty, and expense. On occasion, too, conflict among millowners led to cooperative arrangements from which all benefited. In short, a motive power that in many respects was indeed a welcome gift of nature was often attended by a lack of independence in use and management and an expense in money and emotional energy from which steam-power users were entirely free.

The Geographical Determinants

Waterpower and steam power were in their origins alike dependent upon the energy radiated from the sun, but between the rainfall, runoff, and streamflow, on which the former depended, and the fossil fuel, which became the primary source of steam power, there was a world of difference. Coal, the product of exuberant plant life in remote geological ages, could be mined, transported, and stored for use as need and occasion required. Its supply was rarely interrupted by natural causes. The streamflow on which waterpower is dependent is a product of the ceaseless process, energized by the sun, by which the moisture of the atmosphere, precipitated to the earth as rain and snow, serves the life needs of the land, finds its way in greater or lesser part into the streams that drain the land, and at length returns through evaporation and vegetal transpiration to the atmosphere whence it came. At every stage this hydrologic cycle is determined by natural conditions: the distribution of rainfall, temperature, humidity, evaporation, and air currents in their diurnal and seasonal variations; the amount and character of vegetation; the character of surface materials and subsurface formations. Of especial importance for the conversion of potential into available waterpower is the physical pattern of drainage basin and streambed as formed through the ages by geological and hydrological forces.[3]

The complex of physical conditions upon which streamflow depends varies from region to region and from one drainage basin, however large or limited in extent, to another. It varies especially from arid and semiarid country to regions of more or less abundant rainfall. It varies from rugged highlands to alluvial delta, from mountains to plains and prairie,

[3] See Luna B. Leopold and Walter B. Langbein, *A Primer on Water,* U.S. Geological Survey (Washington, D.C., 1960); also, with numerous related articles, W. C. Ackerman et al., "From Ocean to Sky to Land to Ocean," in *Water: The Yearbook of Agriculture, 1955* (U.S. Dept. of Agriculture, Washington, D.C., n.d.), pp. 41 ff.

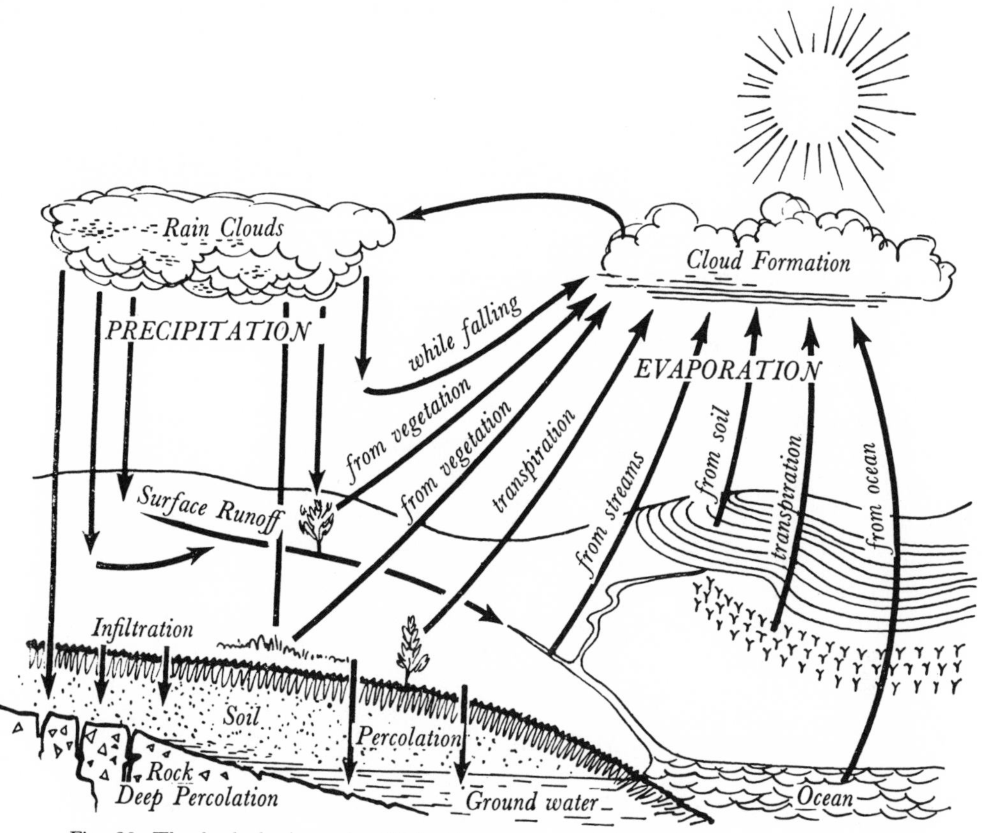

Fig. 30. The hydrologic cycle. (U.S., Department of Agriculture, *Water: The Yearbook of Agriculture, 1955* [Washington, D.C., n.d.].)

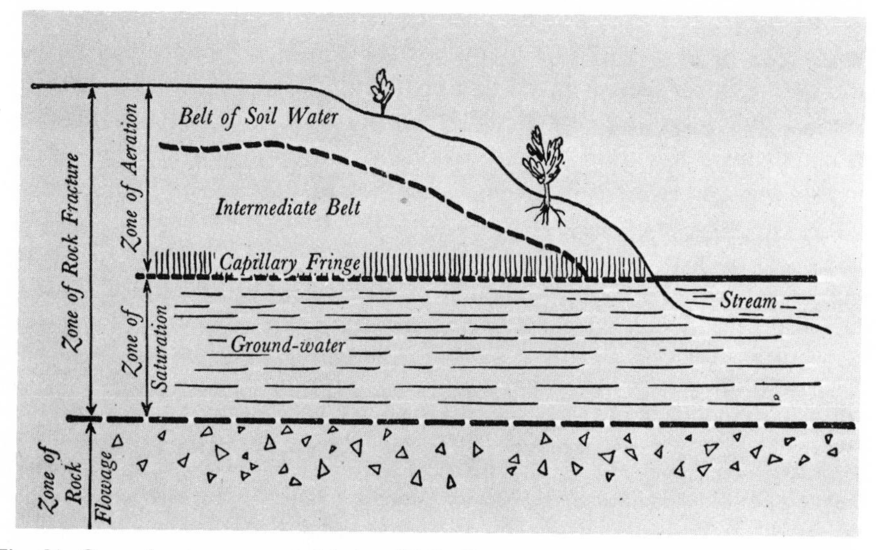

Fig. 31. Groundwater zones and belts. (U.S., Department of Agriculture, *Water: The Yearbook of Agriculture, 1955* [Washington, D.C., n.d.].)

from broken hill country to bottomland. It varies greatly, too, with the nature of soil and subsoil, with the presence or absence of forest or other vegetal cover, with the degree of settlement and the cultivation of the land. Streamflow is affected particularly by the contour and slope of the land, the extent of a drainage basin, the frequency and area of lakes and ponds, and the presence of bogs and swamps.[4] The basic elements of waterpower—volume of flow and amount of fall—are, in short, the product of the varied meteorological, geological, physiographic, and hydrologic conditions. Moreover, the role of these varied factors and conditions and their interrelations were little known and understood to the end of the nineteenth century. The essential dimensional data, especially those relating to rainfall and streamflow and covering extended periods of time, indispensable to understanding and to planning, were almost wholly lacking.[5] Like the mechanical engineer for want of adequate thermodynamic theory, and relevant operating data, the hydraulic engineer hardly less than generations of millwrights before him worked largely in the dark.

Patterns of Millstreams and Waterpowers

The basic pattern of river systems, of the complex of basins they drain and the waterpowers to which they give rise, is the pattern of the tree with its trunk, limbs, and branches. Each river system, small or large, has its main, or trunk, stream and its tributaries; the tributaries their tributaries in turn and so on until in the remotest headwaters within and at the outer fringes of the watershed the system is resolved into a myriad of rivulets and rills. Each stream has its drainage basin, ranging in extent from less than an acre to half a continent. Save for the dead ends of bog and swamp, of lake or pond lacking outlet, and of desert and

[4] See Mead, *Water Power Engineering,* chaps. 4–9.

[5] Most of the basic data presented here, especially with respect to rainfall, runoff, and streamflow, did not exist at the time of the 1880 census survey of waterpower. The civil engineers engaged in that project realized in some measure the importance of long-term, systematic records of streamflow and attempted rough and admittedly inadequate estimates of streamflow for many of the larger streams surveyed. The systematic recording of such data was prompted first by the rapid extension of irrigation in the West and was supported from the 1890s by the rapid spread of hydroelectricity. The systematic stream measurement program of the U.S. Geological Survey had its beginnings in 1889, with periodic reports on progress appearing in the survey's Water-Supply Papers. See C. H. Pierce and R. W. Davenport, *The Relation of Stream Gauging to the Science of Hydraulics,* U.S. Geological Survey, Water-Supply Paper no. 375-C (Washington, D.C., 1915).

arctic waste, no part of the land is without running water, no stream however small without its power potential. For running water is falling water; and falling water has only to be supplied with dams, races, and waterwheels, in some instances by waterwheels alone, to serve the needs of industry and man.

As we have seen, a mill seat, water privilege, or waterpower, in the varying terminology of the early industrial age, was simply a point of marked descent in the bed of a stream where the concentration of fall greatly simplified the capture and harnessing of the flow. It was but an interruption, more or less pronounced, in the otherwise fairly regular and gradual slope of the stream. The falls or rapids that marked such privileges were occasioned by irregularities in the geological formations. Within each drainage basin the distribution of waterpowers, whether numerous or few, simply repeated the trunk-limb-branch, or dendritic, pattern of the river system. The succession of such powers interrupting the streamflow at irregular intervals on any millstream gave a characteristically linear pattern to the industry depending upon these powers. The presence of many millstreams and mill rivers as parts of a larger river system in turn produced a wide scatter pattern of waterpowers and, to the extent utilized, of mills and manufactories.

The frequency and extent of falls and rapids that gave rise to waterpowers reflected the topographic character and geological structure of a river basin. Since in any river system creeks are more numerous than small rivers and the latter in turn outnumber the "limbs" of which they are the "branches," the number of waterpowers generally varied inversely with their size. Whereas country mill seats often fairly blanketed a region, waterpowers capable of driving a textile mill of several thousand spindles or ironworks processing many tons daily were far less common, and those with the capacity to supply the needs of several such establishments clustered in an industrial village were correspondingly less numerous. Only a substantial fall interrupting the course of a river of sizable volume could provide the power base of an industrial city. Of such powers there was no great number, but they gave rise to some of the important early industrial centers in the United States, the like of which were not to be found in Europe.

Topography and Power Potential

As one ascends from the lower to the upper and outer portions of a large river system, tracing the courses first of the main stream and then of the larger and successive groups of lesser tributaries, the slope of streambeds typically increases and the volume of flow diminishes. As the plains or valley bottoms give way to rolling country and then to the broken terrain of hills, piedmont, or mountains as one advances toward headwaters, falls and rapids are apt to occur more frequently and usually become more pronounced, offsetting in varying degree the lessened volume of flow. Whatever uniformity is present in a river system and its drainage basin as a whole, the occurrence, size, and characteristics of individual waterpowers reflect the circumstances of local geological structure and history and offer little regularity or uniformity. Within the framework of its drainage basin, the power *potential* of a river system is a product of aggregate volume of flow and total amount of descent (loss of elevation) of the streambed between headwaters and river mouth.[6] The potential is greatest, of course, where the sources are in hilly uplands, high plateau, or mountains and the outlet is at the sea. The shorter the distance in which this loss of elevation is effected, the greater is the slope, or gradient, and, provided the topography and geological structure are favorable, the greater the amount of fall available for concentration in waterpowers.

Regions and river basins differ widely in these respects. Many New England rivers flowing into the Atlantic have average slopes of five to ten feet per mile. Much of the Atlantic slope of the state of Maine, including that traversed by such major rivers as the Saco, Androscoggin, Kennebec, and Penobscot, is marked by a mean fall of some 1,100 feet in a mean distance of 140 miles, or nearly eight feet per mile. The Androscoggin River in its 180-mile course through New Hampshire and Maine possesses an aggregate fall of some 1,500 feet—nearly as great as that of the Mississippi River in its 2,600-mile length from its headwaters to the Gulf of Mexico.[7] Companion streams of the south Atlantic seaboard have gradients of three to six feet per mile, but west of the Appalachians average slopes are much less. The Ohio River in its nearly thousand-mile course from Pittsburgh to the Mississippi has an average slope of

[6] Tenth Census, *Water-Power,* pt. 1, pp. 18–27. Leopold and Langbein state that the average vertical distance traversed by water in rivers of this country is about 1,650 feet, which, with the total average flow of about 2 million cubic feet per second, gives an aggregate national potential of some 300 million horsepower, of which by 1960 only about 38 million had been developed, 25 million of this amount within the preceding twenty-five years (*Primer,* p. 47). See also Marshall O. Leighton, "Undeveloped Water Powers," in *Papers on the Conservation of Water Resources,* U.S. Geological Survey, Water-Supply Paper no. 234 (Washington, D.C., 1909), pp. 52 ff.

[7] George F. Swain, "Statistics of Water Power Employed in Manufacturing in the U.S.," *PASA*, n.s. 1 (1888–89): 17–18.

less than six inches per mile. Its more important southern tributaries descend no more than twelve inches per mile except in their upper portions.[8]

In general the topography best suited to power development was that intermediate area between steep mountain slopes and the alluvial bottomlands of the lower valleys, where slopes were slight and falls limited in descent and occurrence. Quite apart from the question of accessibility, mountain streams with their marked slopes and steep sides were usually too variable in flow to afford satisfactory powers. At the other extreme the length of dam required by a wide stream and, in the absence of sufficient concentration of descent, the cost of increasing the fall by raising the height of the dam or of building long races discouraged development. Also, increasing the height of a dam could produce excessive flooding of upstream farmlands and result in a heavy assessment for damages. For waterpowers of the kind most in demand in the middle decades of the nineteenth century, the most favorable topography was the broken terrain of hill country and piedmont—in the northeastern United States frequently approaching the coastline—in which the underlying rock formations occasioned fairly abrupt, frequent, and permanent descents in streambeds.[9] For the special case of tidal power sites, see Appendix 5.

[8] Tenth Census, *Water-Power,* pt. 1, tables, pp. 19–26. Louis Bell, an electrical engineer in the early 1900s, suggested the following classification of waterpower streams: (1) mountain rivers, streams with the high heads characteristic of much waterpower in the Far West, ranging from several hundred to a thousand feet. With heads of this height each cubic foot of water per second might be equivalent to as much as 140 horsepower at the waterwheel below. (2) Upland rivers, such streams as were common in New England with descents usually between 20 and 40 feet; without considerable volume of flow the power capacity would be low, since within this range of fall a cubic foot of water per second accounted for two or three horsepower only. (3) Lowland rivers, streams with heads at point of fall of no more than 10 to 15 feet and requiring an immense volume of flow to produce any power of consequence; falls in this class were rather reliable and while affording often a large number of very useful powers of moderate size seldom provided any considerable power development (*Electric Power Transmission,* 4th ed. [New York, 1906], pp. 388 ff.).

[9] Swain, "Statistics," pp. 11–15. Swain develops this theme in his introduction to "Water-power of Eastern New England," in Tenth Census, *Water-Power,* pt. 1, pp. 49 ff. A later source remarked upon the irregularities of riverbed slope "so characteristic of the streams of New England, consisting of a succession of still reaches alternating with falls and rapids" (H. Gannett, *Profiles of Rivers in the United States,* U.S. Geological Survey, Water-Supply Paper no. 44 [Washington, D.C., 1901], p. 11). Hydraulic engineers stress the importance of the hardness of the New England rock formations in giving permanence to the falls. The near-disaster attending development of the Falls of Saint Anthony, with its soft sandstone formation, is described below, chap. 4. At the other extreme in scale was the small stream in West Virginia where a survey-

A good waterpower must offer more than the required potential. At the immediate site conditions must favor the construction of dams and races as well as provide space for the factory buildings and dwellings for workers. Where streams cut their way through light glacial drift or alluvial materials, the provision of secure foundations, especially for the dam, often presented considerable difficulty. The rocky terrain so favorable to falls usually afforded good foundations, although if much excavation in rock was required for races and wheel pits, costs could run high. The pondage facilities important for storing and saving overnight flow during periods of low water also depended upon the character of the terrain. A steep-sided, narrow gorge upstream would provide negligible pondage; at the other extreme, flooding bottomlands suited to farming could be costly in damages.

The very ruggedness of terrain that gave rise to otherwise satisfactory powers could present serious problems of accessibility. Such districts were often remote from centers of population and channels of transportation and trade and were handicapped by high transport charges. Manufacturers were dependent upon access to raw materials, markets, labor, and commercial facilities as well as power; and these factors assumed increasing importance as the scale of industry increased and markets widened. In short, many otherwise excellent waterpowers were unsuited to development because of technical difficulties and costs of construction, site limitations, or general inaccessibility.

Rainfall, Runoff, and Streamflow

If a river system served by its drainage basin and supplemented by numerous hydraulic facilities may be likened to a vast and sprawling engine with many power takeoffs, it was the weather that supplied the dynamic force to set this engine in motion. The most favorable disposition of terrain had no meaning save in relation to volume and regularity of streamflow, as the periodically dry streambeds of arid regions eloquently testify. It was the variations and uncertainties in streamflow that gave millowners and millwrights their greatest problems. The fall generally could be determined once and for all by survey and measurement, although millowners and even millwrights were often ignorant and negligent about such matters. Volume of flow was, by contrast, difficult

ing party about 1900 discovered that a falls used as a landmark in 1790 had "walked upstream" nearly 100 feet (Alvin Johnson, *Pioneer's Progress: An Autobiography* [New York, 1952], p. 133).

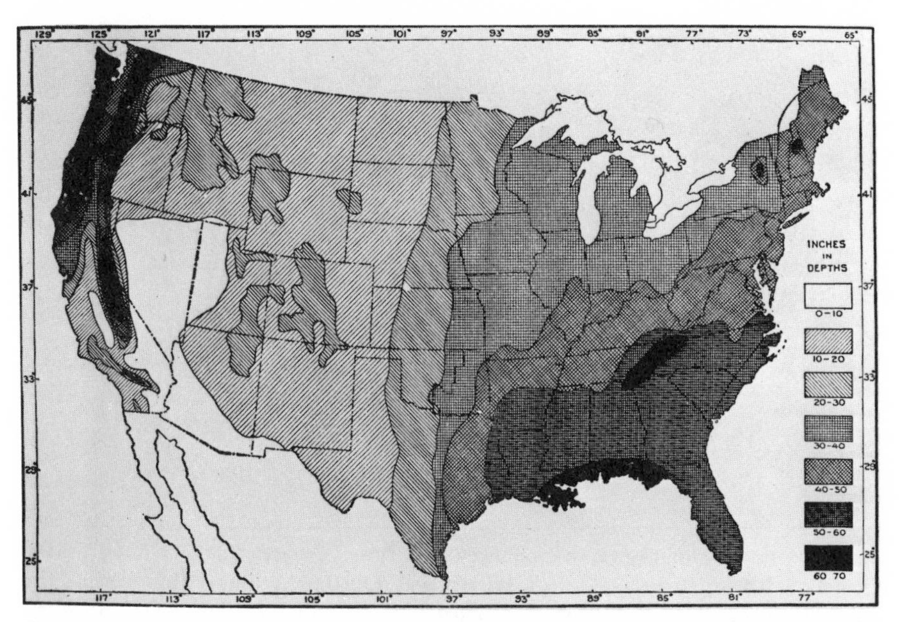

Fig. 32. Mean annual rainfall. (Frederick H. Newell, *Irrigation in the United States,* rev. ed. [New York, 1906].)

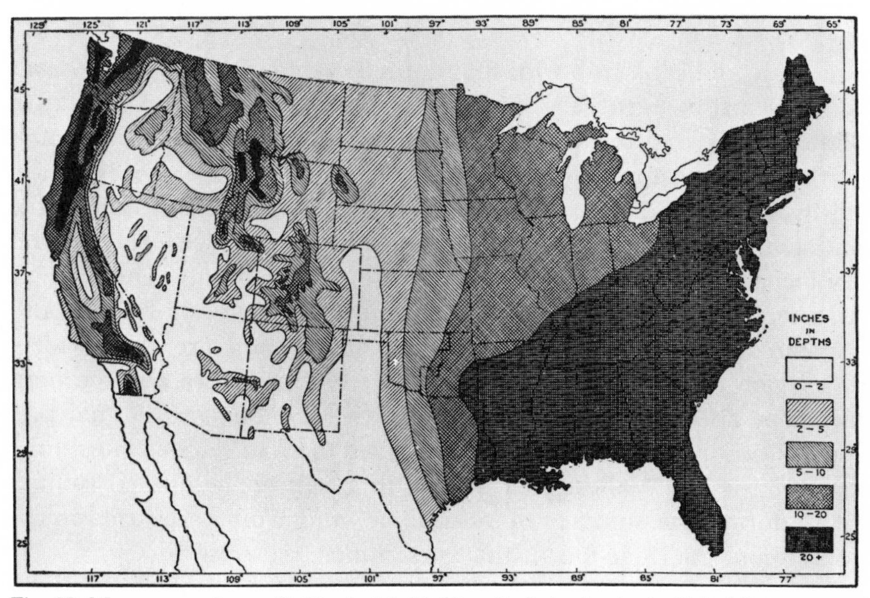

Fig. 33. Mean annual runoff. (Frederick H. Newell, *Irrigation in the United States,* rev. ed. [New York, 1906].)

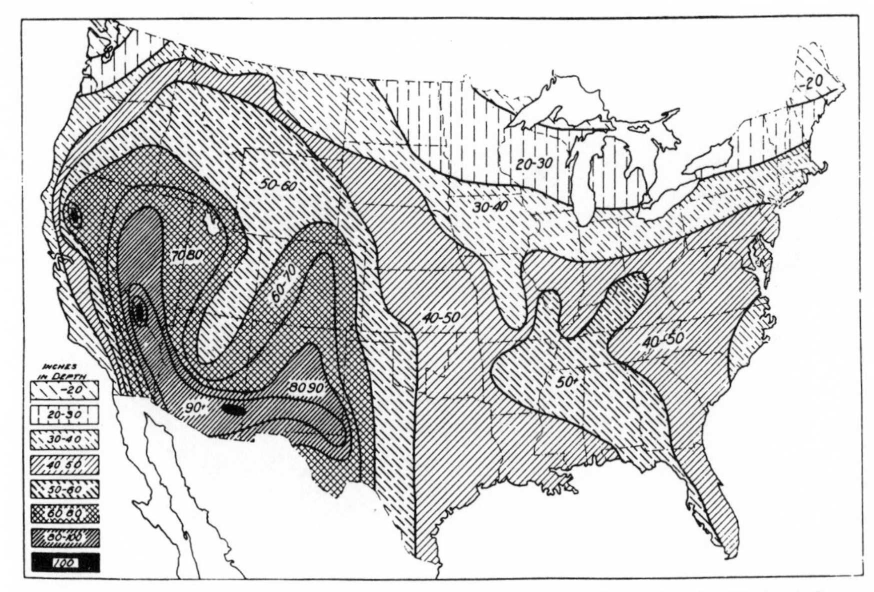

Fig. 34. Relative annual evaporation from free water surfaces in the United States. (Daniel W. Mead, *Hydrology*, 2nd ed. rev. [New York, 1950].) "Other things being the same, the rate of evaporation is nearly proportional to the difference in the temperature, to the velocity of the wind . . . and nearly inversely proportional to the pressure of the atmosphere. Evaporation is usually but not always the source of the greatest loss on a drainage area and commonly other sources of loss are insignificant compared with it." (Daniel W. Mead, *Water Power Engineering* [New York, 1908], pp. 141–44.)

to measure and subject to wide and unpredictable variations, by seasons and by years, owing to a wide range of meteorological and vegetative conditions, the basic elements of which are rainfall and runoff. Rainfall sets the limits of supply by its amount, its form and mode of occurrence, and its distribution by months and seasons. In the eastern half of the continent average annual precipitation, ranging between twenty and fifty inches, is by comparative standards fairly abundant. The amount varies regionally, however, and within any drainage basin from month to month and from year to year. Seasonal distribution also varies widely from region to region, especially as between the northern and southern states, but also between East and West. In New England annual precipitation is distributed fairly evenly among the four seasons, but in the upper Mississippi Valley from a third to a half of the annual rainfall occurs during the summer months alone, and winter precipitation is hardly more than one-fourth that of summer (see Appendix 6).[10]

[10] Tenth Census, *Water-Power*, pt. 1, pp. 49–58, 657; pt. 2, pp. 146, 290–91; Mead, *Water Power Engineering*, chap. 6.

Within the limits set by precipitation, what counted most for streamflow and power potential was runoff, the amount of rainfall actually reaching streams. The losses of water in runoff reach as much as 87 percent of the total in rivers of the plains country in the West. In the northeastern United States the federal survey of 1880 estimated the ratio of runoff to rainfall at more than 60 percent for the upper Connecticut Valley, the Merrimack, the Passaic, and the headwaters of the Hudson River. The average ratio for the rivers of Maine was placed at 40 percent. At the other extreme are the rivers draining the rolling prairies and plains of the Mississippi Valley, where the ratio of runoff to rainfall was estimated at 15 percent for the Missouri Valley and 24 percent for the Ohio Valley.[11]

So variable are the conditions affecting runoff, declared one authority, that it was "impossible to determine any definite relationship between the discharge of streams and annual precipitation."[12] However, major sources of loss could be identified if not measured. Much rainfall is lost to streamflow by absorption in plant and tree growth and the process of transpiration. While most rainfall penetrating subsurface channels later enters streamflow, some reaches streams at too low a point to have value for waterpower or has direct outlet in lake, sea, or estuary. Above all, heavy seasonal losses of water through evaporation, varying with temperature and humidity and the exposure of land and water surface to sun and wind are mainly responsible for the low ebb of streamflow in the summer months in much of the country.[13] In the northern states streamflow is affected in varying measure by the locking-up of moisture in snow, ice, and frozen ground and its release with the coming of warm weather.[14]

[11] Tenth Census, *Water-Power,* pt. 1, p. 175.

[12] Ibid., pt. 2, p. 21.

[13] Mead, *Water Power Engineering,* chap. 7.

[14] In the north, streamflow was often maintained in winter from frost-free groundwater sources, especially springs. Further, the surface ice as it thickened provided an insulating layer protecting from frost the water below on the larger streams, ponds, and lakes. Samuel McElroy pointed out that the mean winter temperature of the southern two-thirds of Maine was 19 degrees. Although "there are a large snowfall and deposit," he stated, "its under stratum thaw maintains winter stream flow, while spring freshets are delayed until the ice, as a rule is brittle, and does not make the dangerous freshets of milder climates. . . . upper pond ice sheets tend strongly to prevent sludge ice and maintain full power head on the dams" ("Water-Power of Caratunk Falls, Kennebec River, Maine," *TASME* 17 [1895–96]:59). In large installations effective countermeasures were feasible, as Francis pointed out: "The extreme rigor of the New England winter renders it necessary to afford to water-wheels of all descriptions complete protection from the cold. The result is, that less interruption from frost is experienced, than in many milder climates. The wheel house in which these turbines

Thus even under the most favorable conditions streamflow fluctuates widely from one season to the next and from year to year. Even if the extreme variations of flow are omitted as typically of quite short duration, the ratio of maximum to minimum flow by monthly averages, as estimated by Joseph P. Frizell for streams of the northeastern United States, was ten to one, as the following tabulation of the volume of streamflow shows (see also Appendix 7).[15]

	Percent of annual total		*Percent of annual total*
January	10	July	2
February	14	August	3
March	20	September	3
April	15	October	5
May	10	November	6
June	4	December	8

Even on the Merrimack River with its elaborate system of reservoirs the ratio of maximum to minimum flow by months, averaged for 1893-1905, was nine to one.[16] The pattern of streamflow, too, fluctuates from year to year and often widely.

Natural Regulators

The variations in streamflow to which power users must conform as best they may would be far more aggravated but for the influence of several natural elements that serve to regulate runoff, in effect functioning as reservoirs to accumulate and release water. The primary regulators of streamflow include not only forest, woodland, and other vegetation with their litter, roots, and accumulated humus, but the soil itself, the subsurface strata, and that basic subterranean reservoir, ground water, whose upper surface forms the water table. These agents of varying capacity absorb rain in its intermittent and sporadic fall, reduce direct surface runoff and evaporation, and, save for the large amounts lost in plant growth and transpiration, eventually release much if not most of

---

are placed is a substantial brick building, well warmed in winter by steam" (James B. Francis, *Lowell Hydraulic Experiments,* 4th ed. [New York, 1883], p. 7). The general practice of such heating is described in David Craik, *The Practical American Millwright and Miller* (Philadelphia, 1870), pp. 99–100.

[15] Frizell, *Water-Power,* 3d ed. (1903), p. 4. The latitude is that of Boston.

[16] Mead, *Water Power Engineering,* table 16, pp. 173–74.

this water to streamflow. To the extent that these natural regulators are lacking or deficient in capacity, surface runoff is more rapid and variations in streamflow greater. Although some of the water passing underground is lost to streamflow, much of it, released over periods ranging from days to weeks and even months, eventually joins the running waters of the river system. Thus the peaks and valleys of rainfall are to some extent moderated and variations in streamflow reduced.[17]

A second class of natural regulators that serve to moderate the variations in streamflow is less widely distributed but of comparable importance. These are lakes, ponds, swamps, and bogs, which find outlet in streamflow directly or by underground channels, such as springs. By accumulating water during periods of rain and melting snow and ice and gradually releasing it later, they serve as natural reservoirs. Lakes and ponds often offered millowners and operators an additional advantage: dams and gates built at their outlets at relatively small expense could increase their storage capacity, often very materially, and regulate their outflow in accord with the needs of the mills below. Lakes and ponds are found in large number throughout the glaciated regions of New England and the Old Northwest, especially in Maine, Wisconsin, and Minnesota.[18] In other parts of the eastern United States they are relatively few in number. The Middle and South Atlantic watersheds south of the Susquehanna basin and the Ohio and Missouri river valleys are largely lacking in lakes and ponds, much to the disadvantage of the waterpower

[17] Ibid. The precise functioning of these regulators and the extent of their influence were not readily observed and measured and continued to be subjects of investigation, and in some instances, controversy. For example, forests were long believed to have a powerful regulative influence upon streamflow; their destruction was deplored as reducing streamflow and increasing the range of seasonal fluctuations. See observations of Peter Kalm in mid-eighteenth century, cited in Stevenson W. Fletcher, *Pennsylvania Agriculture and Country Life, 1640–1840,* 2 vols. (Harrisburg, 1950–55), 1:324–25; Zadock Thompson, *A Gazetteer of the State of Vermont* (Montpelier, 1824), pp. 11–12; Joseph K. Angell, *A Treatise on the Common Law, in Relation to Water-Courses,* 5th ed. rev. (Boston, 1854), p. 185n; and numerous references in Tenth Census, *Water-Power.* For the growing skepticism of such influence by forests, see W. G. Hoyt and H. C. Troxell, "Forests and Stream Flow," *TASCE* 99 (1934):1–111.

[18] Not counting the multitude of small ponds scattered through the state in such profusion that "almost every neighborhood and school district has one," Maine was reported in 1869 to have 1,620 lakes. The Androscoggin, Kennebec, and Penobscot lake systems comprised 148, 311, and 467 lakes, respectively (Maine, Hydrographic Survey, *Water-Power of Maine,* pp. 29–30). According to land office maps there were 4,920 lakes and ponds in Minnesota and 2,465 in Wisconsin; the number in eastern Dakota was but 133 (Tenth Census, *Water-Power,* pt. 2, pp. 23–24; also pt. 1, pp. 347, 422, 523).

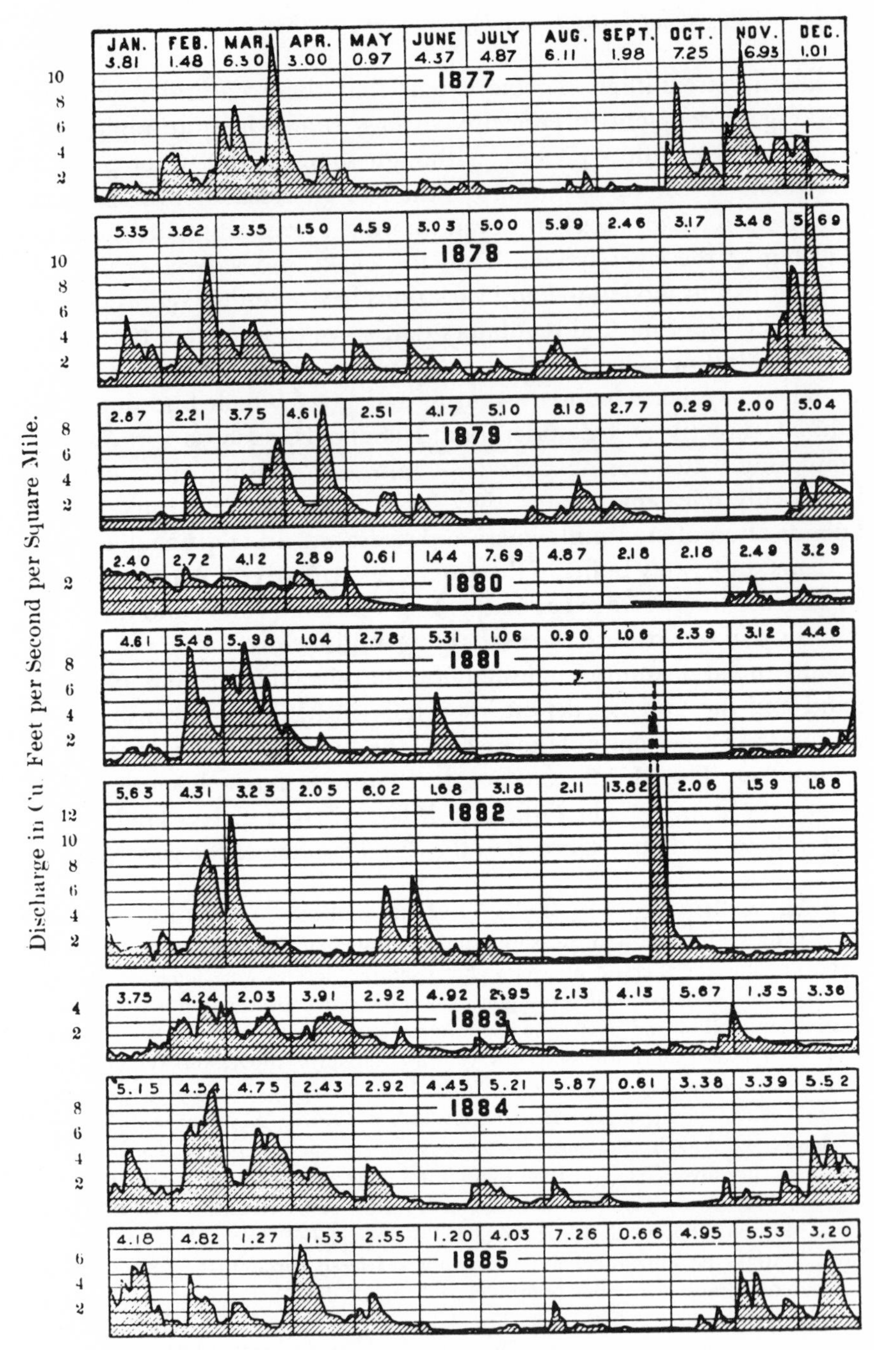

Figures near top of each diagram show total monthly rainfall.

Daily flow of Passaic River, Little Falls, N.J.

Figs. 35-36. Hydrographs of selected streams in the U.S. "The information of primary importance in a water power project is the amount and variation in the run-off of the

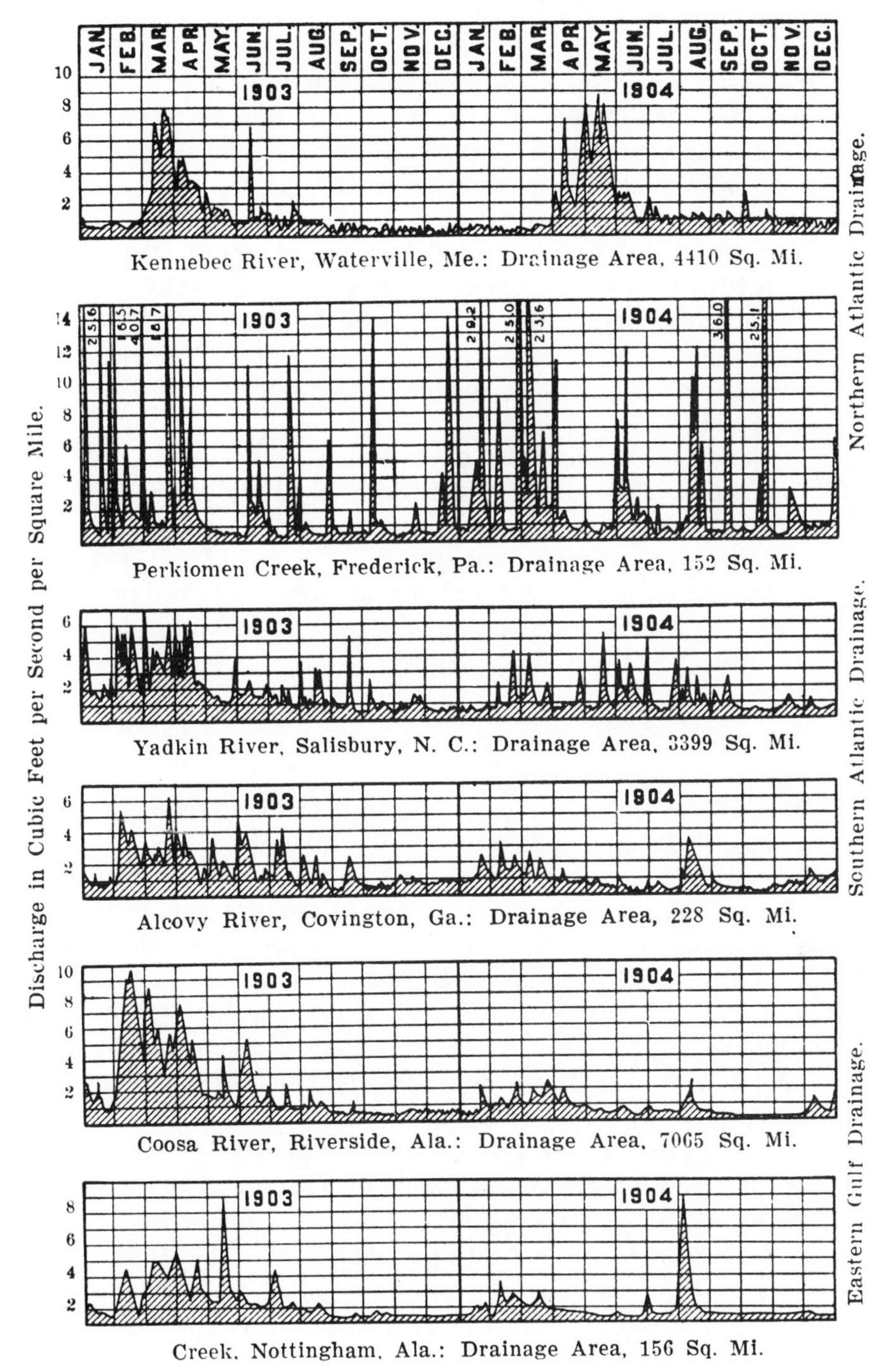

stream itself. . . . a simple gauging of the stream is of little or no value. . . . such hydrographs must be available for a considerable term of years . . . [covering] all extremes of rainfall and drought." (Daniel W. Mead, *Water Power Engineering* [New York, 1908].)

resources of these regions in the view of nineteenth-century hydraulic engineers.[19] George F. Swain in 1888 summed up the regional conditions favorable to waterpower development. In general, he pointed out, waterpower resources would be the more valuable:

1. The greater the slope of streams
2. The more concentrated the fall of streams of definite points, with the fall "neither too great nor too small for economical development"
3. The more permanent the falls, that is, the harder the rocks that comprise them
4. The nearer the falls to navigable waters and the better the rail transport facilities
5. The larger the average flow of streams
6. The more uniform the flow of streams as favored by
    a. the lesser severity of the winters
    b. the greater extent of forests
    c. the larger the number and area of lakes or artificial reservoirs
    d. the less steep and rocky the drainage basin (within limits)
    e. the larger the drainage basin[20]

Regional Power Potentials

New England was the region above all others endowed by nature with waterpowers of the kind best suited to the requirements of our early industrial development. Its predominantly hilly and broken terrain gives rise to a multitude of streams of high average slopes, interrupted in their courses by numerous strata of hard metamorphic rock, occasioning abrupt, numerous, and durable falls. Since virtually no coastal plain exists from New Brunswick, Canada, to New Jersey, many substantial powers lay within practicable transport distance of seaports and coastal waterways. Rainfall is relatively abundant and well distributed throughout the year. The handicap of long and hard winters was measurably offset by the favorable temperatures and evaporative rates of the summer months.[21] Not only was the greater part of the region well forested to the believed advantage of regular streamflow, but after the early nineteenth century neither lumbering nor farming made significant inroads upon

[19] Tenth Census, *Water-Power,* pt. 1, pp. 52–53, 523, 669; pt. 2, pp. 23, 288, 437. For differing views of the influence of swamps, see ibid., pt. 1, p. 183; pt. 2, pp. 29, 155, 437.

[20] Swain, "Statistics," pp. 13–14.

[21] Tenth Census, *Water-Power,* pt. 1, pp. 41–160, on the streams of eastern New England, and pp. 161–333 on the region tributary to Long Island Sound.

this protective mantle of the land. The literally innumerable lakes and ponds distributed throughout the greater part of New England, a large proportion of which found outlet in streams, served as natural reservoirs for storing and releasing water to streamflow. Often, too, such reservoirs could be improved in capacity and regulative value to the advantage of power users by relatively small outlays for the purchase of land ill suited to other uses or for the compensation of its riparian owners.

This important natural resource was exploited to great advantage. As of 1880 the six states of New England possessed 33 percent of the developed waterpower of the United States, although occupying but 2 percent of the nation's land area. Within the region this resource was widely if unevenly distributed. Maine displayed in outstanding measure the physical characteristics favorable to power, although deficiencies in other resources and distance from markets limited their industrial exploitation. Excluding brooks and rivulets and taking only those bodies of water appearing on maps (as of the late 1860s), the state possessed over 1,600 lakes and more than 5,000 streams, including among its major river systems the Saco, the Androscoggin, the Kennebec, and the Penobscot. A careful estimate reported one square mile of lake surface to every fourteen square miles of area in the state.[22]

Outstanding in central and western New England were the Merrimack and Connecticut river systems. Before the hydroelectric age these two systems with their numerous and widely distributed powers, in capacities ranging from the requirements of country mills to those of large industrial centers, presented without doubt the largest complexes of developed waterpower in the world. Of far less extent but proportionally impressive were the Thames and Housatonic river systems of south-central Massachusetts and Connecticut. Of considerable industrial importance, too, were the otherwise minor streams along the seacoast, such as the Salmon Falls River in New Hampshire; the Charles, Taunton, and Quequechan (Fall) rivers in Massachusetts; the Pawcatuck and Blackstone rivers in Rhode Island, and the Quinnipiac River in Connecticut. Fifty miles or

[22] Maine, Hydrographic Survey, *Water-Power of Maine,* pp. 29–30. Wells's estimates and appraisals were amply confirmed by the U.S. Geological Survey, Water-Supply Paper no. 69, published some thirty years later (Henry A. Pressey, *Water Powers of the State of Maine* [Washington, D.C., 1902], pp. 16–20). Despite their omnipresence, the vast number of inland waters comprising the natural drainage systems of any large region are beyond the comprehension of all but the specialist in hydrology. The six million acres (some 9,500 square miles) of upstate New York's Adirondack Park, about the size of the state of Vermont, contain 2,300 lakes, 6,000 miles of rivers, and 30,000 miles of brooks and streams (*New York Times,* 20 May 1973).

less in length, with drainage basins comprising only two or three hundred square miles and with slopes ranging from five to ten feet per mile, these streams offered many fine waterpowers of moderate capacity that came to be quite extensively developed. As of 1880 the Blackstone led all rivers in the United States in utilized horsepower per square mile of drainage basin. Eleven of the first twenty waterpower streams in the United States, ranked in order of utilized power per square mile of drainage basin, were in New England.[23]

Southward of New England the physical conditions favorable to waterpowers and their utilization gradually declined. The upper Hudson River system and the streams draining into Lake Ontario form an exception, however, presenting many of the characteristics of topography, climate, and accessibility prevailing in New England. The Hudson River above the head of navigation at Troy and the Mohawk and Hoosic rivers with their tributaries offered in the aggregate an impressive power potential. So, too, on a somewhat lesser scale did the Genesee, Oswego, Black, and other rivers flowing into Lake Ontario.

South of the small but well-endowed streams of the Raritan and Passaic basins in New Jersey are the numerous and extensive river systems of the Middle and South Atlantic seaboard states, especially the Delaware, Susquehanna, Potomac, James, Roanoke, Santee, and Savannah. These river systems, and their Gulf Coast companions, principally the Alabama and Apalachicola, fell short in power potential of those of the North Atlantic seaboard. Average slopes generally are lower, good falls less frequent, rainfall less favorably distributed, runoff less gradual, streamflow more variable, and lakes serving as natural storage facilities far less abundant than in New England, New Jersey, and New York. Conditions vary from river to river and tend to become less favorable for power development as one advances southward. Falls are either infrequent in the more accessible portions of the important rivers or take the form of extended shoals and rapids, often combined with wide and shallow streambeds that would require long and costly dams and races for development. On the Susquehanna proper, the largest river of the Atlantic seaboard, there was as late as 1880 not a single utilized power, owing to the lack of sites suited to economical development.[24]

[23] Tenth Census, *Water-Power,* pt. 1, p. 39.

[24] See reports by George F. Swain on the waterpower of the Middle Atlantic and South Atlantic watersheds ibid., pp. 513 ff., 661 ff. See also Frizell's summary of the differences between northern and southern streams and streamflow in *Water-Power,* pp. 3–4, and Swain, "Statistics," p. 15. Swain also noted that whereas numerous lakes in the upper basins of the Delaware and Susquehanna rivers served as natural reser-

Finally, and of great practical importance well into the closing decades of the century, the coastal plain that is nonexistent or very narrow in New England and the Middle Atlantic states steadily widens to the south, reaching a breadth of 150 miles or more in the Carolinas and Georgia. The fall line, which marks its western limit and the beginning of the waterpower resources of the Piedmont, moves gradually inland and away from the water transport facilities of the seacoast. Crossing the Delaware River at Trenton and the Potomac River a few miles above Washington, the fall line farther south is marked by such inland cities as Richmond, Fayetteville, Columbia, Augusta, and Columbus. In short, until the greatly delayed extension of the railway network through much of this region, the waterpower resources were least accessible where most abundant, providing an additional element in the complex of causes that retarded the advance of industry in the South.[25]

Across the Appalachians lay the vast inland basin drained by the Ohio, Mississippi, and Missouri rivers. Here with notable exceptions nature was unfavorable to the development of waterpower on an industrially significant scale. Nor did the regions drained by rivers flowing into the Great Lakes or the Gulf, with certain exceptions, offer much waterpower potential above the requirements of country mills. The western slopes of the Appalachian system were in the main difficult of access; and the abruptness of the descent of the Appalachian plateau concentrated much of the available fall within a relatively short distance. The southern tributaries of the Ohio River, especially the Cumberland and the Tennessee, were distinguished by navigability over extensive portions of their long courses, to the disadvantage of power. The few noteworthy power sites occurred at extended shoals requiring forbiddingly high development expense. The northern tributaries of the Ohio, with limited drainage basins and low aggregate slope, offered greater accessibility; and in a number of regions the broken topography gave rise to power sites of a medium class often suited to development.

Yet north of the Ohio River there were few large waterpowers and none on a scale comparable to scores of powers in New England. Only the Miami River and, at a few points, the Allegheny and Beaver rivers could provide support for industry other than grain, lumber, and other small mills. The power situation for the northern half of the Ohio Valley

voirs contributing to streamflow, there were none to the southward through Maryland and Virginia, and thence through the South Atlantic watershed there were likewise virtually none (Tenth Census, *Water-Power,* pt. 1, pp. 523, 669).

[25] See Swain's "Concluding Remarks," Tenth Census, *Water-Power,* pt. 1, pp. 823–24.

was summarized, circa 1880, as follows: streamflow small relative to area drained; low runoff relative to rainfall; inadequate water in summer and fall; trouble from freshets and ice flows; no falls of significant drop; and difficulty in securing foundations for dams.[26] Similarly the watershed draining northward into Lake Erie was too limited in extent and aggregate fall to afford power facilities for other than gristmills, flour mills, and sawmills. The rivers of lower Michigan likewise afforded no waterpowers of any importance.

The far more extensive western portion of the Mississippi River basin, excluding the upper Mississippi Valley, was served by the great western affluents: the Missouri, the Arkansas, and the Red. With their headwaters on the slopes of the distant Rockies, these rivers trace courses ranging from 1,200 to nearly 3,000 miles across the vast expanses of the West and have aggregate falls, outside of the mountains proper, of thousands of feet. The power deficiencies of these river systems lay less in their slopes, which for extensive portions (especially in their larger tributaries) averaged several feet per mile, than in low annual rainfall, high rates of evaporation, wide fluctuations in volume, and the rarity of abrupt falls. Although the vast deposits of loess, sand, and gravel comprising most of the plains country were generally ill suited to the construction of hydraulic facilities, the development of limited waterpower was often practicable, usually with dams of brush, logs, and earth. Power could be developed, however, only in amounts that offered little support for manufacturing requiring more than 25 to 50 horsepower, even had other circumstances been more favorable to industrial growth.[27]

The one region in the continent's great inland basin offering powers comparable to those of the North Atlantic seaboard was a substantial portion of the upper Mississippi Valley and the adjoining drainage basins with outlet in Lakes Superior and Michigan. Included principally in this power-abundant region were Wisconsin and Minnesota, eastern Iowa, northwestern Illinois, and the upper peninsula of Michigan. The northern half of this extensive region was rugged, broken country, deeply covered with glacial drift and overlain with thick forests. Southward lay the rolling prairie country whose numerous streams often possessed considerable slope although extensive falls were uncommon. The most important rivers for power potential were the upper waters of the Mississippi itself, culminating in the spectacular Falls of Saint Anthony at

[26] Ibid., pt. 2, pp. 435–37.

[27] Ibid., pp. 277 ff.; also A. G. Allan, "Water Powers of the Western States," *Engineering* 9 (1895):17 ff.

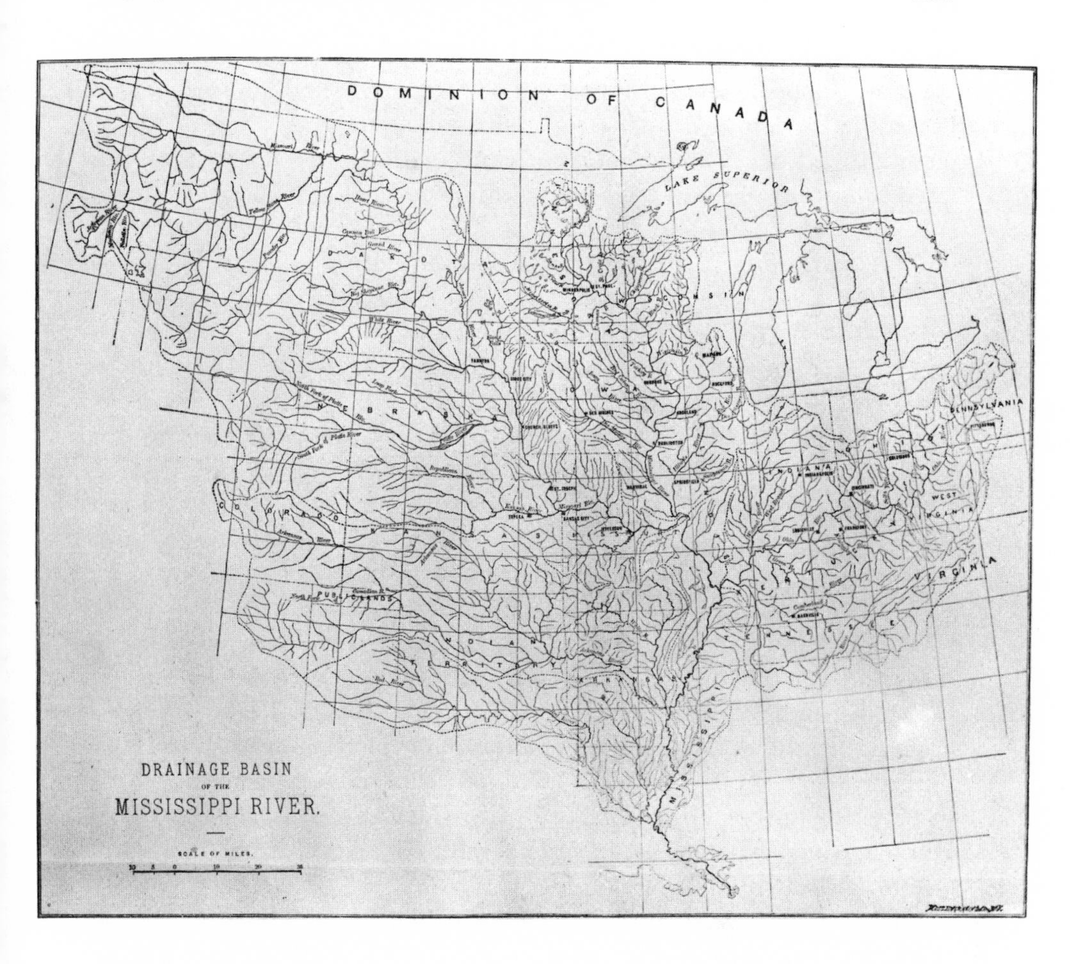

Fig. 37. Drainage basin of the Mississippi River. The vastness of the extent of this drainage basin contrasts sharply with its relatively minor power potential adapted to direct-drive development in years before hydroelectricity. (U.S., Census Office, Tenth Census, 1880, *Reports on the Water-Power of the United States,* pt. 2 [Washington, D.C., 1887].)

Minneapolis, and major tributaries such as the Minnesota, Saint Croix, Chippewa, and Wisconsin. The streams entering Lake Michigan were unimpressive with one exception: the Lower Fox River, less than forty miles long from its origin in Lake Winnebago to its outlet in Green Bay. Yet the Lower Fox, served by an extensive drainage basin, possessed at Appleton and Grand Kaukana, Wisconsin, an aggregate power potential equal to but less developed than that of the combined great Merrimack powers at Manchester, New Hampshire, and Lowell and Lawrence,

Massachusetts. To the south, arising in southern Wisconsin and extending through northwestern Illinois to the Mississippi at points above Saint Louis, the Rock and Illinois rivers, with moderate slopes and falls and proximity to lines of communications and markets, were to occasion the rise of a number of small but vigorous industrial centers. In eastern Iowa only the Des Moines and Iowa rivers offered waterpowers sufficient to support even quite small industrial communities.[28]

Most of the foregoing region had been subject to severe glaciation, which was reflected not only in the courses and profiles of the streambeds but in the almost innumerable lakes dotting the countryside in Wisconsin, Minnesota, and adjoining areas of South Dakota. These lakes, together with the swamplands of the timber country to the north, did much to offset the effects of the severe winters of the region in restricting streamflow, promoting a degree of uniformity of flow comparable to that characteristic of New England rivers. With notable exceptions, however, such as the Falls of Saint Anthony, the Lower Fox River, Rock River, and the upper Illinois and affluents, the abundant waterpower resources of the Old Northwest were to remain without significant industrial development to the end of the direct-drive waterpower age. In the greater portion of the region economic development before 1900 did not advance much beyond the exploitation of the rich agricultural and timber resources. Effective utilization of its waterpowers, as well as those in much of the region east of the Mississippi and south of the Ohio-Potomac line, awaited the practical industrial application of hydroelectric power.

The same was even more true of the vast waterpower potential of the mountain regions of the Far West, in the Rocky Mountain and Pacific Coast states. Apart from important local, and in some areas more widespread, applications in the mining and milling of minerals dating from the 1850s, the development of this potential was to depend upon hydroelectricity, which found its first significant development here. The controlling geographic considerations, as these became known, stood in marked contrast to those prevailing in the northeastern United States: rugged and lofty mountainous terrain, steep slopes and very high heads or amounts of fall, and an arid or semiarid climate in the greater part of this vast region. For direct-drive waterpower at mines and mills the low annual precipitation and its wide seasonal variations were offset by the very high heads to be had on a number of the major watersheds and by the moderate costs of developing waterpower in the limited amounts required.[29]

[28] Tenth Census, *Water-Power,* pt. 2, pp. 1–67. See also pt. 1, pp. 30–32.

No comprehensive view of waterpower resources of the Far West was available before 1900. The elaborate survey of the nation's waterpower undertaken by the federal census of 1880 gave only perfunctory attention to watersheds in the western half of the continent. The 1880 census report on power used in manufactures showed that nearly 66 percent of the waterpower in use in the United States was located in the North Atlantic states, chiefly in New England and the state of New York. If the North Central states are added, the proportion becomes more than 80 percent. If the returns for the remaining parts of the country are not too greatly in error, the South Atlantic states accounted for 12 percent of the total—and the South Central and western states contributed some 4 and 2 percent, respectively. These figures, however, understate the degree of regional concentration of waterpower used in manufactures in 1880, for they do not take into account the size of the several regions, as does the following tabulation showing developed waterpower in horsepower per square mile of area.[30]

Region	*Area in square miles*	*Horsepower per square mile*
North Atlantic	162,065	4.81
South Atlantic	268,620	0.54
North Central	753,550	0.30
South Central	540,385	0.09
Western	1,175,550	0.02
Total United States	2,900,170	0.42

In relation to the size of a region, the concentration of water horsepower in use in manufactures was nearly twelve times as great in the northeastern United States as in the country at large; and nine times as great as in the region next in degree of concentration, the South Atlantic states. This was, of course, a reflection at least as much of the marked concentration of industry in this older, settled part of the United States as of the abundance of waterpower in the Northeast.

In sharp contrast to the pattern of use in 1880 was the distribution of waterpower potential as estimated some forty years later by Gilbert and Pogue. Only 8 percent of the national total was assigned to the North Atlantic states and over 70 percent to the Rocky Mountain and Pacific Coast states.[31] Much happened during the intervening years,

[29] Swain, "Statistics," pp. 21–22; Titus F. Cronise, *The Natural Wealth of California* (San Francisco, 1868), pp. 596–97.

[30] Tenth Census, *Water-Power,* pt. 1, p. 16.

[31] Chester G. Gilbert and Joseph E. Pogue, *The Energy Resources of the United States: A*

above all the revolution in the generation and transmission of power made possible by electricity. This revolution, highlighted in the early 1890s by the striking development of Niagara power, gave new importance and greatly enhanced value to waterpower. It led to intensive exploration of hitherto little-known regions in the eastern and western United States alike and to the discovery of power resources far greater than anything envisaged as of practical importance during the earlier age of direct-drive waterpower.

George F. Swain of the Massachusetts Institute of Technology reviewed briefly the position of this country in 1880 in respect to waterpower development and use in his introduction to the massive Tenth Census survey of waterpower east of the Rockies. There was no doubt, he remarked, "that in no other country in the world is an equal amount of water power utilized, and that, not only in regard to the aggregate power employed, but in regard also to the number and importance of its large improved powers, this country stands preeminent."[32] Furthermore, he noted in another place, in the engineering problems of developing and managing very large powers this country had long taken the lead. American engineers were frequently called into consultation on power projects abroad. The total theoretical capacity of all running waters in streams large and small from their sources in highland and mountain regions to their outlets in the sea was estimated at more than 200 million horsepower. The vast volume of flow of the lower Mississippi River from its point of juncture with the Ohio River to the Gulf—despite an average fall per mile of less than a foot—accounted for only 6 percent of the national total. The total net horsepower in actual use in manufactures as

---

Field for Reconstruction, Smithsonian Institution, Bulletin no. 102 (Washington, D.C., 1919), pp. 23–24; *Report of the Commissioner of Corporations on Water Power Development in the United States* (Washington, D.C., 1912), pp. 53 ff. For a later and more sophisticated survey of hydroelectric-power development, potential and actual, in the United States, see W. B. Langbein, "An American Survey," *Water Power* (Nov.–Dec. 1950): table 2.

[32] Tenth Census, *Water-Power,* pt. 1, pp. 11–12. The estimate of potential capacity adapted to commercial development is from Swain, "Statistics," p. 8. James B. Francis of Lowell had much earlier called attention to the preeminent position of this nation in waterpower, potentially and in use. The utilization he attributed especially to "the active and inventive genius of the American people," combined with the very high price of labor. Citing a recent estimate of the total useful effect of waterpower employed in France as about 20,000 horsepower, Francis remarked that power far in excess of this amount was "already derived from the Merrimack river and its branches, in Massachusetts and New Hampshire" (*Lowell Hydraulic Experiments,* 2d ed. [New York, 1868], p. ix).

reported by the 1880 census seemed slight by comparison, the 1,225,000-horsepower total comprising less than one percent of the theoretical aggregate. In what was to prove a very conservative estimate Swain placed the amount that might in time be brought into commercial development and use at some 4 million net horsepower, available throughout the driest year. By 1920 the aggregate installed hydroelectric capacity of industry and utilities combined was well over 6 million horsepower, a figure to be quadrupled by 1955.[33]

The Law of Watercourses

To the limitations on waterpower imposed by nature were added others arising from competing or conflicting stream uses and reflected in the intricate body of judicial doctrine known variously as the law of watercourses or of water rights.[34] For centuries streams had served a variety of human needs and economic purposes: navigation and commerce; lumbering and mining; fishing and farming; and, in time, urban water supply. The several uses were often incompatible and in conflict, leading to interference, one with the other, and resulting only too frequently in friction, controversy, and litigation. Simply to present in the most summary digest the results of such litigation as reached courts of record—the visible portions of the "iceberg"—required nearly 700 pages in a mid-nineteenth-century American treatise on the subject, Joseph K. Angell's *Treatise on the Law of Water Courses.* The bulk of this litigation was the product of disputes arising from the use of streams for waterpower.[35]

[33] U.S., Bureau of the Census, *Historical Statistics of the United States, Colonial Times to 1957* (Washington, D.C., 1960), p. 509.

[34] The purpose of this section is not to trace the evolution of legal doctrines of water rights in their theoretical aspects and as they have borne upon economic development and the conflicts of interest during the period under consideration. These matters are explicated with much illumination in two studies of recent years: James Willard Hurst, *Law and Economic Growth: The Legal History of the Lumber Industry in Wisconsin, 1836–1915* (Cambridge, Mass., 1964); and Morton Horlitz, *The Transformation of American Law, 1780–1860* (Cambridge, Mass., 1977). In the first, chapter three is the most directly relevant; in the second, "Water Rights and Economic Development," pp. 34–54. The purpose of the present brief survey is simply to call attention to and briefly review the principal doctrines of water rights as they bore upon the everyday management and operation of water mills, giving frequent rise to controversy and conflict in the districts of waterpower's most intensive use, and serving as still another of the problems associated with a source of power which in so many respects was so advantageous and so widely relied upon by industry before the Civil War.

There were two reasons for the large number of disputes arising from this use. First, the milldams almost invariably employed for harnessing streamflow interfered with most other uses, arousing the opposition of those adversely affected. Dams obstructed not only the passage of boats, rafts, and logs alike but also the seasonal movement of fish to and from upstream spawning grounds. More serious in the long run, by raising the level of water upstream and creating the ponds so important for water storage, dams typically caused the flowage of bordering lands, depriving the owners of their use and usually lowering, where it did not destroy, the land's value. This was a matter of particular concern where good farmland was involved, and the farming interest was often strongly opposed to the mill interest. In the second place, among the varied uses of watercourses, navigation only excepted, the development of power held a special and favored position in the eyes of government. This favoritism reflected the early and persistent view that the importance of mills to the community gave them a public character and merited special aid and encouragement. We noted in chapter 1 the forms of assistance given pioneer mills by local communities, especially in New England, and, later, the mill acts giving millowners a preferred position over other riparians, for the most part farmers. But it was not simply with other user-interests in watercourses that millowners found it difficult to live peaceably. They eventually came to be their own worst enemies, interfering frequently and at times seriously with each other's operations through the manner of constructing and managing hydraulic works.

The principles governing the use of watercourses rested upon the common law of England, whose roots went back to Roman law. Under the common law, owners of the land bordering either side of a stream were deemed riparian proprietors, with ownership in the case of non-navigable streams extending to the midpoint of the streambed itself. Streams regularly navigable by vessels were in the public domain and

[35] The importance of Angell's *Treatise* is suggested by the number of editions published. References below are to the 5th ed. rev. (Boston, 1854). This comment is based on a section-by-section check of Angell; see also his prefatory comments, dated Providence, June 25, 1850, p. vi. For a recent discussion, see C. M. Haar and B. Gordon, "Legislative Change of Water Law in Massachusetts," in D. Haber and S. W. Bergen, eds., *The Law of Water Allocation in the Eastern United States* (New York, 1958), p. 8. For similar conflicts of interest in Great Britain, see J. S. Clutterbuck, "The Perennial and Flood Waters of the Upper Thames," *PICE* 22 (1862–63):336 ff.; R. Manning, "On the Results of a Series of Observations on the Flow of Water . . .," ibid. 25 (1865–66): 458–79; Fred S. Thacker, *The Thames Highway,* 2 vols. (1914–20; reprint ed., New York, 1968), 1:3; Thomas S. Willan, *River Navigation in England, 1600–1750* (1964; reprint ed., New York, 1965), pp. 47–52.

might not be obstructed. Mill privileges, the waterpowers to which falls in the streambed gave rise, went with the riparian land bordering either or both sides of a stream, and, unless specifically excluded, were transferred with the land when it was sold. Ownership of stream bank and bed, however, did not carry with it ownership of the stream itself, "moving and transient by nature," but only the temporary use of the stream. This right of use, moreover, had to be exercised with due regard to the rights and interests of other riparians, above and below. The riparian might not unreasonably detain the water in its passage nor diminish its amount (save for certain natural uses such as domestic supply and watering livestock). If diverted somewhat from its natural course, as for example in millraces, the stream must on leaving the riparian's property be returned to its natural bed. To cause water to overflow the land of another and to encroach upon the waterwheels of the next privilege upstream by raising a dam unduly high were actionable injuries that could be stopped by the injured party and damages recovered by law.

Such were the more important of the common-law doctrines bearing upon water rights of riparian owners. Even when modified variously by the legislation of the several states after independence, they remained deceptively simple in statement yet were far from simple in judicial interpretation and application. It has been said that the most certain thing respecting the interpretation of the law of watercourses by the courts was the uncertainty of the conclusions reached in any specific situation, an understandable consequence of the interdependence of the several stream uses and of what came to be known as the doctrine of reasonable use. In effect, the last held that since each use in some measure interfered with other uses, such interference was not in itself actionable if held within reasonable bounds, a doctrine that lent itself to considerable latitude of interpretation.[36]

[36] See discussion of this intricate subject in Angell, *Treatise;* H. P. Farnam, *The Law of Waters and Water Rights,* 3 vols. (Rochester, N.Y., 1904); A. A. Schmid, "Water and Law in Wisconsin," *Wisconsin Magazine of History* 45 (1962); Lincoln Smith, *The Power Policy of Maine* (Berkeley, Calif., 1951), pp. 29 ff.; John A. Fairlie, *Public Regulation of Water Power in the United States and Europe* (Ann Arbor, Mich., 1911); and articles on the law of water use in Massachusetts by Haar and Gordon, in North Carolina by H. H. Ellis, and in Michigan by Arens, in Haber and Bergen, eds., *Law of Water Allocation.* For the quite different law of water rights in the western half of the continent, reflecting the needs of irrigation agriculture and placer mining and stressing prior appropriation rather than riparian interests, see Farnam, *Law of Waters,* 3:chap. 22, and Gregory Yale, *Legal Titles to Mining Claims and Water Rights in California* (San Francisco, 1867).

Navigation Rights and the Milling Interests

The least controversial portion of the law of watercourses related to navigation. On streams of sufficient size to bear regular traffic of vessels a fair portion of the year, navigation rights carried priority. In such watercourses the riverbed was in the public domain and the rights of riparian owners were distinctly subordinate to those of the public in navigation. The common-law definition of navigable as limited to waters affected by the tide was modified by many states in this country to include all watercourses that were in fact used in commerce or were capable of such use.[37] The erection of milldams on such streams commonly required legislative authorization. In a number of states, especially in the South, the erection of any dam required authorization by county authority, and in some states such dams were specifically forbidden by statute.[38] Since rivers suited to commercial navigation were usually ill suited to power development, save at sites of falls, which themselves interrupted navigation, this conflict did not often occur. Yet lesser streams not regularly navigable in commerce were often boatable for brief flood seasons and were often prized as affording the only practicable routes to market for remote settlements in the upper portions of a drainage basin. In such circumstances milldams were often strongly opposed by the seasonal boating interests commonly employing the one-way, one-time-used flatboats of the pioneer years.[39]

Another and typically more formidable navigation interest, particu-

[37] Angell, *Treatise,* chap. 13; Farnam, *Law of Waters,* 1:chap. 3; Haber and Bergen, eds., *Law of Water Allocation,* pp. 12 ff., 246 ff., 402 ff. The emphasis varied with the time and the place. Peter Mathias remarked that "medieval England was more concerned to exploit the smaller rivers as sources of power than as highways, so that conflicts abounded between millers defending their weirs and water rights against bargees and merchants demanding a 'flash' of water through the weirs to get boats past the obstruction" (*The First Industrial Nation: An Economic History of Britain, 1700–1914* [New York, 1969], p. 109).

[38] A Pennsylvania statute of 1771, for example, declared the Lehigh and Delaware rivers to be common highways for the purposes of navigation and fixed a £20 penalty for the erection of dams impeding navigation. In time even wing dams—extending into the stream from one shore but not across—erected on the Delaware required legislative authorization. As late as 1880 no dam could be built entirely across the Delaware, owing chiefly to the opposition of the rafting and fishing interests (Tenth Census, *Water-Power,* pt. 1, pp. 611–12).

[39] See, for example, Ausburn Towner, *Our Country and Its People: A History of the Valley and County of Chemung* [New York] (Syracuse, N.Y., 1892), pp. 118–19; Alta H. Heiser, *Hamilton in the Making* (Oxford, Ohio, 1941), pp. 27–29; James W. Hurst, *Law and Economic Growth,* esp. chaps. 3, 9.

larly prominent in the timber states from Maine to Minnesota, was the logging and lumber industries. Their dependence on watercourses as often the only practicable means of getting timber in log or raft form to sawmills or markets led in some states to the concept of floatable, as distinguished from boatable, streams.[40] Lumbermen and loggers, tempered by the hardships and hazards of life in the woods, were not usually inclined to accept passively the milldams that broke up their rafts and blocked their log drives, although during the spring floods both might pass the submerged dams without much damage. On occasion they tore down the offending obstructions.[41] In a number of the major lumbering areas from Maine westward the lumbering interests reversed the dilemma, building their own booms and dams (with gates) for accumulating logs or rafts along with water and then flushing them downstream on artificial floods, embarrassing millowners below with alternating drought and flood.[42] "At every session from 1850 until the disappearance of large scale lumbering," runs an account of this system in Wisconsin, where it was introduced from New England, "the legislature was besieged with demands for charters empowering individuals or companies to build dams, construct booms or improve rivers. Between 1850 and 1873 the legislature passed more than a hundred boom acts and more than two hundred dam acts"—and this in the face of opposition from millowners.[43]

In his history of the lumber industry of Wisconsin, James Willard Hurst neatly reviewed the simplicity of the issues and the complexity of the relationships:

Law responds to the intricacy of men's relations. On the state's inland waters shippers of logs jostled shippers of lumber, both contended with those who would improve a stream for tolls, all of these with men who would develop a stream for power, and all of these with riparian landowners who complained of flowage and trespass. . . . The lumber industry was interested in Wisconsin's inland waters first for navigation, and second for power. The power was relatively uncomplicated; the single most common type of issue here turned on the right to flow others' lands and the terms of the compensation therefor. . . . the considerable

[40] See the sources just named and Angell, *Treatise,* pp. 599 ff.; Farnam, *Law of Waters,* 1:121 ff.; William G. Rector, *Log Transportation in the Lake States Lumber Industry, 1840–1915* (Glendale, Calif., 1953), the chapter on "Law of the Waters"; *Report of the Wisconsin Commission on Water Powers* (Madison, Wis., 1911), p. 314.

[41] Tenth Census, *Water-Power,* pt. 1, p. 530; pt. 2, pp. 205, 214.

[42] Ibid., pt. 1, pp. 530, 587; pt. 2, pp. 205, 214. Schmid, "Water and Law in Wisconsin," pp. 203 ff.

[43] Robert F. Fries, *Empire in Pine: The Story of Lumbering in Wisconsin, 1830–1900* (Madison, Wis., 1951), pp. 48 ff. See also Pressey, *Water Powers of Maine,* 20–21, 40–41; Rector, *Log Transportation in Lake States,* pp. 101–3.

quantity and complexity of legislation and litigation which attended the building of power dams in the forest area reflected more the importance of accommodating power use to navigation use than any issues peculiar to water power development as such.[44]

Milldams and Fisheries

Fishing was yet another interest that millowners in some regions had to contend with. Fish were an important source of food at many inland points, all the more welcome to relieve the monotony and meagerness of a frontier diet. This bounty of nature extended well beyond frontier days and was shared by communities scores of miles upstream from the seacoast.[45] Any obstacle to the free passage of fish along watercourses quite understandably was viewed with hostility by the deprived communities. This was particularly true of the seasonal, usually annual, migration of such fish as herring, shad, alewives, and salmon to upstream spawning grounds in streams having outlet in the sea. In the spring, "when the shad ran up the river to spawn," according to an account of an eastern Pennsylvania community, "every family in Hanover had at least one of its members down at the river [where] the shad could be caught in immense quantities. Seines were used by some, but the shad could be caught by any one with a hook and line. They needed no bait—only just throw in and pull out. . . . The whole country around came there and caught all they wanted."[46] "What greater boon could be bestowed on the poor of Lowell [Massachusetts] than a cheap and abundant supply of wholesome fish?" inquired another local historian. "As late as 1835, it is estimated that more than sixty-five thousand shad and over eight hundred salmon were taken from the Merrimack in Lowell alone."[47]

[44] Hurst, *Law and Economic Growth,* pp. 140–47.

[45] That this was no passing problem is indicated by the devotion of eight pages to the subject of fishways by a standard treatise on waterpower in the opening years of this century. See Frizell, *Water-Power,* pp. 150–57. In 1973 a newspaper in the upper Connecticut Valley published a communication on the "Battle of Wilder Dam," charging that hydroelectric stream management was making fish migration up the river almost impossible and demanding provision of adequate fish ladders (*Connecticut Valley Reporter* [Lebanon, N.H.], 31 July 1973).

[46] Henry B. Plumb, *History of Hanover Township and of Wyoming Valley in Luzerne County, Pennsylvania* (Wilkes-Barre, 1885), pp. 231–32; also *History of Bedford, New-Hampshire* (Boston, 1851), pp. 204–5.

[47] Charles Cowley, *History of Lowell.* 2d ed. rev. (Boston, 1868), pp. 129–31. A Massachusetts law of 1687 forbade erection of dams preventing the passage of alewives (Margaret T. Parker, "Factors in the Development of the Cities of the Merrimac Valley in Massachusetts" [M.A. thesis, Wellesley College, 1921], p. 33).

By comparison with these annual upstream floods of seafood, the biblical miracle of loaves and fishes becomes an almost paltry affair. Understandably the communities benefited were not to be deprived of this gift of nature by the selfishness of millowners with their milldam barricades. They found support not only in common-law doctrines but in legislative concern and intervention. "In this country," declared Angell, "the statute books of almost all the states shew the solicitude of the Legislature to preserve a free passage for the fish, especially in those rivers which are annually visited by fish from the ocean."[48] In time fishery commissions were established in more than thirty states with the assigned duty, among others, of enforcing state requirements for the construction and maintenance of fishways for passage upstream, a requirement that millowners and manufacturers found burdensome as well as annoying.[49]

The role of fishing rights as an obstacle to milling enterprise is not to be underrated, especially as regards New England. An ambitious ironworks established at Braintree, Massachusetts, in the 1640s, was for many years idled by the contentious fishermen who destroyed the dam. In the 1820s three distinct efforts were made, reportedly, to build the town of Weymouth, Massachusetts, into a Fall River or a Lowell, only to meet frustration in each instance owing to the townspeople's refusal to approve the sale of the alewife fishery in the town, despite the provision of a satisfactory fishway through the proposed dam and an annual lump-sum payment. The erection of a high dam at the lowest privilege on the Housatonic River in Connecticut was delayed from 1839 to 1864 through fear of injuring the shad fisheries. "Pembroke preferred herring to shovels," one historian reported, citing the instance in which a celebrated shovel and tool manufacturer abandoned thought of locating at this Massachusetts town on hearing that he must open his sluiceways for five or six weeks during the busiest season of the year that the herring might run.[50] When in the mid-1840s the combined opposition of the fishing and carrying trades threatened to block legislative authorization of a reservoir dam needed by the Cooper-Hewitt iron interests to insure a full midsummer water supply for the Trenton, New

[48] Angell, *Treatise*, chaps. 3, 6. See also Farnam, *Law of Waters*, 2:chap. 14; George Minot, ed., *Digest of the Decisions of the Supreme Judicial Court of Massachusetts* (Boston, 1844), pp. 337–39; and H. H. Ellis, "Some Legal Aspects of Water Use in North Carolina," in Haber and Bergen, eds., *Law of Water Allocation*, p. 255.

[49] Frizell, *Water-Power*, pp. 150–57; also Oscar and Mary Handlin, *Commonwealth: A Study of the Role of Government in the American Economy, 1774–1861* (New York, 1947), pp. 76–77.

[50] Orra L. Stone, *History of Massachusetts Industries*, 4 vols. (Boston, 1930), 2:1081, 1134, 1314–15. See also Tenth Census, *Water-Power*, pt. 1, p. 311.

Jersey, waterpower company, money was sent to a representative at Harrisburg "so that he might prosecute the matter energetically," although in the event to no avail.[51] Coleman goes so far as to declare that "Rhode Island industrialized rapidly as compared with Connecticut partly because it acted earlier to relax the laws requiring dams to be kept open to enable fish to reach their spawning grounds. These laws were burdensome, especially in summer, because they forced manufacturers to release water during the period when they needed all the storage capacity they could get."[52]

Farmers versus Flowage Rights

Most enduring and important of all was the conflict of interest between millowners and the most numerous class of riparians, the farmers, whose lands were subject to flowage by millponds created by dams. In narrow valleys with sharply sloping lands the extent of such flowage would be slight and the farming value of the land affected small. But the more accessible and desirable millsites were often those adjacent to the broad and fertile bottomlands most valued for farming and most reluctantly abandoned, despite damage awards by local juries. "Water mills," declared Rees's *Cyclopaedia*, "have long been great nuisances to agriculture, by preventing the use of streams on which they stand, in many cases, in irrigating and flooding the adjoining lands, by which much improvement is kept back that otherwise would take place." A similar view was expressed during an engineering discussion in England in 1876: "The vested rights of a few miserably unprofitable mills should not be allowed to cause the flooding of thousands of acres."[53]

[51] Correspondence between Edward Cooper, Peter Cooper, Charles Hewitt, Abram Hewitt, and J. Redmond, 4, 17, 20 Feb. 1846, Cooper-Hewitt Papers, John Crerar Library, Chicago. Allan Nevins, *Abram S. Hewitt* (New York, 1935), p. 85. On the decline of upstream fisheries with the increasing building of milldams and remedial measures, see Cowley, *History of Lowell,* pp. 130–31; Frizell, *Water-Power,* pp. 151 ff. Years after the Civil War, James B. Francis of the Locks and Canals at Lowell pleaded with the Massachusetts Fish Commissioner to relax the fishway regulations, permitting closing the fishway at the dam in the event of very low water. The fishway was kept open "but it was with much sighing and grief. No possibility of the fish going up or down. Coal, which we use as a substitute for the water running over said fishway, $9 per ton" (JBF to Commissioner Lyman, 12 June, 26 Sept. 1871, LCP, vol. DB-6, 253–327. See also 23 and 29 June, 19 August 1880, vol. DB-10, 276, 281, 334).

[52] Peter J. Coleman, *The Transformation of Rhode Island, 1790–1860* (Providence, 1963), p. 76n.

In the United States, moreover, unlike the situations characteristic of other interests competing for stream use, the conflict between millowners and farmers became more rather than less acute with the passage of time. With the increase of population and growth of settlement, good farmland became scarcer and more valuable while power requirements grew ever larger.[54] As the expanding operations of mills demanded more power, dams were raised in height—and often again and again—in part to secure greater head but more commonly to secure greater upstream pondage for storing the nighttime flow for use in working hours. At the same time, the clearing of trees and draining of swamps, which improved land for farming, usually had the opposite effect upon power supply by increasing the fluctuations of streamflow.

"Milldams have always grown higher, year by year, as every man knows who owns land on any stream, till now the farmers, who desire to improve their lands by drainage, are beginning to enquire whether their business, of supplying food to the community is not of public interest," argued a proponent of repeal of an act regulating flowage along the Concord and Sudbury rivers.[55] In the Northeast especially, farmers affected by the flowage of their lands had no choice but to acquiesce, accepting such damages as juries might award. Damages were usually granted on an annual basis, although some states eventually gave landowners a choice between annual or gross damages. The compensation at best covered only the material loss suffered, affording no balm where sentiment and pride of proprietorship were injured.

[53] Abraham Rees, ed. *The Cyclopaedia,* 1st Am. ed. (Philadelphia, 1810–24), s.v. "Mills"; George J. Symons, "On the Floods in England and Wales during 1875," *PICE* 45 (1875–76): 9–10; also R. B. Grantham, "On Arterial Drainage and Outfalls," ibid. 19 (1859–60): 53, 76, 87. One discussant declared that "there were cases in which mills not worth £50 a year, were damaging large districts to the extent of hundreds of pounds" (ibid., p. 89).

[54] See the opinion of Justice Parker in *Stowell* v. *Flagg,* 11 Mass. R. 364, cited in Angell, *Treatise,* pp. 560–61.

[55] *Argument of Hon. Henry F. French, of Boston, on the First Day of March, 1861, before the Joint Committee of the Legislature of Massachusetts on Petition for Repeal of an Act re Flowage of Meadows on Concord and Sudbury Rivers* (Lowell, 1861), pp. 9–11. In the timber country of Michigan, Wisconsin, and Minnesota the widespread practice of damming streams for log-driving purposes at times aroused bitter opposition by riparian farmers (Rector, *Log Transportation in Lake States,* pp. 174–81; also, R. G. Lillard, *The Great Forest* [New York, 1947], pp. 233–36).

The Public Character of Water Mills

The opposition encountered by millowners from these varied interests was substantially mitigated with time by a variety of circumstances: the concentration of lumbering in the more remote, thinly settled, and agriculturally and industrially unattractive backwoods area, together with its cut-out-and-get-out nature; the confinement of flatboating on most streams to the pioneering years; and the limited agricultural value of much, even most, of the land subject to flowage on many important millstreams. Yet the main strength of the millowners' position vis-à-vis competing stream users probably lay in the widespread and early manifested eagerness to obtain and increase mill industries. "The man who first erects a mill in a new country," reads an early nineteenth-century court decision in Pennsylvania, "is considered as a public benefactor, and no subject ought to be treated with more tenderness, no possession more respected."[56] The application of water to the working of mills and machinery, declared Angell, is "a use profitable to the owner, and vastly beneficial to the public." As the Handlins have pointed out, "Manufacturing had always borne a public aspect from the necessity of utilizing mill sites for power." Opponents to milldams, wrote a nineteenth-century historian of the Chemung Valley in New York, "formed a powerful and numerous class, but people must have flour and boards."[57]

Community aid to mill industry, as noted earlier, dated almost from the beginnings of colonial settlement and in the form of aid and encouragement to gristmills and sawmills was repeated on virtually every frontier until the opening decades of the nineteenth century. Public assistance was raised to a new level and generalized when the Province of Massachusetts in the Mill Act of 1714 modified the common-law doctrine to give millowners a decided advantage over other riparians. This act was confirmed and extended after Independence.

Under the common law a riparian did not have to accept passively the flowage of his land incident to the building of a dam and creation of a millpond. Such flowage without the landowner's consent comprised a nuisance; by repeated suits the landowner could so harass the millowner as to discourage all but the most determined.[58] In effect the Mill Act of 1714 and subsequent legislation extended the right of eminent

[56] *Strickler* v. *Todd,* 10 Sergeant and Rawle (Pa. 1823), pp. 63, 68.

[57] Angell, *Treatise,* p. 93; Handlin and Handlin, *Commonwealth,* p. 110; Towner, *Valley and County of Chemung,* pp. 118–19.

[58] Angell, *Treatise,* pp. 555–59; Minot, ed., *Digest*, 481–86.

domain to millowners. Now the riparian whose land was flowed "had no legal means of preventing such flowage; he could only attempt to collect just compensation." Moreover, the owner could not recover at once the full value of the land, the use of which he was deprived, but only the amount of the yearly or gross damage as determined by a jury.[59]

The principle of the mill acts, with variations in detail, was adopted not only in other parts of New England but in many of the northern states east of the Mississippi River.[60] Originally applicable only to grist- and sawmills, they were in time extended to manufacturing generally, as much—perhaps more—through judicial interpretation and public acquiescence as through legislative action. This extension was early viewed with misgiving by some and long remained a subject of judicial deliberation and concern.[61] Yet even though the principle had been reduced to an anachronism by the wide introduction of steam power, it was not repudiated. Clearly the larger manufacturing interest as the beneficiary was not disposed to alter the status quo and the affected farm interest was usually not in a position successfully to challenge the traditional doctrine.[62]

On occasion the subordination of the farm interest could arouse bitter protest, as in 1861 in the celebrated case of the flowed meadows of the Concord and Sudbury rivers. Counsel for the relief-seeking farmers declared:

> The Flowage Act, by which you cover the land of the farmer with water, and convert his smiling meadows into pestilent swamps and millponds, can only be defended on the ground of the public good. . . . The interest of Manufacturers is considered a public interest, and the land of the farmer is appropriated to the public use of manufacturing cotton or woolen. This flowage of the best lands of the State, under the Mill Act, has come in this Commonwealth, as in England,

[59] Smith, *Power Policy of Maine,* pp. 29 ff. In the words of one judge, the act and its successors were in derogation of the common law and rather arbitrary in their operation. The mill acts "authorize one man to take from another the beneficial use of his property without his consent, and at a price to be fixed by strangers" (*Snell* v. *Bridgewater Manufacturing Co.,* 24 Pick. [Mass., 1840], p. 290). In some instances the deprivation of use was only partial. See Smith, *Power Policy of Maine,* chap. 2; Angell, *Treatise,* chap. 12; Handlin and Handlin, *Commonwealth,* pp. 76 ff.

[60] Angell, *Treatise,* pp. 551 ff. Smith states that the mill act was in effect in at least twenty-nine states (*Power Policy of Maine,* p. 40). See Schmid, "Water and Law in Wisconsin," pp. 203 ff.

[61] Farnam, *Law of Waters,* 3:chap. 23, esp. pp. 2134–43.

[62] See Angell, *Treatise,* pp. 560–65; Smith, *Power Policy of Maine,* chap. 2, esp. pp. 31–41; Handlin and Handlin, *Commonwealth,* pp. 78, 136.

to be a great evil. The organized companies of manufacturers have always been an overmatch for the scattered agriculturalists.[63]

In evident reference to this same situation the *Scientific American*, 19 September 1863, declared that the milldam did more damage to the "interests of humanity and agriculture . . . than would furnish all the mills with steam engines, engineers and fuel," and characterized "the attendant waste of rich land as a living disgrace to the State." Later legislation in some states modified the relationship in behalf of the farm interest; and with the coming of hydroelectricity, greatly extending the requirements for flowage over riparian farmlands, there were complaints that legislative favoritism had shifted from the power interests to riparians.[64]

In sum, among the various beneficiaries of watercourses, millowners from the beginning enjoyed a preferred status. This preference reflected the long-prevalent view that, navigable rivers apart, the propulsion of the machinery of industry was the most important object to which watercourses might be applied. In frontier communities this view was anticipated during periods of settlement when, as we have seen, in their eagerness to obtain milling facilities individuals willing and able to establish mills were not only granted land adjacent to mill seats but given other forms of aid and encouragement. The acceptance of a certain subordination of other riparian rights to the milling interest continued through the rise of the more varied small manufactories serving the immediate environs. With the growth of manufacturing serving more distant markets the traditional hospitality of the farm interest underwent erosion. This was especially the case where, as in the agricultural intervales along the rivers and in the extension of cultivation to lands earlier covered with timber, farmlands took on increased utility and value and the rights of flowage became more burdensome. Damage awards in themselves with many farmers proved inadequate compensation for loss of control of their property. Understandably, too, the swelling manufacturing interest, drawing support from the broader conception of manufacturing as in the national interest, resisted the efforts to change the situation to their disadvantage. After all, power was indispensable to industry and through general acceptance and practice there was no

[63] *Argument of Henry F. French*, p. 9.

[64] Angell, *Treatise*, pp. 586–87; Handlin and Handlin, *Commonwealth*, p. 222; G. B. Leighton, *Report of the [New Hampshire] Commission on Water Conservation and Water Power, 1917–1918* (Concord, 1919), pp. 24–26, 47. On the flowage problem as seen from the viewpoint of a large waterpower corporation, see letters of JBF to J. U. Parker, 22 Mar. 1851, and to T. Wentworth, 19 Apr. 1851, LCP. See also Horlitz, *The Transformation of American Law*, pp. 34–54.

acceptable alternative to waterpower save in situations where it was lacking.[65]

Conflict of Interests: Mills versus Mills

Despite the perennial altercations between millowners and others with a stake in the use of watercourses, the main source of harassment in the more fully developed industrial areas was the millowners themselves. The spread and growth of manufacturing brought about the progressive occupation and development of eligible millsites on the more suitable streams. Millowners in time began to crowd each other, both at privileges large enough to support several establishments and on streams presenting a succession of eligible waterpowers. It became increasingly difficult to meet expanding power requirements without interfering to some extent with the needs and operations of other mills on the same stream. Questions of the apportionment of water and priority of its use among several claimants at the same privilege had to be resolved. With a succession of privileges on the same stream, the lower mills were dependent upon the upper ones for the release and regular supply of water, especially during the low-water seasons. Where privileges were close together or the fall between them slight, the building of a dam or the raising of the height of an existing one might result in the condition known as "backwater," impeding the waterwheels of the privilege above.[66]

There were many other circumstances favorable to dispute and litigation among the common users of a millstream: the division of a privilege between riparians on opposite banks; multiple ownership of a large privilege and the division of streamflow among the several owners; contractual priorities of use and assignment of flow among the several users; arrangements for sharing repair and maintenance costs of dams, races, and reservoirs used in common. A troublesome issue at times was the matter of prescriptive rights under which the undisturbed exercise of a practice for twenty years freed the millowner from the customary restraints of the common law. Differences among owners over such

[65] See Harold Faber, "Dam *vs.* Farmland the Issue," *New York Times,* 9 Apr. 1972. Here the conflict was heightened by critical power shortages in New York, city and state, and the dependence of increased hydroelectric capacity on storage dams and reservoirs of great extent and capacity. In the specific situation in controversy, involving 4,716 acres of rich bottomland, the farming interests secured the desired protection.

[66] See Tenth Census, *Water-Power,* pt. 1, pp. 246–47, 431, 447, 580; pt. 2, pp. 107, 332–33; Frizell, *Water-Power,* p. 619 ff.

matters rested at bottom in the real-property character and physical circumstances of hydraulic power and in the law of watercourses inherited from medieval England, in the absence of contrary local legislation. A mode of motive power incident to ownership of land was hedged about with a variety of conditions of disposition and use from which prime movers in the form of personal property—animals and steam engines, for example—were free.

The physical conditions attending the use of waterpower were inescapable. To repeat an earlier figure of speech, each millstream used by a number of establishments may be likened to a power transmission line upon which each mill privilege in turn was entitled to draw, somewhat after the manner of a lineshaft to which one's machinery and equipment are belted. But whereas the lineshaft distributed its energy simultaneously, impartially, and at the same speed to all machines along its length, the energy of the millstream became available only seriatim as the power passed from one mill privilege to the next down the line. The extent to which the water was held, diverted, or released at any privilege, especially during periods of low water and power shortage, determined the extent and time of its availability at the next privilege downstream.

Water wasting over the dam ground no meal. Each mill operator in turn husbanded the available supply reaching him to meet his own requirements. The extent to which the water could be detained at any point depended upon the storage capacity of the millpond and this in turn upon the character of the terrain, the distance and slope to the next privilege above, and the value of the land flowed and damages paid. The time and rate of release depended also on operational considerations—whether power was used continuously or intermittently, the length and hours of the working day, the extent of nighttime operations, the frequency and duration of shutdowns for repairs or for local holidays.

To a degree, of course, the facts of stream behavior and the operational requirements of mills were recognized in the law of watercourses. As noted above, the millowner, like other riparians, had use of the stream only while it was passing over or along his land. It could be diverted from its bed through races and penstocks and over waterwheels but had to be returned to its natural course below these diversions without undue delay or substantial diminution. The stream could not be used so as to impair unreasonably the rights and interests of millowners upstream or down, as by alternately flooding or withholding water to the mill below or by setting water back (creating backwater) upon the wheels of the mill above. As noted earlier, the test came to be one of "reasonable use," interpreted variously as any use that was beneficial and lawful and not inconsistent with "the reasonable use" by other riparians.

Yet the more favorably circumstanced millstream came to present a fairly continuous succession of occupied millsites, and even during periods of ample flow expanding enterprises were faced with power shortages. The most common method of obtaining relief was by raising the height of the milldam, either structurally by adding a new upper course of timber or masonry or by the temporary but much less costly device of "flashboards," common boards held in place by light iron rods. Under the pressure of rising flood the rods would bend or give way, lowering or releasing the boards and reducing proportionately the height of the water and the degree of flooding above the dam.[67]

If in this situation the dam was raised unduly the water might be set back upon the wheels of the mill at the privilege immediately above, reducing their power and in extreme cases bringing them virtually to a halt. Backwater created in this manner was the cause of frequent controversy, innumerable suits, and much bitterness.[68] Farnam noted that the litigation arising out of such "encroachments" of a lower mill upon the one above it were so numerous as to lead him to introduce a separate chapter on the subject in his treatise *Law of Waters.*[69] However injurious the effect of a mill's backwater, this interference was at least limited to the next privilege above. It was otherwise with the manipulation of streamflow at an upstream privilege in disregard of the effects upon other mill privileges downstream. Chiefly during low-water periods, nighttime flow was often insufficient to fill millponds, and the ability of lower mills to operate depended upon the release of water by the operation of the strategically located mill or mills above. At what hours of the day—or night—such release might be vouchsafed and whether the controlling millowner would exhaust the available supply in a short period of full operation or a longer one on a diminished scale was dependent upon the advantage and convenience or of the consideration shown others by the millowner in question. Only if the lower mill privileges had storage

[67] The consequences of using flashboards to raise the height of a main dam at Lowell by one or two inches on a total fall of thirty-three feet was a matter of serious consideration by Superintendent Francis of the Locks and Canals. He requested legal opinion on the consequences, vis-à-vis upstream riparians, of five alternative arrangements (JBF to A. P. Booney, 15 Feb., 4 Mar. 1875, LCP, vol. DB-8, 41–42, 49, 74 ff.). On flashboards, see Frizell, *Water-Power,* 3d ed., pp. 142–43.

[68] Angell, *Treatise,* chap. 9; James Emerson, *Treatise Relative to the Testing of Water-Wheels and Machinery,* 2d ed. (Springfield, Mass., 1878), pp. 20 ff.; "Backwater Suits," in Minot, ed., *Digest,* pp. 481–86; Tenth Census, *Water-Power,* pt. 1, pp. 247–48.

[69] Farnam, *Law of Waters,* 2: chap. 20.

capacities independent of the main stream and owing at least partly to the terrain were their operators free from the policies of the mills above in releasing or holding back water. Actionable injury, especially in the case of upstream privileges more than once removed, was less readily established than with backwater. Since the courts usually took into account a variety of qualifying circumstances, the withholding of water by an upstream mill evidently had to assume a rather flagrant form before judicial intervention was likely. The decision of a New York court remarked in part: "The common use of the water of a stream, by persons having mills above, is frequently, if not generally, attended with damage and loss to the mills below; but that is incident to that common use, and for the most part, unavoidable."[70]

The professional turbine tester James Emerson cited the hypothetical case of a dozen mills located in close succession on the same extended privilege and served by the same millpond at the head of the stream with ample water to meet the needs of all provided each used efficient wheels. However, if the uppermost mill had a wheel so inefficient as to require twice the water of the wheels at the succession of mills below, it would empty the pond by noon and all the mills below would have to close down for the remainder of the day. "Such cases are very common."[71] In 1820 a flour mill was plaintiff in a suit against an ironworks on Wynant's Kill, near Troy, New York. The plaintiff claimed that the defendant "during the time of heating a quantity of iron, which occasioned more than an hour, entirely stopped the water on their dam, and when they afterwards let it out, it ran out in such torrent that it was wasted by running over the plaintiff's dams."[72]

There were similar complaints of long standing against the federal armory at Springfield, Massachusetts, by the Ames Company and other owners located on the same stream below. In a communication of 13 December 1824 to the superintendent of the armory the lower owners declared themselves to be

[70] Justice Wordsworth in *Merritt et al.* v. *Brinkerhoff et al.,* 17 Johns. (N.Y. 1820), p. 306. See also Angell, *Treatise,* 1st ed. (Boston, 1824), pp. 125 ff. Sometimes a mill operator's practices drew fire from both directions—for lettting the water rise too high and letting it down too fast (Massachusetts, General Court, *Report of the Joint Special Committee upon the Subject of Flowage of Meadows on the Concord and Sudbury Rivers* [Boston, 1860], p. 284).

[71] Emerson, *Treatise Relative to the Testing of Water-Wheels,* pp. 20–22.

[72] *Merritt* v. *Brinkerhoff.* Related cases are *Platt* v. *Johnson,* 15 Johns. (N.Y. 1818), p. 213; *Twiss* v. *Baldwin,* 9 Conn. (1832), p. 291; Haar and Gordon, "Legislative Change of Water Law in Massachusetts," pp. 8 ff.

constrained from a sense of duty to the public as well as themselves to request that you would direct that more proper measures are taken in letting down the water from the upper shop dam, there being times when we have scarcely any—and then without any addition to the stream there is nearly or double the quantity run over our dams more than can be used and is lost to the public—ourselves—and numbers of the workmen employed in the armory. We are persuaded if you was to know how much time would be saved by an equal letting down the water in a constant regular manner day by day, we should be relieved at once—and might then be able to calculate on some regular supply—but in the present situation we know not at what time of the day, nor how much, nor how little, we are to have.[73]

Nearly ten years later the same plea was repeated to another armory superintendent in a petition by, among others, some of the same millowners downstream. Noting the failure of earlier assurances of relief, they urged consideration of their needs: "and cause the water in the Upper Dam to be let down more equally so that we can have the natural stream, rather than to have some days twice as much as our business requires, and other days, and even part of the same day, not one-half what we had a few hours before, by which means you will perceive we are daily disapointed in our business, and wholly unable to calculate before hand what we can do."[74]

Difficulties in Coordination

With forebearance and good will such conflicts between mills on immediately adjacent privileges could be and doubtless usually were avoided or reconciled. With mill privileges scattered along a stream at considerable intervals, on the other hand, it was hardly practicable for the millowner to take into account the effects of his water usage upon mills at considerable and varying distances below, making allowance for the time for water to move downstream. Upstream mills were evidently not expected, for example, to forgo badly needed pondage buildup over Sundays in order that mills some distance downstream would have

[73] David Ames et al. to Roswell Lee, 13 Dec. 1824, Springfield Armory Records, National Archives.

[74] David Ames et al. to John Robb, 4 Feb. 1834, Springfield Armory Records. Another issue in this protracted controversy was the backwater created by the dams of each party obstructing the wheels of the other. This was taken first to referees and then to court. The circuit court decision in *United States* v. *Ames,* Fed. Case (no. 14, 441), 1845, was for the plaintiff on the ground that armory property was not subject to the jurisdiction of state legislation respecting watercourses.

sufficient water to operate on Mondays. We read of instances—which may not have been uncommon—where during some periods mills were hampered as late as Wednesday by weekend shutdowns at mill centers some distance above.[75] Doubtless there were many who, like the owner of the privilege at the outlet of Lake George, New York, endeavored "to pass down uniformly [to the privileges below] what he considers to be the natural flow."[76] The control of upstream reservoirs in the interest of mill centers far below, as on the Merrimack River, frequently and at some places commonly operated to the disadvantage of mills at intermediate points. With the best of will such inequities could not wholly be avoided. Yet in a competitive economy it is not surprising to learn that when a large and nationally celebrated edge-tool manufactory in Connecticut closed down several days to take inventory, it should conserve so far as possible streamflow to the extent of its storage capacity, and so compel mills in an industrial community below to shut down also.[77] Eventually the pressure of necessity brought millowners on many streams together in joint programs of upstream storage reservoirs and stream control. Yet even in the twilight of the waterpower age many communities suffered from an irregular power supply stemming from the control of streamflow elsewhere on the river. However, the combined pondage and reservoir capacities of many mills often served to increase the aggregate yearly flow to the mutual benefit of all.

In still other ways the exercise of water rights created problems of relationships between mills and led to disputation and in some instances to legislation. When different parties owned opposite stream banks, they had to reach some agreement respecting the development and use of the privilege.[78] The common practice of joint ownership and use of a

[75] See references to the Farmington River in Connecticut and the Rock River in Illinois in Tenth Census, *Water-Power,* pt. 1, p. 243; pt. 2, pp. 36, 236.

[76] Ibid., pt. 1, p. 406.

[77] Ibid., p. 247. This was the Collins Company, about 1880 employing some 600 hands and 800 to 1,000 horsepower; the six establishments at Unionville below employed some 450 hands and 750 to 1,000 horsepower. On occasion the mills at Nashua on the Merrimack River closed down for some local holiday, materially affecting the power supply of the mills at Lowell below.

[78] Angell, *Treatise,* chaps. 5, 6. Cf. the complexity of joint ownership at Saco, Maine, where owners of the sawmills that originally controlled the water rights divided them into rights to use a mill so many hours each day or days in each month or year, as described in George S. Gibb: "These rights were widely sold, and further split by inheritance into fractions sometimes as small as minutes." Clear titles to some properties required search for owners and efforts to acquire rights by purchase extending over generations (*The Saco-Lowell Shops: Textile Machinery Building in New England, 1813–1949* [Cambridge, Mass., 1950], p. 118).

large privilege occasioned similar problems, centering not only in maintenance and repair but in the difficult business of measurement and control of flow to insure the rights of each proprietor. James B. Francis, superintendent for the Proprietors of the Locks and Canals on the Merrimack at Lowell, defended the elaborate and costly system of water measurement employed there "as the most practicable mode of preventing those ruinous controversies, which experience shows, almost invariably grow up in the course of time, with water privileges used by parties having separate rights to water."[79] The disposal of water rights subject to special reservations and restrictive conditions, such as priorities of use, which were of great importance at times of water scarcity, was common and led at times to disagreement and hostility. In times of abundant streamflow there was no problem.

Frequenters of casebooks and legal treatises are likely to receive an exaggerated and one-sided view of human behavior, understandably enough. Yet such competent and perspicacious observers as Samuel Slater and Zachariah Allen, and later James B. Francis and Joseph P. Frizell, remarked upon the financial burden of litigation over water rights. The massive 1880 census survey of waterpower referred to "the well-nigh universal litigation arising from conflicting water power interests."[80] So provocative of disagreement and litigation was the common practice of joint ownership and use of waterpower that even at Lowell, where the parties concerned were great textile corporations, it was difficult to maintain harmony. One of the achievements of James B. Francis singled out for high praise was his successful role as "chief of

[79] Statement of 21 Nov. 1857, LCP, DA-4, 230 ff.

[80] *Journal of the American Institute* 1 (1836): 525–26; Zachariah Allen, *The Science of Mechanics* (Providence, 1829), p. 199; Tenth Census, *Water-Power*, pt. 1, pp. 428, 451; pt. 2, pp. 166, 258, 378, 388, 470. That this contentiousness was not peculiarly American is indicated by the following quotation from Leslie Syson: "There were constant disputes over the water supply to the mill. Generally these were caused by some hindrance to navigation or the prevention of fish passing up the river. Many were the result of the flooding of adjacent land or of work which diminished the flow of water to mills further down the stream" (*British Water-Mills* [London, 1965], pp. 42–45). Syson quotes Sutcliffe, *Designing and Building Water Corn Mills* (1816): "In consequence of so many water-mills, the country is never free from litigation and vexations, lawsuits respecting erecting, repairing or raising weirs, by which the peace and harmony of neighbors and friends are often destroyed. . . . The constant bickering over water may have been one reason why some millers chose to build a windmill nearby in order to have a reserve power." See also John Rodgers, *English Rivers* (London, 1948), pp. 11–12, and Louis C. Hunter, "Waterpower in the Century of the Steam Engine," in Brooke Hindle, ed., *America's Wooden Age: Aspects of Early Technology* (Tarrytown, N.Y., 1975), pp. 168–70.

police of water." For half a century Francis administered successfully the distribution of water for power among the corporate users of the power, who were also its owners, effecting economy in water use while "preserving amiable relations and cutting off interminable and expensive law suits."[81]

Contentiousness was but one side of the coin. For all the rugged individualism and independence attributed to the age, millowners frequently did succeed in working in harmony together. On innumerable occasions they subordinated individualist tendencies to cooperative arrangements, reaching informal agreements or setting up formal associations for the common management of hydraulic facilities, including the maintenance and repair of dams and races, and for the resolution of differences arising from the joint use of water privileges. Of particular importance was the practice of joint arrangements for upstream reservoir capacity, typically beyond the resources of individual millowners, as a means of at least mitigating the seasonal shortages of power supply and avoiding the suspension of operations attending extreme summer droughts.[82] Yet even in an area marked by much constructive achievement, a prominent hydraulic engineer declared that such activity as the development of common storage reservoirs "presumes a concert of action and a spirit of mutual concession and accommodation among the parties [millowners] very rarely met with."[83]

The basic fact nonetheless remains: the millowner depending upon waterpower, as the vast majority of enterprises did down to the 1850s, found himself caught up in a complex network of institutional relationships that left him considerably less than a free agent in matters relating to power supply. A source of energy based upon streamflow subject to legal doctrines and institutions derived from medieval England was communal in nature and interdependent in usage.

[81] William E. Worthen, "Life and Works of James B. Francis," *Contributions of the Old Residents Historical Association* (Lowell) 5 (1892–93): 227 ff.

[82] Some of the more important examples of the cooperative arrangements, formally or informally adopted, by a number of mill or factory owners sharing in the utilization of a waterpower are noted in the census of 1880. See Tenth Census, *Water-Power,* pt. 1, pp. 52, 58, 62, 65, 89, 100, 245–47, 349, 353–54, 356, 466–67; pt. 2, pp. 108–10, 114, 229-301, 403–4, 501.

[83] Joseph P. Frizell, "Storage and Pondage of Water," *TASCE* 31 (1894):54.

4

Mill and Factory Villages
Grass Roots Industrialization

FALLING WATER WAS the chief source of stationary power at all levels, in most branches of industry, and throughout the greater part of the United States before the 1860s. Here was a situation and an experience markedly different from that of Britain earlier in laying the foundations of modern industrialism. The American reliance on waterpower is best understood against the background of British experience and the practice of both seen in the light of the treatment of industrial power by historians. The role of waterpower since the eighteenth century, whether viewed in strictly technological terms or within the broader economic context, has fared poorly at the hands of the historian. Although the origins of waterpower have been carefully traced back to ancient times and the increase and spread of water mills through Europe during medieval and later times described, interest in waterpower falls off rapidly with the emergence of steam power in eighteenth-century Britain.

The traditional view of the revolutionary role of steam power, while not without a certain validity, has been accepted by historians almost without challenge and with little qualification until recent years. In its treatment of the Industrial Revolution a well-known survey of the history of technology by Thomas K. Derry and Trevor L. Williams dismisses waterpower, except for some incidental references, with a single page at the opening of a long chapter on the steam engine. A more recent and somewhat more comprehensive interpretation of technology in Western civilization by Melvin Kranzberg and Carroll W. Purcell, Jr., opens a chapter on "The Prerequisites of Industrialization" (which makes no reference to waterwheels) with this statement: "When we think of the Industrial Revolution we usually think of the steam engine, the railway locomotive and the factory system." More than a century earlier Marx and Engels in their *Communist Manifesto* had given wide currency to this concept in declaring that "steam and machinery revolutionized industrial production."[1]

All things considered, the nineteenth century *was* the century of the

[1] Thomas K. Derry and Trevor L. Williams, *A Short History of Technology from the Earliest Times to A.D. 1900* (New York, 1961), Melvin Kranzberg and Carroll W. Purcell, Jr., eds., *Technology in Western Civilization,* 2 vols. (New York, 1967).

Fig. 38. Waterpower on the Black River at Watertown, N.Y. (*Scientific American,* Aug. 27, 1881.)

steam engine. Yet the role of stationary steam power before 1850 has been exaggerated and that of waterpower underrated, although more by inference and implication than by direct statement. The central and striking performance of steam power in Britain's Industrial Revolution and in the American transportation revolution before the Civil War has tended to identify steam power with progress and waterpower with obsolescence. These are oversimplifications that need correction.[2]

[2] See Eugene S. Ferguson's brief reference to waterpower in his chapter "The Steam Engine before 1830" in Kranzberg and Purcell, eds., *Technology in Western Civilization.* See also Alfred R. Ubbelohde, *Man and Energy,* rev. ed. (Baltimore 1963), pp. 15, 42 ff. On the role of steam power in industrialization on the Continent, see W. O. Henderson, *Britain and Industrial Europe, 1750–1870* (Liverpool, 1954); Rondo F. Cameron, *France and the Economic Development of Europe, 1800–1914* (Princeton, N.J., 1961); Arthur L. Dunham, *The Industrial Revolution in France, 1815–1848* (New York, 1955); Thomas C. Banfield, *Industry on the Rhine,* ser. 2, *Manufactures* (1848; reprint ed., New York, 1969).

Power Sources in Perspective

The key innovation in the Industrial Revolution was the mechanization of hand operations and the use of mechanical power to drive the machinery. The kind of power adopted was a secondary consideration, the choice depending chiefly upon such basic factors as availability and cost. Yet for historians as for contemporary observers the commonplace has been overshadowed by the novel. When Americans, like their cousins in Victorian England, celebrated the power source of the emerging industrial age, it was rarely the great breast wheels of Lowell or the newly introduced water turbine closely identified with the mills of Holyoke that they had in mind—nor indeed the common mill steam engine, of whose growing presence few were aware. It was rather the spectacular and highly visible applications of steam power in steam vessels and railway locomotives that captured the imagination and offered the exhilarating experience of speed.[3] Yet when the first steps toward industrialization were taken here, as in Britain, in the manufacture of textiles and later when Waltham, Lowell, and the great successor mill towns pioneered the integrated production of cotton goods with impressive results, the power base was falling water.

Although waterpowers were widely distributed and utilized in Britain, they stood in marked contrast with those of America in two particulars: limitations in aggregate capacity and a paucity of powers of the larger capacities. Both were a consequence of the small size of the island, the very limited extent of drainage basins, and the infrequency of the marked descents—falls—in streambeds essential for good waterpowers. Britain's abundant and seasonally well-distributed rainfall could not offset these limitations.[4] As late as 1880 the largest

These matters are discussed in some detail in Louis C. Hunter, "Waterpower in the Century of the Steam Engine," in Brooke Hindle, ed., *America's Wooden Age; Aspects of Early Technology* (Tarrytown, N.Y., 1975).

[3] See Herbert L. Sussman, *Victorians and the Machine: The Literary Response to Technology* (Cambridge, Mass., 1968), and Leo Marx, *The Machine in the Garden: Technology and the Pastoral Ideal in America* (New York, 1964).

[4] The historian interested in stationary power in Britain is plagued by meager census data on the subject during much of the nineteenth century. Statistical data on power in manufacturing are largely limited to the textile industries, where their collection and publication were long incidental to enforcement of the factory acts from the 1830s. See *Encyclopaedia Britannica,* 11th ed., s.v. "Labour Legislation." According to Laurence D. Stamp and Stanley H. Beaver, as late as the 1930s no comprehensive survey of fresh water resources of the British Isles had been made (*The British Isles: A Geographic and Economic Survey,* 2d ed. rev. [London, 1937]). Works by J. H. Clapham, H. D. Fong, M. Blaug, and Andrew Ure contain statistical tables on motive power,

developed waterpowers of Britain were in Scotland; they would have attracted little attention in the United States. They included the celebrated Catrine textile mills in Ayrshire, built 1825–27, with two large breast wheels together developing 240 horsepower on a forty-eight-foot fall. The Catrine mills were designed and built by William Fairbairn. Another, later development was the Deanston mills on the Teach; they were driven by four thirty-six-foot wheels with an aggregate rating of 375 horsepower. The largest single waterpower development in Britain in 1880 was the widely known Shaw's waterworks at Greenock on the Firth of Clyde. Here a skillfully engineered system based on an artificial reservoir in the hills created a fall of 512 feet and supplied a succession of thirty-odd hillside mills with an aggregate of 1,500 to 2,000 horsepower. In 1890 a paper read before the Institution of Civil Engineers on the installation of Britain's first hydroelectric central station (driven by a 50-horsepower turbine!) occasioned the remark that forty years had passed since a paper on waterpower machinery had been presented and "no Paper on the subject of water power generally, and the best ways of utilizing it, had ever come before the Institution."[5] In sum, while waterpower may have supplied as much as 50 percent of Britain's industrial power needs as late as 1850, from the early years of the nineteenth century the power requirements of establishments using 100 horsepower or so could be met only with steam power and coal.[6] The

but nearly all of these are limited to the textile industries. Phyllis Deane declares that not until the 1850s and 1860s did textile factories "turn over to steam-horsepower on any scale" and that "outside the textile factories and the mines, iron-works and railways, steam power was still a rarity in the mid-nineteenth century," fifty years after the lapse of the Watt patent and the opening of the field of engine-building to all comers (*The First Industrial Revolution* [Cambridge, 1965], p. 273). See also Peter Mathias, *The First Industrial Nation: An Economic History of Britain, 1700–1914* (New York, 1969), pp. 132–33.

[5] H. D. Pearsal in discussion of William P. J. Fawcus and Edward W. Cowan, "The Keswick Water-Power Electric-Light Station," *PICE* 102 (1890):180.

[6] See also *Library of Universal Knowledge* (1880 ed. of *Chambers's Encyclopaedia*), s.v. "Water Power." Fairbairn describes the waterpower plant at the Catrine mills in *Treatise on Mills and Millwork*, 2 pts. (London, 1861–63), pt. 1, pp. 126 ff. The most revealing evidence of Britain's progressive reliance on steam power in industry is supplied by a statistic of 1922–23. Of the electricity produced in this year only 0.2 percent was generated by waterpower (Frederick Brown, "Significant Trends in the Development and Utilization of Power Resources: Great Britain," *Transactions of the Third World Power Conference, 1936*, 10 vols. [Washington, D.C., 1938], 2:600). This figure included all industrial and central station establishments with more than ten employees. Total waterpower capacity was estimated at about one million kilowatts, or 1.34 million horsepower. By 1933–34 the proportion had increased to 2.4 percent. See also ibid., Table 10, p. 224.

abundance of coal in the Midlands and the north of England as elsewhere in Scotland and Wales resolved the fuel and power problem for a century to come.

The fact of the New England textile industry's long-continued reliance on waterpower is well known, yet the conventional view of steam power's superiority, found in most general discussions of industrial power, persists. Waterpower was fixed in location, inelastic in capacity, and unreliable in supply, subject to the vagaries of weather and streamflow. Steam power, on the other hand, could be produced virtually anywhere and in whatever quantity desired, subject only to the availability of fuel; in competent hands it was reliable and at command in all seasons. However accurate these observations, they fail to illuminate adequately the different power practices of Britain and the United States during the early stages of industrialization.

Conditions Controlling the Choice of Power

In the development of American industry before 1860 the controlling conditions differed widely from those in Great Britain. American society was predominantly rural and thinly distributed over the eastern half of the continent. The people were largely engaged in wresting a living directly from the land and in pushing outward the frontiers of settlement. A large part of their energies was absorbed in the creation of capital by clearing the land, building farms, and providing roads and more advanced transportation facilities. Except where favored by a fertile resource base, as with lumber and grain milling, shipbuilding, and primary ironworks, or by mechanical and inventive ingenuity in certain branches of metal- and woodworking, American industry lagged well behind that of Britain. So too did technology. Yet the stirrings of manufacturing in the opening decades of the nineteenth century in the Middle Atlantic and New England states revealed an extraordinary potential for industrial development. In a nation of continental proportions, little more than the land and the forests had begun to be exploited. The mineral resources, except for widely distributed surface deposits chiefly of iron ore and coal, lay largely untouched as late as 1840. Far more important both for the agricultural population and commercial development were the vast timber resources of the eastern United States. Although a major obstacle in land clearance and farm building, the woodlands were an indispensable source of building and fabricating materials and of fuel for cooking, heating, and agricultural processing.

The river systems, large and small, that blanketed the land served as lines of communication and advance, even when otherwise unused, and in many areas were early employed during seasonal freshets as a means of reaching markets with rural produce. Such streams were a main reliance in the commercial development of the lumber industry as it moved inland from the seaboard and thence across the continent.

The reverse side of this coin of abundance had a direct bearing upon the development of manufacturing in relation to power. The inexhaustible abundance of land for farming combined with virtually giveaway government land policies and the activities of land operators eager to promote land values by encouraging settlement served not only to advance settlement but to spread population thinly over the land. Settlers, inclined to see ever greener meadows just ahead, bypassed the less fertile and less favorably situated lands. Settlement thus proceeded well in advance of the means of transportation and communication and out of reach of markets and supplies of much needed goods. Before the rapid spread of the railway network from the 1850s on, this wide distribution of population meant that regions beyond several days' travel time by ox- or horse-drawn vehicles from main transportation routes following inland waterways were thrown almost wholly on their own resources, the home and fireside industries of the subsistence farmstead and in time local water mills. Replacement of such indispensable items as gunpowder and shot, metal tools and ironware, and salt required seasonal trips to distant markets over difficult trails and roads. Substantial relief from the hardships of self-sufficiency usually depended on the gradual emergence locally of manufacturing in a small way. Specialized production, at first largely on a craft and workshop basis, eventually was supplemented by mills and small factories serving the immediate locality often within a radius of but a few miles. At first exchange was chiefly on a barter basis, the produce or raw materials of the countryside supplied in return for the products of blacksmithies, forges, and small blast furnaces, of tanneries, woodworking shops, and small yarn mills—developments that will be considered more fully later.[7]

The two controlling conditions, then, which distinguished so much of American manufacturing from that of Britain were the thin distribution of population over expanding areas of settlement and the burdensome

[7] As noted above, manuscript schedules of federal censuses (National Archives) from 1820 on supply much information, as do state gazetteers of the period. In a report to Congress by Louis McLane, descriptive accounts of townships and counties frequently refer to the limited extent of market areas (U.S., Secretary of the Treasury, *Documents Relative to the Manufactures in the United States,* House Ex. Doc. 308, 22d Cong., 1st sess., 2 vols. [Washington, D.C., 1833]).

means of overland transportation by wagon and cart on common roads, on which most communities long depended and by which they were largely excluded from outside markets. In 1840 the total land area of the United States was twenty times the size of England and Wales combined, but in population the two entities were approximately equal, with sixteen and seventeen million people, respectively. In area England and Wales were smaller than New England by one-eighth, but in density of population they exceeded it by 276 people per square mile to a mere 33 for New England. In New York, New Jersey, and Pennsylvania population density was somewhat greater, with 44 persons per square mile. West of the Appalachians, in the five states north of the Ohio and east of the Mississippi, the figure fell to 12.[8]

As for transportation facilities in relation to population served, the contrast between the old country and the new was hardly less marked. Even before 1750 the English and Welsh rivers were extensively navigated by horse-drawn or man-propelled barges of varying capacity. The greater part of the land was within fifteen miles of navigable water.[9] In the United States more than a century later, in 1840, steam navigation was well advanced on the coastal waters and tidal rivers of the seaboard and on the trunklines and largest tributaries of the Ohio and Mississippi rivers and on the Great Lakes. On hundreds of the smaller rivers and larger creeks in much of the settled country, downstream navigation by flatboats and rafts was seasonably available for short periods, but upstream traffic by keel or pole boats was in most instances negligible owing to the difficulties and expense of relying on men as a source of power.[10] By

[8] Figures cited are from U.S., Bureau of the Census, *Historical Statistics of the United States, 1789–1945* (Washington, D.C., 1949), data B 1–30, p. 25; for England and Wales, *Encyclopaedia Britannica,* 11th ed., 9:404.

[9] Thomas S. Willan, *River Navigation in England, 1600–1750* (1964; reprint ed., New York, 1965), maps, app. 2. The great advantage of barge over wagon transportation is suggested by A. W. Skempton's comparative figures of the load capacity of a single horse pulling a wagon over soft (dirt) roads, macadamized roads, and iron rails: 0.625, 2.0, and 8.0 tons, respectively; and pulling a barge on canal or river, 30 tons (cited in Joseph Needham, *Science and Civilization in China,* 4 vols. [Cambridge, 1971], 4:pt. 3, p. 216).

[10] Louis C. Hunter, *Steamboats on the Western Rivers: An Economic and Technological History* (Cambridge, Mass., 1949), chap. 1. For the flatboat, keelboat, and rafting trade in the Ohio and Mississippi valleys, see ibid., pp. 52–60, 574–82, and Leland D. Baldwin, *The Keelboat Age on the Western Waters* (Pittsburgh, 1941). A comprehensive treatment of the subject for the Atlantic seaboard states is wanting, but frequent references to boating and rafting on creeks and small rivers as well as large appear in state gazetteers of the antebellum years: e.g., "Natural Inland Navigation" and "Canals," in Horatio G. Spafford, *A Gazetteer of the State of New York* (Albany, 1824), pp. 594 ff., and Thomas F. Gordon, *A Gazetteer of the State of New Jersey* (Trenton, 1834).

1840 railroad transport was just getting under way, but the total mileage of canals and surfaced roads was impressive and, joined with steam navigation, made possible a regional interchange of products of great importance to the national economy.

Yet the direct impact of such domestic commerce upon a large proportion of the agricultural population is, perhaps, best described as marginal, owing especially to an isolation enforced by the poor condition much of the year of the common country roads on which most farmers were immediately dependent. The literature of pioneering gives much attention to the problems of roads as a basic condition of rural isolation and of a subsistence economy.[11] The *Scientific American*, 8 April 1854, noting that "good common roads have been held up as an evidence of a country's civilization," went on to comment: "Throughout the rural districts, in almost every part of our country, the people suffer great inconvenience from bad roads—especially during the spring thaw when the frost is leaving the ground, and during the long periods of wet weather. . . . Even very near our cities, the common roads are also often rendered almost impassable." Even toward the close of the nineteenth century the situation was not greatly improved in many parts of the country. Our common roads, declared one commentator in 1891, "are inferior to those of any other civilized country in the world." Our home markets, noted a federal official in 1898, "are restricted by difficulties in rural distribution which not infrequently clog all the channels of transportation, trade, and finance."[12]

[11] For early references to common roads, see Stevenson W. Fletcher, *Pennsylvania Agriculture and Country Life, 1640–1840,* 2 vols. (Philadelphia, 1950–55), pp 246 ff.; Timothy Dwight, *Travels in New-England and New-York,* 4 vols. (New Haven, 1821–22), p. 300; Alan Conway, ed., *The Welsh in America: Letters from Immigrants* (Minneapolis, 1961), pp. 66, 69, 92; and J. A. Durrenberger, *Turnpikes: A Study of the Toll Road Movement in the Middle Atlantic States and Maryland* (Valdosta, Ga., 1931), chaps. 1, 2. See also Louis C. Hunter, *Studies in the Economic History of the Ohio Valley* (Northampton, Mass., 1934), pp. 5–32.

[12] I. P. Potter, "The Common Roads of Europe and America," *Engineering Magazine* 1 (1891):624–25; M. O. Eldridge, "Construction of Good Country Roads," *Yearbook of the U.S. Department of Agriculture, 1898* (Washington, D.C., 1898), pp. 317–24. See also the strong indictment by Max J. Becker, president of the American Society of Civil Engineers, 1889: "During the spring and fall we struggle through the mud manfully, as best we can, and when winter comes, and the bottom literally drops out of the roads we quietly compose ourselves and contently stay at home" (*TASCE* 20 [1889]: 240–41. An aged English miller, quoted by William Coles-Finch, stated that the "one great regret" of his father's life as a miller was "that he had subscribed to the making of hard roads." As Finch noted: "Before the advent of hard roads each village was self-confined and catered for its own needs. Every parish had its own watermill or

The prevailing conditions of transportation thus kept regional and national markets outside the reach of a majority of the predominantly rural population. With occasional exceptions in such industries as textiles, metalworking, and machinery building, manufacturing in 1840 was carried on largely in hamlets, villages, and small towns, widely distributed over the land, save on the outer frontiers. Most of these workshops, mills, and factories depended upon waterpower; yet except in an incidental sense the presence of waterpower was not a determinant of industrial location. Manufacturing establishments had to be near the market served and where the people were; waterpower typically was not far off and existed in a range of capacities adequate to most needs. The rise of small local manufactories not only provided the means for advancing the community's standard of living but afforded an important outlet for ambition and enterprise for those dissatisfied with living on and by the land.[13] Here were found the grass roots of industrial growth.

The United States in the 1790s was hardly on the verge of an industrial revolution, but the winds of change from across the Atlantic were reaching the new republic and finding congenial conditions and responsive minds.[14] From England came the first repercussive effects of momentous

---

windmill and competition in the milling trade was almost unknown." Hard roads brought competition and the concentration of milling in the larger mills with better service and lower prices (*Watermills and Windmills* [London, 1933], p. 53). As late as the years before World War I the roads of rural Illinois, as the author recalls from his boyhood, were often passable only with difficulty during the weeks of spring thaw and rains. Under the worst conditions with wagon and buggy traffic churning the country roads into morasses of felloe-to-hub depth, a heavy strain was imposed on horses and light loads were often necessary. At country schools the spring holidays were a movable feast, known locally as "muddy roads vacation." More than a half century later in rural Vermont the period of the spring thaw is still known among old-timers as "mud-time," when "everybody with any sense stays at home."

[13] The account of the pronounced localism of seventeenth-century English economic life in the early chapters of Harold J. Dyos and D. H. Aldcroft, *British Transport: An Economic Survey from the Seventeenth Century to the Twentieth* (Leicester, 1969), suggests what might be done with the preindustrial stage of development in the United States. Samuel Hollander offers a similar analysis, stressing the intimate relationship between local communities and the countryside—the exchange of materials and subsistence for manufactured goods—that permitted specialization in both country and town. Adam Smith declared of the American colonies that "no manufactures for distant sale have ever yet been established in any of their towns since American industry was confined to coarse and household manufactures by each family for its own use or for some neighbors." Hollander notes that Smith was concerned "less with the smith, the wheelwright and the carpenter than with the major categories of manufacturing for distant sale" (*The Economics of Adam Smith* [Toronto, 1973], pp. 98 ff.).

[14] For the purpose of this study the most useful work covering economic develop-

technological and industrial changes already in progress there. In successive decades the new textile methods and machinery, steam power on the Boulton and Watt plan in both stationary and mobile applications, and major innovations in the refining and shaping of iron secured a foothold on these shores. The mechanical spinning of yarn in small textile factories gradually spread from the first mills in Rhode Island to neighboring states and under the stimulus of embargo and war years had a rapid increase and extension through the country. On the western slopes of the Allegheny the ground was in preparation at Pittsburgh for the new iron-making techniques as ironmasters from central Pennsylvania relocated at a site favored both by expanding western markets and the protective barrier of distance from the seaboard. The first indigenous stirring of a talent for mechanization that was to make Americans famous was reflected in the successful development in New England during the 1790s of mechanical methods for making nails and the hand and machine cards so important in the manufacture of textiles, whether at home or in factories. Despite the efforts of England to prevent the spread of the new techniques beyond its shores, ideas and men were too slippery to be contained. The emigration of English artisans and English methods strengthened American industry in both the traditional and newer branches.

The stream of events was paralleled by the evolution of a supporting rationale, in which the leading voice was that of Alexander Hamilton, reinforced and in some degree anticipated by the work of numerous societies for the promotion of "useful" manufactures and of a host of pamphleteers. The enthusiasm for the cause of manufactures reflected both the experience with wartime shortages and the recollection of colonial handicaps. Sharing in many respects the views of Washington, under whom he had served in the war, Hamilton in his 1791 report to Congress on manufactures marshaled an elaborate body of argument and evidence in support of manufacturing as a national interest deserving and requiring national support.[15] Manufactures not only would provide

ment during the first decades of American independence is Curtis P. Nettels, *The Emergence of a National Economy, 1775–1815,* vol. 2, *The Economic History of the United States* (New York, 1962). Volumes 3 and 4 in this series are equally valuable in their areas: Paul W. Gates, *The Farmer's Age: Agriculture, 1815–1860* (New York, 1960), and George R. Taylor, *The Transportation Revolution, 1815–1860* (New York, 1951).

[15] Hamilton's *Report on Manufactures,* prepared at the request of the House of Representatives, 1790, together with preliminary drafts and useful editorial comment, appears in Alexander Hamilton, *The Papers of Alexander Hamilton,* ed. H. C. Syrett (New York, 1961–), vol. 10. See also Louis M. Hacker, *Alexander Hamilton in the American Tradition* (New York, 1957), chap. 9; Arthur H. Cole, ed., *Industrial and Commercial*

a new and more reliable domestic market for surplus products than foreign demands could supply, but also would create a demand hitherto lacking for various raw materials, give employment to presumptively underemployed persons such as women and children, encourage immigration, and greatly enlarge opportunities for a diversity of talents and enterprise. Hamilton particularly stressed the advantages for productivity of the division of labor and the use of machinery and power the manufacturing industry offered. In machinery we had "an artificial force brought in aid of the natural forces of man, unencumbered by the expense of maintaining the laborer." This country, moreover, was particularly favored "from the uncommon variety and greater cheapness of situations adapted to mill-seats, with which different parts of the United States abound."[16]

Manufacturing as a Condition of Growth

Hamilton's case for manufactures rested in part upon information respecting industry gathered through personal correspondence and by Treasury Department agents at his bidding. More significantly, it reflected the rationale of the ascendant school of political economy in Europe, reacting against the mercantilist philosophy of a now passing age. In his analysis of the 1810 census returns on manufactures Tench Coxe, onetime assistant and coadjutor of Hamilton, drew directly upon the facts of American experience.[17] He found the roots of American concern with manufactures in the struggle for existence on the frontiers of settlement. Manufactures, he declared with added emphasis, *"facilitate the first struggles of the American settlers for decent comforts, thrifty profits, and farming establishment."* Coxe cited the equipment and products of several newly settled counties in trans-Appalachian Pennsylvania, where "the mere presence of a few sheep and cattle, supplying wool, hides, skins, and horns, and tallow and other fat . . . (that is to say, the presence of raw materials) occasions the corresponding manufactures. In such places, profit, comfort and necessity, appear to invite, or rather to compel, the farmers and their families to that mode of industry."[18]

Correspondence of Alexander Hamilton, Anticipating His Report on Manufactures (Chicago, 1928).

[16] *Papers of Alexander Hamilton*, 10:249 ff., 272 ff.

[17] Tench Coxe, "Digest of Manufactures," *American State Papers, Finance*, 2:666–812. See also Albert Gallatin's earlier report, "Manufactures," submitted April 19, 1810, ibid., pp. 425 ff.

[18] Coxe, "Digest of Manufactures," pp. 667 ff.

A similar if soberer analysis appeared in the introduction to John Farmer and Jacob Moore's *Gazetteer of New Hampshire* (Concord, 1823): "New Hampshire is emphatically an agricultural state. Manufactures and commerce engross the attention of a comparatively small portion of its citizenry." In household manufactures, however, its citizens already excel. "To be independent we must manufacture for ourselves. . . . We manufacture comparatively little for exportation: most of our products are required at home." In sum, as Coxe viewed the American scene, from Vermont to the Carolinas and from Pennsylvania and Ohio through Kentucky and Tennessee, agriculture and manufactures, land and livestock, mill seats and raw materials were mutually supporting and in fact inseparable. If the role of mechanical power, especially waterpower, is to be seen in correct perspective, the native roots of manufacturing, which made so deep an impression on Coxe, must be recognized as hardly less significant for the future course of the American economy than the more dramatic industrial advances during the same years making their way here from Great Britain.

In retrospect it is clear that the water mills so eagerly sought by the pioneer communities were but the first rung upward in the struggle to escape the tyranny of a regressive self-sufficiency imposed by isolation on frontier settlements. This isolation was attacked from two directions. At the regional and national level the major forces were the improvements in main- or trunkline transportation on the navigable rivers and lakes and the canal systems joining them and on the turnpikes and railroads. During the half century preceding the Civil War transport systems and facilities occupied the center of public attention and with the varying contributions of private initiative and public assistance laid the foundations for a national economy through the medium of interregional trade. Far less publicized but more directly meaningful to the thousands of communities lacking access to main avenues of trade were the changes at work on the local level. Here, too, not only transportation and trade but manufacturing were seen as the keys to progress, the essential means of escape from the hardships and deprivations of frontier living. They were the means as well for attaining something of the amenities of living as known, if not always experienced, in the older and developed regions from which so many had come.[19]

[19] Richard Hofstadter takes issue with the traditional literary view of rural life, contending: "The farmer himself, in most cases, was in fact inspired to make money, and such self-sufficiency as he actually had was usually forced upon him by a lack of transportation or markets, or by the necessity to save cash to expand his operations" (*The Age of Reform: From Bryan to F. D. R.* [New York, 1955], p. 23). Zadock Thompson

The absence during the first years of anything but trails or frequently impassable roads limited trade largely to exchange within the community—of labor for goods, of one commodity for another, or for a service supplied by craftsman or mill. On many frontiers the first slender link of a new settlement with the outside market economy consisted of annual or semiannual trips by packhorse, sled, or flatboat to a market town with such surplus farm produce and various "country work" as might with much labor be accumulated. Such journeys, lasting up to several weeks and often made in winter when ice and snow made roads passable and fords and ferries unnecessary, were often the only means of obtaining such necessities as salt, shot and powder, edge tools, and ironware, and remembered luxuries such as tea, coffee, chinaware, and dress goods.[20] Sooner or later the setting up of a local store with its meager stock in trade reduced the need of such trips and was received with much the same enthusiasm as the first mill, near which it was often located.

As surpluses of produce increased and improved means of communication permitted, the local mills took on an increasingly commercial role; and as opportunity offered, their operations were enlarged. The carding machine or fulling machinery evolved into a small woolen mill making a cheap satinet with its cotton warp or linsey-woolsey with locally grown flax. To the rough boards of the common sawmill were added shingles

---

reviewed the past of his state, one of the less advantaged at the time, with its rugged terrain and isolated position: "The manufactures carried on in Vermont were, for many years, such only as the immediate wants of the people rendered indispensable, and in general each family were their own manufacturers. . . . By persevering industry they were soon enabled to raise a little flax and wool, which they spun and wove and colored and made into clothing. . . . The only trades which were then deemed indispensable, were those of the shoemaker and blacksmith, and these were, for the most part, carried on by persons who labored a portion of their time upon their farms. . . . As the condition of the people improved, they by degrees, extended their desires beyond the necessaries of life; first to its conveniences and then to its elegancies. This produced new wants; and to supply these, mechanics more numerous and more skilful were required, till at length the cabinet maker, the tailor, the jeweler, the milliner, and a host of others came to be regarded as indispensable" *(Geography and Geology of Vermont* [Burlington, 1848], pp. 112–13).

[20] Local histories dealing with the period 1790–1860 recount with pardonable pride the hardships of the early years. See, e.g., Abby M. Hemenway, ed., *The Vermont Historical Gazetteer,* 5 vols. (Burlington, 1868–91), 1:250; "The Sturdy Pioneers of Van Buren and Cass Counties," *Hist. Collections of Michigan Pioneer and Hist. Society* 38 (1912):641–43; E. Lakin Brown, "Autobiographical Notes," ibid. 20 (1906):433; Fletcher, *Pennsylvania Agriculture,* 1:76; Samuel W. Kercheval, *A History of the Valley of Virginia,* 3d ed. rev. (Woodstock, 1902), pp. 261 ff.

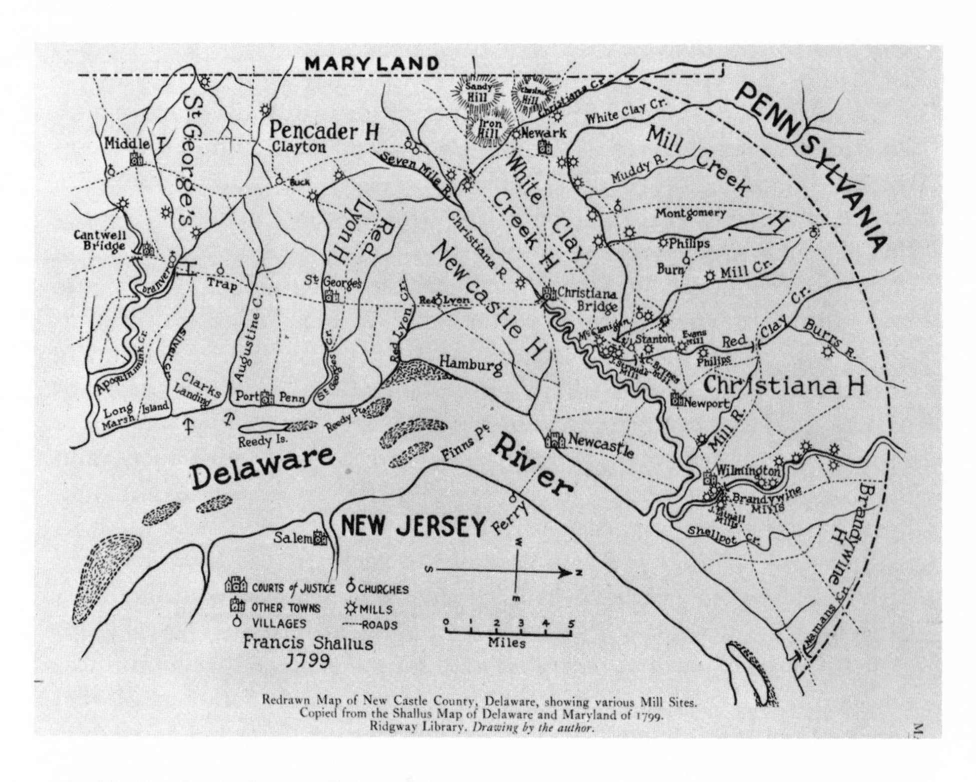

Fig. 39. New Castle County, Delaware, its stream pattern and mill seats in 1799. (Greville and Dorothy Bathe, *Oliver Evans* [Philadelphia, 1935].)

or staves, as we noted in an earlier chapter, with later perhaps a planing mill to supply the finished woodwork for sash, casings, and doors in the frame houses that marked the passing of the raw frontier years. Other workshops and more specialized mill industries gathered about a favorably situated water privilege as far as its capacity permitted. Local materials meeting local or outside market needs led to stone quarries, brickyards, pottery or paint works, manufactories of edge tools and other hardware, or some of the endless variety of wood products in the machine fabrication of which American mechanics demonstrated much ingenuity.

Mill Seats as a Condition of Manufacturing

The development of such small-scale, local manufactures was not dependent upon the initiative and capital of "outside" enterprise; nor was it, in the later phase, imposed upon or "sold" to lethargic and tradition-bound countryfolk, although itinerant peddlers selling "notions," hardware, and dry goods helped stimulate as well as gratify desire. It was a response rather to the same basic drive that brought most settlers to the frontier regions in the first instance, an improvement in the condition of their lives that made the establishment of the basic water mills an important objective. With agriculture and farm building providing the broad base of their mode of livelihood and life, the trend was toward a gradually diversifying economy. The evidence of an abundant body of records demonstrates that within the northern states and especially the northeast there was, as Tench Coxe observed, widespread concern with the development of manufactures. As manufacturing was a condition of economic growth, so waterpower was an indispensable requirement for the advance of manufacturing beyond the workshop level. The emphasis given in gazetteers and local histories to millstreams and mill seats was second only to that given the agricultural resources and activities predominant in most regions.[21] Repeatedly we meet references to "an abundance of mill seats," "the great facilities for mills," "the great advantages for hydraulic works," and the presence of "water machinery of various kinds." Although "mill rivers" were common enough, the millstreams most frequently mentioned were quite small, often creeks or brooks. We read again and again, as in Spafford's *Gazetteer of the State of New York* (1824), that a certain town or township "is abundantly supplied with mill-seats although all its streams are small"; that a certain creek "and its numerous small streams afford an abundance of fine mill-seats and mills"; that a town with "none but small streams is well supplied with mill-seats and mills"; or is "well-watered by springs, brooks and rivulets, the latter of a good size for mills." Many localities were not, of course, so fortunately endowed. Of a Vermont town first settled in the 1780s it was reported that for want of "valuable water power, manufacturing establishments or central place of business the occupation of the people has been exclusively confined to agriculture." In another instance there were "no streams of consequence and consequently no mills or mill privileges," a condition not without its silver lining, "for the people are subjected to no expense for bridges, nor loss by inunda-

[21] Among the more useful for the older states are: Zadock Thompson for Vermont, 1824; John Farmer and Jacob B. Moore for New Hampshire, 1823; Horace G. Stafford for New York, 1813 and 1824, and for Massachusetts, 1828; John Hayward for Massachusetts, 1846 and 1849; Thomas F. Gordon for Pennsylvania, 1832, for New Jersey, 1834, and for New York, 1836; Warren Jenkins for Ohio, 1837; John Blois for Michigan, 1838. The "Township Histories" in Andrew W. Young, *History of Wayne Country, Indiana* (Cincinnati, 1872), pp. 144 ff., typically begin with an account of early settlement followed by a section headed "Mills and Machinery," "Mills, Machinery and Merchants," "Mills, Mechanics, Physicians, and Merchants," and so on.

tions."[22] A Maine village "lacked the elements of permanent growth, in that it had no waterpower and . . . no important manufactures could be there established." Unfortunate was the district with "a scarcity," "a scanty," or an "indifferent" supply of mill seats and where "the want of small streams . . . forbids the introduction of manufactures, except in the household way."[23]

A Factor in the Growth of Towns

The waterpower that played so contributory a role in pioneering and in agricultural communities and was so essential to the growth of manufactures came early to be identified with the rise of villages and towns. As one early nineteenth-century traveler put it, with specific reference to southern Maine:

> The objects of first importance in these settlements are *mills*. If farms are cultivated, then mills are wanted, to make flour of the grain; but if, (as frequently was and is the case, in the countries that we are now entering,) the *lumber-trade* precedes farming, by a period of many years, then mills are indispensable, for sawing boards and plank. The place, therefore, at which a village begins, is either a sea-harbour or other *landing*, where country-produce is exchanged for foreign merchandise, or it is a cataract on a river, or some situation capable of affording a *mill-seat*. In such a situation, the first fabric that is raised is a solitary saw-mill. To this mill, the surrounding *lumberers,* or fellers of timber bring their logs, and either sell them, or procure them to be sawed into boards or into plank, paying for the work in logs. The owner of the saw-mill becomes a rich man; builds a large wooden house, opens a shop, denominated a *store,* erects a still, and exchanges rum, molasses, flour and pork, for logs. As the country has by this time begun to be cleared, a flour-mill is erected near the saw-mill. Sheep being brought upon the farms, a carding machine and fulling-mill follow. . . . the *mills*

[22] Hemenway, ed., *The Vermont Historical Gazetteer,* 1:127, 431.

[23] William B. Lapham, *History of Bethel . . . Maine, 1768–1890* (Augusta, 1891), p. 407. Before completion of the middle section of the Erie Canal, "a much larger business was done in New Hartford village than in Utica; and to its extensive water power—the lack of which has ever been severely felt in the latter place—was this extent of business attributable," wrote Pomroy Jones in *Annals and Recollections of Oneida County* (Rome, N.Y., 1851), p. 274. Of Lee Center, Jones wrote: "A quiet country village, isolated from the bustle of canals, rail or plank roads, yet its water power makes it a place of some importance and considerable business" (pp. 234–35). Despite the advantages of courthouse, jail, meeting house, and pioneer settlement at Bennington Center, Vermont, its location on a hill "without the advantages of water power, a large portion of the business which formerly centered here has passed to more favored locations, on the streams" (Hemenway, ed., *The Vermont Historical Gazetteer,* 3:139).

becoming every day more and more a point of attraction, a blacksmith, a shoemaker, a taylor, and various other artisans and artificers, successively assemble. . . . But, as the advantage of living near the mills is great, even where there is not (as in numerous instances there is) a navigable stream below the cataract—where it is a cataract that supplies the mill-seat—so a settlement, not only of artisans, but of farmers, is progressively formed in the vicinity; this settlement constitutes itself a society or parish; and, a church being erected, the village, larger or smaller, is complete.[24]

Some thirty years earlier a traveler passing through the Cape Fear region of North Carolina reported the same phenomenon. Enterprising men, remarking the occurrence of mill seats on a branch of the North West (now Cape Fear) River, "built mills, which draw people to the place, and these observing eligible situations for other profitable improvement, bought lots and erected tenements, where they exercised mechanic arts, as smiths, wheelwrights, carpenters, coopers, tanners &c. And at length merchants were encouraged to adventure and settle: in short, within eight or ten years from a grist-mill, saw-mill, smith-shop and tavern, arose a flourishing commercial town, the seat of government of the county of Cumberland."[25]

The role of waterpower and mill seats in determining village and town sites has been widely accepted, and indeed at times exaggerated, in the literature of pioneering. Bolles declared that water privileges had "given rise to a myriad of flourishing cities and villages in different parts of the country."[26] Not only did colonial villages, declared Clark, originate in mills but the location of most up-river towns was determined by waterpower. "The happy accident of waterpower in river or stream," according to Hedrick, "determined the location of most of the towns of inland New York as a thousand names attest." A New York gazetteer proclaimed, "It should be borne in mind by everybody, (excepting only the makers of towns-on-paper,) that water power is indispensably necessary, in a thin farming population, to the making of a respectable village." An agricultural history of Pennsylvania states, "Water power for turning mills determined the location of most rural hamlets."[27] A historian of the Old

[24] Edward A. Kendall, *Travels through the Northern Parts of the United States in the Years 1808 and 1809,* 3 vols. (New York, 1809), 3:33–34.

[25] *The Travels of William Bartram,* ed. Mark Van Doren (New York, 1928), p. 377.

[26] Albert S. Bolles, *Industrial History of the United States* (Norwich, Conn., 1878), pp. 302–4. In 1855 a frontier-bound migrant passing through Indiana remarked: "A settlement consists of one log cabin and an acre of cleared land; and a town, of a cabin, blacksmith's shop, and sawmill" (Seth K. Humphrey, *Following the Prairie Frontier* [Minneapolis, 1931], p. 6).

[27] Victor S. Clark, *History of Manufactures in the United States, 1607–1860,* 3 vols. (New

Northwest noted in respect to the gristmills that followed settlement, "A good millsite sometimes boasted a sawmill also; round about would spring up a blacksmith shop, general store and probably a village."[28] There was contemporary witness, too. "Every village aspired to be an important manufacturing center," declared an early historian of a county in Michigan, "even though the only natural advantage was hardwood forests."[29] "While agriculture has always been the leading pursuit of the people of Swanzey [New Hampshire], the manufacturing interests, in some respects, have been hardly less important. The streams of water are well adapted to furnish the propelling power and the pine forests have always supplied abundant material for the manufacture of wooden ware."[30]

York, 1929), 1:187. Ulysses P. Hedrick, *A History of Agriculture in the State of New York* (Albany, 1933), p. 146. Spafford, *Gazetteer of the State of New York,* pp. 424–25. Fletcher, *Pennsylvania Agriculture,* 1:323. "Along almost every brook," wrote L. D. Stilwell of Vermont, "there developed a mill-village, containing grist mills, sawmills, cidermills, tanneries, iron-forges, carding machines, linseed-oil mills, and every other sort of eighteenth century apparatus worked by water-power" (*Migration from Vermont, 1776–1860* [Montpelier, 1937], p. 102).

[28] R. Carlyle Buley, *The Old Northwest: Pioneer Period, 1815–1840,* 2 vols. (Indianapolis, 1950), 1:226 ff. See Stephen S. Visher, "The Location of Indiana Towns and Cities," *Indiana Magazine of History* 51 (1955):341 ff.; Lenora P. Miller, "Princeton—an Early Frontier Village in the Hoosier Pocket," ibid., p. 49. "It is mainly at the local concentrations of descent," stated a description of the Illinois River, "that the inviting water-powers are located, and consequently the prominent river towns are located" (U.S., Census Office, Tenth Census, 1880, vols. 16–17, *Reports on the Water-Power of the United States* [Washington, D.C., 1885, 1887], pt. 2, p. 225). Merle Curti, in a letter to the writer, 9 June 1954, said of Trempeleau County, Wisconsin: "Almost all the villages of the county were platted at points deemed good for mill-sites." Professor A. L. Throckmorton believes that practically all of the towns of early Oregon will be found to have been located with reference to waterpower (letter to writer, 17 Oct. 1962); for Texas, see R. N. Richardson, *Frontier of Northwest Texas, 1846–1876* (Glendale, Calif., 1963), pp. 291–92.

[29] E. W. Barber, "The Past and the Present," *Hist. Collections of Michigan Pioneer and Hist. Society* 29 (1901):634.

[30] B. Read, *History of Swanzey, New Hampshire, 1783 to 1892* (Salem, Mass., 1892), p. 220. Materials for pioneering history at the local level are abundant for New England and its westward extension. They are far less full and detailed for states south of the Potomac-Ohio river line. Common water mills were widespread in the southern states, but manufacturing villages, numerous in the northern states, were for generally familiar reasons infrequent in the South. For instance, Robert Mills, *Statistics of South Carolina* (Charleston, 1826), covers the state county by county and reports little but flour, grist-, and sawmills and household manufactures. Frederick B. Kegley, *Kegley's Virginia Frontier: The Beginning of the Southwest* (Roanoke, 1938), covers in elaborate detail land grants, land acts, and, in less detail, community settlement, but makes infrequent reference to water mills. For data on the hundreds of flour and grist

Aspiration and achievement are different things. With the probable exception of southern New England the generalizations on the influence of waterpower on village and town location are much too sweeping. Of some 320 village communities in Ohio as reported in Warren Jenkins's *Ohio Gazetteer and Traveller's Guide* only one-fourth contained one or more mills or other manufactories using waterpower. Yet of the thirty inland towns having a population of 1,000 or more, twenty-three were located on streams and three on canals from which power was obtained in amounts that were important during at least the industrial beginnings of these communities—among them Canton, Columbus, Dayton, Hamilton, Mansfield, Massillon, Newkirk, Springfield, Warren, and Zanesville.[31]

In the neighboring state of Michigan active settlement did not get under way until the 1830s, a generation later than in Ohio, when there had developed a more general appreciation of the importance of manufacturing in community welfare and of waterpower as a condition of its development. Townsite speculators and lumber operators joined with settlers in exploiting the abundant resources of land, timber, and numerous millstreams. Favorable gradients combined with frequent rapids to provide mill seats of the common sort in considerable number on the tributary streams. Waterpowers adequate to the larger purposes of manufacturing were found at rapids and falls interrupting the main streams. Settlement typically began in the open places about the river mouths along the lake and thence followed the course of the streams, we are told, "in search of open land and mill sites." Enterprising land operators encouraged and aided the establishment of mills and stores as a means not only of promoting townsites but of stimulating settlement and land values in the agricultural environs. "To be near a suitable millsite was frequently a motive of settlement and the 'sawmill town' usually had its gristmill, distillery, and tannery from the same cheap supply of power." The larger streams affording waterpower ranked with the presence of older settlement and good roads in determining the distribution of settlement and the location of centers of trade and manufacturing.[32]

---

mills reported for Georgia at the middle and end of the nineteenth century, see M. T. Thomson, "The Gristmill in Georgia," *Georgia Review* 7(1953):14–15.

[31] Warren Jenkins, *The Ohio Gazetteer and Traveller's Guide of the State of Ohio,* 1st ed. rev. (Columbus, 1837). Of more than forty villages along the Hudson River named in Allen Corey, *Gazetteer of the County of Washington, New York* (Schuylerville, 1850), about half contained grist- and sawmills and about a quarter of these had other manufactories, chiefly cotton and woolen mills. The only reference to manufactures in the remainder are to shops of mechanics.

[32] George N. Fuller, "An Introduction to the Settlement of Southern Michigan

Waterpower and the Industrial Village

Industrial villages in the proper sense, as distinguished from crossroads villages with a miscellany of mills, workshops, and stores, are of much greater interest and importance. The manufacturing interests of this country, declared Zachariah Allen in 1829, "are all carried on in little hamlets, which often appear to spring up in the bosom of some forest, gathering around the water fall which serves to turn the mill wheel." These villages "were scattered over a vast extent of territory, from Indiana to the Atlantic, and from Maine to North Carolina, instead of being collected together, as they are in England, into great manufacturing districts."[33] According to Percy Bidwell, it would have been difficult to find 50 out of the 479 townships in Massachusetts, Connecticut, and Rhode Island in 1840 "which did not have at least one manufacturing village clustered around a cotton or woolen mill, an iron furnace, a chair factory or a carriage shop" or similar establishment.[34] New England was preeminently the land of the factory village; but in upstate New York, in New Jersey and Delaware, and in portions of Pennsylvania and Maryland much the same pattern of industry is found. Whether we leaf through such gazetteers as Hayward for Massachusetts and New England, Spafford or French for New York, or Gordon for Pennsylvania, New York, and New Jersey, and in less degree Jenkins for Ohio, and Blois and later Clark for Michigan, we come upon many references to, or brief accounts of, factory and mill villages. For the more fully developed districts, the references and descriptions are too numerous to count, as witness the federal waterpower survey of 1880. These small industrial communities in important respects prepared the way for the larger industrial developments ahead; they were the next upward rung on the ladder of industrial advance.

Two examples from western Massachusetts illustrate the mill village phase of industrial growth, the town of Middlefield in the Berkshires and the valley of the Mill River, which had issue in the Connecticut River

---

from 1815 to 1835," *Hist. Collections of Michigan Pioneer and Hist. Society* 38 (1912):578 ff.; idem, *Economic and Social Beginnings of Michigan, 1805–1839* (Lansing, 1916). The latter pioneering monograph is of unusual interest and emphasizes the role of waterpower in land settlement. Related material, especially on the linkage of townsite plats and speculation to waterpowers, appears in J. J. Hogaboam, *The Bean Creek Valley* (Hudson, Mich., 1876), and Lakin Brown, "Autobiographical Notes," pp. 424 ff.

[33] Zachariah Allen, *The Science of Mechanics* (Providence, 1829), p. 352.

[34] Perry W. Bidwell, "Rural Economy in New England at the Beginning of the Nineteenth Century," *Transactions of the Connecticut Academy of Arts and Sciences* 20 (1916): 241 ff.

at Northampton.[35] To the usual complement of early water mills at Middlefield, a settlement dating from the 1780s, two fulling and carding mills were added about 1815. The enterprising owners successfully extended these activities into the making of woolen cloth of different grades, sending the products by wagon to Connecticut River market towns thirty miles distant. By 1840 Middlefield had a small but sturdy "Factory Village" located upon and taking its power from the Factory Branch of the small Westfield River. Some forty workers were accommodated by several "tenements" and a two-story boarding house. The coming of a railroad in 1841, resisted by the predominant farming element, provided access to the larger market of Boston. The Civil War brought a boom in woolen textiles and raised the Factory Village to the full flower of its industrial achievement; its population in 1870 accounted for about one-fourth the town's 1,700 inhabitants. Competition from much larger and more favorably situated woolen mills, the gradual replacement of woolens by lighter and more popular worsteds, and serious losses by fire in the 1870s brought industrial decline. The coup de grace ironically was given by the Factory Branch, which in two disastrous floods in 1874 and 1901 took away all that this millstream over the years had contributed to building up. The fifty-seven families of Factory Village in 1860 fell to thirty-eight in 1880 and nineteen in 1900.

The experience of an isolated factory village as at Middlefield was replicated innumerable times within New England and elsewhere in the industrial North. Under favoring conditions there arose along many millstreams the less numerous but more significant phenomenon, a succession of mill villages. Expanding industrial opportunities and the demand for power often resulted in the conversion of a millstream into a fairly continuous string of millponds with dams and mills at frequent intervals. Such was the case with the Mill River, a minor Massachusetts tributary of the Connecticut River, which from Northampton, near its mouth, was gradually brought into full development at a dozen privileges with an aggregate capacity of some 500 horsepower.[36] Eventually more than fifty mills and factories were strung along its sixteen-mile length. Development got under way much as at Middlefield with the usual assortment of seventeenth- and eighteenth-century water mills, later followed by textile mills, a large tannery, and various small shop enter-

[35] E. C. Smith and P. M. Smith, *A History of the Town of Middlefield, Massachusetts* (Menasha, Wis., 1924). The Westfield River basin, of which Middlefield's "Factory Branch" is a part, is described in Tenth Census, *Water-Power,* pt. 1, pp. 253 ff.

[36] Agnes Hannay, *A Chronicle of Industry on the Mill River* (Northampton, Mass., 1935–36).

prises. In time the profits accumulated from successful ventures by a closely knit group of local businessmen were invested in new fields. The list of products came to include not only buttons, cutlery, edge tools, and paper but eventually sewing machines and oil stoves. During the period 1815-45 factory and prefactory activities went on side by side, but with the factory gradually taking over most production. In pursuit of additional power sites, manufacturing pushed up the valley and several new villages were created, from Leeds and Haydenville to Florence and Williamsburg. Not until the late 1850s was it necessary to turn to steam for additional power; the previous decade the railroad had resolved marketing and supply problems. Owing to its favorable situation in the heart of the Connecticut Valley with convenient access to both Boston and New York City, this district enjoyed a vigorous prosperity and expansion into the twentieth century.

Commercial manufacturing as conducted in mills and small factories was by no means confined to factory villages of the Middlefield and Mill River types. Gristmills and sawmills and mechanics' workshops apart, many a predominantly agricultural community in New England included small isolated industrial establishments. One such Vermont community, familiar to the writer through long summer residence, followed much the same pattern as Middlefield, lacking only a "Factory Village." The town of Thetford in the upper Connecticut Valley, whose river it borders for some ten miles, was in 1840 near the peak of its population and agricultural growth. With a population of 2,065 living on the farms and in several small villages, as reported in the Sixth Census, Thetford had two tanneries, nine other leatherworking establishments, eight men making carriages and wagons or furniture, thirteen sawmills, four gristmills, and one oil mill. The total value of their products was $27,435.[37]

There was an intimacy between farming communities and manufacturing that persisted in many parts of the northeastern United States into the closing decades of the nineteenth century. In Massachusetts in the early 1870s all but perhaps a score of the more remote and thinly peopled towns nourished some small amount of commercial manufac-

[37] Charles Latham, *A Short History of Thetford, Vermont, 1761–1870* (n.p., 1972), pp. 23–41. Orange County, in which Thetford is located, was reported in the 1840 census to have the following number of persons employed: agriculture, 6,558; commerce, 132; manufactures and trades, 893; navigation of canals, lakes, and rivers, 34; learned professions and engineers, 166 (U.S., Census Office, Sixth Census, 1840, *Compendium of the Enumeration of the Inhabitants and Statistics of the United States* [Washington, D.C., 1841], pp. 18–19).

turing, that is, production for outside sale. Elias Nason, in *A Gazetteer of the State of Massachusetts* (Boston, 1874), declared, no doubt with the sales-minded exaggeration of the later gazetteer compilers, that there was "hardly a village in the Commonwealth whose activities are not quickened . . . by some establishment for the manufacture" of articles ranging, in his listing, from bonnets, boots, and boxes, chairs and cutlery, matches and machinery, through pencils, pianofortes, and pocketbooks.

The Ubiquity and Variety of Waterpowers

All things considered, it is difficult to conceive a source of motive power more admirably adapted to the requirements of an early and developing industrialism than the waterpowers of the northeastern United States. If manufacturing was to advance beyond handicraft and workshop levels, the only practicable alternative was the horse- and ox-powered sweep and tread mills so commonly adopted when other sources of power were lacking. Steam engines were ruled out not only by their cost and in most districts unavailability but by problems of attendance, maintenance, and repair.[38] As late as 1840, as will be shown in a later volume, facilities for their production were limited mainly to the larger urban communities. Where water transport was lacking, the cost of carriage for engine and boiler components weighing several tons was prohibitive except for limited distances. The much acclaimed mobility of steam power was slow to be realized.

Waterpower, in marked contrast, was to be had almost everywhere in the northeastern United States, where other conditions for industrial advance were most favorable. The whole land was an intricate yet orderly system of streams and drainage basins, the larger of which had outlet in the sea. As we noted in chapter 3, a large proportion of their affluents, especially in New England, had their sources in ponds and lakes, natural

[38] Robert H. Thurston, an authority on steam power and a United States commissioner to the Vienna International Exhibition of 1873, wrote in his "Report on Machinery and Manufactures" of that exhibition: "Formerly, when steam-engines of even the best construction consumed ten pounds of coal or more per horse-power per hour, when the cost of manufacture was vastly greater than today, and when transportation and repairs were matters of vastly greater expense than now, water power was so much less expensive that the location of mills and manufactories was determined by the availability of mill-privileges, rather than by proximity either to the source of supplies or to the market" (*Reports of the Commissioners to the International Exhibition Held at Vienna, 1873,* ed. Robert H. Thurston, 4 vols. [Washington, D.C., 1876], 3:175). See also volume on steam power, in preparation, and below, chap. 10.

reservoirs serving as useful regulators of streamflow. The creeks and brooks comprising the smallest class of millstreams were not only the most numerous and most widely distributed but also were the most readily developed.[39] By and large as one moved from the main river to its principal tributaries and to the affluents of these in turn, once, twice, or thrice removed, the number of waterpowers increased and their average capacity decreased in something like geometric progression. In the northeastern United States there was only one Niagara, a few score powers with capacities of 1,000 to 20,000 horsepower, and tens of thousands of mill seats capable of supplying the power required to drive a common water mill.

In short, the arrangements of nature were in substantial accord with the requirements of a slowly emerging industrial economy, beginning with numerous and widely distributed small establishments. In response to the growing density of population, improvement in the facilities and cost of transportation and marketing, the extension of mechanization, and the widening recognition of the advantages of industrial centralization, the trend was toward progressively larger establishments and attendant economies of scale. Power requirements with the shift from hand to machine operations increased in even higher ratio than size and led to a searching out and development of powers capable of meeting the larger demands, culminating in the development of such great waterpowers as those on the Merrimack and Connecticut rivers. Within the range of available choice in any district, mill seats varied widely in respect to such important considerations as accessibility, the difficulty and expense of development, and the volume and regularity of flow.[40] The advantage of firstcomers in the pick of mill seats was often roundly paid for in overly sanguine estimates of streamflow unchecked by record or experience. Yet by the early 1800s the more experienced of millwrights and owners, with generations of water-milling lore behind them, seem to have developed a shrewd intuitive sense in the selection of sites and the manner of their development to best advantage. They managed to reach tolerable judgments on volume and regularity of

[39] See the discussion of the drainage basins of the small brooks (termed "brook areas") emptying directly into the Mississippi River. The aggregate extent of such brook areas for the entire Mississippi River was estimated at some 32,000 square miles, of which the upper Mississippi accounted for nearly 18,000 (Tenth Census, *Water-Power,* pt. 2, p. 145).

[40] Oddly enough, many of these matters received little or no attention in the many editions of the standard handbook of the time, Oliver Evans's *The Young Mill-wright and Miller's Guide.* Comprising the commonplace, they evidently called for no remark.

streamflow based on bits and pieces of evidence, avoiding an undue reliance on hearsay. In this last and vital matter latecomers enjoyed a margin of advantage over the pioneers.

The immobility of waterpower—to become a major deficiency in the years after the Civil War—was substantially mitigated in the earlier decades by the widespread occurrence of mill seats. The wide scattering of settlement in small rural communities for many years corresponded to the availability of small powers on the creeks and brooks everywhere close at hand. The problem of industrial location arose as opportunities for an increased scale of operations were gradually extended by improvements in transportation and marketing facilities. When power needs rose above the few horsepower adequate for most common water mills, the range of choice was narrowed, especially since accessibility to markets and materials demanded increasing consideration. As industrialization gathered headway in any district or region, the capabilities and limitations of millstreams were not long in the testing. Initially the usual practice was to select powers of an estimated capacity not greatly in excess of anticipated requirements in order to minimize the difficulties and costs of development, for capital was scarce and interest rates high. In contrast with what was to prove the case with steam power, initial outlay per horsepower capacity tended to increase rather than decline with the size of installation.[41]

With the increase in manufacturing, questions of full or part ownership, or lease, of a waterpower arose. Ownership of land on both sides of a stream gave full and independent control of the manner of its development and use. Sharing possession with the owner of the opposite bank or of the entire privilege with several others reduced initial outlay and maintenance expense but presented difficulties in the division and measurement of streamflow and the sharing of maintenance costs, too often leading to controversy and litigation. Such difficulties might be out-

[41] George F. Swain, ranking civil engineer in the 1880 census survey of waterpower, noted in another publication: "While steam power costs much more in small than in large quantities, the reverse is frequently the case with water power, so that to compare the cost of large quantities of power is—so far as concerns the ordinary conditions through-out the country—much more unfavorable to water power than would be the comparison of the cost of small quantities" ("Statistics of Water Power Employed in Manufacturing in the U.S.," *PASA*, n.s. 1 [1888–89]: 30–31). The same point was emphasized by a leading hydraulic engineer in textile manufacturing in New England: Samuel Webber, "Water Power in New England," *Engineering Magazine* 1 (1891):528–30. The explanation evidently is that as one moved from smaller to larger powers, the engineering difficulties and structural requirements increased faster for waterpower than for steam power.

weighed by the other advantages of a situation, such as transportation facilities and the growth prospects of a thriving industrial community.

Many of the larger and more flourishing industrial villages of the mid-nineteenth century had their origin in broken or hilly country that gave rise to a succession of falls on the same stream at intervals of some hundreds of feet or yards. Even if the total outlay for development of a succession of mills was not materially different, the engineering difficulties were usually much less than with a single waterpower of corresponding capacity. Space for races and millsites could be provided more easily, and the difficulties of joint ownership and use would be mitigated if not wholly avoided. Situations of this kind occasioned many of the larger industrial villages, some of which became important manufacturing centers. An early example was Vergennes, Vermont, where in 1824 some twenty mills and factories were gathered about a series of three falls in Otter Creek and supported a village of some 1,000 people. At Seneca Falls, New York, a community of about 3,000 in 1836, the Seneca River fell forty-seven feet within three-quarters of a mile; four successive dams diverted the stream to serve some thirty manufacturing establishments. At Watertown, New York, the Black River a dozen miles above its mouth at the eastern extremity of Lake Ontario, fell 120 feet within three and a half miles divided among eight privileges. In 1880 its incompletely used power supplied many mills and factories with an aggregate of 4,675 horsepower. During the same period Fitchburg, Massachusetts, harnessed the rapidly falling Nashua River with eleven dams within two miles, laying the basis for vigorous growth in the manufacture of paper, textiles, and machinery.[42]

Other industrial communities taking advantage of readily developed multiple falls on streams of no great size were Willimantic, Moodus, and Rockville in Connecticut; Fisherville, Franklin Falls, Claremont, and Tilton in New Hampshire; Athol and Fall River in Massachusetts; Schaghticoke, Troy, and Watertown in New York; and Cuyahoga Falls in Ohio. On a far larger scale and requiring much greater resources to develop were the waterpowers at Cohoes and Rochester, New York; Richmond, Virginia; Appleton, Wisconsin; and the great waterpower at Minneapolis. In each instance the division of a large fall on a river of some size among a succession of privileges simplified and encouraged the task of development and use.

[42] The examples cited are from state gazetteers and Tenth Census, *Water-Power.* This phenomenon was noted years later by a prominent hydraulic engineer who began his career in Lawrence, Massachusetts: John R. Freeman, "General Review of Current Practice in Water Power Production in America," in *Transactions of the First World Power Conference, 1924,* 5 vols. (London, [1925]), 2:382–83.

River Systems and Patterns of Industrial Location

By 1850, except in the more developed portions of southern New England —in the absence of relevant census data on both villages and waterpower one can only speculate—much the more numerous type of industrial community was the relatively isolated factory village of the type in Middlefield, Massachusetts.[43] Such villages in the aggregate presented at best a loose scatter pattern without developmental sequence. Of greater interest and significance in the emerging industrial economy based on waterpower was the cluster pattern of villages centering in and more or less dependent upon a coastal or inland town of some commercial importance. Here were marketing facilities and outlets, supplies, and, usually, financial resources, which for want of motive power could not be invested in local industries. From Boston to Baltimore clusters of mills and mill villages sprang up in widening circles around the leading commercial centers. In the Middle Atlantic states at Wilmington, Philadelphia, and Baltimore the grain trade with the backcountry gave rise to a flourishing flour-milling industry. In time a similar gathering of flour and other mills occurred at such inland points as Chambersburg and Lancaster in Pennsylvania; Albany, Troy, and Ithaca in New York; and Circleville and Mount Pleasant in Ohio.[44] The most striking example of the mill-

[43] It is probable that most of the small one-industry or one- or two-mill villages of the Middlefield type suffered the same fate. An important exception is treated comprehensively in John B. Armstrong, *Factory under the Elms: A History of Harrisville, New Hampshire, 1774–1969* (Cambridge, Mass., 1969), an outstanding volume in the same class as the Smiths' study of Middlefield.

[44] See discussion of the Mill Dam tidal power project in Boston in the companion volume on steam power, in preparation. For a Connecticut cluster, see C. E. Chandler, "The Manufacturing Interests of Norwich," in William T. Davis, ed., *The New England States,* 4 vols. (Boston, 1897), 2:1003 ff. In Philadelphia and its immediate vicinity in 1823 there were "upward of 30 cotton factories, most of them on an extensive scale" (J. Thomas Scharf and Thompson Westcott, *History of Philadelphia,* 3 vols. [Philadelphia, 1884], 3:2318). On Wilmington, Del., and Lancaster, Pa., see Clark, *History of Manufactures,* 1:185–87, and Fletcher, *Pennsylvania Agriculture,* 1:324–25. *Hunt's Merchants' Magazine* 23 (1850):50–51, states that in Baltimore and its immediate suburbs there were 22 flour mills with about 75 pairs of millstones; within a circuit of 20 miles there were 10 millstreams, each with an aggregate fall of 106 to 326 feet and over 80,000 horsepower altogether. Israel D. Rupp reported that in Chambersburg, and within a radius of five miles, there was waterpower with a capacity of 100 pairs of millstones (*The History and Topography of Dauphin, Cumberland . . . Counties . . .* [Lancaster, Pa., 1846], p. 462). On Troy and Albany, New York, see *Hunt's Merchants' Magazine* 14 (1846):521–22, 21 (1849):51. For Ithaca, see Thomas F. Gordon, *A Gazetteer of the State of New York* (Philadelphia, 1836), p. 730, which reported 160 mills within 12 miles, a number which had increased to 299 by 1832—presumably all types

village cluster pattern was that of the textile industry in southern New England centering at Providence. As late as 1826 there was only a single cotton mill (with less than a thousand spindles) in the city, as Zachariah Allen, a leading manufacturer, noted in the *Science of Mechanics* (1829), "whilst several hundred thousand [spindles] were in operation on the mill streams in the country adjacent." Providence, according to Peter J. Coleman, "supplied most of the capital, the managers, and the technical knowledge."[45]

With time and growth it became necessary in the more active industrial regions to move out ever further in search of unoccupied mill seats, and a new and eventually dominant pattern took form wherever industry flourished, as illustrated on a small scale by the mill river development described above. Mills and mill villages followed the course of the millstreams, eventually replicating in the more fully developed areas the trunk-limb-branch configuration of the millstreams. An early example of this locational phenomenon was described in 1832 by Samuel Slater, the man who contributed so large an impulse to the progress described. The manufactures of Rhode Island, he declared in response to the McLane inquiry into the state of manufacturing, were found chiefly along the waters of five streams: the Blackstone, Moshassuck, Wonasquatucket, Pautuxet, and Pawcatuck rivers and their tributary streams.

> The most considerable of them would, in the western country, be designated as a creek. Its course, from its head waters to its place of embouchere, does not, perhaps, exceed sixty miles, and the volume of water which it discharges would be inconsiderable in comparison with that discharged by what is called a creek; yet these streams, more steady in their volumes than those of the western country, and descending, in their short courses, an elevation of from two to four hundred and fifty feet to the tide waters of the bay and sound, furnish, with their tributaries, innumerable cascades, and a power of propelling machinery almost incalculable in amount. . . . Many streams which, with their natural flow, would drive the works upon them only ten or eleven months of the year, have been made [by artificial reservoirs upstream] to operate them throughout the year.[46]

---

of water mills. "The great hydraulic powers furnished by Walnut, Darby and Deer creeks and the [Ohio] canal" drove "the large number of flouring mills in operation in the neighborhood" of Circleville, Ohio (Jenkins, *Ohio Gazetteer,* pp. 118–19). For similar though less striking concentrations, see ibid., pp. 201 ff. (Germantown), p. 302 (Millbrook), pp. 316–17 (Mount Pleasant and Mount Vernon), p. 414 (Springfield).

[45] Peter J. Coleman, *The Transformation of Rhode Island, 1790–1860* (Providence, 1963), p. 83.

[46] "General summary for Rhode Island," in *Documents Relating to the Manufactures in the United States,* 1:927–28.

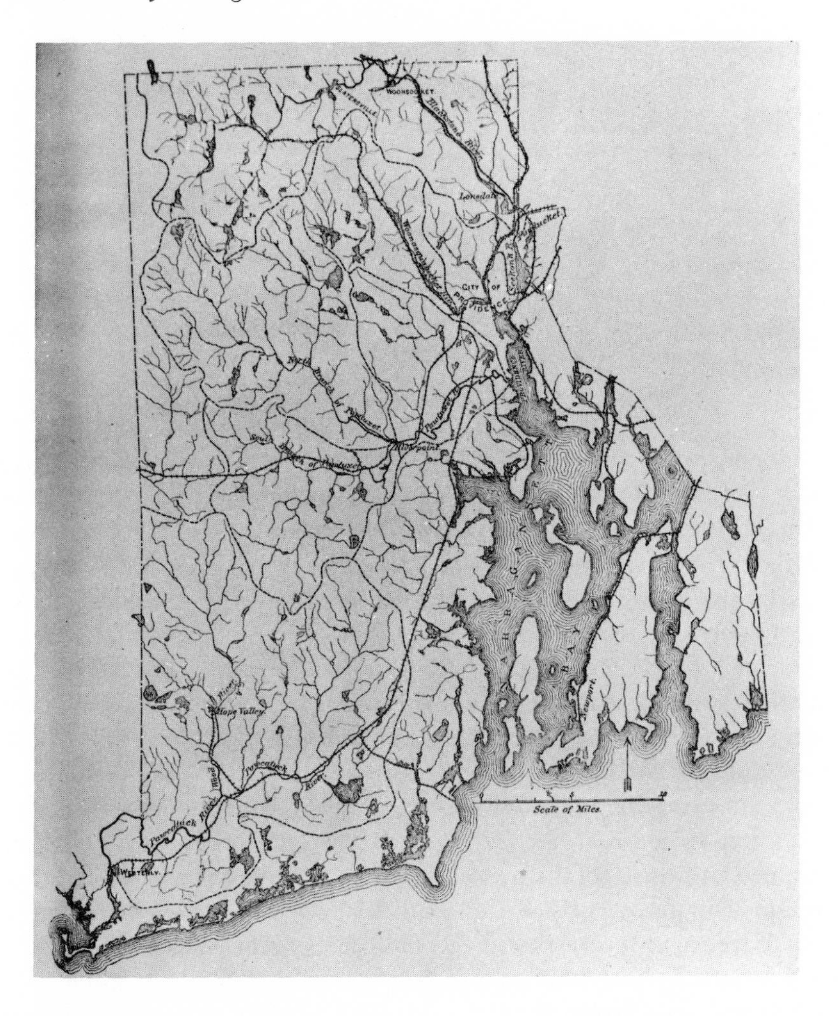

Fig. 40. The drainage basins of Rhode Island. (U.S., Census Office, Tenth Census, 1880, *Reports on the Water-Power of the United States*, pt. 1 [Washington, D.C., 1885].)

When nearly a half century later a national survey of waterpower was carried out under the direction of the federal Census Office, there were few parts of the more industrialized portions of New England in which variations of the Rhode Island pattern described by Slater were not to be found. As the following tabulation giving water and steam horsepower in use per square mile in 1880 shows, concentration of waterpower in northern New England was much less than in southern New England and declined as we move southward.

	Horsepower per square mile			*Horsepower per square mile*	
	Water	*Steam*		*Water*	*Steam*
Rhode Island	20.5	38.10	New York	4.61	4.93
Massachusetts	17.21	21.32	New Jersey	3.63	9.76
Connecticut	12.63	11.77	Pennsylvania	2.45	8.94
New Hampshire	7.68	2.06	Delaware	2.44	5.43
Vermont	5.72	1.21	Maryland	1.83	3.37
Maine	2.67	0.69	Virginia	0.93	0.49

To the southward and westward of the states listed, save in isolated river basins, it fell off rapidly, in no state attaining as much as one horsepower per square mile, more commonly less than one-half of one horsepower.[47]

Although by 1880 steam power had come to supply over 60 percent of the motive power used in manufacturing in the United States and was progressively extending its domain, the greater part of industrial America outside of the larger urban centers still relied on waterpower. The distribution of waterpower by regions in this year gave to the six New England states 35 percent of the developed waterpower nationwide. The Middle Atlantic states, here taken to cover the seaboard from New York through Maryland—with half again the area of New England and twice the population—accounted for 32 percent, giving to these two regions combined two-thirds of the waterpower in use in the entire United States. The much more recently settled and developed North Central states—north of Kentucky and Arkansas on both sides of the Mississippi and east of the Montana-New Mexico line—with four times the area and a somewhat larger population than the Middle Atlantic and New England states combined, had an aggregate waterpower equal to one-half of the waterpower in New England alone. Thus, owing both to differences in physiography and climate and in levels of economic development there were in 1880 wide differences in the availability and use of waterpower across the continent. The extreme range in horsepower per square mile by region was from 0.30 for the North Central states to 4.81 for the Middle Atlantic and New England states combined.

[47] Tenth Census, *Water-Power,* pt. 1, pp. xiv, xv.

The River Basin: Basic Unit of Power Supply

The distribution of waterpower in the eastern half of the United States is presented in definitive and often exhaustive detail in the Tenth Census *Reports of Water Power of the United States.* The men who planned and carried out the 1880 survey were civil, not mechanical, engineers, drawn from eastern engineering schools. Their training and experience led them to concentrate attention not upon the machines involved, of which, except in a general way, they may have had no special understanding,

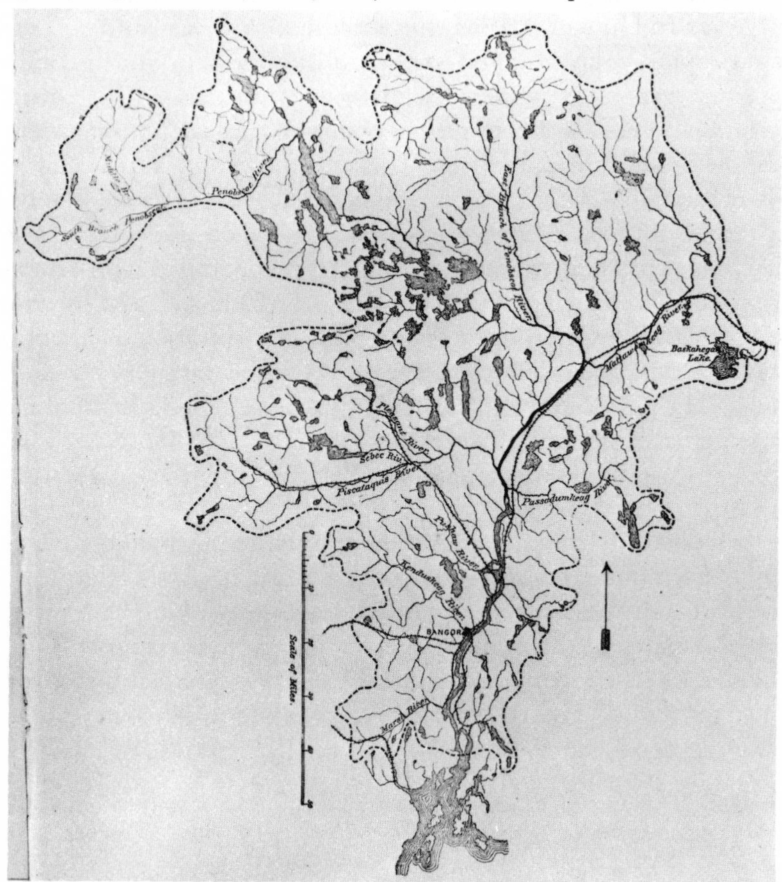

Fig. 41. Drainage basin of the Penobscot River. Of the 9,200 horsepower in use in this area—about one horsepower per square mile—7,100 horsepower were used in sawmills and 830 horsepower in flour and grist mills. (U.S., Census Office, Tenth Census, 1880, *Reports on the Water-Power of the United States*, pt. 1 [Washington, D.C., 1885].)

but upon topography, geology, stream gradients, and streamflow, which they did. This bias was fortunate, for over the centuries much had been learned about waterwheels but little was known about the hydrology upon whose principles their most effective utilization depended. The focus of the waterpower reports of the 1880 census, unlike that of steam-engineering literature and of the history of technology, was not upon prime movers, to which only occasional and incidental attention was given, but upon the means by which a "head" of pressure necessary to place prime movers in motion was obtained and maintained. River drainage, or catchment, basins, surface runoff and streamflow, waterfalls and rapids, supplemented at the millsite by dam and races, were the pressure generators analogous to the hot, dirty, and little-studied boiler-furnace units of steam-power plants. Basic calculations were of streamflow volume in cubic feet of water per second and began with records of rainfall in inches per year, distributed over the days and months of the year with seasonal and unseasonal variations, per square mile of the catchment basin, or watershed area. They concluded with estimates of the proportion of such precipitation, including snowfall and snowmelt, entering streamflow after losses due to plant transpiration, evaporation, and the like had been deducted, eventually to be carefully checked in test areas. The far more reliable calculations based upon periodic on-site measurements of streamflow, obtained through a national network of hydrographic stations and accumulating over the years, lay well in the future.[48]

By extended personal exploration and reconnaissance and with materials assembled from many sources, local and professional, the young engineers of the 1880 survey recorded with painstaking thoroughness the basic features of the hydrological system of power supply and distribution upon which American industry had principally relied for more than two centuries.[49] Here was described, with all the deficiencies of what was more reconnaissance than survey, at the high plateau of its develop-

[48] Methods of measuring or gauging streamflow in order to estimate waterpower capacities are described in *Operations at River Stations, 1900,* pt. 1, *East of Mississippi River,* U.S. Geological Survey, Water-Supply and Irrigation Paper no. 64 (Washington, D.C., 1901). See also E. C. Murphy et al., *Hydrographic Manual of the United States Geological Survey,* Water-Supply and Irrigation Paper no. 94 (Washington, D.C., 1904), or, more concisely, William A. Liddell, *Stream Gaging* (New York, 1927). A comprehensive treatment is Daniel W. Mead, *Hydrology: The Fundamental Basis of Hydraulic Engineering,* 2d ed. rev. (New York, 1950). A concise summary appears in Robert W. Abbett, ed., *American Civil Engineering Practice,* 3 vols. (New York, 1956–57), 2:sec. 13, pp. 2–75.

[49] See App. 4.

ment the energy base of an industrialism hardly less impressive than that of England with its long reliance primarily on steam power. Here were recorded the results of the gradual extension of manufacturing along the steamflow lines of an increasingly obsolete mode of power supply. Using the unwieldy but well-organized quarto volumes of the survey as our guide, we can observe the structure and distribution of manufacturing industries in their geographic setting, as reported by these civil engineer "special agents" of the Census Office, region by region, river basin by river basin, trunkline, limb, and branch. The distinctive characteristics of individual millstreams and mill privileges and the number and kind of manufacturing establishments and the amount of power used were reported with varying detail and summarized in tabular statements covering the larger drainage basins. Despite the meager and often dubious data at their disposal the engineers presented carefully qualified estimates of streamflow for the larger rivers.

Representative Drainage Basins in New England

In the ten river drainage basins in the country with the greatest concentration of waterpower in use per square mile, the range was from 7 to 44 horsepower. Eight of the ten were in New England. They included some of the largest—the Merrimack and Connecticut river basins with 4,916 and 10,924 square miles, respectively—and some of the smallest—the Charles, Taunton, and Blackstone rivers with 281, 337, and 458 square miles, respectively. In addition to the Blackstone, the Merrimack, and the Connecticut rivers, a fourth, the Thames River in Connecticut and the Hudson River in New York provide illuminating examples of industrial regions long based on waterpower.

The five millstreams cited by Samuel Slater in 1832 as the power base of Rhode Island industry had during the succeeding half century amply fulfilled their early promise. These streams in 1880 drove some 350 mills, supplying more than 30,000 horsepower. The forty-mile-long Blackstone River, with its sources near Worcester, Massachusetts, and its mouth at Providence, Rhode Island, together with its tributaries provided nearly two-thirds of the total power, supplying 177 mills with 19,989 horsepower. Sustained during the dry seasons by a large number of ponds, reservoirs, and nearly as many millponds as mills (and eventually supplemented by steam power), the Blackstone led all streams in the country in horsepower per square mile, twice that of the "number two" stream, the Thames River basin. With an average fall of ten feet per mile, the Blackstone provided the classic example of the fully developed millstream, a succession of millponds occupying nearly all the river between headwaters and mouth. Thirty-one establishments on the main stream, chiefly large textile mills, employed more than half the power used in the entire basin, with 358 horsepower per mill. The numerous tributaries supported nearly five times as many mills, averaging 61 horsepower each and including a variety of manufactures, including numerous small textile mills.[50]

Some fifty miles southwest of the Blackstone Valley lay the Thames River, a twenty-mile-long navigable tidal river, formed by the junction of the Yantic and Shetucket rivers at Norwich, Connecticut. Its drainage basin of 1,450 square miles was three times the extent of the Blackstone's but with half the latter's horsepower per square mile.[51] With its affluent streams the Thames revealed clearly the dendritic pattern of streamflow and power distribution, as shown in Table 6.

In an area less than forty miles square we have a cross section of industrial America in the nineteenth century, lacking only heavy industry. The main strength of the Thames basin industry was cotton and woolen textiles, some 150 mills small and large. There was a wide scattering of metal and wood products, pulp and paper mills, and some machinery. Flouring mills, gristmills, and sawmills here as elsewhere led in numbers. By 1880 auxiliary steam power had been introduced widely, but nearly three-fourths of the motive power was supplied through the intricate system of millstreams flowing into a short tidal river. Not counting the smaller brooks, most of these streams ranged in length between fifteen and thirty miles; few were wider than 50 to 75 feet for any distance. Few, too, were without at least one upstream reservoir, and many had several of these artificial or converted ponds. Well over a score of industrial communities were supported by this power system. They ranged in size from Norwich, with Greeneville, its industrial suburb of about 15,000 population, to a number of towns with several thousand inhabitants or less, not counting the hamlets along the lesser streams such as the five-mile-long East Branch of the Yantic with "each mill clustering around it a little village of its operatives." The names themselves offer the reader little to conjure

[50] Tenth Census, *Water-Power,* pt. 1, pp. 61 ff., 68 ff. Similar but less striking examples on smaller streams may be cited in the various river basins of eastern New England and tributary to Long Island Sound, such as the Charles, Great Pond, North Branch of the Andover, Suncook, East Branch of the Willimantic, the Moodus, Whitestone Brook, and Hookanum (ibid., pp. 66 ff., 89 ff., 195 ff., 236 ff.).

[51] Ibid., pp. 187 ff. Much of the land had little agricultural value; the principal industry was manufacturing.

Table 6. Utilized waterpower in the Thames River drainage basin, 1880

	Number of mills	Total horsepower	Average horsepower
Thames River: union of Yantic and Shetucket rivers	0	0	0
Tributaries below Norwich			
Alewife Brook	4	90	22
Oxoboxo Brook	14	612	44
Others	21	763	36
Subtotal	39	1,465	38
Yantic River	11	2,025	184
Tributaries	8	155	20
Subtotal	19	2,180	115
Shetucket River: union of Natchaug and Willimantic rivers	8	4,230	529
Sundry small tributaries	28	1,139	41
Subtotal:	36	5,369	149
Natchaug and tributaries	52	1,312	25
Willimantic and tributaries	70	3,705	53
Quinebaug (principal tributary)	34	6,454	190
French River	44	3,585	81
Other tributaries	126	6,378	51
Subtotal	204	16,417	80
Total, Thames Basin	420	30,448	72

SOURCE: U.S., Census Office, Tenth Census, 1880, vols. 16–17, *Reports on the Water-Power of the United States* (Washington, D.C., 1885, 1887), pt. 1, pp. 187 ff.

with: Taftville, Hopeville, Fitchville, Eagleville, Bozraville, and Almyville, as well as Occum, Baltic, Willimantic, Scotland Station, Mansfield Hollow, Bean Hill, and Stafford Springs. Most of these communities by this date had convenient access to rail service; some teamed supplies and products a few miles at no heavy outlay.

The Thames basin was but a fragment, although a brilliant one, of a vast industrial panorama taking its physical pattern from nature's distribution of a centuries-old source of motive power, now on the eve of eclipse, and covering much of New England, many portions of the Middle Atlantic states, and occasional modest and scattered projections to the southward and west of the Appalachians. This middle period of American industrial advance, in round numbers, let us say, 1825-75, has been largely ignored except for those features directly and obviously related to the industrial economy that came to full flood in the closing years of the century. Yet the journeyman years represented by the widespread factory villages and the rudimentary stages of mechanization preceding the rise of the heavy metal and engineering industries have their own special character deserving of exploration and analysis.

The Merrimack, Connecticut, and Hudson River Valleys

The trunk-limb-branch, that is, dendritic, pattern of industrial location illustrated by the Thames River system was repeated in other parts of the Northeast, most notably in the Merrimack, Connecticut, and Hudson river basins. The Merrimack was described by Special Agent George F. Swain as "the most noted water power river in the world."[52] Forty years later it was called by a prominent engineer whose apprentice training was obtained there "the hardest worked river in the world."[53] In horsepower per square mile the Merrimack basin ranked third in the nation, approaching closely the level of concentration in the Thames basin but having an area more than three times as great. The Merrimack River system supplied some 900 mills with nearly 80,000 horsepower. The fame of this celebrated river rested primarily on the three leading textile cities situated at intervals of some fifteen to thirty miles along the lower half of the Merrimack's 110-mile southward and eastward course through New Hampshire and Massachusetts. The 29,000 horsepower, divided not too unequally between Lowell, Lawrence, and Manchester, marked the use of power on a very different order of magnitude from that characteristic of the vast majority of industrial establishments in New England. The three Merrimack River centers contained twenty-one companies, counting only the leading textile corporations, employing an average of more than 1,000 horsepower each. These companies had installed equipment representing more than 25,000 of the 29,000 water horsepower used in the three textile centers. Nearly 27,000 horsepower were in use on the fifteen leading tributaries of the Merrimack, but this total was divided among some 380 mills, giving an average of 70 horsepower. The tributaries of the direct tributaries and their affluents in turn supplied more than 500 mills with some 23,000 horsepower, an average rating of 45 horsepower. Several hundred

[52] Ibid., pp. 71–110.

[53] John R. Freeman, "The Fundamental Problems of Hydroelectric Development," *TASME* 45 (1923):529. The census survey tabulation for the Merrimack gives no totals, either for the system as a whole or for individual streams (pp.104–10). Totals cited are those of the author.

mills on the lesser streams, lumped under "other tributaries," had an average rating of 31 horsepower. In sum, the Merrimack basin, most celebrated of the "natural" systems for gathering and distributing power in the form of streamflow, provided average mill capacities ranging from some 1,200 horsepower for the great textile mills at Lowell, Lawrence, and Manchester on the river itself to some 30 horsepower for the miscellany of small industrial establishments located on minor tributaries.

In its 300-mile-long and forty- to fifty-mile-wide course to Long Island Sound through central New England, the Connecticut River system supplied 118,000 horsepower to 3,000 mills (see Table 7). It was simply the Merrimack writ large. The three largest powers on its trunkline—at Holyoke and Turners Falls, Massachusetts, and at Bellows Falls, Vermont—supplied 22 percent of the total power of the Connecticut River system, as compared with 37 percent for the three Merrimack River textile centers. The nine largest direct tributaries of the Connecticut River supplied about 50 percent of the aggregate number of mills and horsepower on the entire system. The average horsepower per mill in this group was 50 horsepower, almost identical with that of the entire system. For the 456 mills on the "sundry small tributaries" of the Connecticut the average rating was 31 horsepower, a little less than the average flour and grist- or sawmill in this region. Except for the "super textile mills" at Lowell and its peers on the Merrimack and a comparable dominance of paper mills at Holyoke, Turners Falls, and Bellows Falls on the Connecticut, these two leading power systems of New England served much the same kind and scale of manufactories.[54]

Historically the Hudson River owes its fame to its role as a great highway of trade, migration, and travel not only between the upper and lower valley within New York State but between the seaboard and the trans-Appalachian West by way of river and roads, the Erie Canal, and, in time, the New York Central and connecting railroads.[55] Between Albany and Troy on the main river and Buffalo on Lake Erie this highway followed the "water level" route provided by the Mohawk Valley through the

[54] Tenth Census, *Water-Power,* pt. 1, pp. 216–305. The tabulation on pp. 290–303 covers the utilized powers on the Connecticut and its tributaries.

[55] The chief source for the Hudson system is ibid., pp. 335–401. This section of the survey was prepared by Dwight Porter, instructor of civil engineering, Massachusetts Institute of Technology. Valuable contemporary sources include Spafford's *Gazetteer* (1824) and John French's *Gazetteer of the State of New York* (1861); Jones, *Annals . . . of Oneida County* (1851). Jones is particularly useful for waterpowers along headwater streams of the upper Mohawk Valley.

Table 7. Utilized waterpower on the Connecticut River and its tributaries

[With a few unimportant exceptions this table is based upon the returns of the census enumerators, and represents the power in use in 1880. The

	River.	COTTON-MILLS.			SILK-MILLS.			WOOLEN-MILLS. (*a*)			PAPER-MILLS.		
		Number of mills.	Water-power utilized.	Auxiliary steam-power.	Number of mills.	Water-power utilized.	Auxiliary steam-power.	Number of mills.	Water-power utilized.	Auxiliary steam-power.	Number of mills.	Water-power utilized.	Auxiliary steam-power.
			H. P.	*H. P.*		*H. P.*	*H. P.*		*H. P.*	*H. P.*		*H. P.*	*H. P.*
1	Connecticut river (main stream)	8	3,441	65	2	110		6	1,213		30	14,981	23
2	Salmon river and tributaries	13	683	85	1						3	140+	
3	Hockanum river and tributaries	11	725+	525+	2	77	758	10	1,125	635	16	1,499	797
4	Farmington river and tributaries	3	1,270	100	1	100		3	260		10	1,259	60
5	Scantic river and tributaries							7	333	420	1	42	
6	Mill river (Hampden county, Massachusetts)	2	65	25	1	75		1	60				
7	Westfield river and tributaries	3	350+					3	118	120	13	1,978	650
8	Chicopee river and tributaries	23	7,610	1,390				20	1,640	800	1	513	
9	Mill river (Hampshire county, Massachusetts)	2	185	125	3	100+	85	1	115	50	1		
10	Deerfield river and tributaries	4	571	405				1	50	40			
11	Miller's river and tributaries	4	187+		1	30		6	553	95	3	468	80
12	Ashuelot river and tributaries	4						12	956	315	2	244	
13	Sugar river and tributaries	1	300					5	410		3	255	50
14	Mascomy river and tributaries							1			1	50	33
15	Lower Ammonoosuc river and tributaries												
16	John's river												
17	Israel's river and tributaries							1	10		1	25	
18	Upper Ammonoosuc river and tributaries												
19	Mohawk river							1	12				
20	West river and tributaries							1	20				
21	Saxton's river							2	95	58			
22	Williams river and tributaries												
23	Black river and tributaries	2	121					5	340				
24	Ottaquechee river and tributaries							4	590	50			
25	White river and tributaries							3	67				
26	Ompomponoosuc river							1	40				
27	Wait's river and tributaries												
28	Wells river and tributaries							1	20		1	70	
29	Passumpsic river and tributaries							2	58		2	1,090	
30	Nulhegan river and tributaries												
31	Sundry small tributaries of Connecticut river	8	926	1,650	2	25	8	10	293	170	16	924	341
	Total, Connecticut river and all tributaries	88+	16,434+	4,370+	13	517+	851	107	8,378	2,753	104	23,538+	2,034

a Including also worsted-mills.

b Comprising blacksmithing shops, lock- and gun-smithing shops, brass and iron founderies, and establishments for the manufacture of agricultural implements, iron bolts and nuts, machinery, needles and pins, plated and britannia ware, pumps, saws, screws, scales and balances, sewing-machines and sewing-machine

c Comprising carpentering, cooperage, wheelwrighting, and wood turning and carving shops; planing-mills, and establishments for the manufacture of billiard carriages and sleds, coffins and undertakers' goods, excelsior, furniture, general house-furnishing goods, matches, models and patterns, picture-molding, piano

d Comprising bleaching and calendering, dyeing and cleaning, lithographing, marble and stone, calico printing, printing and publishing, soapstone, and wool and hose, boots and shoes, boot- and shoe-findings, bricks and tiles, brooms and brushes, buttons, carpet yarns, crashes, twines and bagging, cigars, cordage, drugs whetstones, hosiery, horse-blankets, kaolin and ground earths, leather board, linen, mattresses and spring beds, mosquito- and fly-nets, mucilage and paste, musical starch, stationery goods, tape, toys and games, upholstering materials, vinegar, whips and lashes, whip materials, wood-pulp, and wool extract.

e Power used mainly in the manufacture of wood-pulp.

NOTE.—In considering the results furnished by the above table, it should be borne in mind that while saw-mills stand first in number very small period, altogether, of the year. In reality, paper-mills far outrank all the other classes mentioned in the aggregate of water- Reckoning upon the same basis of working hours common among other mills, or, say, from ten to twelve hours, the total power utilized aggregate has since been increased in a very considerable degree, as may be seen, for example, from the accounts of Holyoke and Bellows Falls.]

Flouring- and grist-mills.			Saw-mills.			Various metal-working establishments. (b)			Various wood-working establishments. (c)			Sundry other establishments. (d)			Total.			
Number of mills.	Water-power utilized.	Auxiliary steam-power.	Number of mills.	Water-power utilized.	Auxiliary steam-power.	Number of mills.	Water-power utilized.	Auxiliary steam-power.	Number of mills.	Water-power utilized.	Auxiliary steam-power.	Number of mills.	Water-power utilized.	Auxiliary steam-power.	Number of mills.	Water-power utilized.	Auxiliary steam-power.	
	H. P.	H. P.		H. P.	H. P.		H. P.	H. P.		H. P.	H. P.		H. P.	H. P.		H. P.	H. P.	
5	373		6	845		22	1,628	170	6	110		13	665		98	23,366	258	1
4	71		6	139		10	154	40	5	245	30				42	1,432	155	2
4	130		4	95		2	18	8	2	38		3	135	6	54	3,842	2,729	3
15	432	8	53	1,429		54	3,110	990	26	679	35	13	313	110	178	8,852	1,303	4
7	310		6	125								2	312	240	23	1,122	660	5
2	100					2	255	100							8	555	125	6
10	393		29	771	65	13	226	5	26	504	8	33	380		130	4,720	848	7
28	910	150	68	2,045	60	13	674	77	17	263	60	12	949	45	182	14,604	2,582	8
2	67		1	33		5	407	300				6	154	95	21	1,061	655	9
20	528		58	1,896	20	6	559	30	22	660	60	6	88	30	117	4,352	585	10
13	462		58	2,194	80	17	662	57	76	2,863	810	10	153	6	188	7,572	1,128	11
5	158		48	1,310	45	8	139	2	32	814	99	10	173	45	121	3,794	506	12
6	689		26	836		5	135		15	367		8	184		69	3,176	50	13
7	320		17	599		6	120		8	321		8	166		48	1,576	33	14
5	228		18	890		2	145		11	343	40	9	920		45	2,526	40	15
			2	150	310				2	18					4	168	310	16
2	140		14	661	40	1	20		5	90		2	28		26	974	40	17
2	65		10	480	150	1	3								13	548	150	18
1	70		2	60		2	18		5	33	23	5	40		16	233	23	19
6	129		44	1,370	118				23	352	12	4	96		78	1,967	130	20
4	148		4	100					5	92		3	145	8	18	580	66	21
4	181		3	110	15				4	40		1	20	15	12	351	30	22
13	441		18	646		5	66	15	14	427		5	91		62	2,132	15	23
4	215		11	501		4	125	10	9	232		2	60	25	34	1,723	85	24
18	846		53	1,875	25	12	243		18	509		4	78	30	108	3,618	55	25
1	30		7	245					2	40	25	2	50		13	405	25	26
6	131		13	341		2	10+		8	82		1	32		30	596		27
4	168		10	345		1			1	25					18	628		28
14	713		37	1,348		9	367	175	14	569		7	e 2,840		85	6,985	175	29
			1	240											1	240		30
91	3,131	120	167	5,515	365	54	1,463	138	57	1,126	152	51	925	342	456	14,328	3,286	31
303	11,579	278	794	27,194	1,293	256	10,547+	2,117	413	10.842	1,354	220	8,997	997	2,298	118,026	16,047	

bells, bits and gimlets, brass-ware, bronze statuary, clocks, coffin-trimmings, cutlery and edge-tools, files, fire-arms, general hardware, hooks and eyes, iron forgings, materials, springs, steam fitting and heating apparatus, stencils and brands, swords, tin-, copper-, and sheet-iron ware, watch and clock materials, wire, and wirework.
and bagatelle tables, cues, and materials, cigar- and packing-boxes, bobbins, carriages and wagons, carriage and wagon materials, chairs, chair-stock, children's materials, rules, sashes, doors and blinds, shoe-pegs, spools, washing-machines and clothes-wringers, wheelbarrows, wooden handles, and wooden ware.
grading and scouring works; tanneries, watch- and clock-repairing shops, and establishments for the manufacture of baskets, rattan- and willow-ware, leather belting and chemicals, emery-wheels, explosives and fireworks, fancy and paper boxes and other fancy articles, fertilizers, gloves and mittens, gunpowder, hones and instruments and materials, patent medicines and compounds, preserves and sauces, rubber and elastic goods, shoddy, soap and candles, spectacles, sporting goods,

and in the aggregate horse-power of wheels, a large proportion of them are not operated continuously, while many run during only a power actually employed, since, under ordinary circumstances, they are run day and night and continuously through the year. by paper-mills would probably correspond, in round numbers, to at least from 40,000 to 50,000 horse-power.

SOURCE: Tenth Census, *Water-Power,* pt. 1, pp. 304-5.

Appalachian barrier. Although in utilized power the Hudson River basin approached the Merrimack in importance, a drainage area nearly three times as large resulted in an average horsepower per square mile but one-third that of the New England river. Its largest developed power in 1880, the impressive Cohoes Falls on the Mohawk River a short distance above its entrance to the Hudson, measured some 6,500 horsepower and was outranked by at least a half-dozen powers in New England, including the Merrimack Valley textile centers and Holyoke and Bellows Falls on the Connecticut. Apart from trunkline powers of such magnitude, the Hudson basin shows a marked similarity to the river systems of New England in respect to average horsepower per mill on the main tributaries and those at second and third remove. Yet the Hudson provides a useful reminder that New England was not the only important industrial region to rise to prominence on the basis of waterpower and that terrain so inhospitable as to discourage farming and drive ambitious men to distant frontiers or the cities was not a necessary condition of industrial development. The greater part of the Hudson drainage basin, excepting chiefly the Adirondacks, was marked by a relatively high density of population, a prosperous agriculture, and a flourishing commercial and industrial life—in short, a more balanced diversification of the economy than was to be found in New England.[56] To a fuller extent, too, than in any other drainage basin outside of New England, the Hudson basin displays to advantage the role of stream flow as a system of power supply. This aspect was reflected not only in the large number of millstreams and water powers distributed throughout the river basin but in the wide range of capacities adapted to the needs of most kinds and classes of industrial establishments.

Table 8 condenses and rearranges the tabular data in the 1880 census survey and distinguishes between trunkline or main river and primary and secondary tributaries of the Hudson River system. The primary tributaries are designated as "major" or "minor" if they show totals of

[56] See description in Tenth Census, *Water-Power,* pt. 1, pp. 343–45. Of a tributary of the upper Hudson, Pomroy Jones wrote: "Fish Creek has a course of ten miles in this town [Annsville] with a fall of from thirty to one hundred feet per mile. Indeed, so far as fall is concerned it is believed that its whole waters can be used every hundred yards in that distance. Its tributaries furnish almost as much power as the main stream." Although hardly one-twentieth of the power was reportedly in use, it supplied the following establishments: 2 flour and grist mills, 21 sawmills, 12 shingle mills, 4 lath and 2 stave mills, 4 turning lathes, 1 woolen manufactory, 1 blast and 2 cupola furnaces, 2 tanneries. Other machinery was being erected (Jones, *Annals . . . of Oneida County,* pp. 66–67). For Sauquoit Creek, see ibid., pp. 299 ff.; *Scientific American* 8 (30 Apr. 1853):261.

Table 8. Utilized waterpower in the Hudson River drainage basin, 1880

	Primary tributaries			Secondary tributaries		
	No. of mills	Total HP	Average HP	No. of mills	Total HP	Average HP
Hudson River trunkline	47	11,445	244			
Major tributaries of the Hudson						
Rondout Creek and tributaries	137	4,098	30			
Kinderhook Creek	14	1,352	34	86	2,953	34
Hoosic River	47	6,371	136	86	3,292	38
Subtotal	198	11,821	60	172	6,245	36
Minor tributaries of the Hudson						
Croton River and tributaries	34	825	24			
Murderer's Creek and tributaries	14	766	55			
Jansen's Kill and tributaries	15	455	30			
Poesten's Kill and tributaries	27	1,638	61			
Schroon River and tributaries	28	1,037	37			
Fishkill Creek	12	1,813	151	9	239	26
Wappinger Creek	17	666	39	17	503	30
Batten Kill	35	1,834	52	45	1,240	28
Sacondaga River	8	758	95	79	2,041	26
Fish Creek	7	1,445	206	42	2,278	54
Esopus Creek	29	1,480	51	33	897	27
Catskill Creek	14	681	49	54	1,219	23
Subtotal	240	13,398	56	279	8,417	30
All other tributaries to the Hudson	189	6,454	34	451	14,662	33
Total	674	43,118	64			
Mohawk River trunkline	86	9,157	106			
Tributaries of the Mohawk						
Oriskany Creek and tributaries	25	1,301	52			
Schoharie Creek and tributaries	126	3,424	27			
Cayadutta Creek and tributaries	23	782	34	2	80	40
East Canada Creek	11	440	40	31	1,019	33
West Canada Creek	19	708	37	48	872	18
Sauquoit Creek	18	1,954	109	7	56	8
Subtotal	222	8,609	39	88	2,027	23
Minor tributaries of the Mohawk	150	5,162	34			
Total	458	22,928	50			

Table 8. (cont.)

	Primary tributaries			Secondary tributaries		
	No. of mills	Total HP	Average HP	No. of mills	Total HP	Average HP
Totals for Mohawk and Hudson	1,132	66,046	58	539	16,689	31
Grand total of primary and secondary tributaries	1,671	82,735	50			

SOURCE: U.S., Census Office, Tenth Census, 1880, vols. 16–17, *Reports on the Water-Power of the United States* (Washington, D.C., 1885, 1887), pt. 1, pp. 342–99.

more or less than 4,000 horsepower. The Mohawk River is freed from its tributary status to the Hudson and for convenience is treated as an independent system, which in effect it virtually was. The trunk-limb-branch pattern of streams and mills represented in this table differs from that illustrated by the Thames River, and on a much larger scale by the Connecticut River, chiefly in the somewhat wider dispersion of manufactures and mills, the smaller size of the industrial communities, and the less fully occupied and developed condition of the millstreams. As we move from the river trunklines to the primary and secondary tributaries—those at still further remove are merged and submerged in such phrases as "and tributaries," "Minor," and "all other" tributaries—there is a progressive reduction in the average size of mill as measured by the horsepower employed. Throughout the Hudson Basin the range of choice was heavily weighted on the side of the small powers, as the accompanying tabulation makes clear.

	No. of streams	*No. of mills*	*Total HP in use*	*Average HP per mill*
100 HP and over	6	217	30,565	141
50–99 HP	8	194	11,407	59
Under 50 HP	24	1,260	40,729	32
Total	38	1,671	82,701	—

Three-fourths of the developed waterpowers in the basin, but only one-half of the aggregate horsepower, were concentrated in establishments requiring no more power than the typical flour, grist-, or sawmills of the time. This largest class included nearly one thousand such mills, with the remainder accounted for by several hundred metal- and wood-working establishments and tanneries. Cotton, woolen, and paper mills accounted for the greater part of the power used on streams averaging more than 100 horsepower per mill. The five streams supplying much the greater part of capacity in these three branches of industry were the Mohawk and Hudson rivers and Kinderhook Creek for textiles, and the Hudson, Mohawk, and Hoosic rivers and Kinderhook and Fish creeks for paper.

Regional Variations in Waterpower Supply and Use

In 1880 the industrial regions of the Northeast stood well apart from the remainder of the country in the active development and use of waterpower. The lesser resources in waterpower elsewhere are only part of the explanation. Even at this high plateau in the use of direct-drive waterpower in the United States, its employment throughout the South and in the Northeast's extension into the trans-Appalachian West could not begin to bear comparison with its use in the states of the north Atlantic seaboard. In utilized waterpower per square mile the leading river basins from Maryland to Florida rarely exceeded one horsepower, compared with 5 to 40 horsepower for comparable river basins to the north. More significantly, the percentage of waterpower used in flour, grist-, and sawmills—61 percent of the total waterpower in all manufacturing in 1880—was far higher in the South and in the trans-Appalachian West than in the Northeast. In most of the leading river basins between Pennsylvania and Maine these traditional bedrock industries employed between 15 and 50 percent of the aggregate waterpower in use. On the leading streams of the south Atlantic seaboard the range was from 73 to 93 percent. In the Ohio Valley only two minor streams, the Big Beaver River in western Pennsylvania and the Miami River in southern Ohio employed appreciably less than 90 percent of their utilized power—74 percent and 66 percent, respectively—in these basic industries.[57]

Throughout the vast Mississippi River basin, in regions as yet little penetrated by manufacturing, 90 percent or more of the available waterpower was employed in flour, grist-, and sawmills. The use of the rich waterpower resources of the upper Mississippi Valley, especially in Minnesota and Wisconsin, awaited the coming of hydroelectricity in

[57] See Tenth Census, *Water-Power,* pt. 1, for reports on the Middle and South Atlantic "watersheds" and the Eastern Gulf "slope," and pt. 2 for reports on the major drainage basins of the Mississippi Valley. Many useful tables appear in the "General Introduction," pt. 1. The computations of horsepower in this and the following paragraphs are mine, since summary tables by industry and stream are not given for the drainage basins here considered.

the years ahead. A few industrial communities on the Rock and Wisconsin rivers, chiefly in Illinois and within the economic orbit of Chicago, were sustained by waterpower. A beginning had been made in industrial exploitation of the abundant power of the Lower Fox River in eastern Wisconsin, in 1880, chiefly at Appleton. In the upper Mississippi Valley, the Falls of Saint Anthony at Minneapolis stood out like a monadnock above a plain. With some 13,000 horsepower in use in 1880, the falls at Minneapolis led even the largest developed powers in New England in size, providing the power base for the mass conversion of the Northwest's leading products, timber and wheat.

The limitations of river systems as sources of power supply were clearly revealed by the 1880 survey. With exceptions principally in some of the more congested industrial areas of southern New England, we find a combination of underutilization and shortages of power on the same streams. On the one hand attractive waterpowers often remained undeveloped, and developed powers were not used to fullest advantage through carelessness, ignorance, or considerations of cost. Yet there were relatively few streams on which the supply of water was adequate so far as reliable year-round operations were concerned. The extent to which auxiliary steam power was necessary to relieve this condition varied from stream to stream, as also did the readiness of mill and factory owners to install steam capacity to meet seasonal deficiencies. In 1880 auxiliary steam power amounted to about 12 percent of the aggregate power used on the Connecticut River system and to 20 percent on the Thames and Hudson river systems.[58]

By 1880 the tide had turned and was moving fast against the traditional source of motive power. During the 1880s steam power was to provide almost the entire basis for the extraordinary industrial expansion taking place in the United States. The wide distribution of small- to medium-scale powers, which continued to be the chief source of waterpower's strength, was entirely inadequate, for reasons to be discussed elsewhere, to the requirements of an industrial production increasingly concentrated in urban communities and ever larger industrial establishments.[59] The innumerable hamlets "gathering around the water fall which serves to turn the mill wheel" upon which Zachariah Allen had remarked in 1829 had in the intervening half century grown frequently into large and vigorous villages. Yet by 1880 the days of the flourishing industrial village and of streams that were a continuous succes-

[58] Tenth Census, *Water-Power,* pt. 1, pp. 214–16, 401.

[59] Adna F. Weber, *The Growth of Cities in the Nineteenth Century* (New York, 1899), pp. 187 ff.

sion of mills and mill villages were numbered. During the eighties the increase in waterpower in manufactures, nationwide, slowed to a virtual standstill; in these and succeeding years the decline of villages generally in this country became a matter of wide comment and some social concern. Despite the mounting evidence, the deeply rooted belief that manufactures and waterpower were inseparable companions was slow to yield; trends clear in retrospect were discernible to few at the time. Absorbed in their demanding tasks, the civil engineers engaged in the 1880 survey continued to envisage industrial growth in the traditional terms. Reporting on the waterpowers of the south Atlantic watershed and referring specifically to the Santee River basin, George F. Swain wrote with enthusiasm of the prospects of the region:

> It would be difficult to select another stream of equal drainage area which can offer so large a number of excellent powers, from the smallest to the largest. From the great falls of the Catawba, with a fall of 173 feet, to the numberless fine small powers on the smaller streams in western South Carolina, the range is large, and offers powers of all scales of magnitude; and as the manufacturing interest in the South develops, there is no doubt that many of the fine powers now lying idle will be turned to account. Hand in hand with this development will go the construction of railroads, until the southern streams become, like many of the northern ones, a succession of mill-ponds, with all kinds of manufactures on their banks, and the country becomes threaded with a network of railroads and studded with factory villages.[60]

Waterpower was to play a large role in the industrial development of the South, but this did not come until many years later and through the quite different agency of hydroelectricity.

[60] Tenth Census, *Water-Power,* pt. 1, p. 782; see also, pp. 186–87.

5
The First Industrial Cities

IN THE quarter century following the War of 1812 industrialization in New England based on waterpower entered upon a far more advanced and sophisticated stage than hitherto described. The new stage had its beginnings in Massachusetts, spreading in time to the southern portions of New Hampshire and Maine. Its origins lay in the commercial and financial strength of Boston and in the shift of capital and entrepreneurial energy from overseas commerce to manufacturing, above all to textiles, especially cotton. In the process waterpower emerged from a craft shell going back to ancient times and within two decades became a branch of engineering in which Americans led the world.

The exploitation of large-scale privileges of several thousand or more horsepower located within practicable transport distance of the larger seaports between Boston and Portland was favored by a combination of factors. A new and highly profitable method of conducting the manufacture of textiles led quickly to a scale of operations and a demand for powers of a capacity much greater than those hitherto in use.[1] In addition there was a growing awareness of the speculative and investment opportunities offered by well-situated powers of large capacity. The striking success of the pioneer textile complexes at Waltham on the Charles River and Lowell on the Merrimack, within ten and twenty-five miles of Boston, not only led to similar projects elsewhere but resulted eventually in the commercial development of large waterpowers not for use by the projectors but for lease or sale. By mid-century companies engaged in the development and sale of waterpower, usually

[1] The outstanding monographs on the industrial developments discussed here are Caroline F. Ware, *The Early New England Cotton Manufacture in the United States* (Boston, 1931); George S. Gibb, *The Saco-Lowell Shops: Textile Machinery Building in New England, 1813–1949* (Cambridge, Mass., 1950); Constance McLaughlin Green, *Holyoke, Massachusetts: A Case Study of the Industrial Revolution in America* (New Haven, 1939); Margaret T. Parker, *Lowell: A Study of Industrial Development* (New York, 1940); Vera Shlakman, *The Economic History of a Factory Town: Chicopee* (Northampton, Mass., 1963). A valuable account is Edward Stanwood, "Cotton Manufacture in New England," in William T. Davis, ed., *The New England States,* 4 vols. (Boston, 1897), 1:117–87. See also Committee of the American Society of Civil Engineers, "American Engineering as Illustrated at the Paris Exposition of 1878," *TASCE* 7 (1878): 370–71.

providing building sites as well, had become a distinctive part of the American industrial scene.[2] Such projects often provided speculative and investment opportunities, but in numerous instances they received community encouragement and support as a means of promoting industrial growth and urban prosperity.

Breaking New Ground at Lowell

The founding of Lowell in 1822 inaugurated the new phase in the history of industrial power. Lowell not only established a new scale and concentration of factory production but it pioneered in the design and execution of a waterpower on a scale that dwarfed all previous enterprises. The exploitation of water privileges, heretofore carried out on creeks and streams with the rule-of-thumb methods of the millwright, here demanded engineering competence of a high order and a heavy investment of capital. The massive structures of timber and masonry required to control the powerful hydraulic forces of a river at flood dwarfed the largest of earlier projects of the same kind. In strictly hydraulic terms the construction of the Erie Canal, a major training school for civil and hydraulic engineers at the time, was, save in sheer magnitude, a relatively minor venture. The engineering and business success at Lowell from its relatively modest beginnings in the early twenties—full development was spread over some thirty years—led to the creation in New England during the next generation of new and wholly industrial towns, engaged in the manufacture of textiles, together with supply and service industries. Lowell and its brilliantly successful prototype, Waltham, paved the way for Manchester, Nashua, and Lawrence, Biddeford and Saco, Chicopee and Holyoke. The 1880 federal census survey reported that New England possessed "at least ten developed powers of 10,000 theoretical horsepower or over during working hours . . . at least eighteen of over 3,000 horsepower continuously, and at about twenty of over 2,000 horsepower continuously."[3] The development and management of most of these larger powers was influenced by, if not patterned upon, the Lowell prototype. Within a generation the cotton

[2] See U.S., Census Office, Tenth Census, 1880, vols. 16–17, *Reports on the Water-Power of the United States* (Washington, D.C., 1885, 1887), pt. 1, pp. xxv–xxxvii, for a list of and data on large waterpower companies. See also George F. Swain, "Statistics of Water Power Employed in Manufacturing in the U.S.," *PASA*, n.s. 1 (1888–89), esp. table 13.

[3] Tenth Census, *Water-Power*, pt. 1, p. xxx.

textile manufacture was transformed from an industry in which the typical mill produced yarn only with a thousand spindles or less to one in which the largest establishments contained 25,000 spindles or more and performed all operations, starting with the bales of raw cotton and ending with cloth ready for finishing or use. The striking success of the textile mills at Lowell was felt throughout the country, kindling the hopes of large power-site owners and townsite promoters.

Industrial ventures of the magnitude of the new textile centers required a power base vastly greater in size and complexity than any previously known in this country or in Europe. They called for hydraulic works conceived on a truly magnificent scale and carried out with little guidance from the experience of the past. There was inevitably much improvisation at Lowell and the recasting of major facilities as the demands for power grew beyond all anticipations. There were engineering blunders, as at Holyoke when the main dam gave way within hours of its completion, and near disasters, as when carelessness led to the undermining of the Falls of Saint Anthony and the threat of a complete washout.[4] At a time when it was unusual to harness more than 100 horsepower in a single installation, projects were initiated for the control of powers with capacities measured in thousands of horsepower. In place of the traditional millstream, often no more than a brook or creek, rivers hundreds of feet wide, with volumes of flow measured in tens of thousands of cubic feet per second, had to be harnessed to the tasks of industrial production within great mill structures. Such in New England were the Saco, Kennebec, Androscoggin, Salmon Falls, and, above all, the Merrimack and Connecticut rivers. The country millwright's methods of calculation and construction, which had hitherto sufficed, were inadequate to the new conditions and requirements. They gave way to the techniques of applied science and engineering as these were gradually developed by an evolving profession of civil and hydraulic engineers. Large rivers, seasonally swollen by floods, had to be brought under control by dams and guard gates, channeled through millraces larger than navigation canals to the penstocks and waterwheels of mill and factory, and eventually returned, all but empty of energy, to the riverbed below. To secure adequate space for the factories, mill yards, roads, and other supporting facilities of a large industrial community called for extensive systems of hydraulic "canals," constructed often in two and sometimes three levels, equipped with regulating gates and wasteways, proportioned in dimensions, profile, and

[4] On the Holyoke dam, see ibid., p. 221; the near disaster at the Falls of Saint Anthony is discussed below, pp. 238–39.

slope to the volume of water to be handled and the requirements of the mills and machinery to be served. In a large hydraulic installation of this kind efficiency of water use was hardly less dependent upon the design and construction of these great raceways as on the waterwheels by which streamflow was converted to power on the wheel shafts. The distributive canal network at Lowell, probably the most extensive of its kind, had an aggregate length of about five miles.[5]

Engineering and Management Tasks

In addition to the initial design and construction of such major waterpowers there were the continuing engineering and management tasks of maintaining, operating, and, with time, enlarging the hydraulic facilities to meet growing demands. These responsibilities included adjusting the at times conflicting requirements of different power users, of devising methods and equipment for measuring and monitoring the use of water, of discouraging its waste, and, if need be, of rationing its use during seasonal shortages and unseasonal droughts. In growing industrial communities there arose in time the pressing problem of meeting ever greater power requirements both by eliminating waste and by improving efficiency in every component of the system. The latter included particularly the penstocks, gates and waterwheels of the millowners, and the components of the system under the direct control of the power corporation: dams, canals, millpond, and the like. In important instances, too, the full development of a waterpower required the extension of engineering control from the immediate hydraulic works to the supporting river system itself, chiefly by the establishment and management of storage reservoirs on its upper waters. In the creation of these early nineteenth-century "new towns" the provision of housing and related facilities was usually undertaken or assisted by the corporations developing the waterpower, directly, or, in effect, as agents of the mills. Mill machinery apart, which was commonly produced in a "machine shop" established for the purpose, the establish-

[5] Accounts of the hydraulic facilities of the larger developed waterpowers as of 1880 appear ibid., pts. 1 and 2. For Lowell, see pt. 1, pp. 78–83. Only Holyoke developed a canal power system approaching that of Lowell. The Holyoke canals were arranged on three levels and were extended in length as required. The total length in 1880 was 3.5 miles; later the total reached 4.5 miles. The second level in 1880 was 9,100 feet long (ibid., p. 222).

ment of supply, equipment, and service industries was usually left to independent and typically small enterprise.

The development of large-scale, direct-drive waterpowers—as distinguished from their hydroelectric successors from the 1890s on—had no counterpart in the industrial countries of western Europe. Only one can be cited for Great Britain, that at Greenock, near Glasgow. The few installations on the Continent, chiefly in Switzerland and adjacent portions of France, were of too small a capacity to bear comparison, although they were imaginative in design and development and broadly social in their utility.[6] In the United States, with one major exception at Minneapolis on the upper Mississippi, the largest class of such waterpowers was confined to the region east of the Appalachians.

Some thirty years before Lowell a similar venture on a far less ambitious scale had been undertaken at the site of Paterson, New Jersey. It was backed by prominent men of affairs, chiefly from New York City, and led by Alexander Hamilton. The Society for the Establishment of Useful Manufactures (S.U.M.), launched in 1791 with a corporate charter of wide and liberal powers, was a bold attempt to combine an industrial project of national importance with the pursuit of private profit. The aim was the establishment of a large industrial community, centering in cotton textiles and based upon waterpower to be developed at the Great Falls of the Passaic River, with its sixty-five-foot drop and 2,000-horsepower potential. The celebrated William Duer was the first general manager of S.U.M., and no less a designer and engineer than Pierre L'Enfant was engaged to plan the town and direct construction of the hydraulic facilities. Persons of some prominence were placed in other positions of responsibility, and men skilled in the manufacture of cotton were somehow gotten from England. A beginning was made in attracting other enterprises to the site. But the project

[6] See William C. Unwin, *On the Development and Transmission of Power from Central Stations* (London, 1894), a valuable series of lectures by a leading British engineer. Unwin stated: "The method in which water power is distributed in America to a number of consumers is almost peculiar to that country. . . . the water itself is distributed in canals to consumers, at a level permitting the creation of a waterfall at the mil, or factory" (pp. 80 ff.). He mentioned only four waterpower companies of any size in Europe, those at Schaffhausen, Freiburg, Zurich, and Bellegarde. These powers were exceeded in capacity, often manyfold, by at least a score in the United States. See also the brief summary of American waterpower development in "Address of James Bicheno Francis, President, ASCE," *TASCE* 10 (1881):187 ff. On the imaginative and skillfully engineered Shaw's Waterworks of Greenock, Scotland, see p. 00, and Loammi Baldwin, *Report on the Subject of Introducing Pure Water into the City of Boston* (Boston, 1834), pp. 18–20; and letter of James B. Francis, 24 Nov. 1849, in *Hunt's Merchants' Magazine* 22 (1850):32.

proved overambitious, and within five years of S.U.M.'s creation it was virtually abandoned. The site was well chosen, the plan comprehensive, the goal by no means visionary, the waterpower more than adequate. But insufficient capital, incompetence and irresponsibility among the directing personnel, and, most important, the want of practical industrial and business experience among the leaders of the enterprise brought its early collapse. In time industrial Paterson arose from the wreckage of the early plans, with S.U.M. continuing in the limited role of a somewhat litigious landlord and lessor of water rights. Paterson's excellent waterpower, on which the manufacturing of the community depended almost wholly before 1860, provided the basis for a solid if, for some years, unspectacular growth.[7]

It remained for Lowell, Massachusetts, to demonstrate first the potentialities of a waterpower capable of supporting a truly large industrial center. But Lowell's success rested on far more than a magnificent power site, with a capacity of some 10,000 horsepower, at Pawtucket Falls on the lower Merrimack River with five times the capacity of the Passaic Falls selected by S.U.M. It was a product not only of the imagination and boldness of its promoters but of their large financial resources and extended business experience. Lowell's success rested also, as the S.U.M. enterprise could not, upon a generation of operating experience in the making of cotton textiles, dating from the pioneering work of Almy, Slater, and Brown in Rhode Island during the 1790s and culminating in the striking success of the Boston Manufacturing Company at Waltham two decades later. Indeed, the Lowell enterprise was in fact virtually an extension and enlargement of the Waltham enterprise, made necessary by the exhaustion of the latter's power base. Lowell drew from Waltham technical, engineering, and managerial talent and the right to use the spinning and weaving machinery developed so successfully there. It took its name as well; Francis Lowell, member of a wealthy Boston mercantile family, was the gifted leader of the Waltham Company. Although almost wholly mercantile in business experience, the score or more of Boston and associated capitalists who backed the Lowell

[7] On S.U.M. and beginnings of Paterson, see Joseph S. Davis, *Essays in the Earlier History of American Corporations,* 2 vols. (Cambridge, Mass., 1917), 1:essay 3; *Scientific American,* n.s. 1 (29 Oct. and 5, 12, 19 Nov. 1859); James B. Kenyon, *Industrial Location and Metropolitan Growth: The Paterson-Passaic District* (Chicago, 1960), esp. pp. 26 ff, 51 ff. On revival of the dormant S.U.M. in 1814 and restriction of its activities to management of real estate properties and waterpower rights, see *John Kean* v. *R. N. Colt et al., New Jersey Equity Reports,* 5:365 ff.; also *S.U.M.* v. *Morris Canal Company,* ibid., 1:157. For the Passaic waterpower, its development and the establishments using the power in 1880, see Tenth Census, *Water-Power,* pt. 1, pp. 647–50.

project and the later ventures patterned upon it displayed a singular gift for dealing with the technical as well as the financial and management problems and for selecting men of unusual managerial and engineering talents. The capabilities of these men were hardly less marked in the development of the power bases of the industrial communities created than in the direction of manufacturing operations proper.[8]

The magnitude of the operations of this group, sometimes termed the Boston Associates, is suggested by the fact that over a period of some twenty-five years they organized and placed in operation some thirty large textile and associated industrial enterprises, most of them with the very large initial capital for the time of some half-million dollars. These establishments were concentrated in a half-dozen major industrial communities for whose existence these mercantile capitalists were chiefly responsible: Lowell, Manchester, Lawrence, Saco-Biddeford, Chicopee, and Holyoke. The last two were in the upper Connecticut Valley; one adjoined two communities of coastal Maine; the remaining three were located in the lower valley of what came to be termed, in anticipation of the age of commercial hyperbole ahead, the most noted waterpower stream in the world.[9]

The amount of power required to support manufacturing operations on so unprecedented a scale is suggested by the energy unit that quite early became the standard, not only at these industrial centers but widely throughout the textile industry. Replacing the conventional horsepower, which originated with James Watt and was the equivalent of the sustained labor of eight to ten men, this new unit was the "mill power," the amount of power required to drive a completely integrated cotton mill of the kind pioneered at Waltham. It was based on the power required to drive the 3,584 spindles, together with preparatory machinery and looms for weaving the spun yarn, of the second mill erected

[8] Ware, *Early New England Cotton Manufacture,* and Gibb, *Saco-Lowell Shops,* are the best sources for the Waltham enterprise and its relation to the Lowell project which followed. For a brief but excellent biographical account of Francis Cabot Lowell (1775–1817) see *Dictionary of American Biography.* See also William R. Bagnall, *Sketches of the Manufacturing Establishments in New York City, and of Textile Establishments in the Eastern States,* ed. Victor S. Clark, 4 vols. (Washington, D.C., 1908), 3:2150–51; *Correspondence between Nathan Appleton and John A. Lowell in Relation to the Early History of the City of Lowell* (Boston, 1848). For descriptions of the early textile processes, machinery, and equipment, a still useful account is Carroll D. Wright, *The Industrial Evolution of the United States* (Meadville, Pa., 1895), chaps. 3, 4, and 10.

[9] Tenth Census, *Water-Power,* pt. 1, p. 71. See also table in Shlakman, *Economic History of a Factory Town,* pp. 39–42.

by the Boston Manufacturing Company at Waltham in 1817, one of the largest mills of the time. Although the definition of the mill power in terms of spindles and appurtenant machinery long remained in nominal use, at a very early date a far more precise definition was substituted, describing a mill power as twenty-five cubic feet of water per second falling thirty feet, but with the amount of water varying according to the actual distance of the fall, whether more or less than thirty feet. By what was long termed the "Lowell standard," the mill power was defined as equal to 62.5 net or 85.2 gross horsepower. By the mid-1880s Superintendent James B. Francis of the Locks and Canals company at Lowell estimated the mill power, with the turbines of superior efficiency then in use, at about 68 horsepower.[10]

The waterpower at Lowell, as developed to 1845, was estimated at 90 mill powers, available the year round, an amount increased by later improvements in the hydraulic facilities to some 140 mill powers, equivalent with waterwheels of the types in use here at mid-century to about 9,000 horsepower. Table 9 provides additional data for 1882. In time, as we shall see in chapter 6, there became available during the greater part of the year—at rates varying with management policy for discouraging or encouraging use—"surplus" power amounting to much more than the foregoing "permanent" power.[11]

Power Supply: Evolution of Control

The founding of Lowell began in 1821 with the selection of its site at Pawtucket Falls on the lower Merrimack, slightly over twenty miles from Boston, with which there was direct communication via the Middlesex Canal, completed a generation earlier. Here the river fell over thirty feet in a little over a mile. The promoters bought out the Proprietors

[10] JBF to O. Guthrie, 2 Sept. 1886, LCP, DF-1; Maine, Hydrographic Survey, *The Water-Power of Maine, by Walter Wells* (Augusta, 1869), p. 155. In his extensive correspondence with textile manufacturers and engineers in other parts of the United States, Francis made frequent references to the mill power, with explanations of its origin. See also Nathan Appleton, *The Introduction of the Power Loom, and Origin of Lowell* (Lowell, 1858), p. 28; William Kent, *The Mechanical Engineers' Pocket-Book,* 9th ed. (New York, 1916), p. 766.

[11] JBF to T. G. Carey, 4 Dec. 1858, LCP, citing figures for 1845. The power owned by the several Lowell mills in 1882 totaled over 139 mill powers, equivalent to 11,845 gross horsepower. In the early 1880s the amount of surplus water used, flood- and backwater conditions excepted, ranged from about 30 to 40 percent (Tenth Census, *Water-Power,* pt. 1, pp. 81–82).

Table 9. Mills and power at Lowell, Massachusetts, 1882

Name of corporation.	Designation on map.	Goods manufactured.	Number of spindles. (a)	Number of looms. (a)	Water taken from—	Water discharged into—	Head and fall in feet. (b)	Power owned.		Steam-power in horse-power. (a)	Quantity of water in cubic feet per second.	No. shares in Proprietors of Locks and Canals.
								Mill-powers.	Gross horse-power.			
Merrimack Manufacturing Company.	A.	Cotton	153,552	4,267	Merrimack canal	Merrimack river	30 (33.5)	$24\frac{20}{30}$	2,097	6,000	616.667	740
Hamilton Manufacturing Company.	B.	do	59,816	1,597	Hamilton canal	Lower level of Pawtucket canal.	13 (14)	16	1,360	1,200	968.000	480
Appleton Company	C.	do	45,000	1,228	do	do	13 (14)	$8\frac{16}{30}$	725	750	516.267	256
Lowell Manufacturing Company	D.	Cotton and woolen	22,750 (worsted and wool). 2,000 (cotton).	317 (power carpet). 75 (lasting).	Merrimack canal	do	13 (14)	$8\frac{12}{30}$	714	1,040	508.200	252
Lowell Machine Shop (c)	E.				Merrimack canal and Machine-shop basin.	do	13 (14)	$3\frac{9}{30}$	280	375	199.650	99
Middlesex Company	F.	Woolen	18,640	250 (broad).	Lower level of Pawtucket canal.	Concord river	17 (17)	$5\frac{23}{30}$	490	125	262.383	173
Boott Cotton Mills	G.	Cotton	127,000	3,600	Eastern canal	Merrimack river	17 (19)	$17\frac{26}{30}$	1,519	1,000	812.933	536
Massachusetts Cotton Mills	H.	do	119,528	3,658	do	Merrimack and Concord rivers.	17 (19)	$24\frac{16}{30}$	2,085	950	1116.267	736
Tremont and Suffolk Mills.	I.	do	94,000	2,700	Northern canal	Lawrence basin	13 (14)	13	1,105	1,500	786.500	390
Lawrence Manufacturing Company.	J.	do	100,000	2,360	Lawrence basin	Merrimack river	17 (21)	$17\frac{9}{30}$	1,470	1,000	787,150	519
Total			742,286	20,052				$139\frac{11}{30}$	11,845	13,940		4,181

a These data are from *Annual Statistics of Manufactures in Lowell and Neighboring Towns, January*, 1882. Published by the Lowell *Vox Populi.*
b Figures in parentheses are actual falls (see page 32).
c This establishment consumes per year 1,160 tons wrought-iron, 8,500 tons cast-iron, and 200 tons steel—in all 9,800 tons of metal.

SOURCE: U.S., Census Office, Tenth Census, 1880, vols. 16–17, *Reports on the Water-Power of the United States* (Washington, D.C., 1885, 1887), pt. 1, p. 81.

of the Locks and Canals on the Merrimack River (incorporated 1792), the company owning and operating the navigation canal joining the river above and below the falls, and proceeded quietly to purchase some 700 acres of land adjacent to the river, thereby completing their control of the waterpower at the falls. Transferring these properties to the Merrimack Manufacturing Company, incorporated in 1822, the promoters proceeded with such vigor to advance their plans that a temporary dam and the necessary enlargement and extension of the navigation canal, to convey water to the mills, were completed and the first cotton mill equipped and placed in motion in September 1823.[12] Additional mills were soon under construction and in succeeding years

[12] In addition to the accounts of the founding of Lowell by Ware and Gibb, see memorandum headed "Pawtucket Canal," presumably by JBF, in LCP, DB-8, 392–94. Other relevant JBF correspondence is found in DB-3, 233 ff., 565 ff.; DB-7, 445 ff.; DB-9, 474. See also the letters of James F. Baldwin and Timo. Williams to L. Baldwin, 7, 10 Dec. 1821, Loammi Baldwin Papers, John Crerar Library, Chicago. In an account of the Merrimack Manufacturing Company, *JFI* 2 (1826):365–66, citing the Boston *Commercial Gazette,* stated that five mills with 4,000 spindles each had been erected; three supplied with proper machinery were in operation.

new corporations were organized to erect and operate still others. Between 1821 and 1844, one machinery-building and ten textile corporations were established at Lowell. They operated mills that by 1850 contained in the aggregate some 275,000 spindles, rising to 380,000 in 1862, together with preparatory and weaving machinery, and occupied sites on a network of hydraulic canals some five miles in extent.[13]

At the outset of the enterprise it was the evident intent at Lowell, as in the case of S.U.M. at Paterson, to have the original corporation, the Merrimack Manufacturing Company, responsible for the management of the entire affair: development of power facilities; construction of mills, tenements, and machinery; management of the extensive landholdings; and the actual conduct of textile-manufacturing operations.[14] This arrangement turned out to be "unwieldy," and in 1826 the properties and operations not concerned directly with the making of textiles were transferred to a reorganized Locks and Canals company "for the convenience of management," the "real parties in interest remaining the same."[15] Thereafter for some eighteen years the Locks and Canals, as the company was commonly known, devoted itself vigorously and profitably to the development, disposal, and management of the waterpower, the building of machinery and the sale and management of real estate. Despite some interlocking ownership between the Locks and Canals and the textile corporations organized after 1826, the "real parties of interest" were no longer quite the same. Each new textile corporation made its bargain with Lowell's master corporation.[16]

So far as power alone was concerned, the bargaining of the Locks and Canals could scarcely be called hard. With some exceptions, water was delivered at the mill penstocks at a fairly uniform rate of approximately $12,500 a mill power, plus an annual "rent" of $300.[17]

[13] Gibb, *Saco-Lowell Shops,* pp. 102–3; Memorandum for Deposition of JBF in *Great* Falls Manufacturing Company v. *The United States,* LCP, DB-3, 233 ff.

[14] Gibb, *Saco-Lowell Shops,* pp. 66–68; JBF, memorandum for A. Gilman, Dec. 1878, LCP, DB-9, 474. On the legal, technical, and substantive aspects of the water-power system at Lowell, see "Legal Opinion of Payson Tucker," 15 Jan. 1884, ibid., A-17.

[15] JBF to A. J. Rogers, 4 Dec. 1874, LCP, DB-7, 445; to B. Saunders, 21 Oct. 1864, DB-3, 565. A Massachusetts act of 1825 authorized the Locks and Canals to buy and hold these properties. See Bagnall, *Sketches of Manufacturing Establishments,* 3:2072; and "Legal Opinion of Payson Tucker."

[16] Correspondence of Kirk Boott, LCP, File 145, A.29; and of Boott's successor, Patrick Tracy Jackson, DA-1, 69 ff., DA-2A, 29 ff.

[17] JBF to Abbott Lawrence, 2 July 1859, ibid., DB-2; to T. G. Carey, 4 Dec. 1858, DA-5, 58 ff. Some variations in the terms of the contracts other than price were eliminated by the adoption in 1853 of a uniformly worded contract for all lessees. In a report to the City Council of Austin in 1890, Joseph P. Frizell stated that the total

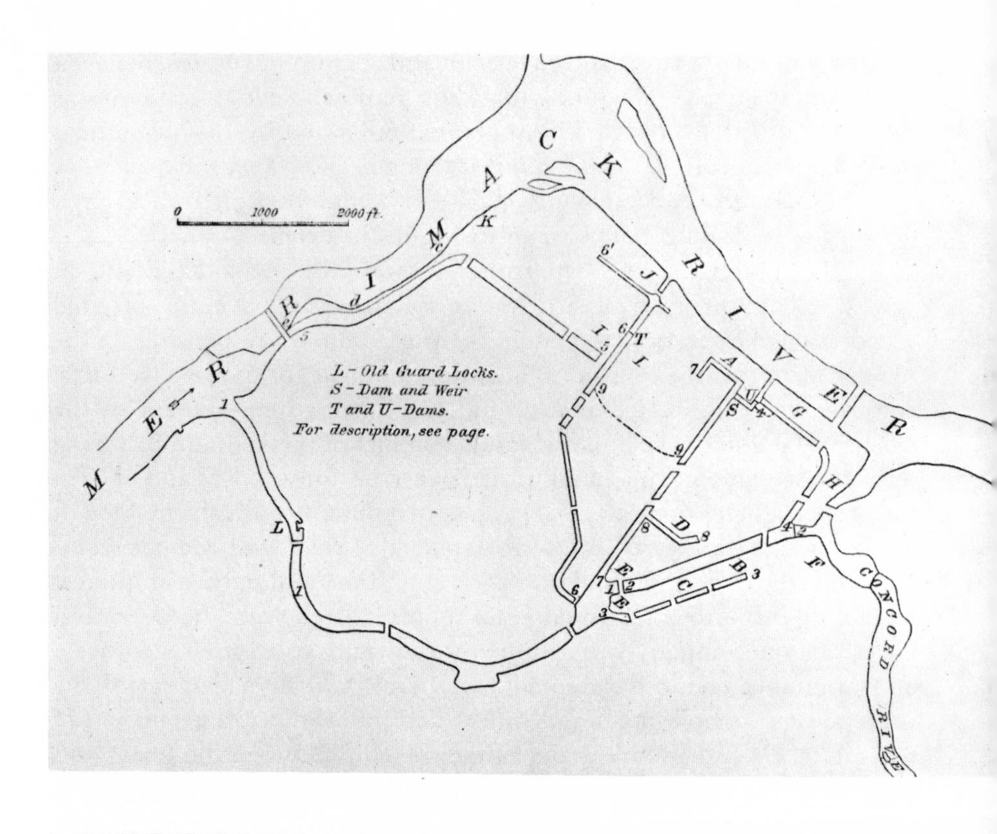

Canals at Lowell, Massachusetts. (*a*)

Name.	Designation on map.	Section.	Length (approximate).	Depth (approximate).	Width (approximate).	On w[...] lev[...]
Pawtucket:			*Feet.*	*Feet.*	*Feet.*	
Upper level	1–1	Irregular	6, 500	10	Irregular.	Uppe[...]
Lower level	2–2	Rectangular	2, 200	7	100	Lowe[...]
Hamilton	3–3	....do	1, 600	10	50	Uppe[...]
Eastern	4–4	....do	2, 000	8	42 to 66	Lowe[...]
Northern	5–5	.. do	4, 100	16 to 21	100	Uppe[...]
Western	6–6	....do	3, 500	10	15 to 115	Do
Do	6–6′	....do	1, 000	8	30 to 80	Lowe[...]
Merrimack	7–7	Rectangular and irregular	3, 400	10	50	Uppe[...]
Lowell	8–8	Rectangular	500	10	30	Do
Moody-street feeder	9–9	Rectangular and arched over.	1, 390	*b* 10	30	Do

a Total length of canals, about 5 miles. Bottom of lower level, at about 9 feet on scale. River behind mills, at low wa[...] about 2 feet on scale. Water surface, lower level, at about 17 feet on scale. Bottom of upper level, at about 22 feet o[...] Water surface, upper level, at about 32 feet on scale.

b In middle of arches.

Figs. 42–43. Map of Lowell, Mass., showing location of the mills and canals, and summary list describing the canals. (U.S., Census Office, Tenth Census, 1880, *Reports on the Water-Power of the United States,* pt. 1 [Washington, D.C., 1885].)

Taking account of the land for millsites and tenements, ceded in most instances before 1846 with the power, and the charges on waterwheel installations borne by the mills, the aggregate cost of power to the textile corporations was probably not much in excess of $25 per horsepower per year, a very moderate figure for the time. Francis estimated the cost of the Lowell wheel installations at approximately $100 per horsepower, a figure he believed "was probably double what was usually expended . . . but it has been the object here to make the very best of the water power and no expense has been spared."[18] If, as was contended, leases at such annual rates brought little or no profits to the Locks and Canals in view of the amount of its investment, the ability to grant or withhold power rights to new companies provided the company with the leverage to secure contracts for building mill machinery and equipment on very rewarding terms.[19] There were further handsome returns from the sale of real property held in what became the heart of the city of Lowell. According to George Gibb, "the heavy first cost of water power had been paid for largely by the sale of city land." Gibb, author of the most careful and detailed study of this enterprise, declared: "No enterprise could locate advantageously in Lowell except on Locks and Canals land, and no wheels could turn except by means of Locks and Canals water. More advantageous circumstances

---

permanent power of the Merrimack River at Lowell and Lawrence and Manchester was not over 11,000 horsepower, on an average during working hours, "the power that can be supplied without interruption." "It is not easy to state," he continued, "the rental received for water power . . . [at such centers] as grants of water are usually covered with grants of land, the water being regarded as an easement of the land. A round sum was paid for the land and a nominal rent for the water, which was intended as a fund for the maintenance of the appliances of the water power. When manufacturers draw in excess of their grant they are charged all the way from $3 to $12 a day for mill power for water terminable at will. My opinion is that $1,200 per annum fairly represents the value of mill power" (quoted by Thomas U. Taylor, *The Austin Dam,* U.S. Geological Survey, Water-Supply and Irrigation Paper no. 40 [Washington, D.C., 1900], p. 15).

[18] JBF to Washington Hunt, 20 Jan. 1858, LCP, A-30; to Walter Wells, 19 Nov. 1868, DB-5, 337. The figure of $100 was also taken by Samuel Webber, a well-known hydraulic engineer familiar with installations at Lowell and other New England textile towns ("Water-Power–Its Generation and Transmission," *TASME* 17 [1895–96]: 51–52).

[19] Gibb, *Saco-Lowell Shops,* pp. 102–3. JBF to J. P. Frizell, 7 Dec. 1883, LCP, DB-12, 274–75. See also JBF to B. Saunders, 21 Oct. 1864, DB-3, 565 ff.; and to Washington Hunt, 20 Jan. 1858, A-30; and James Montgomery, *A Practical Detail of the Cotton Manufacture of the United States of America* (Glasgow, 1840), p. 43.

for this company's operations could hardly be imagined."[20] Here is one of the complications attending the confusion of land as property with riparian rights to and control over streamflow, having their origin in the evolution of the common law in Britain. Whatever the net return from the investment in the hydraulic facilities alone, the operations of the Locks and Canals as a whole were unquestionably very profitable. When in 1845 the company disposed of its machinery-building plant, its corporate stock, and remaining town lots, enough was realized to divide among the stockholders more than three times the par value of their stock.[21]

This last action was preliminary to a reorganization that marked the passing of the independent status and authority of the Locks and Canals and its takeover by the manufacturing corporations. The stock of the new Locks and Canals, its property now largely reduced to the vital water rights and hydraulic facilities about which Lowell was built, was taken up by the textile corporations in direct proportion to the power leased by each. Master and servants changed places: the lessees of power became joint owners of the leasing corporation, a change that should not be exaggerated in view of the considerable interlocking of ownership previously existing. The purpose of the reorganization was clear: to protect the interests of the textile corporations by preventing the further lease of power to outside parties, thereby jeopardizing the expansion of their own operations. According to a later account, it was "deemed wise" that the stock "should no longer be held by persons whose only interest was to make money in the furnishing of power but by the manufacturing companies themselves."[22]

Thereafter the Locks and Canals, now managed by a board of directors consisting of the treasurers of the textile corporations, was operated as a service organization in the joint interest of the textile corporations as lessees and owners. As explained by Francis some years later, under the new ownership the Locks and Canals became practically "trustees for the management of the water power of the manufacturing companies using the same."[23] Heretofore the whip hand over Lowell's industrial development had been exercised by the Locks and Canals in such

[20] Gibb, *Saco-Lowell Shops,* p. 71. See also John P. Coolidge, *Mill and Mansion: A Study of Architecture and Society in Lowell, Massachusetts, 1820–1865* (New York, 1942), chap. 3.

[21] JBF to B. Saunders, 21 Oct. 1864, LCP, DB-3, 565; John A. Lowell, *Memoir of Patrick Tracy Jackson* (Boston, 1848).

[22] "Legal Opinion of Payson Tucker," LCP, A-17.

[23] JBF to A. Gilman, Dec. 1878, ibid., DB-9, 474; to J. Goforth, 11 July 1874, DB-7, 329.

manner as to admit only large corporations to the industrial fellowship of the city. Now that the textile companies themselves were directly in the saddle, there were no additions to their number during the half century that followed the reorganization. Theirs was literally a watertight monopoly.

Small Industry: The Outsiders

The use of steam power was open to all, but it was a burden borne perforce by the manufacturer of small means at Lowell because he had no other choice.[24] In time, as the supply of waterpower was exhausted, the large textile corporations were themselves obliged to adopt auxiliary steam power, an action reluctantly taken and for ample reason. In the 1850s James B. Francis, superintendent of the Locks and Canals, estimated the annual cost of waterpower to the using corporations as not over $25 per horsepower, as was remarked earlier. This estimate covered both the cost of water supplied to the mill by the Locks and Canals and the cost, including interest, amortization, and repairs, of the wheel installations supplied by the power user. In 1854 Francis, using data supplied by the nation's leading steam engine builder, Corliss and Nightingale of Providence, carefully estimated the cost of steam power. Allowing for interest, depreciation, attendance, consumption of anthracite coal at the low rate of 2.5 to 5.0 pounds per horsepower per hour, Francis found the cost of steam power per horsepower year ranged from $64.24 to $90.34.[25]

The wide differences in the cost of steam and water power largely explain the determination of the Locks and Canals company under Francis's direction to employ all possible means of promoting efficiency and economy in the use of the available supply of streamflow. These included not only internal measures for improving the hydraulic facilities themselves and for promoting efficient practices of water use by the manufacturing corporations, to be considered in the following chapter, but the complete exclusion by means direct or indirect of small

[24] Writing to T. Lyman, 13 July 1865, Francis referred to "a considerable number of small establishments on the Concord River [which joined the Merrimack in Lowell] and a large number of little mills, shops, &c., in the city driven by steam" (LCP, DB-3, 658). The limitations of steam power for the small user will be discussed in a work in preparation.

[25] JBF to Corliss and Nightingale, 6 Jan. 1854, LCP, A-2. Over a century after the founding of Lowell, waterpower continued to maintain a small cost advantage over steam power (Parker, *Lowell,* chap. 4).

industry from the benefits of cheap waterpower. Except for several hundred horsepower available from the Concord River, which joined the Merrimack at Lowell, the Locks and Canals through its ownership of the land bordering the Merrimack on either side of Pawtucket Falls possessed the unchallenged control over the runoff of the 4,100-square-mile drainage basin in its passage through Lowell. Although Francis was fully aware of the power problem of the small manufacturer and mill-owner, and, his official duties apart, was not unsympathetic in the matter, company policy reflected no recognition of the important, even indispensable role of numerous small supply and service industries in the prosperity and growth of the dominant textile industry.[26] The policy of limiting the disposal of waterpower to large corporations, declared a historian of Lowell in 1868, was based partly "on a love of methodicity and an unreasoning attachment to incorporated forms of industry, and partly on the selfish desire to have the whole body of the people of Lowell subject to their sway."[27]

Over the years the power problems of the small user—for which steam power, cost apart, did not provide a satisfactory answer—kept recurring. The more obvious solutions were either ignored, considered and rejected, or determinedly opposed by the Locks and Canals, the controlling authority in what was in basic respects a "company town." A system of distributing power in the form of water under a pressure head—through a system of open canals as in this country or in closed pipes as practiced effectively in some places in Europe[28]—to a multitude of small users presented practical and commercial difficulties requiring for their resolution more imagination and motivation than were present at Lowell. A simpler and quite practical and widely adopted plan of serving small users in this country was that of power buildings in which space equipped with power from a central plant was rented. Employed to some extent in such similarly situated industrial com-

[26] A later inquiry by Francis into the cost and operating expense of small gas engines remarked upon the "great call for small powers" at Lowell (JBF to Schleicher, Schumm and Co., Nov. 1879, LCP, DB-10, 130–42). See also JBF to Hamilton Smith, Jr., 23 Jan. 1884, on the need for fractional horsepower motors, DB-12, 306. The power-supply problems of small users will be treated in a forthcoming volume.

[27] Charles Cooley, *History of Lowell* (Boston, 1868), pp. 60 ff. This history was in substance the second edition of Cooley's *Handbook of Business in Lowell* (Lowell, 1856), a work highly laudatory of the founders and corporations of Lowell. In the *History* praise for the latter was confined to the early years of Lowell's growth; there was strong criticism of the corporations' Civil War policy of largely closing down production and dismissing some 10,000 operatives "in cold blood."

[28] See Unwin, *Power from Central Stations,* pp. 80 ff.

munities as Manchester, New Hampshire, Fall River and Holyoke, Massachusetts, and Rochester, New York, this method was briefly considered in 1851 at Lowell.[29] With a building site available and a considerable surplus of water during much of the year, the Locks and Canals prepared detailed cost and income estimates for building space to be equipped with auxiliary steam power for use during the low-water periods. The proposal called for the eventual erection of two buildings, each 720 feet by 42 feet wide and four stories high, not including basement and attic. The cost of the initial structure with power was estimated at $60,607 and the net income on the investment at over 9 percent. The plan was to rent the building in its entirety to a middle party who would sublet space and power on such terms as he pleased.[30] For whatever reasons the scheme was dropped.

Another means of relieving the power needs of small users was also denied at Lowell. This process, which became widely available in the sixties and seventies with the rapid spread of public water-supply systems in American cities, involved the use of small water motors, mainly in the fractional-horsepower class, operated by water pressure from city mains. The Locks and Canals and the textile corporations determinedly opposed the establishment in Lowell of a public water supply system. When in the 1870s a public system was nonetheless created, a clause was inserted in the city's contract with the Locks and Canals forbidding the use of water taken from the Merrimack River for power, *except* in the form of steam.[31] Repeated efforts of the Lowell Water Board in the late seventies to secure Locks and Canals permission for the use of small water motors were unsuccessful, even when a limit of one-half horsepower per motor was proposed. (Church organs and hospital elevators were exempted from the general prohibition.) Such use, contended the corpora-

[29] JBF to Corliss and Nightingale, 6 Jan. and 7 Nov. 1851, LCP, A-2 and DA-3, 114–15; and to Directors, Locks and Canals, 25 Nov. 1851, A-2.

[30] In 1869 Francis wrote to G. A. Baldwin in Quebec: "Our manufacturing interest, which pays more than half the taxes do not want it and I am collecting information to show that it is not for the interest of the city to go into it at present" (LCP, DB-5, 385). See also JBF to A. P. Roussey, 29 Apr. 1870, DB-6, 39.

[31] This provision had in fact been established by law. See Mass. Statutes, chap. 435 (1855). An agreement between the city of Lowell and the Locks and Canals was reached in 1875 by which the city paid the corporation $50,000 covering "not only the damage of taking water, but also for certain pieces of real estate taken by the city for laying pipes, etc." This sum was greatly exceeded by the more than $140,000 paid by the city to the owners of the waterpower along the lower Concord River, who were deprived of water by the city's supply requirements (George A. Kimball, "Water Power–Its Measurement and Value," JAES 13 (1894):86–88.

tion, in arguments prepared by Francis, was not only inefficient and costly but if allowed would threaten the industrial foundations of the city by a serious depletion of the water supply of the textile mills. Yet by this date the mills were already using more steam power than waterpower.[32]

Lowell was perhaps not the first "company town" in this country, but it was the largest of its day. Its authority rested on the possession of two inseparable factors indispensable in any large industrial enterprise of the period: land and waterpower. "Imperfect though the steam engine may be, it is by no means certain," stated John C. Merriman in 1868, "that water power is cheaper, and there is one disadvantage in the latter that is often overlooked, it is that of monopoly. The rich company who own the water power let it out at their own price, for there is no competition; but with the steam engine it is very different; if the price asked by your neighbor for power is too great, you can readily purchase an engine of just the power you require, and run it independently."[33] Much the same view was expressed by a leading steam engineering authority twenty years later: the great waterpowers were usually controlled, as at Lowell and Holyoke, by a corporation with the legal right to charge such rates as they found most profitable; and "all consumers of power are compelled to deal with this corporation on such terms as it may dictate."[34] Yet the largest among the other waterpower corporations in New England, notably those at Lawrence, Manchester, and, much later, Holyoke, recognized to some extent the shortsightedness and lack of wisdom of Lowell's policy in the selection of consumers. Each made some provision for the needs of small users, in recognition not only of the usefulness and power needs of small establishments but also of the desirability of diversifying the community's industrial base.[35]

[32] See LCP, DB-9, 54, 162, 191, 225. For the argument against water from the city supply system being used for water motors, see the communication to the president of the Lowell Water Board, 2 Dec. 1879, signed by the president and the agent (Francis) of the Locks and Canals Company, DB-10, 132–40. A renewal of the request two years later, limiting water use for power to motors not in excess of one-half horsepower, was rejected with equal firmness (JBF et al. to the Lowell Water Board, 28 June 1881, DB-11, 136–40).

[33] Chap. on "Steam," in *Eighty Years' Progress in the United States* (Hartford, Conn., 1868), esp. p. 271. Articles in this volume were written "by eminent literary men, who have made the subjects of which they have written their special study."

[34] Robert H. Thurston, "Systematic Testing of Turbine Water-Wheels," *TASME* 8 (1886-87):360-61. See also Gibb, *Saco-Lowell Shops,* p. 87. The evidence of widespread underutilization of capacity, as reported in Tenth Census, *Water-Power,* suggests that the great majority of waterpower companies operated in a buyer's market and were hardly in a position to enforce their monopoly position at a given site.

The Lowell Achievement and Influence

If there were flaws, the Lowell achievement was nevertheless a notable one, considered alike from engineering, economic, and business viewpoints. The engineering success of the design and operation of Lowell's hydraulic facilities was emulated throughout the United States. On the economic side the effective and low-cost exploitation of a large-scale energy resource met for half a century here and in similarly situated textile centers most industrial power needs and measurably offset the fuel deficiencies of New England. Those who had the enterprise to undertake so ambitious and daring an innovation or the opportunity to contribute to its capital were rewarded by returns that for the more ably managed corporations were very substantial.

As we shall see in subsequent chapters, there were defects in the execution of the engineering works. The power system was less planned than gradually evolved, as the success of the initial measures required the enlargement of the hydraulic facilities to meet the requirements of ever more mills. The limitations of the original Pawtucket navigation canal, long the sole channel of water supply, and of the hydraulic science and engineering knowledge available in the early years were never fully overcome in later improvements and extensions. Certain irregularities in water supply to the several mills were never entirely eliminated. Nor did the close interlocking of ownership and community of interest among the owning corporations and the strict impartiality of management by Superintendent Francis prevent the friction and ill-feeling that so often accompanied and marred arrangements for the common use of a large waterpower jointly owned. Yet Lowell both demonstrated the rewards of power development as a field for business enterprise and anticipated in a limited but impressive manner the central-station mode of power generation and distribution. What the Erie Canal was to the age of inland waterways, Lowell was to a generation in which the manufacturing industry, large and small, relied chiefly upon falling water for the propulsion of machinery. Lowell stirred the imagination, aroused the hopes, and provided guidance and instruction in varying degree for most of the large waterpower developments in the eastern United States. The early experience and experiments and the advanced practices

[35] The legislative act giving the city of Lawrence the right to take water from the Merrimack above the dam of the Essex Company "contemplated" that the water taken would be used "for domestic supply only, and the city pays the further sum of 5 percent on all receipts for water used for running motors, elevators, etc." (Kimball, "Water Power," pp. 87–88). Tenth Census, *Water-Power,* pt. 1, p. 223.

and equipment in the management of falling water long made this city a leading center in this branch of hydraulic science and engineering in the Western world.[36]

The influence of the Waltham-Lowell system is strikingly and fully illustrated by the several other New England textile centers drawing their inspiration, capital, and direction from much the same group of men and presenting much the same interlocking of power, machinery building, manufacturing, and community development, as well as of ownership and management.[37] Although the idea of the planned industrial community on this scale found little acceptance elsewhere, in many other respects the repercussions of the success of the pioneer industrial center on the Merrimack were widely felt throughout the eastern half of the nation. Lowell's success greatly strengthened the conviction that waterpower was the indispensable natural base for industrial growth. In the mid-1830's the proprietors of "an immense water power" at Grand Rapids on Ohio's Maumee River could well believe that "ere long this will be one of the principal manufacturing points in the West." Some twenty years later the promoters of a power site at Lawrence, on the Des Moines River, Iowa, declared the obvious, that "the first great requisite for building up a manufacturing town is, of course, the abundant supply of water power." When the developers of the substantial waterpower at Hamilton, Ohio, on the Miami River laid out the "Hydraulic Addition" in the summer of 1843, a main street was named Lowell because "the hydraulic would make Hamilton the Lowell of the West." Similar schemes in the late 1840s for developing the large powers at Great Falls on the Potomac River near Washington, D.C., and on the Savannah River at Augusta, Georgia, were stimulated by the great northern example. That at Great Falls was in fact named South Lowell, although it never advanced beyond the town plat stage.[38]

[36] See Swain, "Statistics," p. 5. The Lowell influence is further revealed in the correspondence over several decades of James B. Francis of the Locks and Canals, in reply to requests for information and advice. Francis's *Lowell Hydraulic Experiments* (Boston, 1855) went through several editions and was influential in the practice of hydraulic engineering in this country and abroad. See Desmond Fitz-Gerald et al., "James Bicheno Francis: A Memoir," *JAES* 13 (1894):1–9.

[37] Shlakman, *Economic History of a Factory Town*, pp. 30 ff. Parker noted that, in the 1834 Lowell directory, place of residence was indicated not by streets but by textile corporation (*Lowell*, chap. 3).

[38] Warren Jenkins, *The Ohio Gazetteer, and Traveller's Guide of the State of Ohio*, 1st ed. rev. (Columbus, 1837), p. 202; *Description of the Town of Lawrence, Van Buren County, in the Des Moines Valley, Iowa: Its Hydraulic Power and Manufacturing Facilities* (Keokuk, 1856); Alta H. Heiser, *Hamilton in the Making* (Oxford, Ohio, 1941), pp. 140 ff.; R. H.

As late as 1870 a company with a 2,500-horsepower site on the lower Rock River in Illinois announced its intention "to establish a manufacturing city on a scale unparalleled in the West"; and a New York colonizing company with a large land tract at Big Rapids, Blue River, Kansas, "set to work to develop a manufacturing town" with, unfortunately, but limited success.[39]

In most instances power-development projects of this kind were, of course, privately owned and managed. Typically they were objects of community encouragement and occasionally were launched with the aid of community subscriptions, somewhat after the manner of many early railroads.[40] In a number of instances, chiefly within the southern states, such power projects were undertaken and carried out by municipal authority as a means of promoting industrial development. Outstanding examples were at Manchester, adjoining Richmond, and Lynchburg, Virginia; Columbia, South Carolina; Augusta, Georgia; and, long after the Civil War, Austin, Texas.[41] One of the earliest and much the most impressive of these civic enterprises was that at Augusta, head of navigation on the Savannah River, to be discussed below. The commercial development of most large waterpowers reflected the Lowell influence only in respect to the hydraulic system itself; the usual scale of operations was measured in hundreds rather than thousands of horsepower. Water privileges had early been recognized as a form of real property, subject to sale, rental, and lease, by fractional shares or as a whole, often offering attrac-

---

Shryock, "The Early Industrial Revolution in the Empire State," *Georgia Historical Quarterly* 2 (1927):114, citing *DeBow's Review* 7(1848):370; Alexander Mackay, *The Western World; or, Travels in the United States in 1846–47,* 2 vols. (Philadelphia, 1849), 1:241. According to Mackay, "a few energetic speculators," evidently from New England, got so far as to buy the land at Great Falls and lay it out in land and water lots.

[39] *Report of the Water Power and the Superior Advantages Afforded for Manufacturing and Transportation on the Rock River* (Chicago, 1870), p. 11; Tenth Census, *Water-Power,* pt. 2, pp. 237 ff., 345 ff. For a similar abortive attempt, see ibid., pp. 375–76.

[40] Tenth Census, *Water-Power,* pt. 1, p. 189; pt. 2, p. 375; Allen Corey, *Gazetteer of the County of Washington, New York* (Schuylerville, 1850), p. 140; Thomas F. Gordon, *Gazetteer of the State of New York* (Philadelphia, 1836), p. 489; and John H. French, *Gazetteer of the State of New York* (Syracuse, 1860), p. 484.

[41] See Tenth Census, *Water-Power,* pt. 1, pp. 533–34, 539, 761–62, 787–88; Frank E. Snyder, "The Great Dam at Austin, Texas," *Engineering Magazine* 8 (1894–95):245 ff.; Joseph P. Frizell, *Water-Power: An Outline of the Development and Application of the Energy of Flowing Water,* 3d ed. (New York, 1903), pp. 245 ff. The city of Marseilles, Illinois, on the Illinois River, about 1870 subscribed $60,000 to the stock of the waterpower corporation; the company went bankrupt and the power remained unused (Tenth Census, *Water-Power,* pt. 2, pp. 248–49).

tive opportunities for speculative investment.[42] Waterpower corporations simply represented the organization of such activities on a large and systematic scale. The usual makeshift arrangements for building sites and rude raceways were replaced by far more regular, convenient, and efficient ones. By undertaking to deliver a fairly regular supply of water under stable head at the mill penstocks and assuming responsibility for the maintenance and repair of all facilities, the waterpower company sought to attract customers within the limits of the power's capacity. This more modest method of exploiting medium to large waterpowers was better adapted to the prevailing scale, resources, and organizational capabilities of manufacturing during the decades before the Civil War than the far more ambitious integrated company towns, such as Lowell and Holyoke. Whatever the mode of development, captive or independent, the unified development of large waterpowers, frequently on an impressive scale, was a distinctively American phenomenon. In no other country, declared Professor George F. Swain in the late 1880s, were there so many large manufacturing cities owing their existence to the proximity of power sites or "so many instances of large companies developing power at great cost, and selling it to extensive manufacturing concerns like any other commodity." In resolving the engineering and management problems associated with their development, he added, "this country has taken the lead."[43] "The method by which water power is distributed in America to a number of consumers," declared W. C. Unwin, the British hydraulic authority, confirming the point made by Francis of the Locks and Canals a generation earlier, "is almost peculiar to that country. . . . The water is distributed to mill-owners, who construct the turbines and pay a rental to the Water Power Company proportioned to the amount of power used."[44]

The Lowell concept of a planned industrial center based on a water-

[42] See Frizell, *Water-Power,* 3d ed., chap. 26, "Leases and Rentals of Water-Power"; also William D. Williamson, *The History of the State of Maine,* 2 vols. (Hallowell, 1839), 1:513–14; Blake McKelvey, *Rochester, the Water-Power City: 1812–1854* (Cambridge, Mass., 1945), pp. 19 ff., 33 ff.; *Proceedings of the General Convention of Agriculturalists and Manufacturers . . .* (Harrisburg, Pa., 1827), pp. 59–60; William T. Utter, *The Frontier State, 1803–1825,* vol. 2, *The History of the State of Ohio,* ed. Carl F. Wittke (Columbus, 1942), p. 232; E. L. Brown, "Autobiographical Notes," *Hist. Collections of Michigan Pioneer and Hist. Society* 30 (1906):460–70; Tenth Census, *Water-Power,* pt. 1, pp. 186, 196, 202; *Scientific American* 3 (8 Apr. 1848):229.

[43] Swain, "Statistics," p. 5.

[44] Unwin, *Power from Central Stations,* pp. 84 ff. See also JBF, statement of 19 June 1858, prepared for rev. ed. of D. Stevenson, *Sketch of the Civil Engineering of North America,* at request of John Wale, a London publisher, LCP, A-20.

power of unprecedented capacity was applied by the same group of Boston and associated merchant capitalists, although with varying participation, at a number of other outstanding sites in or adjacent to Massachusetts. The three most impressive examples were the sister textile centers at Manchester, New Hampshire, thirty miles above Lowell on the Merrimack River, dating from 1831, and Lawrence, Massachusetts, twelve miles below Lowell, dating from 1845; and one hundred miles to the westward on the Connecticut River, the paper-milling city of Holyoke, dating from 1847. On a smaller scale, but in varying degrees similarly planned and executed, were the power and industrial developments at Chicopee (1823), adjacent to Springfield; Nashua, New Hampshire (1823), between Lowell and Manchester on the Nashua River; Greeneville (1828), adjacent to Norwich, Connecticut; and Saco-Biddeford (1831) and Lewiston (1837), Maine.[45]

At the three larger centers of Manchester, Lawrence, and Holyoke we find much the same planned industrial community grouped about the hydraulic system with its great dam and upstream millpond, its pattern of supply canals arranged on two or three levels—somewhat less efficient in water use but affording proportionately greater space for millsites than a single level of canal—and all the appurtenances of guard locks, headgates, wasteways, and tailraces. At each center, too, we find much the same plan of leases and rentals of water, with very similar if not identical units and charges, the same differentiation adopted sooner or later between "permanent" and "surplus" power, with regulations controlling the use and rates of each, and much the same problems of measurement and management. The total power available at these three industrial communities was of about the same magnitude as at Lowell, with Holyoke leading the others by about one-fourth. Except at Manchester, where a single corporation long controlled most activities, power development and management—together for a time with operations in land and machinery building—were handled separately from the manufacture of goods, such as textiles or paper.

The story of these celebrated direct-drive waterpower developments

[45] See Shlakman, *Economic History of a Factory Town,* pp. 25, 39–42, and under the appropriate drainage basin in Tenth Census, *Water-Power,* pt. 1, and tabular data pp. xxx–xxxii. See also Gibb, *Saco-Lowell Shops,* pp. 113 ff.; Green, *Holyoke,* pp. 11 ff., 19 ff., 25 ff.; Chandler E. Potter, *The History of Manchester, New Hampshire* (Manchester, 1856); Edward E. Parker, ed., *History of the City of Nashua, New Hampshire* (Nashua, 1897); Frances M. Caulkins, *History of Norwich, Connecticut* (Hartford, 1866); Margaret T. Parker, "Geographic Factors in the Development of the Cities of the Merrimac Valley in Massachusetts," M.A. thesis, Wellesley College, 1921; and Stanwood, "Cotton Manufacture in New England," pp. 117–87.

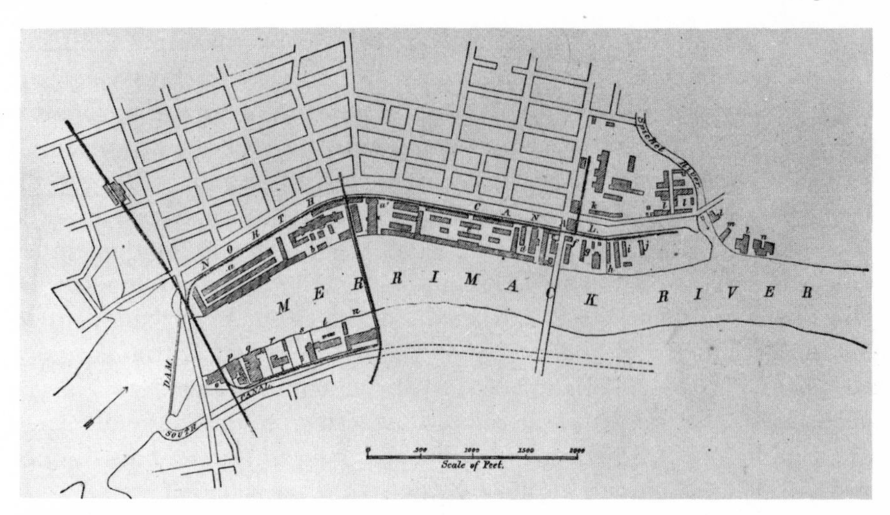

Fig. 44. *Above,* Lawrence, Mass., showing location of the mills and canals; *below,* sections of the North Canal at Lawrence. (U.S., Census Office, Tenth Census, 1880, *Reports on the Water-Power of the United States,* pt. 1 [Washington, D.C., 1885].)

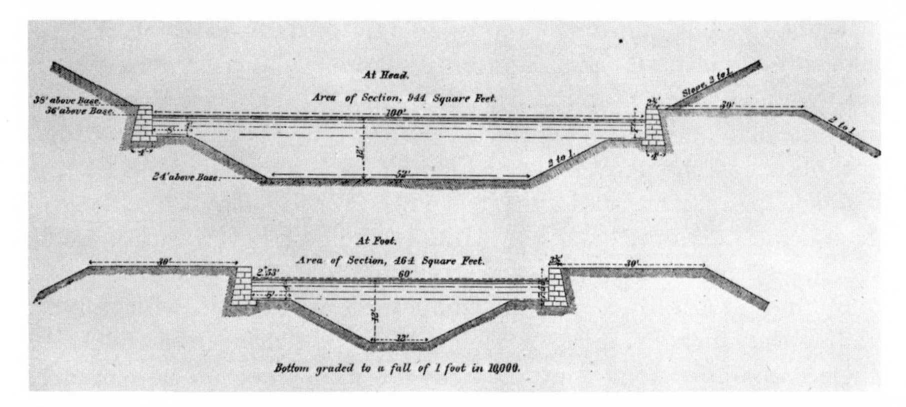

Noteworthy in the above drawings are the diminishing width and cross-section area, and constant low gradient, of the power canal from head to foot as water is drawn off into the penstocks supplying the waterwheels of the mills in this single-level distributing system. The cross-section area decreases slightly more than one-half (944 to 464 square feet) over the length of the canal. The low gradient of the canal bottom of 1 foot in 10,000, or a little more than 6 inches per mile—about the gradient of the Mississippi River in much of its course—minimizes the loss of energy (weight of water x distance of fall) represented by the fall estimated as necessary to deliver the water required by the mills when working at capacity. The low gradient is offset by the ample dimensions of the area section. The South Canal, begun in 1866, twenty years after the North Canal, extended in 1880 only 2,000 feet down the right bank of the Merrimack River, compared with the 5,330 feet of the North Canal. In 1880 only thirteen mill powers (1,108 hp) were in use by the mills on the South Canal as compared with 115 mill powers (9,802 hp) on the North Canal.

was by no means one of uninterrupted successes. As with the cotton textile mills of New England with which they were so closely associated, the business was in time overdone and the anticipated profits not realized. As a distinguished hydraulic engineer, whose career began in New England, later recalled:

> In the great early American power developments at Holyoke, Lewiston, and Cohoes, the original investors are reported to have lost all or the greater part of what they had put into these developments, because of the fixed charges and operating costs accumulating while waiting for customers to absorb the output. I heard the venerable Charles S. Storrow, treasurer and first engineer of the great water-power development at Lawrence, Mass., testify that "they paid one dividend on expectations and then passed thirteen on facts," while waiting to sell that portion of the power left over after supplying the three factories built simultaneously with dam and canal. . . . [In the early 1890s] the water power company at Lawrence still had water rights for factories unsold, while customers were slow in coming.[46]

Waterpower and Paper

The Holyoke enterprise marked the first significant departure from the pattern established at Lowell a generation earlier.[47] Here on the Connecticut River the fall under control of the waterpower company was fifty-six feet (as compared with thirty-three feet on the Merrimack at Lowell) from the top of the dam to the smooth low water below. The drainage basin above the falls was twice that serving Lowell, the volume of flow three times greater, and the gross horsepower double that available at Lowell. Holyoke was virtually the last of the large planned industrial centers based upon waterpower and the first of such centers to have the monopoly control of the waterpower corporation directly challenged and in a measure limited. The original intent to make the new city a tightly knit, closed company town on the Lowell model faltered from

[46] John R. Freeman, "The Fundamental Problems of Hydroelectric Development," *TASME* 45 (1923):541, also p. 532.

[47] The account that follows is based largely on Green's admirable study, *Holyoke*. See esp. pp. 95–97, 152 ff., 158–59. See also Stanwood, "Cotton Manufacture in New England," pp. 147–48, 170–71. An anniversary issue of the Holyoke *Transcript-Telegram,* 14 Sept. 1960, presents a wealth of historical data gathered from various sources, particularly the files of the Holyoke Water Power Company, and deals with technological and economic aspects of the company's history. For comparative technical data on the Holyoke and other large developed waterpowers in New England, see Tenth Census, *Water-Power,* pt. 1, pp. xxviii–xxx, 221 ff. (For Lowell, see esp. pp. 78 ff.).

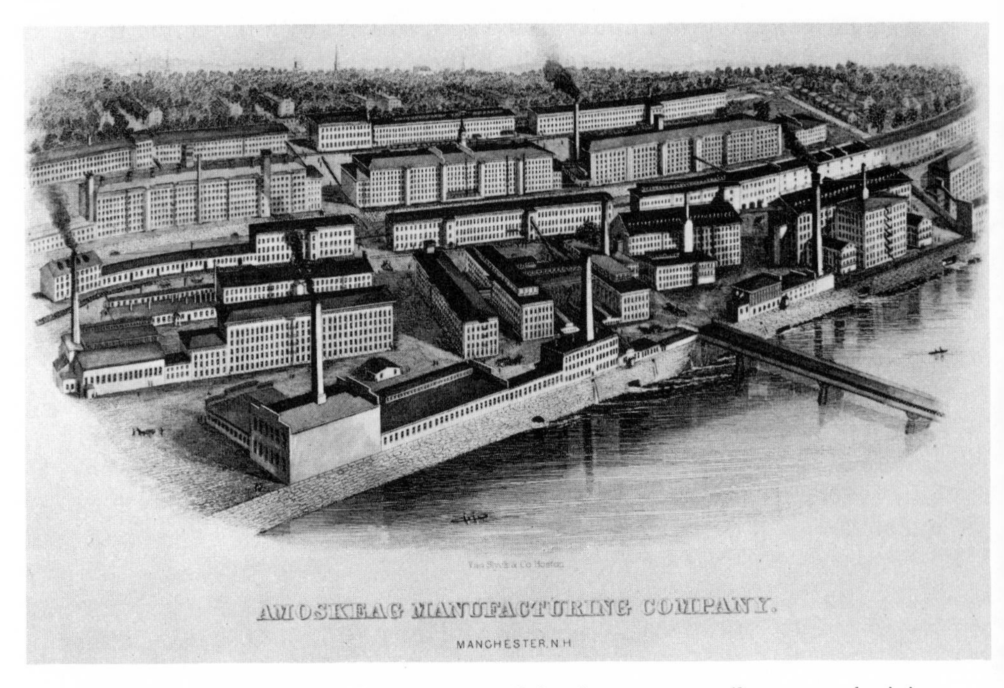

Fig. 45. At Manchester, N.H., uppermost of the three great textile centers obtaining their power from the trunkline of the Merrimack River system, a single corporation, the Amoskeag Manufacturing Co., both controlled and developed the waterpower system and engaged directly in manufacturing, as shown, using nearly one-half of the waterpower in use in 1880. (J. D. Van Slyck, *Representatives of New England Manufacturers,* vol. 1 [Boston, 1879].)

the very beginning. Capacity in textiles by the late 1840s was already overextended in New England, and the peak of the lush profits of the pioneering years well past when the first steps were taken to launch the Holyoke enterprise. Backers of the Hadley Falls Company, the development corporation, lacked something both of the capital and the will to carry on as at the earlier centers. The washout of the first dam within hours of its completion in late 1847 resulted in heavy loss, especially in delay. The first cotton mill did not begin operations until 1850; the second followed in 1852. Others were slow to be established, the financial difficulties of the Hadley Falls Company mounted, and the depression of the late 1850s led to bankruptcy and receivership in 1859 and to the exit of the absentee Boston capitalists who had sought here still another triumph.

The new development corporation, the Holyoke Water Power Company, set itself to the task of extending its operations and enlarging the community's industry within the framework of a buyer's market. The separation of textile manufacturing from power and development functions had taken place in 1854. The new company shortly disposed of its machinery-building division, concentrating thereafter on power and real estate operations. The original intent to make Holyoke exclusively a textile center was gradually abandoned. By the early 1870s the power company had definitely committed itself to the encouragement of diversified industry. The paper industry, despite its round-the-clock operations so disturbing to the conventional practice of accumulating water in the millpond overnight for use the following day, was admitted to Holyoke, with large consequences for the city's industrial future, for by the early 1880s this industry had become the leading consumer of power at Holyoke and employed 84 of the 140 waterwheels in use there. To the attraction of an abundant supply of waterpower required by the central grinding operation in papermaking was added the availability of large quantities of preferably soft river water for use in processing. Water from a municipal supply system would be expensive and from the millowner's own wells might be both difficult and expensive to obtain in sufficient quantity and wanting in the quality of softness.

The power company engaged in the active recruitment of new power customers, frequently extending financial and other assistance. For a time all comers were welcome with one mill power as the minimum lease; but increasingly the Holyoke Water Power Company leaned toward the big, strongly financed companies, preferring, as Constance McLaughlin Green has pointed out, "to see mill sites and power utilized by a few large corporations."

When business recovered in the late 1870s, industrial expansion reached the point where a shortage of power had at last to be faced. To cope with this new problem, the power company gave more rigorous attention to checking upon power consumption, introduced new and more precise methods of water measurement, brought pressure for the installation of the more efficient types of turbines, and took over and perfected the operation of the first commercial turbine-testing flume in the United States. When the last of the "permanent" or year-round powers was disposed of in 1881, the company both tightened its regulatory procedures and raised substantially the price of "surplus" power, available during the "wet" months. Such measures, together with the general strengthening of the company's market position owing to the scarcity of power provoked criticism and protest and led to a nearly successful attempt by the manufacturers' association to obtain control of the company and the power system. Whether or not the power company's long monopoly of the best industrial sites as well as the supply of waterpower hampered the city's growth, as some believed, low-cost waterpower had long been the central feature of and stimulus to the city's industrial life. Only when electric power became widely available for manufacturing was the industrial advantage of Holyoke, based on waterpower, materially weakened.[48]

[48] With the introduction of electricity a feud developed between the city with its publicly owned gas and electric plants and the Holyoke Water Power Company, which entered the field with hydroelectricity (Green, *Holyoke,* pp. 243–45).

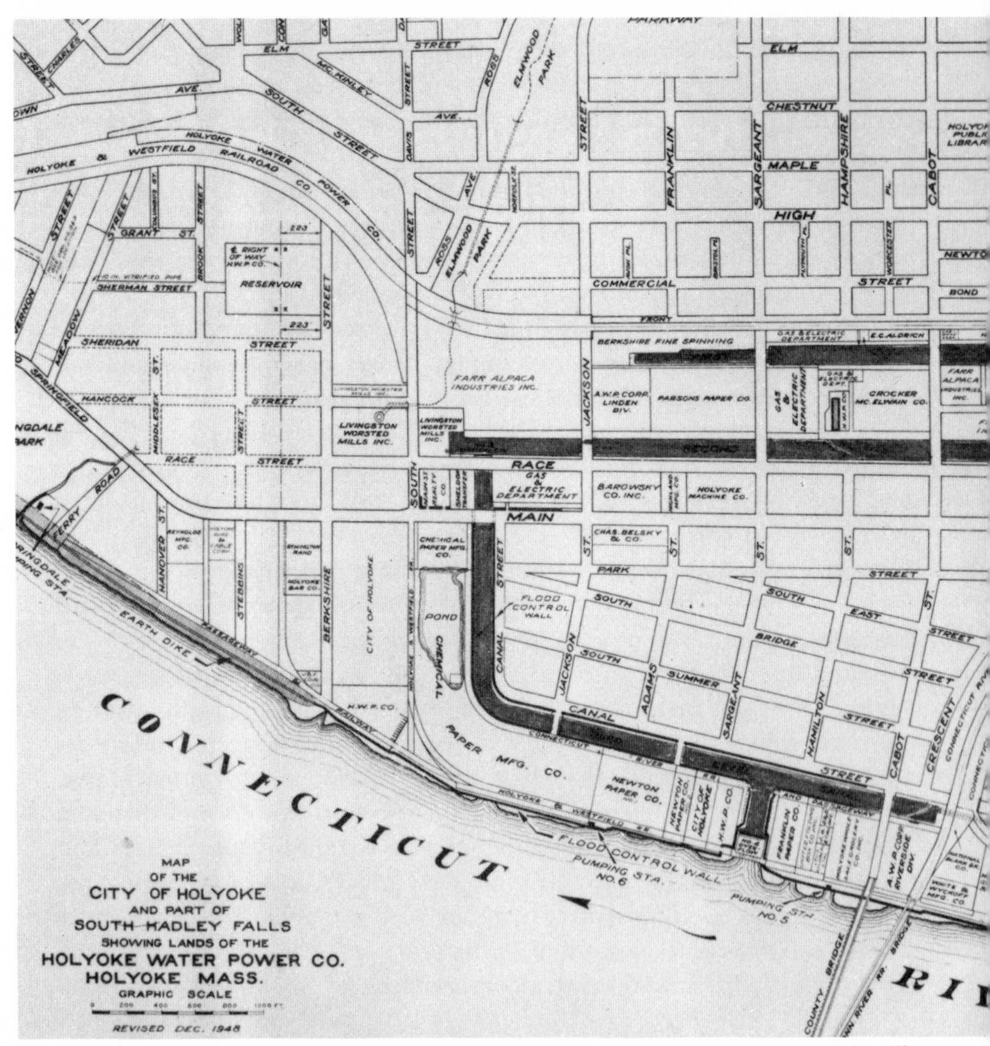

Fig. 46. Plan of the city of Holyoke, Mass., showing the location of canals and mills. (Holyoke Water Power Company.)

Some forty years after the future paper center at Holyoke made its fumbling debut on the upper Connecticut, the last of the great direct-drive installations of New England was founded in western Maine on the Androscoggin River. Sixty miles or so inland from tidewater, the river fell 182 feet within a mile in three successive pitches, the uppermost with a fall of nearly 100 feet, with a minimum estimated potential of 50,000 horsepower. Drawing its incentive from the beginning from paper-making, the Rumford Falls project was carried out in the years marking the first truly large-scale hydroelectric development at Niagara Falls. Behind this development lay the great expansion of the wood pulp and paper industry, chiefly after 1880, for which the extensive spruce timber

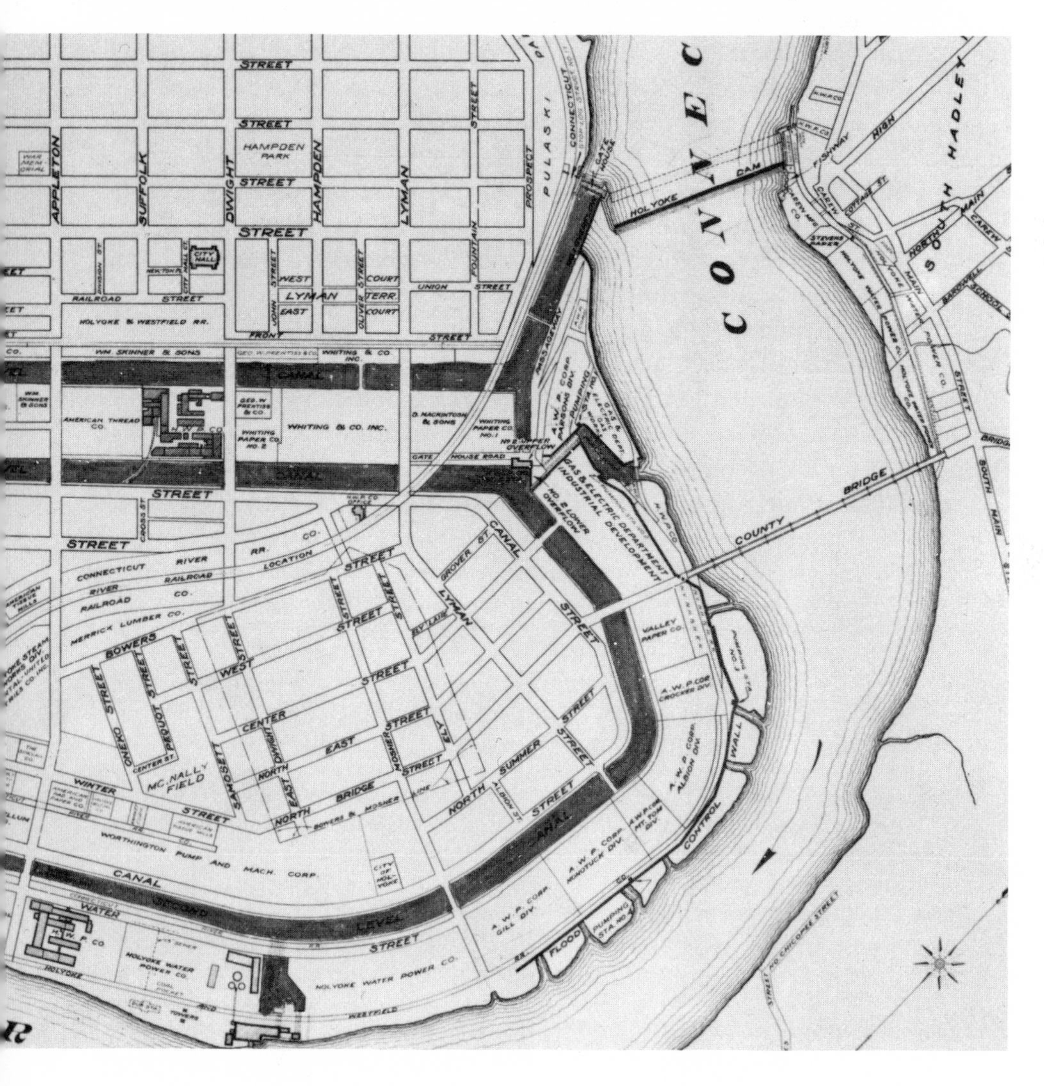

of Maine, largely ignored by the lumber industry, was admirably suited. Active development of the Rumford Falls power followed the incorporation of the Rumford Falls Power Company in 1890 and took much the same course as at the Merrimack River cities. Plans were announced for selling millsites accompanied by perpetual leases of water for power on three projected levels of canals. Actual development and use were concentrated in several large paper companies with interlocking ownership. By 1900 these had been brought under the control of the International Paper Company, one of the leading operators, with timber holdings at that time of more than 300,000 acres in Maine. By this date some 24,000 horsepower were said to be in actual use at this site; and the population of the town of Rumford Falls had increased from a base of some 900 in 1890 to over 6,000.[49]

The Lowell policy toward small enterprise was followed, with some qualification, at Lawrence, at Holyoke, and for a time at Manchester. Eventually Manchester's ruling corporation, the Amoskeag Manufacturing Company, modified its earlier policy, realizing, according to a local historian, the vulnerability of a community dependent upon a single industry. The corporation erected "for the convenience of individual enterprises" a building that came to be known as Mechanics' Row, where room was rented with power; land and shops were also made available to "small corporations."[50] A few fractional mill powers, ranging from some 10 to 50 horsepower, were leased by the Essex Company, the waterpower corporation of Lawrence; and a power building was erected there in 1866 for the accommodation of small users.[51] The long reluctance and eventual submission to a pressing need for small powers by the Holyoke Water Power Company has been noted. The decision of that company in 1881 to erect a large power building to accommodate "infant industries" proved rewarding to the company and community alike.[52] The

[49] A detailed account appears in the *Sixteenth Annual Report of the Bureau of Industrial and Labor Statistics for the State of Maine, 1902* (Augusta, 1903), pp. 116–54. See also James E. Defebaugh, *History of the Lumber Industry of America,* 2 vols. (Chicago, 1907), 2:96–99.

[50] [John B. Clarke], *Manchester: A Brief Record of Its Past and a Picture of Its Present* (Manchester, N.H., 1875), pp. 267 ff., 272 ff.; Potter, *History of Manchester,* p. 756.

[51] Tenth Census, *Water-Power,* pt. 1, table p. 75; Horace A. Wadsworth, comp., *History of Lawrence, Massachusetts* (Lawrence, 1880), pp. 150–51.

[52] Green, *Holyoke,* pp. 94–95, 151 ff.; Tenth Census, *Water-Power,* pt. 1, p. 223. See also the discussion of the Holyoke Water Power Company's "incubator" policy on occasion of its 111th anniversary in 1960: "In the nearly 80 years since that time [1881], the Company has assisted many small firms by providing rental quarters for them in Company-owned buildings. In addition it has built several factory buildings, and sold them to expanding industries" (Holyoke *Transcript-Telegram,* 14 Sept. 1960).

rank and file of "independent" waterpower companies commonly sold or leased power to all comers, subject only to practical limitations in the subdivision of building sites and the provision of the necessary water races. Power buildings were sometimes erected for small users, as at Pittsfield, Massachusetts, and Rochester, New York. As described elsewhere, small enterprises were often accommodated from the surplus power of primary users, conveyed by shafting or cables to the point of use.[53]

Flour, Lumber, and the Falls of Saint Anthony

The Falls of Saint Anthony, head of steamboat navigation on the Mississippi River and site of the twin cities of Minneapolis and Saint Paul, was the last and greatest of the major waterpowers brought under development in the United States. Until it was eclipsed by the far more striking conquest of Niagara Falls, whose development in the early nineties symbolized the beginnings of the hydroelectric age, the massive cataract at Minneapolis led the field.[54] Here a year-round average of some 25,000 tons of water a minute fell some seventy feet within a distance of a mile, representing a gross capacity estimated at 120,000 horsepower. This potential was greater than that of the three major waterpowers on the lower Merrimack combined. By 1880 the round-the-clock industrial use of this power was equivalent to 20,000 horsepower on a daytime basis; within twenty-five years 40,000 horsepower were in use.[55] The importance of a location at the head of steamboat navigation on the Mississippi trunk-

[53] Tenth Census, *Water-Power,* pt. 1, pp. 205, 467–68. In 1874 all the waterpower of the town of Pittsfield, Massachusetts, being occupied, a series of public meetings led to the organization of a company to erect a building with steam power to be leased in portions as needed, "there being a strong desire to extend manufacturing" (J. E. A. Smith, *History of Pittsfield, Massachusetts, 1800–1876* [Springfield, 1876], p. 684). On transmission by shafts and cables, see below, chap. 9.

[54] None of the leading nineteenth-century direct-drive waterpowers has received so comprehensive and authoritative a treatment as that provided in Lucile M. Kane's admirable study *The Waterfall That Built a City: The Falls of St. Anthony in Minneapolis* (Saint Paul, 1966). Unless otherwise noted, the following account is based on this study.

[55] In 1880 the power in use here was developed on an average head of but 25.5 feet of the fall available of some 70.0 feet. See Tenth Census, *Water-Power,* pt. 2, pp. 163–77. For the years after 1900, see *Encyclopaedia Britannica,* 11th ed., s.v. "Minneapolis." An excellent description by an engineer associated with the repair and redevelopment of the Minneapolis waterpower is Joseph P. Frizell, "The Water Power of the Falls of St. Anthony," *TASCE* 12 (1883): 412 ff., and a later account by the same author, *Water-Power,* 2d ed. (1901), pp. 573 ff.

line was greatly enhanced by the rich timber resources and grain-growing potential of the Minnesota-Wisconsin country, which at mid-century was just beginning to be opened up. With an abundance of cheap power and with river and, shortly, rail transportation at hand, the industrial future of the vigorous settlements on either side of the river above and below the falls seemed assured. The event fulfilled the promise. Lumber and flour were the great power-using industries of the nation, consuming in the decade 1870-80 half of the power used in all manufacturing. The average consumption per hand employed by these two industries was nearly three times that of the iron and steel industry.[56] In sum, Minneapolis at the Falls of Saint Anthony was the apotheosis of the country grist- and sawmills. By 1900 it had become the largest flour- and lumber-milling center in the world, as also much the largest industrial center based on waterpower.

The achievement was fairly matched by the difficulties of bringing under effective development a major river, at this point some 1,200 feet wide, with an unstable riverbed and a year-round average daily flow equal to the maximum flood-time flow of the Merrimack at Lowell. From the early 1820s, when a grist- and sawmill had been built to supply the Fort Snelling garrison below the falls, the situation here was a subject of speculative concern and political maneuvering on site and in Washington. By 1856 control of the falls on the two sides of the river, here separated by large islands, had come into the hands of separate waterpower corporations—the Minneapolis Mill Company on the west side and the Saint Anthony Falls Water Power Company on the east.[57] Here manifestly was a site calling for unified development of a great waterpower after the manner of a Lowell or a Manchester. There was no lack of awareness of potential; the industrial pioneers of Minneapolis were from the power-oriented states of Maine, New Hampshire, and Massachusetts; the name of Lowell was even proposed for the community.[58] But development was slow in gathering momentum. There were no established commercial and monied interests here as in Massachusetts ready and eager to invest large

[56] U.S., Census Office, Tenth Census, 1880, vol. 2, *Manufactures* (Washington, D.C., 1883), pp. 491–96, "Steam and Water-Power Used in Manufactures." "Over two-thirds of the entire business of Minneapolis is manufacturing, and of this three-quarters is milling. The industry which ranks next to milling in value is the manufacture of lumber" (*Water-Power,* pt. 2, p. 164).

[57] On the early development of milling and waterpower, see Kane, *Waterfall;* Isaac Atwater, ed., *History of the City of Minneapolis, Minnesota* (New York, 1893), pp. 33 ff., and chap. 15, "Manufactures," by J. T. Wyman; Agnes Larson, *History of the White Pine Industry of Minnesota* (Minneapolis, 1949), pp. 32 ff.

[58] Kane, *Waterfall,* p. 38.

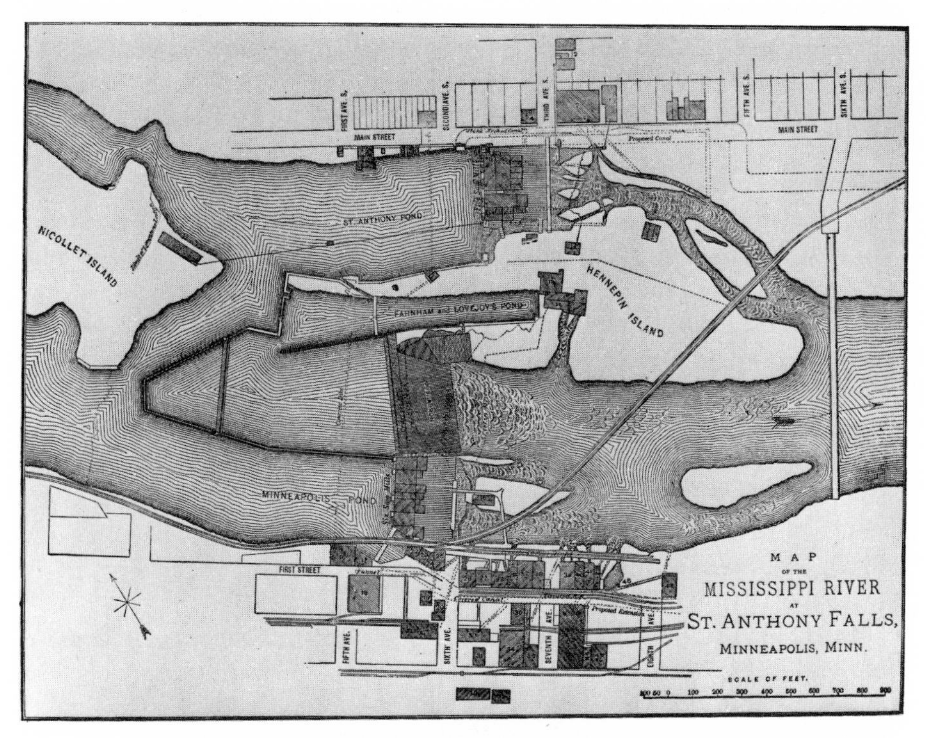

Fig. 47. Developed waterpower at St. Anthony Falls, 1880. (U.S., Census Office, Tenth Census, 1880, *Reports on the Water-Power of the United States,* pt. 2 [Washington, 1887].)

capital in power development and the manufacturing industries upon which it must depend. As late as 1855 the village of Minneapolis was close to the fringe of outer settlement; much of the wheat to supply its first mills was brought upriver from Iowa.[59] Joint action by the two power corporations resulted in a 2,200-foot-long, rock-filled, timber-crib dam, pushed out from either shore and completed in 1857. Thereafter, except for sporadic and limited cooperation on specific projects of improvement and maintenance, each company went its own way, until both were purchased in 1889 by a flour-milling combine.[60]

[59] Atwater, ed., *Minneapolis,* pp. 33 ff. This situation was reversed for the first wheat growers in the vicinity of the falls, who in the early fifties had to ship their grain several hundred miles downstream to Prairie du Chien for conversion into flour (John H. Stevens, *Personal Recollections of Minnesota and Its People, and Early History of Minneapolis* [Minneapolis, 1890], pp. 115–16).

[60] Kane, *Waterfall,* p. 146; John W. Bond, *Minnesota and Its Resources* (New York, 1853), p. 155.

The Minneapolis Mill Company was for many years the more capable and aggressive in its development of the larger power available on the west side of the river, following in important respects the models provided by the Merrimack centers in the East (see Table 10). Consulting engineers, including James B. Francis of Lowell, were brought from Massachusetts, first to prepare an overall development plan for the Minneapolis power and later to appraise the policy and practice of water use as a basis for stricter measures of control to cope with developing shortages. The leasing system adopted by the mill company took over almost verbatim the contract terms and regulations used by the Hadley Falls Company of Holyoke. The eastern practice of appointing a trained hydraulic engineer as agent or superintendent was introduced; two of the more capable engineers had eastern and European experience. Similar methods were taken by the Saint Anthony Falls Water Power Company to raise the level of performance on its side of the river (see Table 11). However, the results of careless practices in the early development of the falls in respect to the location and construction of races and tunnels were not readily corrected.

In part the long-continued, incomplete, and inefficient methods of power development at Minneapolis reflected not only divided ownership and complicated legal relationships but the absence of any strongly felt need for a comprehensive and unified plan for the development and use of the power. There were formidable engineering difficulties as well, owing to the magnitude of the hydraulic forces and the peculiar geological structure of the falls. The massive limestone riverbed, extending upstream from the brink of the falls some 1,200 feet as of 1870, overlay a deep body of soft sandstone. The ease with which the sandstone could be excavated by ordinary pick-and-shovel methods led to the adoption of a shaft-and-tunnel system of conveying water to and from the mill wheels, by this time invariably submerged turbines. The shaft-and-tunnel system was cheaper and gave greater flexibility in the siting of mills than conventional hydraulic installations with their arrangement of supply canals and tailraces limiting the location of buildings. From a main supply canal of large capacity but limited length, canals cut into the limestone bedrock could be carried as needed to the millsite. Here turbines were installed near the bottom of vertical shafts, or wells, sunk through the limestone. Conventional draft tubes carried the discharge from the turbines down to tunnels cut through the underlying sandstone. The tunnels conveyed the discharge to the river below the falls.

An upward extension of the turbine's wheel shaft carried the power through the mills' several stories—as many as seven—to the milling

Table 10. Power controlled by the Minneapolis Mill Company, 1880–81

Kind of mill.	Name of establishment.	Number of horse-power to which entitled (effective).	Average head, in feet.
Flouring	Washburn "C" mill	900	35
Do	Washburn "A" mill	700	37
Do	Crown Roller mill	525	27
Do	Pettit mill	450	37
Do	Standard mill	413	27
Do	Washburn "B" mill	375	35
Do	Northwestern mill	375	35
Do	Pillsbury mill	300	32
Do	Palisade mill	300	45
Do	Anchor mill	300	27
Do	Empire mill	263	30
Do	Excelsior mill	263	32
Do	Minneapolis mill	225	32
Do	Galaxy mill	225	37
Do	Humboldt mill	225	37
Do	Zenith mill	188	37
Do	Model mill	150	27
Do	Dakota mill	150	27
Do	Union mill	113	27
Do	Arctic mill	75	32
Do	Holly mill	75	28
Do	Cataract mill	75	28
Saw-mills	Minneapolis Mill Company's saw-mills. No. 1	300	12
	No. 2	300	12
	No. 3	300	12
	No. 4	225	12
	No. 5	375	12
	No. 6	300	12
Saw-mill		338	27
Paper-mill		225	35
Woolen-mill	North Star woolen-mill	150	27
Machine-shop		150	27
	City water-works	150	27
Wheel running the elevated railroad		75	32
Cotton-mill	Minneapolis cotton-mill	75	20

Source: Tenth Census, *Water-Power,* pt. 2, p. 175.

Table 11. Power controlled by the Saint Anthony Falls Water-Power Company

Nature of manufacture.	Name of factory.	Number of horse-power to which entitled (effective).	Average head, in feet.
Flouring	Pillsbury "A" mill	*a* 750	50
Five saw-mills		1,000	35
General manufactories on Nicollet island		400	35
Iron works; Three machine-shops; Eaves-trough factory; Mattress factory	Union Iron Works	*a* 130	16
Flouring	Phœnix mill	75	25
	North Star mill	75	25
Sash and door factory	Northwestern Fence Works	75	20
Furniture factory		75	25

SOURCE: Tenth Census, *Water-Power,* pt. 2, p. 173.

machinery. The unsatisfactory feature of this otherwise convenient and economical arrangement was the limit placed upon the amount of fall obtained. The structural weakness of the sandstone substrate required that the limestone stratum be used as the roof of the tailrace tunnel or that costly sidewalls be carried up to the limestone level if the tunnel was carried much lower. Since there was a practical limit to the length to which the supply canals could be carried, there resulted a marked concentration of high and narrow mill structures within a milling district that for many years was only a few blocks in length and one block in depth. As a noted engineer has pointed out, this method of developing power at the falls "worked admirably so long as the discharging tunnels have been but little below the lime rock, utilizing about one-half the total fall."[61] Short-term gains were obtained at the expense of long-range advantage. The abuse of the system led to costly disasters. The manner in which several dams were located and built to

[61] See the illustrated account in Tenth Census, *Water-Power,* pt. 2, pp. 163 ff.; F. U. Farquhar, "The Preservation of the Falls of St. Anthony," *TASCE* 12 (1883):393 ff.; Frizell, "The Water-Power of the Falls of St. Anthony," p. 412 ff.; idem, *Water-Power,* 2d ed., pp. 573–74. See also the photographs in Kane, *Waterfall,* and Edward A. Bromley, *Minneapolis Album: A Photographic History of the Early Days in Minneapolis* (Minneapolis, 1890).

raise the head and channel the flow on the two sides of the river narrowed the overflow section of the main dam to less than 500 of its 1,200-foot length.

Unfortunately the concentration of the area of flow accelerated the erosive action of the stream at the falls. As the underlying sandstone exposed to the falling water was eroded, the undermining and caving of the limestone resulted. The normal upstream recession of the falls, some four to five feet a year, doubled during the first decade of active power development and in the years 1866–68 reached the alarming total of two hundred feet. Such was the disturbing situation when in 1869 a misguided project for making power available at Nicollet Island above the falls by means of a long, deep tunnel to serve as a tailrace brought the whole power system to the brink of catastrophe. It caused "such threatening breaks in the limestone rock underlying the falls as to lead to the fear that the whole ledge would be undermined and carried away, leaving but a long irregular series of rapids in place of the falls." Joseph P. Frizell, who was associated with the federal government in reclamation work at the falls, pointed out that the upstream tunneling venture was conducted "in utter disregard of the precautions which ordinary foresight would suggest."[62] Emergency measures of a costly, near-desperate, and community-divisive character by municipal authorities and the milling and power companies staved off the impending disaster. More permanent restorative measures were undertaken during the 1870s by the city government and the power companies, with the federal Corps of Engineers eventually assuming control and direction of the work. To 1880 the project involved expenditures of $940,000. Both city bond issues and federal appropriations for remedial and protective measures had been opposed on the ground that private interests were the chief beneficiaries.[63]

Conflicting interests and divisiveness eventually were followed by a concentration of ownership and control in both the development

[62] Frizell, "The Water Power of the Falls of St. Anthony," pp. 417–18. See also *The Minneapolis Chamber of Commerce, 1881–1903* (Minneapolis, 1903), pp. 22–23; Farquhar, "Preservation of the Falls of St. Anthony," pp. 393 ff.

[63] Kane, *Waterfall,* pp. 67 ff., 76 ff. It was remarked in 1883 of the preservation project that "possibly the Government and private citizens had already spent enough money here to have bought and maintained steam engines sufficient to produce the power." An eastern authority on the cost of power stated that the $340,000 already spent by the city and the waterpower companies and the $600,000 spent by the federal government would have established and maintained indefinitely no more than 1,500 to 1,800 horsepower (C. E. Emery in discussion following Farquhar, "Preservation of the Falls of St. Anthony," p. 410).

and management of the waterpower at the falls and of the industries depending upon it. In addition to the division of control of the power at the falls between the east- and west-bank companies, there was a conflict between the needs of the lumber- and flour-milling operations and also, after the disastrous handling of the falls development was remedied at great cost during the 1870s, a conflict between the power companies and the community, leading to the demand that the city buy out the power companies and place them under a board of public works. Minneapolis, at the head of the falls, and Saint Paul, at the foot, were opposed in respect to measures necessary for developing the unused power at the lower falls and the improvement of navigation to permit vessels to reach Minneapolis.

With its great but unruly power base Minneapolis had grown from a village in the mid-fifties to an industrial city of nearly fifty thousand in 1880, heavily weighted on the side of flour milling in twenty-nine mills with a daily capacity of 18,000 barrels. Ten years later capacity had risen to 40,000 barrels in a city now of 165,000. By 1891 the bulk of milling capacity had been concentrated in the hands of four companies. The division of control of the waterpower, which had seriously hampered its development, came to an end with the merger of the two waterpower companies in 1889, the single organization itself owned by the leading flour-milling interests. Thereafter monopoly ruled at Minneapolis no less than at the New England textile centers. During the 1870s the sawmilling interests were bought out or had their power leases terminated so that the family controlling the Saint Anthony Falls power company, which had chiefly served the lumber industry, could shift its interests to flour milling. During the 1880s fire wiped out the few remaining sawmills at the falls and left the flour-milling industry in full possession.[64]

The increasing efficiency and economy of steam power during these years progressively reduced the control of waterpower over the industrial development of the community, yet that influence persisted. In contrast to the joint action of the three major waterpower corporations on the Merrimack in developing the power potential of a common drainage basin, the Minneapolis flour-milling and waterpower group for years blocked the efforts of businessmen at Saint Paul to exploit the undeveloped

[64] Kane, *Waterfall*, pp. 78, 92 ff., 97, 114 ff., 126, 146 ff. Lumber production in Minneapolis, relying increasingly on steam power (and wood waste as fuel) rose from 118 million board feet in 1870 to a peak of 594 million feet in 1899, thereafter declining until the last lumber mill in the city closed in 1919 (Mildred L. Hartsough, *The Development of the Twin Cities as a Metropolitan Market* [Minneapolis, 1925), pp. 40 ff.).

power of the lower falls. This was done at the sacrifice of the larger interests of Minneapolis, for the associated project of river improvement by which steam navigation would have been advanced to that city was also delayed. With the growing interest in upstream reservoirs as a means of regularizing and thereby enlarging power supply, the Minneapolis power companies had an advantage over those of New England in that federal engineers for years had under consideration reservoirs as a means of improving river navigation and flood control. The power companies gave full support to a movement obtaining its chief drive from a regional interest in a revival of river trade as a means of breaking the railroad monopoly of transportation. The completion of a number of regulating dams at upstream lake outlets during the 1880s added appreciably to river flow in late summer and autumn. Even so, less than a third of the full power potential of the Falls of Saint Anthony, upper and lower falls combined, had been put to effective use by 1890. Further progress was to be made in this respect in the years ahead but without great relevance to industry in this metropolitan community. As manufacturing continued to shift to steam power, Minneapolis "ceased to be a city dependent upon water power for its existence and growth."[65]

Minneapolis, as we noted earlier, was the last and the largest of the great direct-drive waterpowers that played so distinctive a role in the early and middle periods of American industrial development. Development of the Falls of Saint Anthony had much in common with that of the major eastern waterpowers preceding it. It drew much of its inspiration, technology, and entrepreneurial drive from the New England states whence came so many of its leading figures. There were important differences, of which perhaps the least were those related to the physical character of the falls, both in respect to magnitude and geological structure. Two centuries of pioneering settlement and commercial development in New England had preceded the arrival from England of the new mechanized textile processes and the opportunity to take advantage of great powers which the region afforded. A generation after Lowell came Holyoke and the shift from textiles to paper, a further advance beyond the basic requirements of food, shelter, and clothing to meet the cultural demands of a Puritan society with its emphasis on education and enlightenment and the ensuing demands for books and newsprint. At Minneapolis the first significant waves of settlement and the demands for large and low-cost power arrived virtually together. In the upper Mississippi Valley the subsistence and the commercial

[65] Kane, *Waterfall*, pp. 128 ff., 133. See also Tenth Census, *Water-Power*, pt. 2, pp. 130–32.

phases of frontier development were telescoped; promoters of towns and industry advanced side by side with farmers into the retreating frontier zone. Favored by mainline steamboat transportation, extending lines of railways, and the rapid rise of cities, the milling industries were hard pressed to keep pace with the demands, local and for export, for lumber and flour. "We want men to come here with money to invest in producing something," ran an exuberant account of Minnesota in 1853, "in steam and water saw and grist mills, which are now much wanted in all directions. A hundred mills would pay well, if they could at once be located" along the upper Mississippi, Saint Croix, and Minnesota rivers.[66] Steam- and water-powered mills came in together; the machinery, including prime movers of the most improved types, was imported from the East. Yet the margin of cost advantage, while no longer wide, favored waterpower where large power was required and other circumstances were favorable.

Power Developments in the South

The extent to which direct-drive waterpower at the industrial-town level penetrated the agricultural South is illustrated by three power developments in the several thousands of horsepower class and a fourth later and lesser venture, celebrated in engineering annals for its spectacular failure, in Austin, Texas. The northernmost of the group was at Richmond and Manchester, Virginia, located on opposite sides of the James River. The centers to the south were at Augusta, on the Savannah River, and Columbus, on the Chattahoochee River, in Georgia, both within the region having the greatest waterpower potential east of the Rockies, Niagara excepted. The three joined the advantages of power in great abundance with situations at the heads of tidal navigation on their respective rivers. In three of the four cities—Manchester, Augusta, and Austin—the want of private capital and enterprise led municipal authority, prompted in each instance by need of power for city water supply, to undertake the development and provide the capital, the only important instances of the kind before the coming of hydroelectricity.

At Richmond and Manchester the James River dropped eighty-four

[66] Bond, *Minnesota and Its Resources,* pp. 50 ff. This book describes early commercial and industrial conditions and facilities of the upper Mississippi country, with particular reference to the two communities at the Falls of Saint Anthony and the basic milling industries.

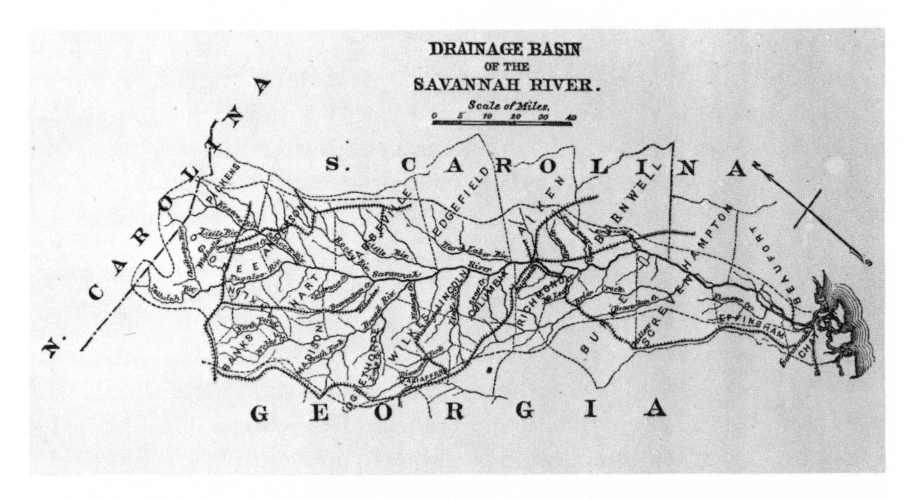

Fig. 48. Drainage basin of the Savannah River. (U.S., Census Office, Tenth Census, 1880, *Reports on the Water-Power of the United States,* pt. 1 [Washington, D.C., 1885].)

feet in three miles. The 4,000 horsepower in use there in 1880 by numerous establishments accounted for but a small fraction of the total potential. This power was developed through a number of dams, irregularly sited with reference to circumstances of terrain and industrial convenience on both sides of the river. Additional power on the north bank was obtained by taking advantage of the fall between the terminal section of the uncompleted and virtually abandoned James River and Kanawha Canal and the river below, which served as a tailrace. Here was the power base of the South's largest antebellum industrial center and the major industrial support of the Confederacy. It was especially strong in wrought- and cast-iron production and in flour milling. The Tredegar Iron Works, long the South's largest industrial establishment, took its power from the canal; it continued in active though dwindling operation until the 1950s, water-powered to the end. In 1880 some 700 to 800 horsepower were employed in rolling mills, machine shops, foundries, and accessories. Other large establishments, chiefly flour and paper mills, raised use of the canal-based supply to about 1,400 horsepower. Across the James in Manchester nearly 1,000 horsepower were disposed of under fifty-year leases to various cotton, flour, and paper mills. The city-owned waterpower system also supplied the city waterworks with 280 horsepower. Still more of the river's power was extracted at points conveniently situated for development in amounts measured in hundreds of horsepower. Development costs were relatively small by New England standards with supply canals and dams representing outlays ranging from several thousand to twenty thousand dollars in the various fragmented enterprises. With a certain understatement the Tenth Census *Report on Water Power* remarked "that the method of using the water-power at Richmond is rather complicated. . . . only a very small proportion of the total available power is at present utilized." With a theoretical capacity estimated to range from 12,400 horsepower at minimum flow to 57,000 "maximum with storage," there seemed "no technical reason why Richmond should not be one of the great manufacturing centers of the Atlantic slope." The contrast with the large integrated power developments of New England's industrial cities is illuminating if not very relevant.

At Augusta, Georgia, municipal enterprise early intervened to promote the city's growth by developing a portion of the Savannah River's

Fig. 49. Dam and bulkhead of Augusta Waterpower. (U.S., Census Office, Tenth Census, 1880, *Reports on the Water-Power of the United States,* pt. 1 [Washington, D.C., 1885].)

large potential to meet the needs of the city-owned waterworks and of the manufacturing industries it was desired to encourage. The system was begun in 1845 with the construction of a seven-mile-long canal extending from above a dam built at the head of an extended rapids; it was considerably enlarged during the 1870s. With three levels of canals, two miles in total length and affording falls ranging from eleven to thirty-eight feet, and 400 acres of land, this power project represented an investment of more than $800,000. In 1880 it supplied well over a dozen establishments. Cotton textiles and flour led by a wide margin a miscellany of other manufactories, employing altogether about 3,600 horsepower. The project was managed as well as owned by the city. The power was leased at the modest rate of $5.50 per horsepower year, one-third or less the rate at some of the leading waterpower centers of New England.

At Columbus, Georgia, said to have ranked next to Richmond as supplier of the Confederate army, the Chattahoochee River within four miles of navigable tidewater fell 120 feet with an estimated potential of more than 20,000 net horsepower. "A fall in the river of 115 feet within a mile of the city furnishes a valuable water-power, which has been utilized for private and public enterprises," stated a 1910 account. The Chattahoochee River was described in the Tenth Census *Report* as offering great industrial prospects with its freedom from ice and drift,

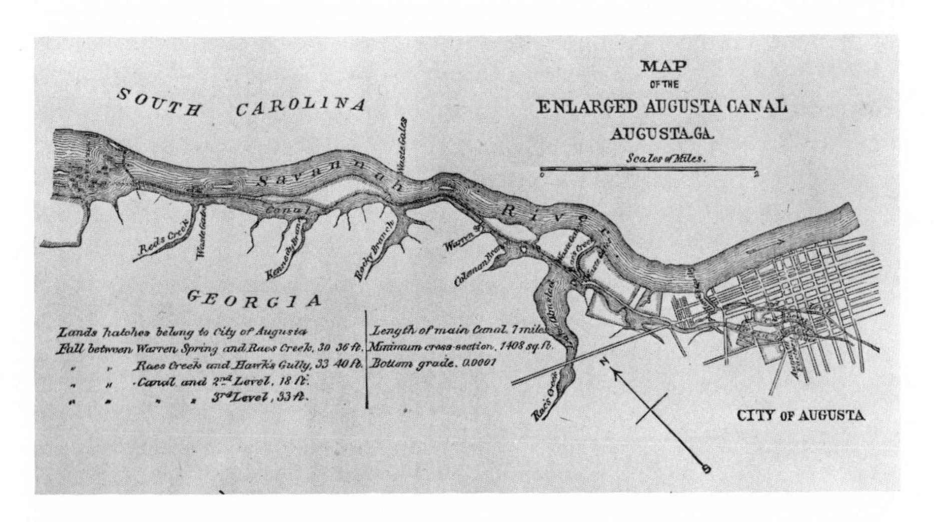

Fig. 50. The enlarged Augusta Canal, Augusta, Ga. (U.S., Census Office, Tenth Census, 1880, *Reports on the Water-Power of the United States,* pt. 1 [Washington, D.C., 1885].)

a "tolerably uniform flow," and great power potential from numerous and impressive privileges. The power in use at Columbus barely exceeded 2,000 horsepower in 1880, concentrated chiefly in a postbellum development, the Eagle and Phoenix Manufacturing Company, whose cotton mills included 44,000 spindles and 1,530 looms. The new main mill alone was equipped with four Swain turbines of 400 horsepower each. The total investment at this date was nearly two million dollars; the employment 1,800 hands.[67]

Far more illuminating, if in the outcome anticlimactic, was the experience of Austin, Texas, 200 miles up the Colorado River from the Gulf of Mexico. Under discussion for some years, a power development project for this thriving commercial and trading center became a leading campaign issue in the mayoralty contest of 1890. In a spirit of enthusiasm and large expectations, construction of the dam and associated facilities as a municipal enterprise was begun almost at once and was completed by 1893. The Austin dam was intended to supply power for the city waterworks, public and private electric lighting, and two street railways—about 2,000 horsepower altogether, leaving a surplus, by one calculation, of 12,000 horsepower as a base for manufacturing enterprise. The dam was described as "an immense structure of granitic masonry, 1200 feet long, 60–70 feet high and 18–66 feet thick." The outcome of the project was disappointing; the volume of streamflow and storage in a lake extending thirty miles upstream fell far below expectations. According to a later account, "the watershed and rainfall were rather accurately ascertained, but the keystone of the whole project, the biggest and controlling factor, namely the minimum flow of the river, was overestimated." The failure to obtain streamflow data over several years, together with errors in design and construction resulting from a confusion of engineering responsibility and contractual irregularities, had results extending well beyond the serious shortfall in capacity, culminating in catastrophe. During the great flood of 1900 what had been termed "one of the largest dams in the world" was undermined and washed out with a loss of several lives and heavy property destruction, a municipal tragedy compounded of many errors. Forty years were to pass before its replacement.[68]

[67] Tenth Census, *Water-Power*, pt. 1, pp. 531–37 (Richmond-Manchester), pp. 786–89 (Augusta), and pp. 860-62 (Columbus). See also Shryock, "The Early Industrial Revolution in the Empire State," pp. 114 ff.

[68] See account by Joseph P. Frizell, the eastern engineer in charge of the initial survey, plans, and early stage in construction, *Water-Power,* 3d ed., pp. 130–32, 245–50. See also Taylor, *The Austin Dam,* pp. 1–20; *Encyclopaedia Britannica,* 11th ed., s.v. "Austin."

Waterpower for Sale: The Overall Dimensions

Many of the larger waterpower corporations or companies were from the outset or shortly became in effect, through some form of interlocking ownership and direction, "captive" corporations. Operationally their most distinctive feature was their management as separate if not independent enterprises. A point was often reached, especially with large consumers of power, where responsibility for meeting power needs was best turned over to an organization set up or existing for the purpose. The locations of some of the most important commercial waterpower sites in the nineteenth century are included in Table 22, Chapter 10, but that list is not exhaustive. Descriptions and maps of commercial waterpowers can be found in the two volumes on waterpower in the census of 1880. East of the Mississippi there were probably a comparable number of small power companies, each with a capacity of no more than 200 to 300 horsepower, developed successfully in emulation of the more celebrated power corporations. Others were of a more speculative character, often identified with townsite-promotion enterprises. The 1880 waterpower survey records several ambitious projects in the trans-Mississippi West at such widely scattered points as Waterloo, Iowa; Sioux Falls, Dakota Territory; Lawrence, Topeka, and Blue Rapids, Kansas; and Arkansas City, Kansas.[69]

To such independent power companies must be added the canal authorities and the slack-water navigation companies in the region east of the Mississippi that found a useful source of income in their surplus flow. Each dam and lock on the many navigation canals and canalized rivers was a potential power site. Each lock marked a change in the level of the canal; usually a surplus of water beyond navigation needs was available during much of the year. Canal legislation usually allowed the renting of surplus water to mills, whose owners often found the advantages of location on a transportation route more than compensation for the priority of water use accorded navigation needs. Water rates were typically low and the amounts of power available usually adequate to the needs of small establishments. There were important exceptions—such as at Manayunk, near Philadelphia; Oswego, Lockport, and Troy, New York; Richmond and Petersburg, Virginia—where surplus water from virtually abandoned canals supported substan-

---

The first Austin dam "was finally replaced by the Tom Miller Dam in 1940" (*Encyclopedia Americana* [1973], s.v. "Austin").

[69] Descriptive accounts of waterpowers on the western tributaries of the Mississippi River, Tenth Census, *Water-Power,* pt. 2, pp. 277 ff.

tial amounts of manufacturing. A total of some 6,600 horsepower was leased on the Ohio State canals in 1880; about half this amount on the Pennsylvania canals; and 2,400 horsepower along the Erie Canal. An aggregate capacity of possibly 15,000 horsepower counted for little in the overall balance, however welcome to the small establishments served.[70]

Table 12 shows that in 1880 the large, developed, and in most instances independently managed waterpower companies accounted for exactly one-fourth of the national total of 1,225,000 water horsepower in manufacturing reported by the Tenth Census. The amount of this power in use was slightly under 50 percent, or one-eighth of the national total. The ratio of utilized to available power as shown here varied widely. In the proportion of utilization to available power the New England and Middle Atlantic states led the field; the leading individual states were Vermont, New Hampshire, Massachusetts, and New York in the East and Minnesota in the West. Understandably the southern states ranked lowest. Three-fourths of the total commercial horsepower of the nation were concentrated in New England and New York. As revealing as the aggregate amount of such waterpower in use was the substantial margin of unused capacity at most of the large developed powers. The most notable exceptions were the three Merrimack textile centers, where the actual deficits in supply were met by resort to auxiliary steam power. Among the larger developed powers only Lewiston, Maine, Bellows Falls, Vermont, and Cohoes, New York, were in or approaching a similar condition. In short, supply in most places where power was developed for sale was more than adequate to need. The remarkable developments in the Northeast were owing not only to the availability of power but to the accumulation of capital during several generations of profitable overseas trade and to the entrepreneurial response to the revolutionary advances in textile technology in Britain.

The impact of leased power was most strongly felt in those industries with large power requirements in which the trend was toward concentration of production in large establishments. These included most importantly paper and wood pulp, cotton textiles, and flour- and gristmill products. During the decade 1869–79, when waterpower in manufacturing nationwide increased only 8 percent while steam power from

[70] Horatio G. Spafford, *A Gazetteer of the State of New York* (Albany, 1824), p. 325. The most useful source is Tenth Census, *Water-Power*, pt. 1, pp. 441 ff., 480 ff., 532 ff., 559 ff., 580 ff., 622 ff.; pt. 2, pp. 491–99. See also New York, Legislature, *Report of the Joint Committee of the Legislature on the Conservation of Water* (Albany, 1912), pp. 18–19.

Table 12. Large developed waterpowers by region, 1880

	Number of powers	Available horsepower	Utilized horsepower	Percent of available power utilized
New England				
Massachusetts	4	45,920	31,757	69
New Hampshire	6	19,292	14,854	77
Maine	10	36,712	17,387	47
Connecticut	1	11,339	1,850	16
Vermont	1	8,004	7,040	88
Subtotal	22	121,267	72,888	60
Middle Atlantic				
New York	18	58,640	34,683	59
New Jersey	3	8,404	2,607	31
Subtotal	21	67,044	37,290	56
South Atlantic				
Virginia	3	14,401	5,775	40
Georgia	2	31,826	5,750	18
Alabama	1	7,257	900	12
Subtotal	6	53,483	12,425	23
Northwest				
Wisconsin	3	18,883	8,520	45
Minnesota	2	44,152	20,500	46
Kansas	1	1,501	300	20
Subtotal	6	64,536	29,320	45
Total	55	306,330	151,923	50

SOURCE: Tenth Census, data in *Water-Power*, pt. 1, xxx-xxxii, supplemented and corrected by data from descriptions of individual installations, ibid., pts. 1 and 2.

NOTE: Horsepower is variously given as gross (weight of water times height of fall per second) and net, effective, and utilized, the last three here taken as synonymous. Gross horsepower figures, chiefly given as available horsepower, have been converted to net horsepower by taking average waterwheel efficiency at 66.7 percent, a conservative figure. Although the compilers used 2,000 gross horsepower as the cutoff for "large powers," many of the utilized powers were but several hundred horsepower. Wheel ratings have been the basis for utilized figures in many instances; estimates have sometimes been used to fill gaps.

nearly the same base rose 80 percent, the use of waterpower in the industries named increased twofold, one-half, and one-seventh, respectively. Not only were the large developed waterpowers, accounting for one-fourth the waterpower nationwide, only half used but an additional eighty large undeveloped powers, with an estimated 750,000 horsepower potential, ran untrammeled to the sea.[71] Not counting some tens of thousands of waterpowers of the smaller classes, the existence of some 900,000 horsepower of unused or undeveloped capacity in large powers as late as 1880 is a reminder that throughout the direct-drive period it was industry that commonly gave meaning to waterpower rather than the converse. The discrepancy between the geographical pattern of power supply and the locational distribution of industrial demand would not be corrected until the development of high-efficiency and low-cost electrical power transmission made unutilized water-power accessible.

[71] Tenth Census, *Water-Power,* pt. 1, p. xvi, and table, pp. xxxiii–xxxv.

6
Management of a Great Waterpower The Proprietors of the Locks and Canals on the Merrimack

LOWELL AND THE two other textile centers on the main stream of the Merrimack, Manchester (established 1831) and Lawrence (established 1845), accounted for nearly one-third of the developed waterpower in the Merrimack River basin, nearly 90,000 horsepower, or over 7 percent of the developed water horsepower in the United States as reported by the census for 1879.[1] Long before large industry took firm roots along its banks in the 1820s and 1830s, the Merrimack River served as an important channel of trade and communication between a large part of New Hampshire and the seaboard. It was early improved by short navigation canals around the larger falls, and its lower reaches were joined directly to Boston by the Middlesex Canal, completed in 1803. In time the river was paralleled by rail service along its length. The unusual combination on the same stream of transportation service with power development was largely owing to the circumstance that the greater part of the fall in the Merrimack's 110-mile length, averaging but 2.5 feet per mile, was concentrated at some half-dozen points. At the sites of Manchester, Lowell, and Lawrence, long the only points where power was developed, the riverbed fell some 49, 35, and 28 feet respectively, in each case over distances of one to two miles. These great waterpowers of the lower Merrimack benefited not only from the runoff of a drainage basin thousands of square miles in extent but also from several lakes at the headwaters that served as natural reservoirs providing a valuable regulation of streamflow. These lakes were later converted to managed

[1] Carroll R. Daugherty, *The Development of Horsepower Equipment in the United States* (reprint; Philadelphia, 1927), p. 49; U.S., Census Office, Tenth Census, 1880, vols. 16–17, *Reports on the Water-Power of the United States* (Washington, D.C., 1885, 1887), pt. 1, pp. 104 ff. See also description of the Merrimack basin and the numerous water-powers on the tributaries, ibid., pp. 71–110. An excellent account of the area's physical characteristics and industrial development is Margaret T. Parker, "Geographic Factors in the Development of the Cities of the Merrimac Valley in Massachusetts," M.A. thesis, Wellesley College, 1921.

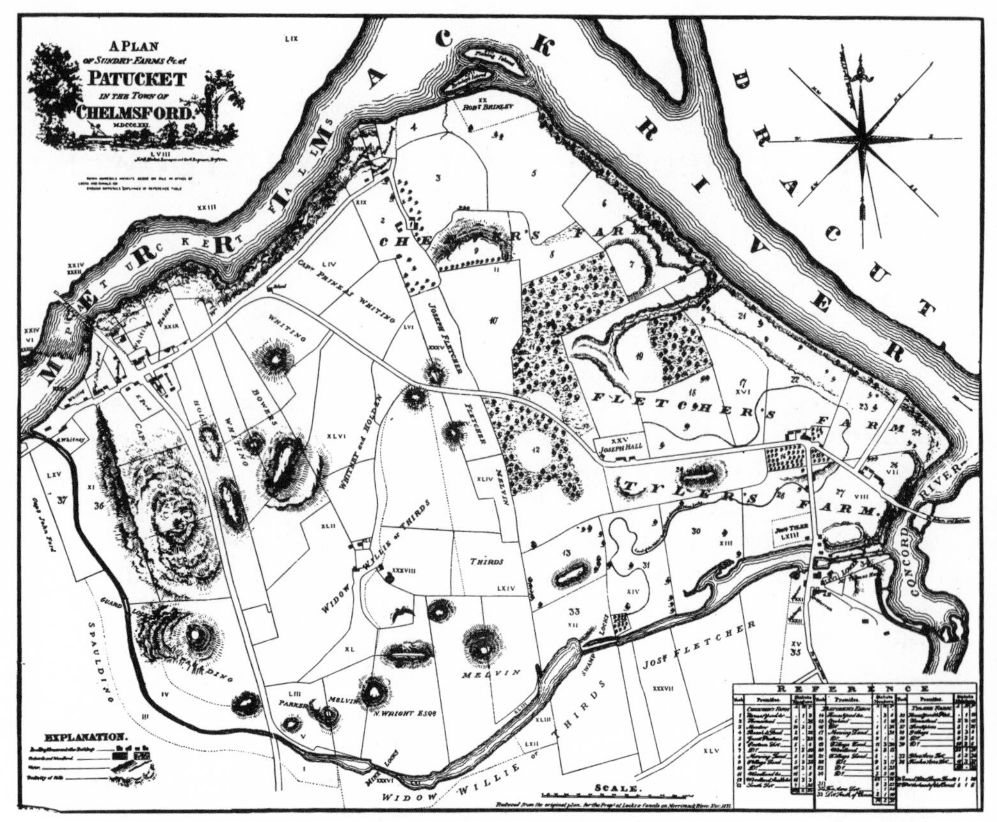

Fig. 51. Lowell before industrial development. (Courtesy of the Lowell Historical Society.)

reservoirs by which the year-round reliable flow was substantially increased. Other circumstances favorable to power development were the comparatively moderate scale of freshets, the high riverbanks affording much protection from floods, and a terrain in the vicinity of the falls well suited to the construction of the necessary dams, canals, and other structures and to the creation of large millponds for storing most if not all of the ordinary nighttime streamflow for daytime use by the mills.[2]

For the great textile centers of Lowell, Manchester, and Lawrence, established in turn at Pawtucket, Amoskeag, and Bodwell's falls on the lower Merrimack River—all within fifty miles of Boston—the waterpowers on which they rested comprised much more than the hydraulic structures and facilities created in the immediate vicinity of the falls (Fig. 54). They included at greater remove the thousands of square miles comprising the drainage basin above each falls, together with the innumerable streams, small and large, whose waters eventually contributed

[2] Tenth Census, *Water-Power,* pt. 1, pp. 71–72.

Fig. 52. Lowell, Mass., in 1876. (Courtesy of the Smithsonian Institution.)

to the volume of flow in the main river. They included, too, those features of terrain that modified and regulated streamflow, from the forests which covered much of the country about the headwaters to the countless ponds and lakes having outlet in the streams.

On a far more limited scale the same points may be made of the many hundreds of lesser waterpowers scattered throughout the Merrimack basin supplying the needs of country mills and the small factories and other mills of many kinds characteristic of the period. Yet between the typical factory and mill privilege and waterpowers of the magnitude of those on the lower Merrimack, the difference in scale was so great as to make a difference in kind. The vast majority of mill privileges, owned, developed, and managed by the establishments using their power, were selected typically because their capacity did not greatly exceed the power requirements of the establishment, which rarely was more, and commonly was much less, than 100 horsepower. In such manner the cost and difficulties of power development were minimized; responsibility for the adequacy of power supply rested almost wholly with the millowners. On millstreams of the magnitude of the Merrimack or Connecticut rivers, on the other hand, where power potentials at major sites were measured in thousands of horsepower, anything like full development and utilization was far beyond the needs and resources of even the largest industrial establishments of the time. The method first devised and applied at Lowell, less in original intent than through early recognition of opportunity and need, was, as we have seen, that of the separate waterpower corporation, developing a major waterpower in its entirety and disposing of the power commercially as an end product.

With thousands of horsepower to be leased or sold, even when the customers were fairly large establishments, generation at a central power plant was out of the question. Even had it been possible, distribution of the power by conventional lines of shafting or moving cables was impractical as well as wasteful, owing to the amounts of power and the distances involved and to heavy losses from friction in mechanical transmission. The method adopted at Lowell and elsewhere in nearly all instances was to distribute power by conveying the water itself to the wheel penstocks of the customer's mills. With the low heads characteristic of the eastern United States, at large waterpowers usually not over 25 to 30 feet such power distribution required canals of substantial size. At Lowell they ranged from 30 to 100 feet wide and from 7 to 21 feet deep.[3] If this method of power distribution was cumbersome and initially expensive, it was nonetheless both effective and efficient, although it presented many difficulties in development and problems in management. Power losses resulting from giving the necessary canals slope to convey the water to the mills probably did not exceed 5 percent on the average, a small fraction of the losses through friction, had conventional mechanical transmission been used.[4] Maintenance and management, as we shall see, presented difficulties and headaches, but operating expenses compared with steam power were low. For waterpowers with a potential of some 5,000 to 25,000 horsepower the Lowell system of power distribution by canals was admirably adapted during the years before hydroelectric power entered the scene. It was employed at nearly all large waterpowers in the eastern United States down to the hydroelectric developments of the late nineteenth century, except at Rochester, New York, where the character of the terrain led to a different arrangement, as did the geological structure of the Falls of Saint Anthony at Minneapolis. The canal method was also introduced at Niagara Falls in a limited way previous to the first large-scale hydroelectric installation there in the early 1890s.

[3] Ibid., p. 79.

[4] Of the total fall in the river at Lowell of 35 feet, 2 feet were lost through the descent of the canals, giving a net fall of 33 feet (ibid., p. 80). See also Margaret T. Parker, *Lowell: A Study of Industrial Development* (New York, 1940).

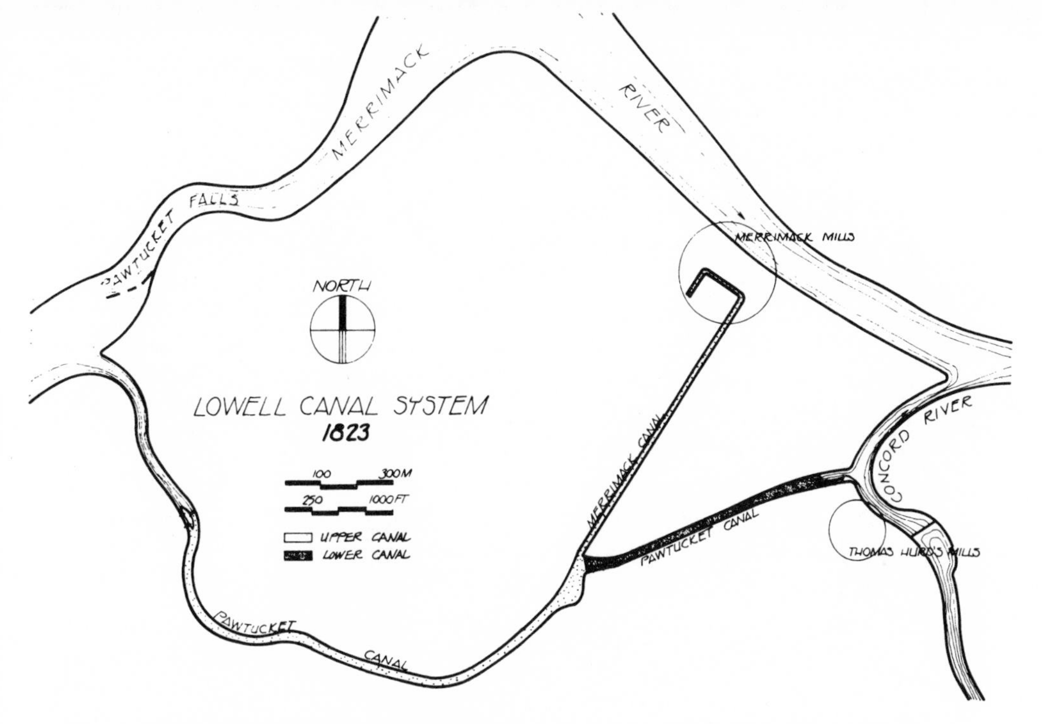

Fig. 53a. The Lowell canal system, 1823. (Drawn by Mark M. Howland, 1975. Historic American Engineering Record, National Park Service.)

The Supply Canal Network

The physical plan of the Lowell power system as it evolved during the first several decades is shown in Figure 53. The industrial community was concentrated in the eastern half of the area outlined by the bend in the Merrimack below the falls and the circuitous course of the Pawtucket Canal, designed originally for the passage of boats around the falls. The initial facilities of temporary dam and an enlarged Pawtucket Canal supplying a branch canal to serve the first group of mills were extended and improved as the number of mills multiplied. New canals were constructed from time to time to supply water to additional building sites: chiefly, the Hamilton Canal (1825–26), the Western Canal (1831–33), the Eastern Canal (1835–36), and largest and most impressive of all, the Northern Canal, built from 1846 to 1848. The dam across the river above the falls, which served to raise the water level and create a millpond and also to divert water into the canal system, was improved in 1826, was raised in height by masonry in 1833 and by flashboards about 1840, and in 1875–76 was replaced by an entirely new masonry structure equipped with flashboards three feet instead of two feet high. The lightly fastened and replaceable flashboards, erected on top of the dam to raise

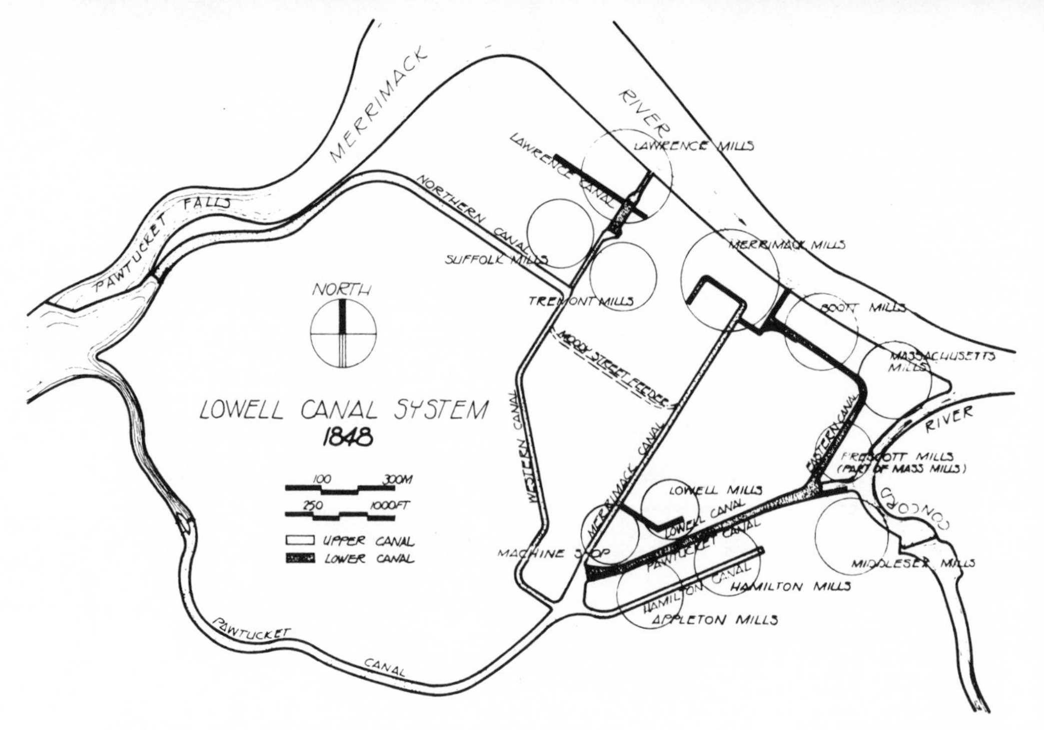

Fig. 53b. The Lowell canal system, 1848. (Drawn by Mark M. Howland, 1975. Historic American Engineering Record, National Park Service.)

the level of the millpond and increase the available fall, were designed to be carried away by flood waters, thus relieving the effects of the flooding above the dam and avoiding costly damage awards to owners of riparian lands.

Although the first group of mills, erected by the Merrimack Manufacturing Company, employed the entire fall of thirty-three feet on its great breast wheels, the need to obtain additional building sites together with the difficulty of building and installing thirty-foot breast wheels led in later canals to a dual-level arrangement, employing falls of seventeen and thirteen feet, an expedient that long remained a source of power waste and managerial embarrassment.[5] In its completed form this cumbrous but effective power-distributing system consisted of ten canals, ranging from 500 to 6,500 feet long and having a combined length of about five miles. Most of the supply canals ranged from fifty to eighty feet wide and were ten feet deep, thus rivaling in size many of the larger navigation canals of the period. The Erie Canal as originally projected was to be forty feet wide and four feet deep—dimensions adopted a few years later for the Chesapeake and Ohio Canal, which even-

[5] Joseph P. Frizell, *Water-Power: An Outline of the Development and Application of the Energy of Flowing Water,* 3d ed. (New York, 1903), p. 377; Samuel Webber, "Water Power in New England," *Engineering Magazine* 1 (1891):526 ff.

tually was built sixty feet wide and six feet deep.[6] The Lowell canals in important instances served also for transportation, providing a convenient means of distribution to the mill warehouses of supplies and materials brought from Boston by the Middlesex Canal. To about 1860 this distributing system, with such appurtenant facilities as guard and head gates, secondary dams, wasteways and waste gates, represented an investment of over $1,500,000, a very large sum for the time.[7] From the great millpond, which raised the level of the river for a distance of some eighteen miles upstream, to the tailraces, by which the largely spent water passed into the river below the falls, these hydraulic facilities comprised a massive power transmission line conveying the greater part of the Merrimack River's average flow to the penstocks of over a score of widely distributed mills. Here the formal responsibilities of the Locks and Canals company, save in supervisory and advisory capacities, ceased. Waterwheel installations were the responsibility of the mills to provide and maintain, with the chief engineer of the Locks and Canals serving on a consulting basis to the textile corporations in respect to these installations and other matters related to power and production.[8]

The ownership and early development of the Lowell waterpower was vested first in the pioneer textile corporation at Lowell, the Merrimack Manufacturing Company. As we have seen, the waterpower was shortly given quasi-independent status under the name of the original canal navigation corporation, The Proprietors of the Locks and Canals on

[6] For Lowell, see table in Tenth Census, *Water-Power,* pt. 1, p. 79. The supply canals at Lawrence and Manchester were less wide but of about the same depth. The canals at Holyoke were probably the largest of the kind in the country, ranging from 100 to 150 feet wide and 8 to 20 feet deep (ibid., pp. 72–73, 81, 222). On dimensions of early navigation canals, see Ronald E. Shaw, *Erie Water West: A History of the Erie Canal, 1792–1854* (Lexington, Ky., 1966), p. 69. The Erie dimensions were based on the Middlesex Canal between Boston and Lowell. When the canal's first enlargement was begun in 1835, its dimensions were nearly doubled to 70 feet wide and 7 feet deep. See also Walter S. Sanderlin, *The Great National Project: A History of the Chesapeake and Ohio Canal* (Baltimore, 1946), pp. 62–63.

[7] "Statement of expenditures, by and on account of, the Proprietors . . . Nov. 1845 to July 31, 1857," LCP, DA-4, 226–233. This statement, prepared by JBF, was described by him as not strictly correct but free from material errors. See also statements of 28 Dec. 1850 and 18 Mar. 1863, DA-3, 44, 285.

[8] The chief source of information on waterpower development at Lowell is the correspondence, memoranda, and other records in LCP. An excellent brief sketch of the evolution of the power system is given in "Legal Opinion of Payson Tucker," 15 Jan. 1884, ibid., A-17; see also George S. Gibb, *The Saco-Lowell Shops: Textile Machinery Building in New England, 1813–1949* (Cambridge, Mass., 1950), pp. 64 ff., 102 ff.

the Merrimack River, which for twenty years combined its basic functions of power development and maintenance with the management of real estate and the building of textile machinery in the Lowell Machine Shop. Then, in 1845, the Locks and Canals was shorn not only of most of its landholdings and the machine shop but of its independence as a waterpower company. Its stock was taken over by the former manufacturer-lessees in proportion to the amounts of power previously leased by each, and the Locks and Canals became the common service agency of the eleven lessee corporations, comprising the main body of Lowell's industry. All but the large machinery-building establishment, the machine shop, were engaged in the production of textiles.

In developing and managing the power-supply system on which rested America's first wholly industrial city, the Locks and Canals faced, on an unprecedented scale, many novel hydraulic engineering problems. Whereas successor organizations at Manchester, Lawrence, and elsewhere could draw upon its hard-won experience in harnessing a large river to the service of industry, the Locks and Canals had to learn its lessons the hard way and at first hand. In the absence of hydraulic science adequate to the massive scale of its operations, the Locks and Canals had to experiment and devise its own rules of thumb. Whereas later comers usually enjoyed the opportunity and advantage of starting with a more or less comprehensive development plan, the Lowell complex grew far beyond the first intentions and early hopes of the founders. Such basic facilities as canals were to prove in important respects ill suited to serve by enlargement and extension the expanding power needs of the industrial community.[9] Some of the original deficiencies were never to be corrected, owing to excessive cost: eventually auxiliary steam power was reluctantly accepted as the simpler and cheaper alternative. Yet in its time the Lowell power system

[9] Loammi Baldwin, in a report to the Amoskeag Manufacturing Company on developing the hydraulic facilities at Manchester, New Hampshire, wrote in 1836: "We have all a very valuable lesson in the mills at Lowell and I should not wish to see another great work on the same river suffer from the want of attention to scientific principles in the outset. Had their canal from the river been constructed upon a proper form they might have had more water about two feet higher, and not have been obliged to raise their dam to get what they now obtain." After visiting the Amoskeag site, Baldwin reviewed some features of the canal layout and design but was prevented by existing engagements from participating actively in the Amoskeag development (Baldwin to Francis C. Lowell, agent of the Amoskeag Manufacturing Company, dated Charlestown, 29 Aug. 1836, Loammi Baldwin Papers, Baker Library, Harvard University, Cambridge).

was a pioneering engineering and managerial achievement of the first order.

The essential function of the power company was simple enough: to maintain year-round the necessary volume of flow at uniform head (depth) throughout the canal system to meet the requirements of mill operations during the eleven-hour working day.[10] The performance of this function brought many difficulties to, and imposed heavy responsibilities upon, the Locks and Canals. The operating requirements of the great integrated textile mills were far more precise and exacting than those characteristic of the typical industrial establishment. The supplying of power was complicated by the interdependence and interaction among the many components of the system, from the various canals to the waterwheels of the individual mills, all drawing upon a common source and each affected in varying degree by the operations of the others. Owing to the large scale of operations at the Lowell mills there was much concern with efficiency and costs; interruptions and irregularities in operations were serious matters. The pressure on management, the superintendents, or "agents," from absentee but nearby Boston owners, and from intercorporate rivalry was always present. Within the Locks and Canals also there was the challenge to the ability and pride of the engineering staff, in itself a major innovation in industry. Much of the smoothness and efficiency with which the score and more large mills operated depended upon the regularity and uniformity with which each was supplied with water for power. A few minutes' delay in the early morning or midday opening of the supply gates; failure by a few inches to maintain the normal supply level; slowness in detecting and countering marked over- or underdrafts on the canals by some mill or mills disturbing to the operations of adjacent mills; poor judgments in imposing or lifting

[10] Original leases carried the right to draw water fifteen hours a day. The working day was usually eleven hours with an hour's intermission at midday ("Memorandum for deposition of James B. Francis in the case of the Great Falls Manufacturing Company v. the United States" [ca. 20–22 Dec. 1862], LCP, DB-3, 233 ff.; also JBF to W. H. Thompson, 28 Jan. 1869, DB-5, 381). Paper mills were the only important exception to the common practice of working but a single shift. Including the time getting under way and stopping during the day, as at noon, the "use and waste," came to twelve hours daily. This total was reduced one hour by the ten-hour law passed in Massachusetts in the middle seventies (JBF to F. B. Crowninshield, 25 Sept. 1874, ibid., DB-7, 390 ff.; to W. A. Knowlton, 12 Feb. 1875, DB-8, 39). Except during emergencies, Lowell mills did not operate at night until the 1890s and then, at first, only a few mills operated part of the night (H. F. Mills, superintendent, to Directors, 24 June 1898, ibid., BC-2, 309–10).

bans on the use of surplus water during the seasons of low waters—such errors and failures were quick to bring complaints and at times angry protests from the injured parties. It was the unhappy lot of the Locks and Canals superintendent to face the conflicting demands of different mill managements and the importunate pressures of mill superintendents quite ready to subordinate the common need to their special interest. The correspondence of James B. Francis as superintendent of the Locks and Canals consists to a surprising extent of efforts to secure compliance with regulations in respect to water use according to the terms of the manufacturers' leases. There appears to have been a great deal of carelessness, indifference, and outright evasiveness by many mill officials.

In sum, a system of power supply so large and cumbrous yet complex and sensitive as that on which the industrial prosperity of Lowell rested required for its development and management not only engineering knowledge and skill and operating wisdom but a combination of tact, even-handedness, and, at times, inflexibility. Many men contributed to this achievement, especially Paul Moody, Patrick Tracy Jackson, and Kirk Boott during the early pioneering years, and, somewhat later, Uriah A. Boyden, for his role in introducing and perfecting the hydraulic turbine. Yet it was James Bicheno Francis (1815–92), for nearly half a century chief engineer and superintendent of the Locks and Canals, to whom chief credit must be given. Coming to this country in 1833 at the age of eighteen, Francis already had several years experience as engineer apprentice on harbor works and canal construction in the south of Wales and England. He entered the service of the Locks and Canals in 1834 and was occupied chiefly with drafting, design, and supervision of a mill under construction. Upon the departure of the engineer in 1837 Francis assumed that position. In 1844 he was made "agent," or superintendent, of the Locks and Canals as well. Francis rose through his own efforts and talent to become one of the leading civil and hydraulic engineers of his day. Virtually his whole professional career was spent in the service of the Locks and Canals, dedicated to his primary responsibility of meeting the power needs of the textile corporations, who were in fact if not in name his employers. His original engineering work on the flow and measurement of water and in the improvement and adoption of the hydraulic turbine, studies that brought him wide recognition and fame, was carried out in the interest of obtaining the greatest possible utility from the waterpower of the Merrimack River. Francis not only presided over the development and use of the Lowell waterpower but, as chief engineer of the Locks and Canals, also served as consultant to the textile corporations on matters of power supply and equipment as well as on other matters.[11]

Enlargement of the Canal System

With the control of the industrial power base firmly secured by the takeover of the Locks and Canals, the textile corporations proceeded to carry out an impressive, as well as urgently needed, program of enlargement and improvement of the hydraulic facilities. In this engineering task the success of the Waltham-based entrepreneurial and technical achievement—the skillful integration of all processes of textile production made possible by Francis Lowell's development of the power loom—was matched by the Francis-guided rationalization of the power-supply system. This took the form of two interdependent advances: the acquisition of upstream storage reservoirs of large capacity—the New Hampshire lakes some eighty miles up the Merrimack Valley—and the enlargement of the hydraulic facilities centering in the construction of an additional supply canal of impressive capacity, the Northern Canal, to handle the increased supply of water obtained from the lake reservoirs to meet rapidly expanding mill needs.[12] The great millpond was generally adequate for overnight and weekend storage, saving water that otherwise would have wasted over the dam and doubling the power available during mill hours. Yet during seasons of drought or frost the diminished flow of the Merrimack was insufficient in working hours to maintain the desired "static head" in the supply canals. At such times the mill wheels slowed and operations had to be curtailed, often to considerable disadvantage. "From morning to afternoon," ran a later account of the situation, "there would be a serious loss of head, sometimes, amounting to as

[11] See William E. Worthen, "Life and Works of James B. Francis," *Contributions of the Old Residents Historical Association* (Lowell), 5 (1892–93):230 ff., and DAB. Of particular value is Desmond FitzGerald et al., "James Bicheno Francis: A Memoir," *JAES* 12 (1894):1–9; also R. S. Greene et al., "Memoir of James Bicheno Francis, Past President and Honorary Member," *PASCE* 19 (1893): 74–88. On 29 July 1885 Francis wrote that "last November I completed fifty years service with the Locks and Canals, beginning as a surveyor less than twenty years old." Resigning in 1884 as agent and engineer, Francis was appointed consulting engineer to serve as needed.

[12] A concise statement of this program of enlargement with costs is given in JBF to Thomas G. Carey, president, 6 Dec. 1858, LCP, DA-5, 58–64. See also Tenth Census, *Water-Power,* pt. 1, pp. 79 ff.

much as 5 feet, resulting in low speed for the rest of the day. All the mills suffered and all complained."[13]

The industrial use of storage reservoirs to augment supply during periods of low water had long been resorted to in some parts of New England. In time their use would become common on the busier streams where conditions were favorable, that is, where ponds, lakes, or terrain suited to their construction could be obtained or controlled at acceptable cost.[14] The achievement of the Locks and Canals at Lowell, joined by the Essex Company, which owned the waterpower at the newly inaugurated textile center of Lawrence, some miles below, consisted in the creation of a reservoir system on a scale corresponding to the magnitude of the great waterpowers of the lower Merrimack. In 1846, when power shortages owing to enlarged demand and use became pressing, an enviable system of natural lake reservoirs at the headwaters of the Merrimack River was acquired at a considerable but amply justified expenditure in virtually a single transaction. The Locks and Canals and the Essex Company purchased control of the textile company possessing riparian rights at the outlet of New Hampshire's largest body of water, Lake Winnipesaukee.[15] Control was later obtained of the outlets of several smaller lakes, among them Squam and Newfound lakes and Smith's Pond. By constructing relatively small dams and gates at the outlets, arranging for the regulation of outflow when and as needed, and paying rather substantial flowage damages to riparian owners adversely affected by the raising and lowering of lake levels, the Lowell and Lawrence textile interests converted these lakes, some eighty to ninety miles distant and with a total area of one hundred square miles, into splendid storage reservoirs. Francis estimated that the dry-season flow from the lake reservoirs was two to three times the volume of the uncontrolled natural flow. Between 1852 and 1865 water was drawn from the lakes during dry periods an average of 52.5 days annually, with a low of fifteen days in 1856 and a high of 117 days in 1859. The lake levels could be drawn

[13] "Legal Opinion of Payson Tucker," 15 Jan. 1884, LCP, A-17. For the ponding facilities at Manchester and Holyoke, see Tenth Census, *Water-Power,* pt. 1, pp. 84, 224.

[14] See Tenth Census, *Water-Power,* pt. 1, pp. 31, 51, 88 ff., 618 ff.; "Extract from Report to Directors of the Locks and Canals, Sept. 21, 1869," LCP, DB-8, 17 ff.

[15] On the development and management of the Lawrence waterpower, see Tenth Census, *Water-Power,* pt. 1, pp. 73 ff. The controlling interest in the Winnipesaukee Lake Cotton and Woolen Manufacturing Company, commonly called the Lake Company, was divided equally between the Locks and Canals and the Essex Company (JBF to G. Atkinson, 30 July 1888, LCP, DB-14, 42–43).

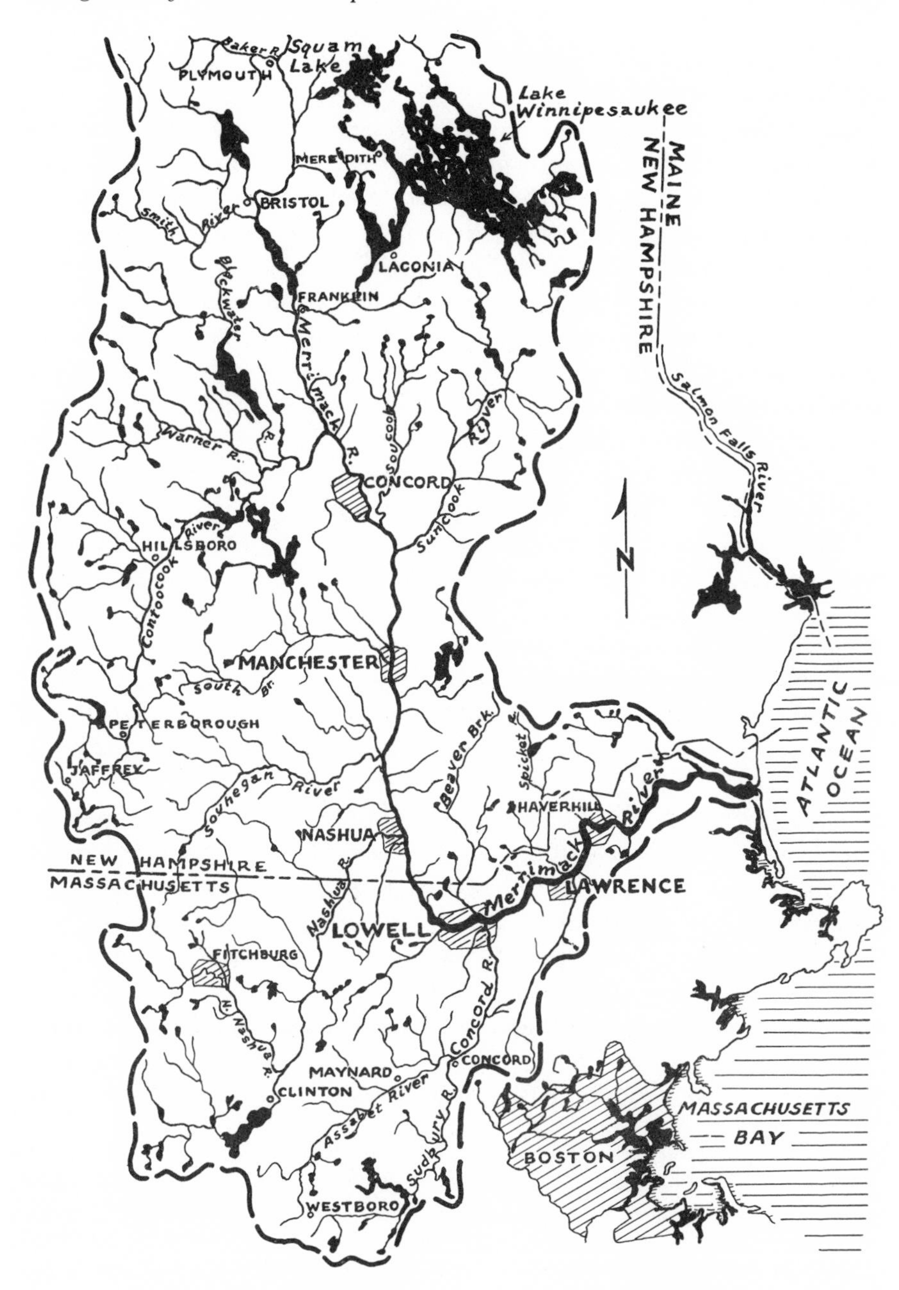

Fig. 54. Merrimack River drainage basin (south of Plymouth, N.H.). (Courtesy of U.S. Army Corps of Engineers and Isabella B. Walker.)

down twelve feet below their full spring height and six feet below the normal summer level.[16]

By means of these New Hampshire reservoirs the floods of wet seasons were moderated, relief from the worst embarrassments and losses of low water secured, and the supply of power for the expansion of mill operations increased fully one-half. Benefits accrued as well to a number of New Hampshire communities. These results were obtained by the Locks and Canals with a capital outlay of some $273,000 during the first ten years of their use, in addition to flowage payments and modest operating costs. Later expenditures brought the total outlay to some $430,000 by 1880; the greater part consisted of flowage damages paid riparian owners on the lakes. The expenditures of the Locks and Canals were more than offset by the income of more than a half-million dollars from the sale of leases resulting from the enlarged capacity of the system and transactions between the Locks and Canals and its owner-clients.[17] As with the earlier takeover of the Locks and Canals stock, the additional power made available by the New Hampshire reservoirs was distributed among the textile corporations in exact proportion to the waterpower leased by each.

The negotiation of the New Hampshire lakes arrangements evidently presented no great difficulties to the shrewd Boston mercantile capitalists in charge of business matters. To take advantage of the greatly improved supply situation, however, presented engineering problems that were as difficult as they were costly in resolution. The additions and improvements included principally the construction of new canals, the enlargement and repair of portions of the old ones, and the provision of supporting facilities in the form of guard gates and wasteways. By far the most important and costly innovation was the Northern Canal, designed by Francis and built in 1846–48, to increase the capacity of the system, heretofore dependent upon the progressively inadequate Pawtucket Canal. The latter was enlarged,

[16] Some of the enlarged volume of dry-season flow in later years was added by the large number of reservoirs on smaller branches of the Merrimack River system, established and operated by millowners situated on the lesser affluents but serving manufacturers on the main river as well (JBF to A. Penfield, 7 June 1867, LCP, DB-5, 200–1; see also JBF to Directors, 10 July 1866, A-17). The area of Lakes Winnipesaukee and Winnisquam was 71.8 square miles, and three smaller lakes brought the total reservoir area to 100 square miles (memorandum, 9 Aug. 1855, ibid., A-11). Samuel Webber, "Water-Power—Its Generation and Transmission," *TASME* 17 (1895–96):42 ff.

[17] Memoranda, 9 Aug. 1855 and later, LCP, A-11, including "Statement of expenditures . . . 1845–1871." See also JBF to C. H. Dalton, 6 Dec. 1879, DB-10, 148.

Fig. 55. Detail from 1876 isometric map of Lowell showing mill area receiving waterpower from the Northern Canal. (Courtesy of the Smithsonian Institution.)

and together the improvements were to relieve although hardly to eliminate the inequities to a number of mills owing to their unfavorable situations on the canals. Over a half-million dollars was expended on the Northern Canal alone. Hundreds of thousands of dollars more were spent on improvements in other portions of the distribution system, including widening certain canals and building the new underground Moody Street feeder, joining the northern and southern portions of the canal network and, in effect, extending the Northern Canal. Additions to the guard gates and wasteways, much less costly, were necessary: the first serving as cutoff valves for both routine and emergency use when portions of the canals required draining and for protection against potential breakthroughs of abnormally high floodwaters into the heart of the city.[18]

[18] Worthen, "Life and Works of James B. Francis," pp. 234–36; Tenth Census, *Water-Power,* pt. 1, pp. 79–80; JBF to A. Penfield, 7 June 1867, LCP, DB-5, 200–1; "Legal Opinion of Payson Tucker," 15 Jan. 1884, A-17. On relation of cross section to

About a thousand feet downstream from the Pawtucket Canal at the site of the huge Pawtucket dam, the Northern Canal, protected by a massive masonry river wall, tapped the Merrimack directly and followed a course at first closely paralleling the river, then bending sharply toward its point of intersection with the existing Western Canal. The new canal—100 feet wide, 16 to 21 feet deep, and 4,100 feet long—overcame the bottleneck of the obsolescent Pawtucket Canal, hitherto the sole source of supply for the many mills along it and several secondary canals. With two supply canals from the river to handle the enlarged volume of water made available by the upstream reservoirs, the whole system was greatly improved. The flow in the two canals was slower, more constant, and therefore more efficient, reducing a source of power loss and inequity among many of the mills.

The Pawtucket Canal was enlarged to sixty feet wide and eight feet deep, but still had a somewhat irregular cross section that was unfavorable to an even flow. The many portions of the system supplied by it continued at a disadvantage, even after the improvements of the 1840s. As new mills continued to be added, the lower ones (in relation to streamflow) were at a disadvantage compared with those above from the loss of head resulting from the canal gradient required for flow. The complexity of the problem is indicated by Joseph P. Frizell in his treatise *Water-power*:

> The head is diminished in time of high water by the rising of the river below the mills more than it rises above, and in time of low water by the lowering of the mill-pond in the afternoon, and the consequent increased velocity in the canals, consuming a large amount of head in giving motion to the water. The loss of head is an evil that aggravates itself. The more the mills draw the greater is the velocity in the canals, and the greater the resulting loss of head. The result is that, unless the draft of water [by the various mills for their operations] is under constant and firm control, the lessees may so diminish the head as to inflict great mutual injury.

One of the most effective measures eventually adopted to increase the carrying capacity of the canals was to line the canal beds and slopes with timber and planking and to smooth the masonry work in side-walls. By such means, Frizell noted, the velocity of flow in rough and irregular canals could often be increased two to three times without change of cross section or gradient.[19]

---

flow, see Frizell, *Water-Power,* pp. 388 ff. The foresight of Francis in preparing against an exceptionally heavy flood in the design of the main guard gates in the Pawtucket Canal is noted in the obituary memoir by Green et al., "James Bischeno Francis," pp. 75–76.

[19] Frizell, *Water-Power,* pp. 621–23.

Fig. 56. Detail from 1876 isometric map showing central mill area of Lowell receiving waterpower from the Pawtucket Canal. (Courtesy of the Smithsonian Institution.)

Another factor contributing to the unequal status of some mills was the shift from old-style breast wheels to turbines, which for technical reasons favored turbine-equipped mills. The introduction for the first time in 1859 of a charge for surplus water added further weight for the grievances of some and to the burden of management, which with the conscientious Francis was a matter of serious concern. The Tucker investigation and opinion rendered in 1884, made in response to demands for legal redress, conceded the need for compensating or offsetting the lower heads—and less power—received by the less favorably situated mills. The difficulties of a system enlarged by increments could hardly be eliminated except at a cost not believed to be warranted; here was a defect to some degree inherent in a system of power distribution by massive volumes of water supplied at low heads. As Dwight Porter remarked of the development of the power at Holyoke:

The development of a great power is a costly enterprise, and can seldom be carried out, in the start, as a completed design; canals are extended and other improvements made only as business and revenues warrant, and so it often happens that some sites are disposed of which are unfavorable to the full utilization of the available power. It is not unlikely that the same fault sometimes arises from too sanguine ideas as to the capacities of a privilege relatively to the demands which will be made upon it.[20]

Other improvements at Lowell carried out under Francis included a new and higher dam, higher flashboards, and the removal of a portion of the riverbed below the main falls—together adding at least 15 percent to the aggregate waterpower.[21] The best measure of these varied and substantial improvements is revealed in the growth of power consumption at Lowell from a base, as of about 1830, estimated at some 50 mill powers (a mill power equals 60 to 65 horsepower). By 1840 power consumption had grown to 65 mill powers, by 1843 to 91 mill powers, by 1852 to 132 mill powers, by 1856 to 163 mill powers, by 1858 to 180 mill powers, and by 1880 to 198 mill powers.[22] These figures show that power consumption at Lowell rose well above the 139.4 mill powers leased following the improvements of 1846–48 and regarded as available on a reliable year-round basis. The increase by 1880 of nearly 60 mill powers represented what came to be known as "surplus power," to be discussed below, resulting from a volume of river flow well above the firm power level much of the year, with shortages taken care of by a steadily increasing resort to steam power.

The Balance Sheet

The cost of the Lowell power system, so far as can be determined by figures that are neither complete nor clearly defined, reached some $1,500,000 in capital outlay, covering dams, canals, reservoirs, and all appurtenances, to 1860. To this figure, covering only the facilities necessary to deliver the water to the mill penstocks, must be added the cost, borne by the textile corporations, of the wheel installations at the

[20] In Tenth Census, *Water-Power,* pt. 1, p. 224; "Legal Opinion of Payson Tucker," 15 Jan. 1884, LCP, A-17.

[21] Worthen, "Life and Works of James B. Francis," p. 5.

[22] JBF to T. G. Carey, 6 Dec. 1858, LCP, DA-5, 58–64; to C. S. Storrow, 16 Dec. 1868, DB-5, 356; to O. Guthrie, 2 Sept. 1886, DF-1, 20. Figures for 1840 and 1843 are for power leased, which may or may not have been the same as power consumed. In 1895 Col. James Francis, son and successor of J. B. Francis as agent and superintendent of the Locks and Canals, reported that as a result of certain improvements in the system the firm, year-round capacity of the waterpower at Lowell had been raised to 12,300 horsepower, or 190 mill-powers (65 hp each) (Webber, "Water-Power," p. 43).

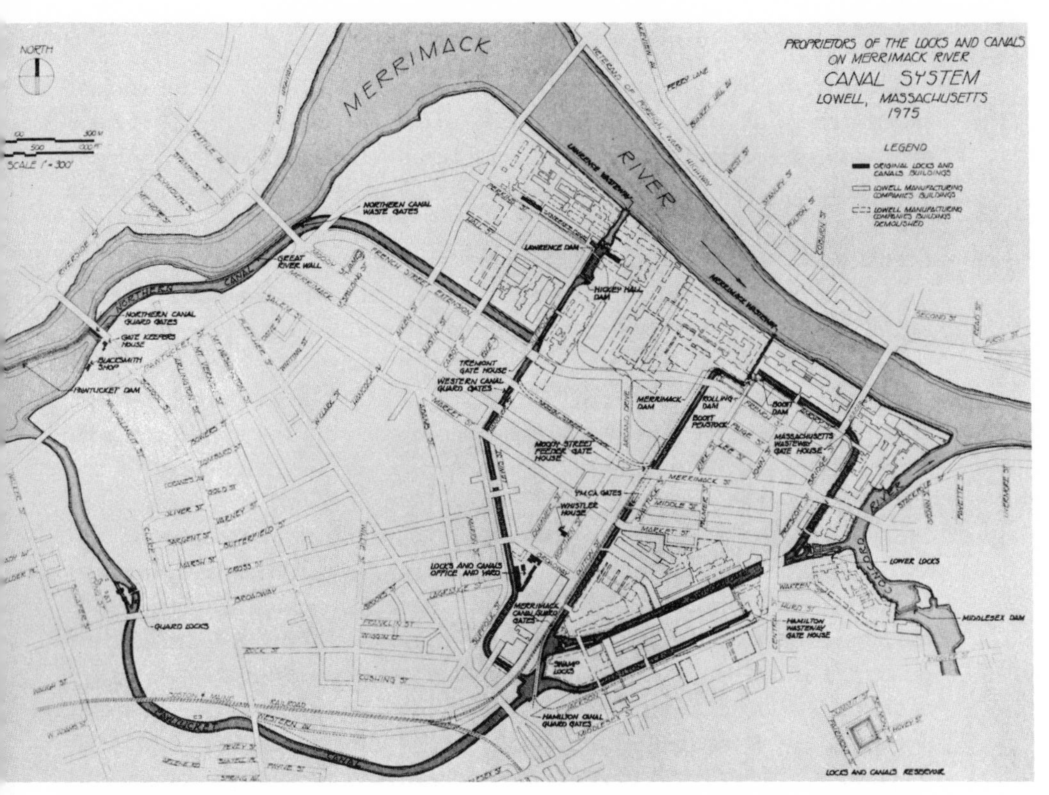

Fig. 57. Canal system at Lowell, 1975. (Drawn by Mark M. Howland, 1975. Historic American Engineering Record, National Park Service.)

mills, extending from the canal of supply to the tailrace and including not only penstocks and waterwheels but the massive gearing to reach the mill shafting and obtain the desired operating speed. Some held this amount to be about as much per horsepower capacity as the cost of the delivery system. It will be estimated here at one-third less, giving the rounded total cost of the Lowell power plant as $2,500,000, or about one-fourth the cost of the original Erie Canal, completed in 1825 at an outlay of some $10 million including not only the 363 miles of canal from Troy on the Hudson to Buffalo on Lake Erie but also 84 locks and hundreds of bridges and aqueducts. More meaningful is the comparison between the cost of the power plant and the aggregate capital of the eleven manufacturing companies it served, reported as of 1855 as $14 million.[23]

[23] Memoranda of JBF, 28 Dec. 1850, 18 Mar. 1863, LCP, DA-3, 44, DB-3, 285; James B. Francis, *Lowell Hydraulic Experiments* (Boston, 1855), p. x. The total capital investment of the ten corporations (nine textile and the Lowell Machine Shop) about 1880 was $13,950,000; total operatives employed, 16,665 (Tenth Census, *Water-Power*, pt. 1, p. 81).

During the twenty years of the Locks and Canals' independent status preceding the takeover by the textile corporations in 1845, the company was responsible not only for the development of the hydraulic facilities but also for the management of the extensive operations of the Lowell Machine Shop. In disposing of the waterpower on perpetual lease, the Locks and Canals set the pattern adopted not only at its sister textile centers on the Merrimack but, with variations, introduced widely elsewhere. During the early years leases were made at rates of $2.50 to $4.00 per spindle, varying with the times and the market, but were eventually stabilized at the latter figure, which amounted to $14,336 per mill power, based on the 3,584 spindles at the second Waltham mill. In lieu of payment of $5,000 of this sum, the lessees were charged an annual "rental" fee of $300, or 6 percent of this amount, an arrangement intended originally as protection to the lessees against nonmaintenance of the hydraulic facilities by the lessor company. If to an outlay for water at these rates of possibly $15 per horsepower per year (assuming the minimum of 60 horsepower per mill power) is added the expense of wheel installations, an aggregate figure of perhaps $25 per horsepower per year may be taken as the approximate cost to the textile corporations at Lowell of their power.[24] Not until the late nineteenth century was steam power generated by large and efficient engines available at figures such as these.

The real estate holdings of the Locks and Canals consisted of some 800 acres of land, originally purchased at an average figure of about $100 an acre by the promoters of Lowell to secure riparian rights to the waterpower and to provide land for millsites and the development of the town. This land was transferred in 1826 to the Locks and Canals, which laid it out in millsites and town lots. These were gradually sold over the years at figures ranging from $200 to as much as $40,000 an acre. In the early grants or leases of the waterpower, one or two acres for building sites were conveyed with each mill power, a practice shortly modified to the right of the lessee to buy some four acres with

[24] This account is based largely on memoranda and letters, written chiefly by Francis, in LCP. On corporate organization and management, see esp. JBF to Benjamin Saunders, 21 Oct. 1864, DB-3, 565 ff.; to A. J. Rogers, n.d., DB-7, 445 ff.; to J. P. Frizell, 7 Dec. 1883, DB-12, 274–75; to W. Hunt, 20 Jan. 1858, A-30; "Hydraulic Experiments and Memoranda, 1849–1863," A-18. See also Nathan Appleton, *The Introduction of the Power Loom, and the Origin of Lowell* (Lowell, 1858), pp. 17 ff.; Gibb, *Saco-Lowell Shops,* pp. 63 ff., 71 ff., 102 ff. George F. Swain gave the cost of water per net effective horsepower at Lowell and Lawrence as $15 and the total cost of power at these cities as $24 ("Statistics of the Water Power Employed in Manufacturing in the U.S.," *PASA,* n.s. 1 (1888-89): table 6).

each mill power. With little competition and with the lease of water-power frequently conditioned upon the purchase of machinery, the Lowell Machine Shop carried on its operations to considerable advantage. According to later reports, little if any profit was realized from the waterpower, the sales of which were frequently made with a view to the profitable employment of the machine shop. Of the overall profitability of the activities of the Locks and Canals there is no doubt. With the transfer of the Locks and Canals stock to the textile corporations in 1845, in exact proportion to the waterpower leased by each, the sale of the machine shop, the town lots, and the stock gave a return to the original Locks and Canals stockholders, as earlier remarked, of more than three times the par value of the stock.[25]

The intimate association of land ownership with power development and use was one of the most distinctive features of water as a source of power. It introduced considerations that were particularly striking in the development of the large textile centers of New England. Joseph P. Frizell, one of the most experienced and knowledgeable hydraulic engineers of the region, expressed his views at some length in his standard treatise on waterpower:

> The great water-power companies of New England made two mistakes in reference to land: 1. They did not secure a sufficient quantity of land. 2. They were in too great haste to dispose of it. The land has rapidly passed out of the hands of the companies and become the property of the general public. A class of population has arisen having aims and purposes entirely different from the manufacturers. The latter generally hold the greater part of the property of the municipality, but the power of taxation is with the former. Taxes are levied and expended with usual municipal extravagance, for purposes in which the manufacturers have little direct concern. The manufacturing interest finds itself compelled to pay the larger half of the taxes, with no direct voice in the assessment of the same. Fierce litigation is continually occurring over taxes and assessments. Such a state of affairs could not arise if the company retained control of the land, allowing it to be held on leasehold tenure, but refusing to alienate. By retaining the fee of the land

[25] "Memorandum for deposition of James B. Francis in the case of Great Falls Manufacturing Company vs United States" (ca. 20–22 Dec. 1862), LCP, DB-3, 236; JBF to C. Estes, 27 Mar. 1884, DB-12, 333-34; JBF to Washington Hunt, 20 Jan. 1858, A-30; to Benjamin Saunders, 21 Oct. 1864, DB-3, 565 ff. In a letter regarding the early history of Lowell to J. P. Frizell, Francis declared, "I think there has been far more profit from the rise in the value of land than from the sale or lease of water power. I recollect hearing P. T. Jackson, who was a leading man in the enterprise [at Lowell] say there was no profit from the water power" (7 Dec. 1883, ibid., DB-3, 565). See also Gibb, *Saco-Lowell Shops,* pp. 101, 103.

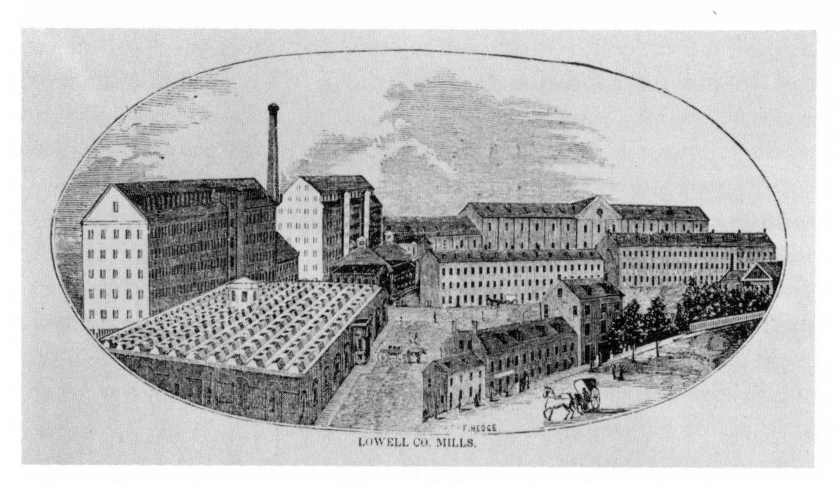

Fig. 58. Lowell Co. mills. (Charles Cowley, *History of Lowell,* 2d ed. rev. [Lowell, 1868].)

Fig. 59. Merrimack Mill No. 6. (Charles Cowley, *History of Lowell,* 2d ed. rev. [Lowell, 1868].)

other than that required for manufacturing, not only could the company control the assessment of taxes, but it could control the liquor traffic, vicious resorts, gambling places, and other matters of vital importance to its working people.[26]

[26] Frizell, *Water-Power,* pp. 624–25.

Maintenance and Management

Freed of its varied business and promotional activities by the 1845 conversion into an agency of the textile corporations, the Locks and Canals thenceforth was concerned solely with the functioning of the waterpower system. With the completion of the basic enlargement of the system by the early 1850s, the duties of the Locks and Canals became largely those of maintenance and management. These duties went well beyond the simple operating routines ordinarily sufficient for a plant having its own waterpower or for the smaller waterpower companies. The typical independent waterpower of no more than a few hundred horsepower, serving a single establishment, required little attention, apart from routine maintenance, save in emergencies. Ordinarily one or two attendants could operate the gates regulating flow, clear out trash racks at the penstock intakes, deal with ice in winter, make minor repairs, and attend to other maintenance duties. By contrast, the management of a waterpower on the scale of Lowell was an exacting and burdensome task, heavy with responsibilities not only to the corporate owners but to the large industrial community served. The great size and complexity of the system, the magnitude of the hydraulic forces to be controlled, the built-in deficiencies that had to be lived with because of the heavy cost of alteration, the large and numerous mills to be served, and the aggressive, competitive, often demanding spirit of management in the textile corporations to which the Locks and Canals was now directly responsible—these varied elements often taxed the managerial and engineering resources of the power company to the utmost, especially during periods of water shortage. The problems of power management must be seen, too, against a background of continued and rapid expansion of production with the value of Lowell's output of cotton textiles multiplying from a little over $4 million in 1845 to nearly $20 million in 1880.[27]

As the demand for power grew at Lowell, the alternative of steam power was accepted slowly and with much reluctance, not only because of its higher cost, but in some measure from force of habit and tradition and quite possibly as a matter of engineering pride in making the most of the power system on which the city's rise to industrial fame had depended. In 1864 Francis estimated the cost of steam

[27] Parker, "Geographic Factors in the Development of the Cities of the Merrimac Valley," pp. 74 ff. At the terminal date there was but a slight difference in wholesale price levels for textiles (U.S., Bureau of the Census, *Historical Statistics of the United States, 1789–1945* [Washington, D.C., 1949], pp. 230–32).

power as twice that of waterpower. Exactly ten years earlier he had estimated that steam power was three times as costly.[28] The margin of cost narrowed as steam engines improved in fuel economy and regularity of motion and were employed in ever larger installations. Further, the increasing requirements for heat in processing textiles—readily supplied by exhaust steam from noncondensing engines—frequently made steam power an inexpensive by-product.[29] The desire to take advantage of very cheap surplus waterpower during some months of the year also encouraged the installation of auxiliary steam power to provide for the remaining months; once installed, the extension of the use of this steam power was chiefly a matter of fuel expense. As early as 1867 one-fourth the power in the textile mills was steam power; within another decade more steam power than waterpower was in use.[30]

The direct-drive system of large-scale water distribution, the use of which Lowell provides the classic example, was in its essential character a gigantic prime mover, in its full scope extending from the distant reaches of the New Hampshire reservoirs to and through the waterwheels of the mills, terminating in the tailrace outlets to the river below the falls. However dwarfed by the hydroelectric systems of a later day, the Lowell plant with its 10,000-horsepower capacity was at mid-century the most striking example of direct-drive waterpower, now approaching its peak years. Its management and maintenance when fully developed—excluding the personnel at the New Hampshire reservoirs—required some twenty hands as a regular working crew, and this was enlarged considerably at times of stress and emergency.[31] This crew included both semiskilled and unskilled laborers

[28] JBF to E. Dwight, 20 Mar. 1864, LCP, DB-3, 449; to Corliss and Nightingale, 6 Jan., 22 July 1854, A-2. See also memorandum, 15 Jan. 1854, ibid., and above, p. 000. For a fuller discussion of costs at the Merrimack textile centers, see Charles T. Main, "Cost of Steam and Water Power," *TASME* 11 (1889-90):108 ff.; Charles E. Emery, "The Cost of Steam Power Produced with Engines of Different Types . . . ," *TAIEE* 10 (1893):144 ff.; and Charles H. Manning, "Comparative Cost of Steam and Water Power," *TASME* 10 (1888–89):499 ff.

[29] See memorandum of JBF, 23 Apr. 1866, LCP, DB-5, 43 ff.; JBF to W. Worthen, 1 Oct. 1866, DB-5; memoranda and other papers in File 104, A-20. See also references to use of steam power in DB-8, 462; DB-9, 384, 445; DB-10, 393.

[30] JBF to W. H. Thompson, 21 Mar. 1881, ibid., DB-11, 81–82; to J. Powell, 3 May 1881, DB-11, 103–4; to J. P. Hutchinson, 25 Jan. 1881, DB–11, 28.

[31] JBF to C. D. Wright, 30 Dec. 1887, ibid., DB-13, 475. Direct-drive waterpower is so termed because of the direct mechanical connection between waterwheel and mill machinery, in contrast to hydroelectric power, in which generators, power lines, transformers, and motors intervene between waterwheel and machinery.

for the routine tasks of operation, maintenance, and repairs of the varied facilities and equipment and an engineering staff—Francis and his assistants, aided by clerical personnel—to supervise and directly take charge of the more critical operating responsibilities. During periods of floodwaters, when hazards attended sheer volume of flow and floating debris and ice, and in drought seasons, when water shortages required close monitoring of use and waste, the engineering staff worked long hours under pressure. For years, too, "apprentice," or student, engineers were recruited for the tedious summer duties of checking the flow and measuring consumption at every operating point in the entire system. Apart from routine maintenance and minor repairs—clearing of trash racks, removal of debris from the canals, breaking up and running off ice as needed, keeping gates, wasteways, and flashboards in order, and the like—operating responsibilities of the Locks and Canals fell mainly into three divisions: management of the water, measurement of the water, and management of the lake reservoirs.

The first two responsibilities accounted together for nearly one-half of the power company's annual budget, which during the 1860s and 1870s averaged about $42,000. The first and principal responsibility was "to see that the mills are properly supplied with water."[32] Each mill had to be supplied, at a minimum, with the amount of water called for by the terms of its lease. This amount had to be provided, as precisely as practical considerations permitted, at a constant level ("static head") throughout working hours, since variations in head were reflected in changes in wheel speed and the motion of the machinery. Canal levels in turn were affected by major changes in mill operations, as when the amount of machinery in use was substantially increased or reduced for operative or emergency reasons during working hours. Despite the growing practice of equipping each waterwheel with its own governor, the canal management had to serve in significant degree as throttle and governor for the entire canal supply system, through main, supplementary, and waste gates. The supply system was "turned on" and shut off at the headgates at the beginning and end of the working day and commonly during the noon period as well. During working hours water levels throughout the system were regulated by numerous supply and waste gates. The management had

[32] JBF to Paul Hill, 30 Apr. 1851, ibid., DA-3, 78–79, and annual statements of "Estimated" Expenditures. Major repairs and nonpower activities such as fire protection were excluded. The extremes were $22,397 in 1868 and $54,000 in 1873; the average, 1867–80, $41,780.

to be ever ready to meet scheduled and unexpected variations in the water requirements of individual mills. The unexpected shutdown or resumption of operations by a large mill would quickly be reflected in the operations of—and complaints from—other mills on the same canal unless the canal management was informed in time to take countermeasures. Other water-management duties included such matters as responding to requests for out-of-hours operations or for draining canals to permit repairs to mill structures, penstocks, or wheel installations.

Shortcomings of the Supply System

The water-management task was complicated, as we have seen, by differences in the hydraulic characteristics and capacities of the several parts of the canal system. The dual-level arrangement of the canals, by which mills on the lower level were dependent for their power supply primarily upon the volume of flow from the tailraces of the upper-level mills, was a continuing source of annoyance and waste during the recurring periods of water shortage. Unless the mills on the different levels required approximately the same quantities of water, the difference had to be obtained, or disposed of, by what were literally called waste gates or wasteways. For example, in 1876, owing to the unequal use of surplus water on the two levels, water to the equivalent of 524 horsepower was drawn from the upper to the lower levels through wasteways and lost. In 1877 the amount of waste was some 630 horsepower. Twenty years later Superintendent H. F. Mills, anticipating the summer stoppage of two weeks by most of the mills, tried to pair the mills on upper and lower canal levels as a means of reducing water wastage during a critically short period.[33]

Another major shortcoming of the distributing system, remarked upon earlier, one never fully remedied owing to cost, was the difference in capacity of the several canals, owing to failure or inability at the time of construction to anticipate future requirements. These differences under conditions of unusual demand, especially during low water, favored some mills over others on the same canal and the mills on canals of large capacity, chiefly those served by the Northern Canal, as against the mills on canals of less capacity, particularly those

[33] JBF to Directors, 27 July 1876, ibid., DB-8, 471, DB-9, 199; H. F. Mills to A. T. Lyman, 1 July 1896, BC-2, 199. See also comment of Frizell, *Water-Power,* pp. 408–9, 416–17; Tenth Census, *Water-Power,* pt. 1, p. 224 (Holyoke).

Fig. 60. Northern Canal. (Charles Cowley, *History of Lowell,* 2d ed. rev. [Lowell, 1868].)

served by the original but outmoded Pawtucket Canal. As the demands on any canal approached and then exceeded its normal capacity, the flow of water was accelerated and the level, or head, drawn down increasingly in relation to the distance from the source of supply. The malady was aggravated when to offset the reduction in head the mills opened their waterwheel gates wider to admit more water to maintain wheel speed and power, thereby further accelerating the flow of the canal. So it was that some mills consistently received less water than was their contract due, placing them at a disadvantage on those occasions when textile markets were favorable and water in short supply. If the inequity was usually not large, it was by no means negligible, and it was admitted by the canal management. It rankled in the minds of mill managers, ruffled intercorporate relations, and was a sore point with the Locks and Canals, frustrated by this continuing flaw in the otherwise progressive improvement of the system and the inability to secure approval for costly remedial measures.[34]

Measurement of water was at once the most tedious and most essential of the labors of canal management. Under conditions of mounting

[34] "Legal Opinion of Payson Tucker," 15 Jan. 1884, LCP, A-17; JBF to R. D. Rogers, 4 Oct. 1875, DB-8, 298–99; Francis to R. S. Fay, 28 July 1881, DB-11, 163 ff.

shortages of water from the 1860s, gaugings of mill consumption were the chief method of checking waste. From the beginning, however, estimates and measurements of water flow were an indispensable basis for engineering calculations and hydraulic control. As the pioneer large-scale hydraulic system, Lowell could obtain little guidance from experience elsewhere. Neither the traditional practice of defining water quantities by the equipment driven, as with a run of millstones, nor formulas from hydraulic treatises based on the flow of water through apertures or over weirs were adequate to the scale of operations.[35] At one of the Lowell mills in 1830 a gauge wheel 19.5 feet in diameter with a fifteen-foot breadth of buckets was placed in the tailrace. In 1840 seven great paddle wheels, each sixteen feet in diameter with a ten-foot breadth of bucket, separated by supporting piers but coupled together, were installed in the canal supplying several mills. With the aid of a counter to record the revolutions, the flow of water through the canal was thus measured. Francis stressed the necessity for a method of measurement that did not interfere, as did the cumbrous and costly equipment cited, with the normal operation of the mills.[36] He devised ways of measurement and recorded the results of their application in the four editions of the *Lowell Hydraulic Experiments* (1855–83), a standard treatise widely consulted in the engineering profession. Reportedly Francis reduced the possible error in gauging water from 10 percent to 2 percent or less. There was no escaping the methodical and careful procedures that during droughts required several gaugings daily at numerous points and the constant employ-

[35] See Frizell, *Water-Power,* p. 619 ff.; Tenth Census, *Water-Power,* pt. 2, pp. 207–13, 229–30, 480 ff.; Zachariah Allen, *The Science of Mechanics* (Providence, 1829), p. 199; Frederick Overman, *Mechanics for the Millwright, Machinist, Engineer, Civil Engineer, Architect & Student* (Philadelphia, 1858), pp. 166 ff., 173 ff.

[36] See James B. Francis and James F. Baldwin, *Report to the Directors . . . of the Locks and Canals on Merrimack River on the Measurement of Water Power Used by Manufacturing Companies at Lowell* (Boston, 1853), prefatory sec.; J. B. Francis, *Lowell Hydraulic Experiments,* 4th ed. (Boston, 1883), pp. 117 ff. Early foreign experiments in measuring the discharge of water from weirs, orifices, and pipes were on too small a scale to supply data capable of generalization for greater discharges. The experiments of Francis "established correct values for the coefficient of discharges of weirs, and his formula is now generally used" (O. Chanute et al., "Engineering Progress in the United States," *TASCE* 9 [1880]: 217–58; "American Engineering as Illustrated at the Paris Exposition of 1878," ibid. 7 [1878]:371–72). Acceptance of this evaluation of Francis's work is noted in George R. Bodmer, *Hydraulic Motors; Turbines and Pressure Engines* (London, 1889), pp. 308 ff., 362, 384. See also JBF to Gustav Ritter Von Wex, 14 Apr. 1888, expressing the belief that results of his experiments were "a nearer approach to the truth than would be given by any formula previously proposed" (LCP, DF-1).

ment of several parties of engineers and assistants.[37] Except during low-water periods occasional checks on consumption were ordinarily confined to mills that "habitually [drew] to excess." The mills' increasing use of "surplus power" after the Civil War, combined with sharply progressive rates intended to discourage its use and, during droughts, strict quantity limitations, gave renewed importance to water measurement. Its cost varied with the extent of the low-water periods but amounted to about one-fourth the total operating cost of the Locks and Canals. In his report to the directors of the Locks and Canals in 1871, Francis stated that owing to the unusual droughts of 1870, costs of water measurement were 50 percent in excess of the estimated figure, absorbing nearly one-half the income from surplus waterpower during the year.[38]

Management of the Upstream Reservoirs

Obtaining control of the New Hampshire lakes and their conversion into reservoirs, although financially self-liquidating through the lease of added capacity, increased materially the operating responsibilities of the Locks and Canals company. The regulation of outflow was readily effected by dams and gate installations of modest size, and attendance and maintenance were minor expenses. The operating problems centered on getting the stored water to the Lowell mills at the times and in the amounts needed, in using the lake water to maximum advantage, and in avoiding injury to local interests in the lake region. Control was also subject to restrictions, legal and extralegal, which hampered freedom of action and became at times a source of considerable harassment. Unlike the more common situation of such

[37] W. R. Bagnall, *Sketches of Manufacturing Establishments in New York City, and of Textile Establishments in the Eastern States,* ed. Victor S. Clark, 4 vols. (Washington, D.C., 1908), 3:2107 ff. Although turbines had largely displaced breast wheels at Lowell by 1871, the convenient gauging arrangement adopted at Holyoke using turbines employed there as water meters, with inspectors daily recording the height of water and opening of each turbine gate, was believed impractical for Lowell. Francis held that the costly methods of measurement at Lowell, requiring "a corps of hydraulic engineers and trained assistants" constantly gauging the quantity of water drawn by each lessee, were required by provisions in the indentures between the Locks and Canals and the lessees adopted to prevent ruinous controversies over water rights (JBF statement, 21 Nov. 1857, LCP, DA-4, 226–233; JBF to Berkey and Gay, 24 June 1878, DB-9, 363; to R. W. Hencker, 29 May 1888, DF-1, 48; to M. Walker, 13 Oct. 1877, DB-9, 274). Measuring methods and equipment at Lowell, Lawrence, and Manchester are described in Tenth Census, *Water-Power,* pt. 1, pp. 82–85.

[38] LCP, 19 Sept. 1871, DB-6, 321.

bodies of water employed as waterpower reservoirs, these lakes, especially Lake Winnipesaukee, were not surrounded by undeveloped lands of slight value but were located in an industrially active part of New Hampshire. In managing the lake reservoirs, the Locks and Canals had to be ever mindful not only of the rights of riparian owners, including summer residents, to whom additional damages must be paid if lake levels were altered beyond a given range, but also of the power requirements of the numerous mills on the twelve-mile-long Winnipesaukee River, through which the lake had outlet. There were, moreover, steamboats engaged in lake transportation whose operations at times were considerably embarrassed by low water caused by large drafts to meet the power needs of the Massachusetts textile cities. The legal obligations in respect to these varied economic interests were not entirely clear, but, as we shall see, a latent hostility in some New Hampshire communities toward out-of-state corporations led the Locks and Canals management to move warily.

The Locks and Canals jointly with the Essex Company of Lawrence had originally obtained control through the purchase by Abbott Lawrence and others, for later transfer to the waterpower companies, of the stock of the Lake Winnipesaukee Cotton and Woolen Manufacturing Company, riparian owner at the lake's outlet. By the late 1860s, if not earlier, the Lake Company became the object of harassment aroused evidently by the exercise of the company's rights under its power leases to restrict water use in local communities, manifestly for the benefit of the "foreign corporations" downriver in Massachusetts. Local communities retaliated by doubling and tripling the valuation for tax purposes of the Lake Company's properties, with the company fighting this action in courts.

Since the textile mills on the Merrimack at Manchester, New Hampshire, largely benefited from the operation of the lake reservoirs, the Amoskeag Manufacturing Company, the dominant corporation there, used its influence to discourage the attacks on the Massachusetts corporations. Back in 1845, however, the desire of the Amoskeag Company to purchase an equal interest in the Lake Company had been rejected by the Massachusetts interests. Now the company's participation in ownership and control of the Lake Company was urged by some as a means of quieting the rising popular and legislative opposition in New Hampshire. C. S. Storrow, treasurer of the Essex Company in the late 1870s, advocated admitting the Amoskeag Company to equal ownership: "If they join, they pay part expenses and have a voice. If they do not join and really want something, they will take it through legislation and pay nothing."[39] In the agreement finally reached, the Amoskeag Company shared operating expenses in return for consideration of their interests in the management of the Lake Company, but received no share in the ownership.

The opposition to out-of-state management of the lake reservoirs continued through the 1870s and eventually led to an investigation by a legislative commission in 1879. Among the charges leading to the appointment of the commission were the assertions that the Lake Company had neglected for thirty years to carry on the manufacturing for which it had been chartered, operating instead simply to develop and sell power; that it had drawn water from Lake Winnipesaukee for use outside of New Hampshire; and it had subordinated manufacturing interests within the state to those in Massachusetts, using its power to crush and oppress manufacturing in various New Hampshire communities. The commission, following hearings in various localities, found no substantial evidence in support of the charges and concluded, to the great relief of the Locks and Canals, that the management of the lake reservoirs was as beneficial to manufacturing interests within the state as to those without.[40]

Drawing upon the lakes for water was complicated by their distance from the cities served, some eighty miles by river from Lowell, and the two or three days required for the released water to reach the mills. It was further complicated by weather conditions affecting the volume of flow throughout the Merrimack basin above Lowell and by dry-season ponding and releasing of water at Manchester, Nashua, and other communities upstream. A sudden coldsnap or a thaw, a heavy rainfall in the watershed, suspension of operation at the Nashua mills to restore the working level of their millpond, shutdown of the Manchester mills on Agricultural Fair Day—such were the frequently unexpected circumstances that might nullify or reverse measures for regulating river flow at Lowell through the reservoirs. In the absence of a comprehensive system of reporting weather and gauging rainfall and volume of flow throughout the river basin, something far in the future, Locks and Canals engineers were frequently confronted

[39] Ibid., n.d. (ca. 1877), A-17; Storrow to T. J. Coolidge, 15 Oct. 1877, ibid. See also George W. Browne, *The Amoskeag Manufacturing Company of Manchester, New Hampshire* (Manchester, 1915), p. 128; correspondence of JBF and other officials, DB-5, 58–59, 171; DB-8, 137; DB-9, 210–11, 276–77, 282; DB-10, 151, 350, 353; DB-12, 2–3; and copy of the *Lake Village Times,* Lake Village, N.H., 5 July 1879, A-11.

[40] See the commission's report in *Lake Village Times,* LCP, A-11.

with changes in the river's condition which they could do nothing to control.

Irregular in time of occurrence, severity, and duration, the late summer and midwinter seasons of drought and low water were anxious times for the Locks and Canals superintendent and periods of sharply increased activity for engineering and maintenance personnel. The use of surplus water was restricted or forbidden; procedures for measuring use and checking waste of water were increased and tightened; gates, penstocks, and waterwheels were inspected for leakage; mill officials were cautioned; and requests for out-of-hours mill operations or for draining canals for repairs were rejected. On the prescience, judgment, and thoroughness of the responsible officials at Lowell largely depended minimal interference with mill operations not only at Lowell but also at Lawrence and Manchester.

As the falling river approached emergency levels, communications between Lowell and the lake control stations mounted in volume, with letters, confirmations, reports, and, from 1856 on, telegrams crowding on one another's heels, requesting and reporting information on levels and trends of lakes and river, weather conditions and forecasts, present needs and forthcoming urgencies at the mills.[41] Code messages for the release of more or less water (usually in hundreds of cubic feet per second) from the lakes were sent, confirmed, and canceled as occasion required, often accompanied by urgent requests for information on weather prospects and lake conditions. From the 1860s the urgency pervading these interchanges evidently reflected the pride and perfectionism of the professional engineer, since the growing use of steam power made the mills progressively less subject to the whims of the weather and the vicissitudes of lake levels. Some typical messages from Lowell during the difficult periods of low water were: "Give us another 200 cubic feet." "Let us have all you safely can." "Water passing over the dam, cut off supplies." "To carry us through next week must have large supply from you." "Give us all you can without endangering next summer's supply." "Unless we have rain or a good deal more water from you we will suffer next week."

Shortage and Surplus

The mid-century decades were periods of rapid expansion in textile operations at the manufacturing cities on the Merrimack, save for the

[41] LCP, DA-4, 150 ff. For the effect of cold weather in reducing streamflow, see JBF to Directors, 21 Dec. 1865, DB-3, 742 ff.

inactive years of the Civil War. Despite formal agreement by the corporations in 1852 to refrain from increased water consumption, there was a steady rise in water use, prompting Superintendent Francis to note in 1858 that ownership of the Locks and Canals company by the textile corporations had "for a time appeared to remove all restraint upon the use of water."[42] The expansion of textile manufacturing was made possible by the enlargement of the waterpower; it was stimulated by the absence, before 1859, of any charge for power in excess of the amounts named in the leases ("surplus power," it was shortly to be termed). As Francis remarked, "Successful manufacturing establishments keep on increasing their product so long as they can obtain power to drive their additional machinery."[43] Mean annual power consumption on the eve of a reapportionment in 1853 was but 132 mill powers, as we have seen, and rose to 163 in 1856 and reached 180 in 1858, over 25 percent in excess of the leased permanent year-round power. In 1860 measurement showed an increase of 22 percent over 1852 in the average quantities of water drawn by one group of mills. On recovery from the sharp slump in production of the Civil War years, the Locks and Canals was faced with a resurgence of the old problem of power shortage, tempered only by the increasing acceptability and declining cost of steam power for use during droughts.[44] Such measures for increasing supply as cutting down leakage and other waste in the hydraulic facilities, raising the height of flashboards on the main dam and prolonging the season of their use, and the progressive replacement of breast wheels by the more efficient turbines brought only temporary relief. Proposals for increasing the overall capacity of the system by such measures as enlarging the canals were to be rejected as much too costly for the anticipated benefits.

Only on rare occasions during extreme periods of drought, and briefly, did the Locks and Canals fail to provide the mills with their leased ("permanent") power. On such unusual occasions, machine speeds fell slowly as the mills lacking reserve wheel capacity were affected first; production declined and with it the earnings and morale of the employees.[45] The real problem centered in the use of power in

[42] JBF to T. G. Carey, 6 Dec. 1858, ibid., DA-5, 58–64; to Directors, 21 Dec. 1858, DB-3.

[43] Ibid., 8 July 1865, DB-3, 651 ff.

[44] "I think mills driven by steam usually run with less power than those run by water, economy of power being more studied in the one case than in the other" (JBF to W. Phillips, 19 Aug. 1856, ibid., A-20).

[45] JBF to T. Lyman, 28 Oct. 1865, ibid., DB-3, 719. Francis wrote W. H. Thompson, 28 Jan. 1869, that only twice in eight years had it been necessary to cut down the mills to "lawful," i.e., leased, quantity (ibid., DB-5, 381).

excess of the leased amount; there was an abundance of surplus power during the greater part of the year. Yet to the Locks and Canals management, responsible for the power supply in all its aspects, there were solid engineering and practical reasons for imposing strict limits on the use of surplus power. As noted above, depending upon the circumstances of their location on the canal system, mills varied in their ability to benefit from the use of surplus flow. In many instances the increased use of water by the more favorably situated mills, through accelerating the flow of water in low-capacity canals, caused a loss of head and of power for other mills, in some circumstances actually encroaching on the permanent power of the latter.[46] In January 1857 the directors of the Locks and Canals directed Francis to order the manufacturing companies to discontinue until further notice the use of any water beyond the permanent power covered by their leases.[47]

With the textile corporations reluctant and slow to introduce steam power, there was inevitably pressure upon Locks and Canals management to meet shortages by drawing upon the lake reservoirs.[48] This Superintendent Francis, ever mindful of the precarious position of the Locks and Canals as a "foreign" corporation in New Hampshire, subject to legal and legislative reprisals, opposed with determination. Upon his urgings the directors of the corporation in 1859 stringently regulated the use of surplus power by introducing charges according to a schedule of rising rates and provision for limitations on the quantities of water used. This decision was also expected to provide additional income to the Locks and Canals of $15,000 to $20,000 per year.[49]

Over the years the principle of increasing rates progressively with the amounts used was continued and strengthened. For example, the initial rates of $3.50 per mill power per day with $7.00 for each mill power in excess of the lessee's permanent power were later raised to $7.00 and $14.00, respectively, and still later were changed to a three-class schedule of $5.00, $10.00, and $15.00. By 1880 the daily rates were $5.00, $10.00, and $20.00, respectively, for less than 40 percent, 40 to 50 percent, and 50 to 60 percent in excess of the permanent power, with the rate of $20.00 applicable to the entire amount of sur-

[46] JBF to T. G. Carey, 17 May 1858, ibid., A-30; to Directors, 21 Dec. 1865, DB-3, 742 ff.; "Legal Opinion of Payson Tucker," 15 Jan. 1884, A-17.

[47] JBF to I. Hinkley, 13 Jan. 1857, ibid., DA-4, 195–96.

[48] JBF to B. Saunders, 21 Oct. 1864, ibid., DB-3, 565 ff.; Gibb, *Saco-Lowell Shops,* pp. 102–3.

[49] JBF to T. G. Carey, 6 Dec. 1858, LCP, DA-5, 58–64; JBF to A. Lawrence, 2 July 1859, DB-2, 57.

plus used in excess of 60 percent.[50] Eventually the income from surplus power became sufficient to make unnecessary the collection of the annual "rental" fees under the leases, the chief source of operating income for the Locks and Canals, bringing a certain advantage to the mills using little or no surplus power.[51] By 1880, the four largest textile corporations leased sixty percent of the permanent power and used 80 percent of the surplus power.[52] Charging for surplus power, Francis remarked, was a kind of police regulation adopted by the owner corporations "for keeping one another within reasonable limits by making them pay roundly when they use too much."[53] William E. Worthen, an engineer at different times associated with the Locks and Canals, in a brief memoir referred to Francis as "the chief of police of water."[54]

As the volume of flow in the river fell in late summer or midwinter, the mills were progressively restricted to smaller amounts of surplus power taken as a percentage of their permanent power, until at length the use of any surplus was prohibited.[55] Conversely, with the river approaching flood stage and the volume of flow well in excess of any possible manufacturing use, the mills in the mid-seventies were free to draw water without limit at a daily charge of $1.00 per mill power, which was but a fraction of the base surplus rate, equivalent in some instances to as little as $5.00 per horsepower per year. In 1888 this rate was increased to $2.00 per mill power per day.[56] From 1875 through 1884, the backwater condition of the river, when surplus water rates were at minimal level and all restrictions upon use were removed, averaged 110.5 days a year; from 1875 through 1888, 122.5 days a year. The working year was considered to be 309 days.[57] The appreciable use of surplus water required additional turbine capacity, which stood

[50] Tabular Statement, 29 Jan. 1866, ibid., DB-5, 11; JBF to J. P. Frizell, 9 Sept. 1869, DB-5, 443; to T. I. Shaw, 10 Oct. 1872, DB-6, 471; to J. A. Dupee, 29 May 1888, DB-13, 50–51.

[51] H. F. Mills to A. S. Covel, 4 May 1896, ibid., BG-2, 191.

[52] Statement, 7 June 1880, ibid., DB-10, 269.

[53] JBF to C. Stott, 26 July 1869, ibid., DB-5, 416.

[54] Worthen, "Life and Works of James B. Francis," p. 234.

[55] Printed forms came into use for giving the mills formal notice of the establishment, the amount, and the withdrawal of restrictions upon the use of power; see examples in LCP, A-13.

[56] Notice from Directors to JBF, 26 Dec. 1876, ibid., DB-7, 479; JBF to J. A. Dupee, 29 May 1888, DB-13, 50–51.

[57] JBF to Treasurer, Mass. Cotton Mills, 24 Oct. 1885, ibid., DB-13, 404; to Directors, Locks and Canals, 18 Mar. 1889, DF-1, 81 ff.

idle when surplus water was not available. Nearly all mills possessed turbine capacity somewhat in excess of their usual requirements, not only to meet occasional above-normal needs but also to meet the situation during backwater in flood seasons. Then the reduced aggregate fall, owing to the greater rise below than above the falls, reduced the power delivered on the wheel shaft and had to be offset by using an increased quantity of water. One student of the Lowell area has reported that by 1884 the wheel installations at the Lowell mills possessed an aggregate capacity of some 23,000 horsepower, or two and one-half times the amount of power covered by leases.[58]

The system of control served the intended purpose; for many years the use of surplus power did not exceed 30 to 40 percent of the permanent power. As steam power was increasingly adopted in the mills, the official pressure to hold down the consumption of surplus power was relaxed. By the mid-nineties the use of surplus power by the Lowell mills at times was as high as 125 percent of the permanent power. Use of surplus power in 1896 ranged from none by two corporations, 18 percent by a third, 85 to 128 percent by six others, and 234 percent by the largest user.[59] Much beyond this point the limiting factor became the limited capacity of the canal system.[60] In 1899 a system of progressive increases of rates on surplus power was reversed, with the charge for all power used above 40 percent of the permanent power leased reduced by 50 percent, or from $3.00 per mill power per day to $1.50.[61]

The declining status of Lowell's waterpower base was reflected in the joint action of the Locks and Canals and the Essex Company about 1890 in selling to New Hampshire interests their ownership of the Lake Company controlling the system of reservoirs. In support of this action Francis gave the following reasons: (1) raising the height

[58] Parker, "Geographic Factors in the Development of the Cities of the Merrimac Valley in Massachusetts," pp. 98–99. The hydraulic engineer Samuel Webber placed the extra waterwheel capacity at Lowell at 19,000 hp (1893) (discussion of paper by Emery, "The Cost of Steam Power Produced with Engines of Different Types," p. 62). Swain estimated the average annual cost of the wheel installations at $9 per horsepower, fixed plus variable costs, or about one-third the aggregate annual cost of the waterpower and one-half the cost of the water used ("Statistics," p. 33, table 6). At Holyoke in 1960 the maximum requirements of water to supply industrial needs occurred but four months a year, leaving much wheel capacity idle the remaining eight months (Holyoke *Transcript-Telegram,* 14 Sept. 1960).

[59] H. F. Mills to Directors, Locks and Canals, 27 Nov. 1896, LCP, BC-2, 215–18.

[60] JBF to R. S. Fay, 28 July 1881, ibid., DB-11, 163–66.

[61] H. F. Mills to all corporation agents, 30 Dec. 1898, ibid., BC-2, 337.

of flashboards on the main dam at Lowell increased the supply of water to the mills during drought seasons; (2) the introduction of steam power enabled the mills to maintain speed during periods of low water; (3) reduction of the working day to ten hours gave more time for water to accumulate in the pond above the dam.[62] Doubtless the decision was influenced by the annual expense of operating the Lake Company, reported for the nine years 1879–88 as averaging $25,623 for the Locks and Canal's half share. This was double the average annual expenditure of $12,681 during the thirty-four years 1845–79.[63]

Steam Power

Steam power had been introduced in the Lowell mills in the middle 1840s; its use increased slowly at first, then more rapidly during the 1860s until it outdistanced waterpower a decade later. By the early 1880s twice as much steam power as waterpower was used, and most mills were supplied with enough steam power to offset waterpower shortages under almost any contingency. In 1881 Francis declared: "Great improvements have been made in the economical use of steam power of late years, and our mills are now, most of them, largely supplied with steam power, which they use more or less, depending on the supply of water in the river, and having the engines and boilers the cost of the power is not materially different, whether they use the surplus power . . . or steam power."[64] A total of 19,703 horsepower in steam power was reported for Lowell in 1882, covering other manufactures as well as the textile mills, whose total steam power at this time was reported as 13,940 horsepower. The Merrimack Manufacturing Company, largest of the textile mills, had eighty-three steam engines with an aggregate of 6,000 horsepower.[65] In contrast, as of 1889 the waterpower used by the Lowell textile mills ranged, according to Francis, from 9,000 to 14,000 horsepower, depending on the season of the year, developed by some eighty-five turbines.[66] Because of

[62] JBF to G. Atkinson, 30 July 1888, ibid., DB-14, 42–43.

[63] John T. Morse to JBF, 8 Aug. 1888, LCP, A-17; JBF to C. H. Dalton, 6 Dec. 1879, DB-10, 148. It is not clear whether these two statements cover the same items of expense or include interest as well as operating costs.

[64] JBF to W. H. Thompson, 21 Mar. 1881, ibid., DB-11, 28.

[65] *Statistics of the Manufacturers in Lowell* (1882), no. 45, cited in Tenth Census, *Water-Power,* pt. 1, p. 81.

[66] JBF, 29 Sept. 1889, LCP, DF-1, 102.

the large requirements for heat in textile processing, supplied conveniently by the exhaust steam of noncondensing engines, the cost of steam power was appreciably reduced.[67]

The rapid expansion of steam power at Lowell from the 1860s pointed up the dilemma of industrial communities dependent upon waterpower. Even at this, one of the best engineered and managed large-scale waterpower systems in the country, much of the water ran to waste during most of the year. Short of a massive and prohibitively costly upstream reservoir system several times the capacity of that developed with the New Hampshire lakes, the limit to the permanent—that is, dependable—year-round capacity of this waterpower was early reached. The system's capacity of some 8,500 horsepower from the extensive enlargement program of the 1840s was raised by 1890 to 9,000 to 14,000 horsepower, according to season. Of surplus power, available when the river ran fairly full, there was a great abundance, but rarely for more than eight months of the twelve, and this supply was hedged about by such management restrictions as materially to reduce its attractiveness.

There were other deficiencies, either inherent in this method of power development or peculiar to the Lowell situation. Some were the result of the gradual development of the power at Pawtucket Falls, the understandable and even inevitable failure correctly to anticipate and plan for future needs, and the believed prohibitive cost of making changes in the system. This shortcoming focused in the limited capacity of a number of the canals and the inequity of service ensuing to certain mills because of disadvantageous location. Yet from the very nature of a large and integrated power system serving many large clients, especially where the hydraulic forces were large, not too well understood, and difficult to control, rules and regulations were set up, elaborated, and administered to insure the effective operation of the system as a whole. In 1845, when the lessees became the owners of the Locks and Canals, this organization in effect became a cooperative body; yet the rules and restrictions imposed on the members were a continuing source of dissatisfaction and complaint to many. Perhaps

[67] As the result of an engineering study made in 1866 by William E. Worthen for the Merrimack Company, exhaust steam from engines of 1,200 total horsepower supplied the process requirements of the print works and reduced the fuel consumption properly chargeable to steam power to not more than one pound of coal per horsepower hour, or a third to a fourth the consumption typical of the best engines of the time. Some twenty years later over one hundred steam cylinders were exhausting into a common main by which all the company's heat requirements were met (William E. Worthen, "Address of the President," *TASCE* 17 (1887): 11–12).

the most disagreeable and exhausting duty of Francis in his role of superintendent was that of judiciously administering the rules and regulations governing water use and of maintaining amiable relations with and among the corporate clients of the Locks and Canals, whose servant he was.[68] Not only were some of the disadvantaged corporations moved to explore legal remedies in the 1880s, but all of the corporations found their independence of action variously restrained by the system of power supply of which they were a part. Steam power, increasingly resorted to of necessity, provided a means of escape and of independence.

In the years of its prime, from the 1830s to the 1860s, the Lowell system was a great engineering and industrial achievement. Yet even before 1870 its force was spent, both as an achievement and as an example. As with waterpower in the American economy as a whole, its technological mission, so to speak, was completed. Although the Lowell waterpower was not the largest of its kind or the most efficiently arranged and managed, it was the most notable installation of the direct drive power age, outstanding for its pioneering role in engineering, development, and management. It anticipated in a crude and cumbrous way the central power station of the future, although in retrospect its canals seem grotesquely elephantine. It helped prepare the way for the hydroelectric age, contributing an important body of hydraulic knowledge regarding the flow of water in canals, the design and construction of such facilities as dams and wasteways, the methods and problems of large-scale water storage, and the essential character of long-range records of rainfall, runoff, and streamflow.

Discontent and Dissent

There was another and less constructive aspect to what has been termed here the "management of a great waterpower." In an agrarian society marked by widespread land ownership and small-scale industrial enterprise, a certain anomaly in the traditional legal doctrine of riparian rights gave effective possession and use of a water privilege of immense capacity to the shrewd purchasers of the land bordering this privilege on either stream side. However suitable as a means for encouraging provision of the small water mills essential for alleviating the hardships of agricultural settlement and growth, the principle

[68] See correspondence and memoranda in the Francis files, LCP. See also Worthen, "Life and Works of James B. Francis," pp. 227 ff. When in the mid-1890s Hiram F. Mills took over as chief engineer, he discovered that his duties were chiefly those of placating and admonishing textile mill officials (LCP, BC-2).

of riparian control in this instance—at the price of a few hundred acres of land—carried with it the disposition of the streamflow of a large river, the product of runoff of some thousands of square miles, comprising in New Hampshire alone a substantial part of the state. The development by the Proprietors of the Locks and Canals of a great waterfall with a potential in excess of 10,000 horsepower was so exercised as to serve the requirements of less than a dozen large industrial corporations, closely linked by interlocking ownership, to the exclusion of all others. The somewhat restricted conception of the responsibilities of the textile corporations to the wider industrial community has been remarked upon above (see chapter 5).[69] It was understandable that having established a water supply system to meet their own needs, the textile corporations should oppose a public waterworks whose cost they must substantially share through taxation. That the city of Lowell should pay the Locks and Canals for the water drawn from the river followed naturally from the doctrine of water rights. Anomaly, however, gave way to irony when the Locks and Canals inserted in the contract for water sale to the city of Lowell a provision forbidding the use of water for power *except in the form of steam,* a power source which because of its cost the corporations themselves had resorted to only when there was no alternative. The scores of small subsidiary industrial and service enterprises necessary to the functioning of this growing textile center were purposely and effectively excluded from the benefits of one of the cheapest and generally most convenient sources of motive power. Since steam power in amounts of less than several horsepower, for reasons discussed in another place, was in most instances impractical, small industry was obliged to resort to the muscular energy of man or beast as applied through such primitive devices as treadwheels, hand cranks, or horse walks if the benefits of machinery were to be enjoyed. Although fully aware and appreciative of the power problems of small users, Francis was not deterred from opposing any measures likely in however limited a way to infringe upon the power needs of the great industry to which his life was professionally dedicated.[70]

[69] On the textile corporations' role in the social and industrial development of Lowell, see the studies of Hannah Josephson, *The Golden Threads: New England's Mill Girls and Magnates* (New York, 1949), and John P. Coolidge, *Mill and Mansion: A Study of Architecture and Society in Lowell, Massachusetts, 1820–1865* (New York, 1942).

[70] The city of Lawrence, Massachusetts, paid the Essex Company a much smaller sum for water for domestic uses than Lowell paid the Locks and Canals. Lawrence could pay an additional 5 percent to use water for water motors, elevators, and the like, long prohibited uses at Lowell (George A. Kimball, "Water Power—Its Measurement and Value," *JAES* 13 [1894]:87–88).

On the lower Merrimack River as elsewhere, time brought increasing demands on the source of power supply that, if anything, diminished with the changing ecological regime of the drainage basin while at the same time seasonal freshets increased in volume and violence. In the years following 1900 the supplies of streamflow from the Nashua and Sudbury tributaries were largely absorbed by the domestic and sanitary requirements of the adjoining communities.[71] Although by this time Lowell and Lawrence were securing over 75 percent of their rising power needs from steam and electric power, the power companies clung to the last cubic foot per second of the Merrimack's flow—despite the mounting indignation of the community interests adversely affected. As early as 1905 there were complaints that over weekends almost no water passed over the power dams of the textile centers. In 1907 it was declared that "sometimes for a month or six weeks continuously no water wastes over the dam, days, nights or Sundays. . . . All that comes down the river is drawn into the canals and used in manufacturing." A United States Army Corps of Engineers report of 1916 on the river cited the complaint by the mayor of Haverhill, the next community downriver from Lawrence, of the selfishness of Lawrence and Lowell, remarking that he knew of no law permitting them "absolutely to stop the flow of this river." During the several previous years, "I have seen the river absolutely cut in two at Lowell and Lawrence; not one drop. The flow absolutely stopped." Except when relieved by the tide the communities below would have each week nothing but mud flats with their terrible stench. To add to the concern and confusion, a movement supported by commercial interests of Lowell and Lawrence proposed the restoration of the navigability of the Merrimack River above tidewater, producing a certain ambivalence of attitude on the part of spokesmen for the Locks and Canals and the Essex Company. The local debate mirrored in a measure the mounting national controversy in respect to dams, rivers, and, increasingly, as dealt with elsewhere, hydroelectric installations. In a congressional debate of 1916 over the general dam law a committee report reflected the swelling sentiments of the Progressive movement: "These immense natural resources . . . should be developed for the real welfare of the whole country and not solely for the benefit of those few individuals who had the shrewdness and foresight to acquire such property rights as may be sufficient to dominate and use mostly for themselves these privileges."[72]

[71] This discussion has been drawn largely from U.S. Army, Chief of Engineers, *Reports on the Preliminary Examination and Survey of the Merrimack River, Massachusetts, from Lowell to the Sea,* 22 Dec. 1916, House Doc. no. 1813, 64th Cong., 2d sess.

[72] Ibid., pp. 25–27.

7
The Hydraulic Turbine

THE FIELD OF applied, as distinguished from theoretical, hydraulics was marked by two major advances in the half century from 1800 to 1850. The countries of origin were France and the United States; the decades of active achievement the 1820s through the 1840s. The American innovation was the success first demonstrated at Lowell in the difficult and costly engineering task of harnessing and managing large-scale hydraulic forces for industrial power supply. The French achievement was the invention and practical perfecting of a new prime mover, a waterwheel known as the "turbine," which in respect to size, cost, efficiency, and operating characteristics represented a marked advance over the best of the traditional waterwheels. The American achievement made possible the exploitation of the hydraulic potential of rivers of substantial size, providing the energy base not only for industrial establishments of the largest capacity but for large industrial centers. The experience of Lowell and its successors during the antebellum years also, in important respects, cleared the way for the age of hydroelectricity. It not only provided important lessons in the control and measurement of streamflow in large quantities; it focused attention on the basic considerations of rainfall, runoff, and streamflow in relation to power supply and dramatized the critical importance of water storage for the effective development of waterpower.[1] The American experience had little relevance to European needs, owing to the scarcity of waterpowers of large capacity in Britain and to the lack of need for, or the inaccessibility of, those existing on the Continent. Nor, as we have seen, was the Lowell system of power development and use appropriate to the needs of the great majority of American industrial

[1] "The engineer, scientifically educated, as distinguished from a mechanic or millwright," declared a New England–based hydraulic engineer of distinction, "may be said to have first fairly come into the water power field about 1835, with the damming of the Merrimack River, and the building of the canals for the systematic distribution of water for power to the great industrial communities at Lowell, Lawrence, Manchester and at Holyoke on the Connecticut River. [Here was found] the highest engineering development of the industrial city epoch" (John R. Freeman, "General Review of Current Water Power Practice in America," in *Transactions of the First World Power Conference, 1924,* 5 vols. [London, (1925)], 2:383).

establishments before the Civil War—and for many years thereafter. These establishments, widely distributed throughout the older and more settled parts of the country, required relatively modest amounts of power.

Waterpowers, as we have noted, were typically selected with a capacity not greatly beyond immediate needs to minimize costs and the difficulties of development. At the same time, growth was a characteristic feature of most districts, communities, and industrial establishments in a young, expanding, and developing country like the United States. Increase in the scale of plant operations was usually accompanied by the extension of mechanization and increased requirements of power. As the more obvious expedients for eliminating waste in and increasing the capacity of the power supply were exhausted, the difficulties and expense of moving to a new site (or in time providing an auxiliary steam plant) made improved waterwheels a welcome form of relief. When to greater efficiency improved operating characteristics and lower costs were added, the appeal of the new types of wheels was often compelling.

Operating Criteria

With the gradual shift from community-service country milling within a subsistence economy to the commercial production of a market economy increasingly concentrated in mill villages and industrial towns, requirements in power supply in waterwheels and supporting facilities alike became more demanding. The frequently intermittent operations of country mills reflected not only the fluctuations in streamflow but the part-time character of the miller's occupation and the subordination of milling to the seasonal labor requirements of farming. Four operating characteristics of waterwheels, functioning within the intractable framework of the seasons, took on an increasing importance and emphasis in wheel design: wheel efficiency in using the available supply of water; higher rotating speeds; adaptability of a wheel to use with a wide range of heads; and ability to operate submerged. It was the need to improve these characteristics, if not completely to remove the limitations of the traditional wheels, that led to the "invention" and development between 1800 and 1850 of two new categories of waterwheels, in important respects quite similar, in both Europe and the United States. Both were a response to essentially the same needs and associated with the same economic and technological stage of the more developed portions of the Western world. In the order of their appearance but in the inverse order of their long-range importance in industrial development, the new prime movers were the reaction wheel and the turbine.

To improve wheel efficiency, stated in terms of the percentage of useful effect obtained from the weight of water falling a given distance, was to raise the power capacity of a mill seat, to extend somewhat the season of operation at a mill seat, or to supply the additional power required by the growth of production. Higher rotating speeds, the level of the prevailing technology permitting—iron as a fabricating material, for example, being more amenable than wood for this purpose—were attended by smaller wheel dimensions and lower cost as well as by economies in space, wheel pits, and foundations and the like. Higher speeds made possible a reduction, or even an elimination, in the amount of gearing required to obtain the desired operating speeds, improvements accompanied by lower maintenance and repair costs—gearing was a major source of millwright worries—and of power loss through friction. Adaptability of a waterwheel for use with a wide range of heads offered advantages in respect to problems of wheel design and construction.

The last of the operating characteristics noted calls for closer and more detailed examination because it is not familiar or well understood in a nonwaterpower age and because it was prominent in waterwheel development in the period of our concern. Most older general accounts of waterpower remark upon the seasonal variations of streamflow but typically only with reference to the reduction or interruption of power supply and mill operations during the months of drought, or ice, and low water. Local histories of early times also often describe the occasional devastating floods which carried away not only dams and bridges but often streamside mills as well. The more commonplace floods of our concern were those that usually marked the end of the seasons of drought in spring and fall and the sporadic ones of unseasonal heavy rains. Yet in the operation of old-style mills too much water was at times as much of a problem as not enough, mitigated somewhat by the circumstance that floodwaters were typically of much shorter duration than seasons of low water, several or more weeks instead of two to four months.

When floodwaters reached a certain height, they actually reduced the available power by raising the level of water below the falls more than above (see Fig. 61), thereby lowering the effective head. This could not be offset by the vastly greater volume of flow since mills could hardly double or treble the capacity of their wheel installations to take

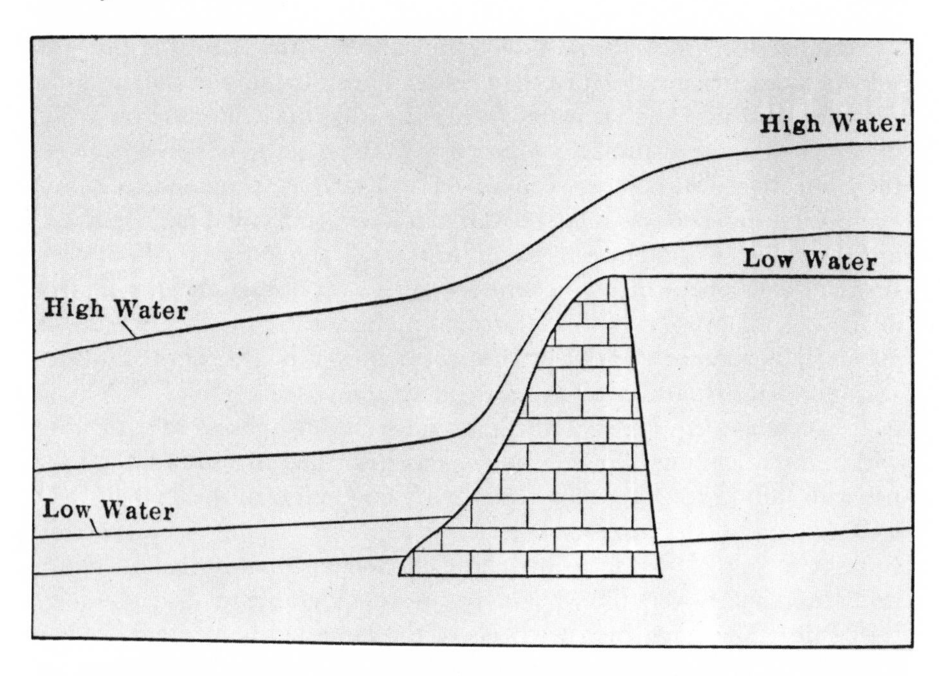

Fig. 61. Changing stream profile of milldam attending wide variations in streamflow. (Robert L. Daugherty, *Hydraulic Turbines*, 2d ed. [New York, 1914].)

advantage of this abundance during a limited portion of the year. To the negative action of loss of head was added the actual interference with the action of the mill wheel in the phenomenon known as "backwater." The rising level of water below the falls reached a point where it backed up the tailrace and, after filling the protective space below the waterwheel provided against this contingency, steadily mounted the wheel itself. The result was a braking action that increased progressively, reducing the wheel's power and in extreme cases compelling the suspension of operations. With country mills, accustomed to intermittent operations, this was perhaps only a matter of inconvenience; with commercial mills, if continued for more than several days, such suspension or even a serious reduction in power might be a matter of serious loss.

Protection against the contingencies of floodwaters, whether of seasonal or irregular occurrence, could be secured by raising the level of the wheel setting above the tailwater, but at the cost of reducing the amount of fall and the power capacity of the mill seat and mill. Millstreams and mill seats varied in their vulnerability to floods and backwater as also did the various types and sizes of waterwheels. With overshot wheels the wheel moved in a direction counter to the tailwater and was subject to greater interference than breast wheels turning in the opposite direction and *with* the tailwater rather than *against* it. Powers on lesser falls were at a disadvantage compared with those on higher ones because the protective distance between wheel and tailrace provided to guard against interference was relatively large in relation to total fall. In an example cited by a middlewestern engineer, it was generally expedient with breast wheels operating on a ten-foot fall to leave two and a half feet of the fall unappropriated. Unfortunately, according to this engineer's estimate, "over nine-tenths of the water power in the United States was found in situations offering no more than 10 feet head."[2] A type of breast wheel equipped with floats rather than buckets and turning within a semicircular wooden breast was described as well adapted for use with falls of four to seven feet, giving good efficiency, but was "not well suited for situations subject to backwater, which very speedily bring it to rest."[3] Owing to their low efficiency, tub wheels and many undershot wheels were the wheel types most vulnerable to obstruction by backwater. The great breast wheels of the large textile centers of New England, under skilled supervision and with the control of water levels in supply canals through headgates and wasteways, presented minimal problems save with the higher floods.[4]

The logic of the situation with floods and backwater called for an implausible solution, considering the character of the impact and gravity wheels in use: a water wheel that would run submerged.[5] Such a wheel was virtually unknown in the United States but in more than

[2] Daniel Livermore, "Remarks upon the Employment of Pressure Engines, as a Substitute for Water Wheels," *JFI* 15 (1835): 166.

[3] *Appleton's Dictionary of Machines, Mechanics, Engine-Work & Engineering,* 2:833; W. W. Tyler, "The Evolution of the American Type of Water Wheel," *JWSE* 3 (1898): 881–82.

[4] It was customary at Lowell to set pitchback or high breast wheels with their bottoms about eighteen inches above the ordinary surface of the tailwater: "These wheels had to be set so that they could obtain a supply of water when the head was lowest as well as at other times—and yet high enough to discharge tolerably well in high water, and subject to stoppage by back water only during short times of maximum flow" ("Legal Opinion of Payson Tucker," 15 Jan. 1884, LCP, A-17). See also "Hydraulics," *Scientific American* 6 (29 Mar. 1851): 224.

[5] Jacob Perkins patented a rather complicated waterwheel designed to overcome backwater caused by high tides and floods. See *The Emporium of the Arts and Sciences* (Philadelphia), n.s. 3 (1814): 273; also George S. Gibb, *The Saco-Lowell Shops: Textile Machinery Building in New England, 1813–1949* (Cambridge, Mass., 1950), p. 25. See also references to backwater in U.S., Census Office, Tenth Census, 1880, vols. 16–17, *Reports on the Water-Power of the United States* (Washington, D.C., 1885, 1887), e.g., pt. 1, pp. 355, 441, 501, 861; pt. 2, pp. 261 ff., 313–14, 375, 453.

a century had been widely in use for gristmilling in the south of France. A small horizontal wheel, in some respects not unlike the American tub wheel, operating within a well or pit and known as the *moulin à cuve*, or pit wheel, was described in some detail by Jean F. d'Aubuisson. The pit wheels were low in efficiency but the simplicity and solidity of their construction occasioned their frequent use, especially in places where there was an abundance of water. They possessed "a remarkable advantage, that of being able to work while submerged, and consequently in the freshets of rivers, so long as there is a marked difference of level between the upper and lower reach." In one form the pit wheels worked entirely submerged. An efficiency of no more than 15 percent, as determined by careful experiment, effectively discouraged its wider use.[6]

The search for a more satisfactory waterwheel was sought alike in France and the United States, stimulated in the former by the interest of scientists and government in applied hydraulics and in this country by the pragmatic inquiries of millwrights and mechanics and by the power needs in the more settled areas by the growing number of small manufactories. By coincidence the year 1823 witnessed significant expressions of the need for a wheel free from the interference of floodwater in both countries. Backwater was one of the impediments that led the Société d'encouragement pour l'industrie nationale to offer in this year a substantial cash award for a wheel free from this defect. All known waterwheels, read their announcement, operate under serious difficulties each time the level of water downstream chokes the wheels and paralyzes their action, often during several weeks running. A solution was necessary for the elimination of this harassing and wasteful condition, during seasonal and sporadic floods alike.[7] In the spring of this year, 1823, the Ordnance Department of the United States Army established a board of three army officers to survey suitable sites for a western armory within the Ohio Valley. The instructions to the board with respect to the waterpower of the armory, some 120 horsepower, specified that "it should be constant and at command, and not liable either to be obstructed *by floods or to fail by droughts*" (Emphasis added). With respect to the contingencies of national defense too much and too little were equally abhorrent. The board's report on western armory sites in 1825 proposed the use of reaction wheels, despite an efficiency but

[6] Jean F. d'Aubuisson, *A Treatise on Hydraulics for the Use of Engineers*, trans. J. Bennett (Boston, 1852), pp. 411–17. The tests on the performance of this wheel were made by Tardy and Piobert and are summarized in the table given on p. 416.

[7] Marcel Crozet-Fourneyron, *Invention de la turbine* (Paris, n.d.), pp. 21–22; M. Burat, cited in *Appelton's Dictionary of Machines*, s.v. "Turbine."

one-half that of gravity wheels such as overshot and breast, at a number of sites under the prevailing conditions of streamflow. They had been adopted "from a conviction that every other kind of wheel, within our knowledge, would be liable to interruption from high water."[8] Thirty years had passed since Oliver Evans in the first edition of the *Young Miller and Mill-wright's Guide* had called attention to James Rumsey's reaction wheel, an improvement on Barker's mill, "said to do well where there is much back water." Nearly twenty more years were to pass before the Fourneyron turbine, possessing this same advantage, reached the United States.

Changing Industrial Requirements

The development of the new types of waterwheels was also a response to the changing character of industrial needs. Both in Europe and the United States the new wheels—reaction wheels and turbines—reflected the first stirrings of industrialization and the rise and multiplication of small manufactories for which the traditional wheels in important respects were ill suited. Turbines were developed and first employed chiefly in northern France and the adjoining piedmont area of Switzerland and western Germany. Although the broad class of what came to be termed, not too satisfactorily, reaction wheels had its origin in Europe, the most prolific breeding ground was the United States. As remarked above, the two classes of wheels shared a number of characteristics in common, indeed have been described as descending from a common ancestor, known in the United States as the tub wheel, the small horizontal wheel that for centuries had been the mainstay of small mills for grinding bread grains. Both reaction wheels and turbines were horizontal wheels with vertical shafts, small in diameter and quick-running. Both differed in motive force from the tub wheel, whose low efficiency was owing to its simple impact operation. Both of the newer wheels employed the reactive force of the water; a common characterization of the turbine described it as taking motion from a combination of pressure and reaction. Several of the better known Amer-

[8] See reports of successive Ordnance Department boards in *American State Papers, Military Affairs,* 2:755, 3:154 ff., 4:497 ff. To the surprise of Ellwood Morris, overshot wheels were removed and replaced by "reaction wheels" at Harpers Ferry Armory ("Remarks on Reaction Wheels Used in the United States and on the Turbine of M. Fourneyron . . . ," *JFI* 34[1842]:227).

ican reaction wheels, as will be seen, took on the character of turbines, however rude and imperfect in design.[9]

The growth of interest in improved waterwheels in this country can be traced in the federal patent office records. Less than thirty wheel patents were issued between 1791 and 1820. The number rose gradually thereafter, from two annually in the 1820s to seven in the 1830s, a level not greatly exceeded until the 1860s, when in an inventive upsurge patents for waterwheels exceeded in number the more than three hundred issued before 1860. "No one class of inventions with the exception of stoves," declared the Commissioner of Patents in 1843, "exhibits such a medley of utility and absurdity as the water wheel."[10] By far the greater number fell within the broad category of reaction wheels.

Reaction Wheels

The better known and more successful reaction wheels, like turbines, were reported to work as successfully when submerged as when placed above water. Builders of reaction wheels often declared they would work equally well in a vertical or horizontal plane, that is, turn on horizontal or vertical shafts; more commonly they were mounted horizontally. A reaction wheel, declared a Franklin Institute committee experimenting with waterpower, was "one propelled by the pressure of the water in the direction of the circular motion of the wheel developed by the discharge of the water in a contrary direction."[11] In contrast, a stream of water struck the blades or floats of impact wheels of the undershot or tub varieties and then splashed off, causing the wheel to operate by impact or impulse. Except for the brief free fall from forebay to wheel, bucket wheels were moved by the weight of the water. To activate the reaction wheels, the water

[9] In tracing the evolution of the American type of turbine, Arthur T. Safford and Edward P. Hamilton take the tub wheel as their point of departure. See their "family tree" diagram in "The American Mixed-Flow Turbine and Its Setting," *TASCE* 85 (1922): 1270. According to J. T. Fanning, "As early as 1819, on the Susquehanna River, water wheels were in use, having vertical shafts and horizontal wheels, each with a row of buckets on its circumference, which discharged water horizontally and outwardly, constituting reaction turbines" ("Progress in Hydraulic Power Development," *Engineering Record* 47 [3 Jan. 1903]:24–25).

[10] U.S., Congress, "Report of the Commissioner of Patents," Sen. Doc. 150, 28th Cong., 1st sess. (serial 433, vol. 3), pp. 325–26.

[11] Quoted in *Scientific American* 3 (27 Nov. 1847):77.

moved *through* the wheel, exerting force through reactive pressure upon wheel passages inappropriately termed "buckets."[12]

Reaction wheels had received much attention on the Continent, beginning in the mid-eighteenth century, from scientists interested in hydraulic phenomena, such as Daniel Bernoulli and Leonhard Euler.[13] The many reaction wheels patented and placed on the market in this country during the years of their prominence, 1820–60, appear to have had their origin in a wheel dating from late seventeenth-century England. This was the wheel devised by one Dr. Barker and known as Barker's mill. Evidently used more commonly to illustrate the principle of reaction than to grind meal, it was pictured in many works on hydraulics.[14] The motor of Barker's mill was a horizontal one, in its simplest form consisting of a hollow vertical shaft carrying water to and through the two horizontal arms branching out from its base (Figs. 62–64). The arms were straight in the Rumsey wheel and curved in an improvement by James Whiteland. The reactive force of the two streams of water leaving the ends of these arms placed the wheel in motion in the manner familiar to all who have observed the common rotary lawn sprinkler in action. Barker's mill was evidently introduced to this country by James Rumsey, who obtained a patent in England in 1791 for such a wheel shortly before his death. Ellwood Morris, a civil engineer who introduced the Fourneyron turbine to America in a series of articles published in 1842, devoted the opening article to reaction wheels in the United States, reporting his own investigations and experiments with wheels in actual use. This very numerous family of wheels Morris described as usually having vertical shafts and curved buckets or vanes against which the impulsive action of the water "*acts indirectly,* or rather *reacts*, thus producing (in reference to the effluent water) a *backward rotary motion,* similar in character and effect to the *forward rotary motion,* produced by direct impulse in the case of undershot wheels." Although these wheels' efficiency, as reported by other studies and as revealed in his own experiments, did not exceed 40 percent, or no

[12] See Morris, "Remarks on Reaction Wheels," pp. 217 ff., 249 ff.

[13] See d'Aubuisson, *Treatise on Hydraulics*, pp. 441 ff; Crozet-Fourneyron, *Invention de la turbine*, pp. 11 ff.

[14] Morris, "Remarks on Reaction Wheels," pp. 220 ff. See illustrated description in *Knight's American Mechanical Dictionary* (1877), 1:231; and in William Waring, "Observations on the Theory of Water Mills, &c.," and "Investigation of the Power of Dr. Barker's Mill, as Improved by James Rumsey, with a Description of the Mill," *Transactions of the American Philosophical Society,* vol. 3, nos. 18, 22 (1792):144 ff., 185 ff. See also James T. Flexner, *Steamboats Come True: American Inventors in Action* (New York, 1944), p. 209.

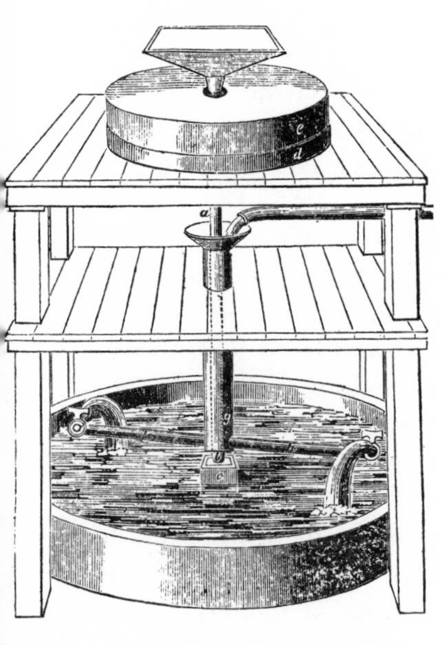

Fig. 62. ("Description of the Island of Ceylon," *JFI* 6 [1828].)

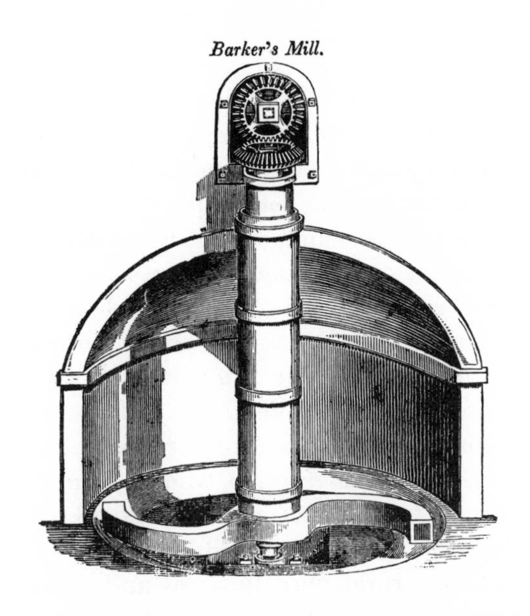

Fig. 63. (James Whiteland, "Suggested Improvements in the Construction of Barker's Mill," *JFI* 10 [1832].)

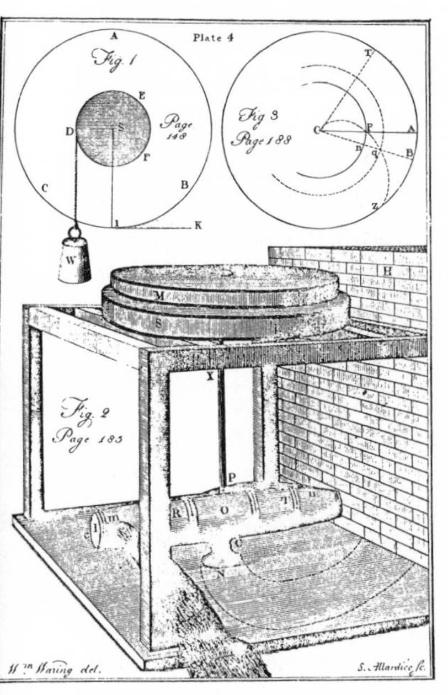

Fig. 64. (William Waring, "Investigation of the Power of Dr. Baker's mill, as Improved by James Rumsey . . .," *Transactions of the American Philosophical Society,* 3 [1792].)

Figs. 62–64. Variations on a 17th-century theme: Dr. Barker's mill.

more than that of well-designed undershot wheels, their ability to run submerged was a great advantage. The additional fall thus obtained added perhaps 20 to 25 percent to their power under average head and somewhat more at low water.[15]

With the reaction wheel, too, waterpower entered the iron age, an important anticipation of things to come. A great merit of the traditional waterwheels was their wooden construction, permitting them to be built on site by the millwright with the wood and hand tools nearly everywhere available. The cost and difficulty of shaping iron limited its use, as elsewhere discussed, to a few parts such as bearing surfaces, as in gudgeons, and reinforcing bands. However, the curved lines and shaped surfaces reaction wheels required for efficient operation were difficult to construct and often failed to retain their shape when made of wood. They were readily obtained in iron castings, which could be had from rural blast furnaces and village foundries at moderate cost. Some of the early reaction wheels were partly built of wood, but all-metal construction was a distinctive and excellent feature of these wheels.[16] Iron gave great strength and durability, made possible greater precision in design and in construction and greater smoothness and efficiency in operation. Its use was essential to provide the higher rotating speeds and greater power increasingly demanded in industry. A communication to the *Journal of the Franklin Institute*, 1826, signed "W. Parkin, Engineer, September 24, 1825," prophetically stated: "In constructing water-wheels, especially those of great power, the introduction of iron is a most essential improvement, and if this metal, and artisans skilled in the working it, could be obtained at reasonable rates, water-wheels might be made wholly of it, and prove ultimately

[15] Morris, "Remarks on Reaction Wheels," pp. 217–20.

[16] T. C. Clarke declared that the reason why the hydraulic turbine had displaced old-style wheels was clear enough: "It is because the age of iron has superseded the age of wood" (discussion of Joseph P. Frizell, "The Old-Time Water-Wheels of America," *TASCE* 28 [1893]:247–48). Both the Parker and Howd wheels, discussed below, used some wood in the early models. A remarkable exception to the rule was the "inclined plane water-wheel," patent no. 168,202, 28 Sept. 1875, invented and built almost entirely of wood with considerable success by Robert Wilson of North Carolina (illustrated communication by Harvey Linton in Safford and Hamilton, "The American Mixed-Flow Turbine," pp. 1334-38). On replacing wood with iron in water-wheels, see *Appleton's Dictionary of Machines,* 2:794 ff.; William C. Hughes, *The American Miller, and Millwright's Assistant* (Philadelphia, 1851), pp. 101–3; and Frederick Overman, *Mechanics for the Millwright, Machinist, Engineer, Civil Engineer, Architect & Student* (Philadelphia, 1858), pp. 309–10.

the cheapest, as if managed with due care, and worked with pure (not salt) water, they would last for centuries."[17]

Hardly less significant, and to many less welcome since it usually required cash outlay by the buyer, the use of iron commercialized production, taking the fabrication of waterwheels out of the hands of local millwrights and carpenters and transferring it to the increasingly numerous and widely distributed blast furnaces and foundries by which the needs of a preindustrial economy for millwork and simple machines were largely met. In the larger ironworking establishments made possible by improved transportation and marketing facilities, shop production came to be supported by sales promotion through advertising, testimonials, and the like. The intricacy and endless modification of design possible with skillful patternmaking and encouraged by the growing demand for improved waterwheels supplied the almost wholly empirical basis for advances in wheel design and performance in the years ahead.

The first reaction wheel to secure more than local acceptance was that of Calvin Wing of Gardiner, Maine. It was patented in 1830 and was described explicitly in the patent as an improvement on Barker's mill, designed to obtain greater economy and efficiency in operation compared with ordinary wheels.[18] The Wing wheel was mounted singly or in pairs on a shaft that was given a vertical or horizontal position to suit the convenience of a particular application (Figs. 65, 66). To obtain greater power than was practicable with a pair of wheels of convenient size, additional wheels could be mounted on the same shaft. The wheel could be made of cast iron plates bolted together, in the manner common with stoves, or could be made as a single casting, according to directions made the subject of a separate patent issued the same day as the one covering the wheel itself. By 1836 eight to ten patents had been issued for various modifications of the Wing wheel. As Morris later remarked, most of the subsequent inventors "have evinced by their production, a perfectly accurate knowledge" of the Wing wheel, "for it is a very difficult matter to distinguish them from it, or from each other."[19]

[17] *JFI* 1 (1826): 103. See also Oliver Evans, *The Young Mill-wright and Miller's Guide,* 13th ed. (1850), pp. 370 ff.

[18] *JFI* 11 (1831): 85–86.

[19] Morris, "Remarks on Reaction Wheels," p. 221. In a characteristic user testimonial a Maine millowner declared that sawmill owners "can save one-half of the water now required; in general cases, two-thirds; and in some instances three-fourths. There are a vast number of water rights which have been abandoned as

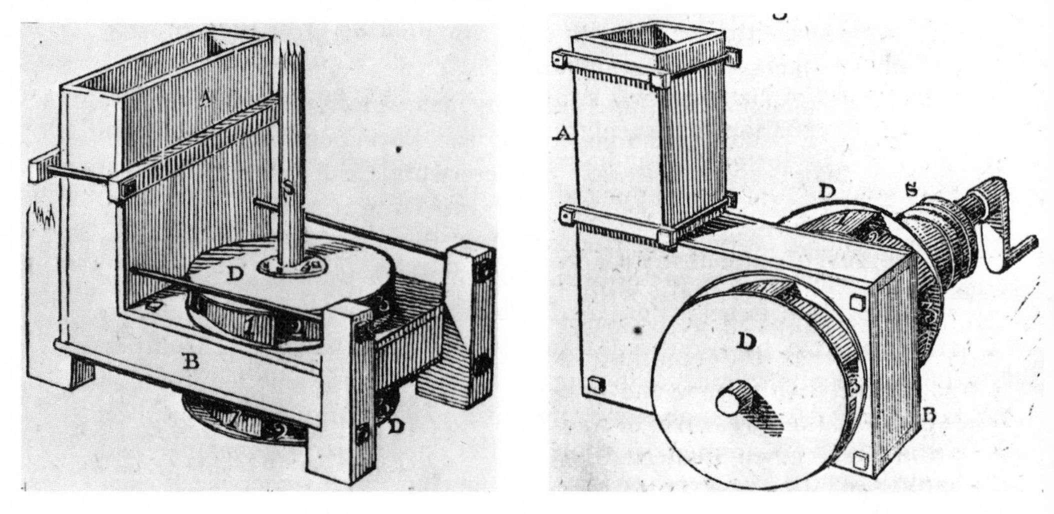

Figs. 65–66. The Calvin Wing reaction wheel. Horizontal wheel on left and vertical wheel on right, the latter with a crank, possibly adapted for use in a sawmill. (*JFI* 11 [1831].)

Durable, compact, quick-running, adapted especially to use with low heads and where backwater was a problem, reaction wheels, from their ability to operate submerged, made a useful contribution to the power needs of their time. Two deficiencies limited their use for the most part to relatively small establishments: small capacity and an efficiency in water use even when well-built and in good order of no more than 30 to 40 percent. Under the low to medium heads characteristic of the eastern United States, reaction wheels probably rarely exceeded 5 to 10 horsepower in capacity, evidently owing to the limited capabilities, equipment, and skills available in the typical local foundry. Their place in the development of waterpower was, overall, a modest though most welcome one and the period of their prominence a passing one, from about 1830 to the Civil War. James B. Francis, writing in the 1850s, summed up the case for the reaction wheel in his *Lowell Hydraulic Experiments.* Remarking upon a certain prejudice against this type of wheel, doubtless among his fellow engineers, he noted that "a great number" had "been used throughout the country, in the smaller mills and with great advantages; for although they usually gave a very small effect in proportion to the quantity of water expended, their

---

valueless to which the application of this wheel will give a power equal to the medium power of mill sites in this country generally" (*JFI,* 11:91–92).

cheapness, the small space required for them, their great velocity, being impeded less by backwater, and not requiring expensive wheelpits of masonry, were very important considerations; and in a country where water power is so much more abundant than capital, the economy of money was generally of greater importance than the saving of water." Citing the more than 300 United States patents issued on reaction wheels, Francis commented upon their manifest improvement. There were several varieties on the market "in which the wheels themselves are of simple forms, and of single pieces of cast iron, giving a useful effect approaching 60 percent of the power expended."[20]

Similar tribute had been paid somewhat earlier by the *Scientific American*: "With a low head and plenty of water, reaction wheels are by far the best, and have been the means of extending all kinds of manufactories throughout our country, by the peculiar adaptation to the propelling of machinery in situations unfavourable to other types of wheels."[21]

Tub Wheel to Turbine

The turbine combined the operating virtues of the reaction wheel with the high efficiency and large capacity of bucket wheels of the overshot-breast type. Within little more than a decade of its first practical introduction, the new motor was being built with a capacity of hundreds of horsepower, aiding the transition to an ever larger scale of industrial production. At the same time, its high efficiency in the use of water—under favorable conditions ranging between 70 and 80 percent—offset for a time the growing shortages of water supply that frequently threatened to check the growth of individual plants and industrial communities

[20] James B. Francis, *Lowell Hydraulic Experiments* (Boston, 1855), p. 2.

[21] *Scientific American* 5 (16 Mar. 1850): 201. In one of a series of articles on hydraulics the same journal declared: "America is the country where the great improvements in such kind of [reaction] wheels have been developed, and more of them are employed in the State of New York alone, than in all the other countries of the world put together" (ibid. 6 [5 July 1851]: 336). This simplistic view of wheel origins can hardly be supported. D'Aubuisson stated: "Towards the middle of the last century, and still later, rotating machines were designed and executed, where the water operated principally by reaction, and which have attracted the attention of mathematicians" (*Treatise on Hydraulics,* pp. 406–7). For d'Aubuisson on early hydraulic works, see ibid., pp. 306–7. An article "Water-Power" in *The Library of Universal Knowledge* (1880 ed. of *Chambers's Encyclopaedia)*, declared of the "reactionary wheel" that "their name is legion and it would take a book to mention them all, or to describe their respective merits."

dependent upon waterpower. In contrast to the typical reaction wheel, assembled from the crude, unmachined castings of the all-purpose local foundry, the early turbines at the Merrimack Valley centers were built in some of the most advanced machine shops of the country. First introduced in the years 1842–44 in the United States, the turbine rapidly gained in use during the next two decades. By the end of the Civil War it had become the leading prime mover in the New England cotton textile industry. In 1880 it was authoritatively described as "the only water wheel in general use . . . having completely displaced all the older forms, whether overshot, undershot, or breast wheels."[22]

The turbine, of course, did not spring like Minerva fully armed from the brow of Jove. Its long history can be traced back to at least the fifteenth century in the thinking and notebooks of engineers like Leonardo and such predecessors as Francesco di Giorgio. The idea of the turbine, Bertrand Gille tells us, was adopted "by all the 16th century engineers."[23] The origin of the turbine lay in the small horizontal waterwheel believed by some to mark the first application of waterpower to industrial tasks, which was, as we have noted, in variant forms widely in use in different parts of Europe from the Middle Ages. In his *Treatise on Hydraulics*, d'Aubuisson remarked that while vertical wheels were most generally used in the north of Europe, "horizontal wheels are most in use in the south; they operate nearly all the [grist] mills in the southern departments of France," employing the most simple mechanism and dispensing "with all gearing and transfer of motion." The shaft with the wheel near its lower end "carries the movable millstone at its upper extremity." Two types of such wheels were noted. Those moved by the impact of an "isolated vein," or course of water, passing down an inclined trough and known as *moulins à trompe* or *à canelle* were widely found in the Alps and the Pyrenees, principally on small streams with great falls. A second class of horizontal wheels, usually found on streams with much water and little fall, were known as pit wheels, or *moulins à cuve*. As remarked above, the wheel was placed in a well, or cylinder of masonry or carpentry, open at both ends; the water in its descent struck first along the side of the well, or pit, before reaching the wheel floats upon which it acted both by impact and weight.[24]

[22] R. H. Thurston, "The Mechanical Engineer—His Work and His Policy," *TASME* 4 (1882–83): 81–83.

[23] Bertrand Gille, "The Growth of Mechanization: The Use of Hydraulic Power," in Maurice Daumas, ed., *A History of Technology and Inventions,* 2 vols. (New York, 1969), 2:49–50.

[24] D'Aubuisson, *Treatise on Hydraulics,* pp. 322 ff., 406–17.

The tub wheel of the United States, most commonly used to drive small gristmills and evidently Scandinavian in immediate origin, fell somewhat between the *moulins à trompe* and the *moulins à cuve*. The American wheel was equipped with flat paddles or floats, inclined or vertical, in contrast with the concave or spoon-shaped floats characteristic of southern and eastern Europe. The wooden hoop within which the wheel turned usually did little more than contain somewhat the splash from the water striking the floats, affording some justification for the name of the wheel and the mill. Tub wheels and reaction wheels each have their place in the genealogy of the turbine.

The turbine was the eventual outcome of the efforts of many generations of men working upon a common problem related to a widespread need. Two fairly distinct lines of influence and development can be traced in the history of the turbine in the United States, one springing from native sources, the other from Europe largely, principally France. On the native side, apart from the tub wheel, the turbine was advanced chiefly by improvements in the reaction wheel, devised by such men as James Rumsey, Calvin Wing, Zebulon Parker, and Samuel B. Howd. The European influences centered in the experimental and inventive activities carried on principally in France during the 1820s and 1830s by Jean Victor Poncelet, Claude Burdin, Arthur Morin, and Benoît Fourneyron, and was developed further in practical application, by Feu Jonval. Turbines of French design were introduced in this country in the 1840s by Ellwood Morris and Emile Geyelin in the Middle Atlantic states and by George Kilburn, Uriah A. Boyden, and James B. Francis in New England, with the last two making significant innovations of their own. The native development was almost wholly pragmatic and empirical, largely ignorant of and indifferent to theoretical considerations. The foreign one was scientific in its environment, stimulation, and methodology, and practical in its objectives.[25] One American authority remarked upon the excitement and discussion aroused among the continental savants by the success of Fourneyron. He noted further that "matters of practical utility are more subjects of interest among scientific men, especially in France, than elsewhere [in Europe], where everything of a technical character is considered to belong exclusively to the workshop."[26]

In view of the vast extent and immense practical importance of

[25] For a systematic treatment, theoretical and applied, of waterwheels in Europe from the traditional wheels through the early turbines, see ibid., chap. 2, pp. 295–410, and Crozet-Fourneyron, *Invention de la turbine,* pt. 1, pp. 7–20.

[26] *Appleton's Dictionary,* s.v. "Water-wheels."

America's waterpowers and the great ingenuity applied to the invention of water wheels, why had Americans made such slight contributions to hydraulic science? So queried a reviewer of d'Aubuisson's *Treatise on Hydraulics* in the *Journal of the Franklin Institute.* The possible answer was "the greater brilliancy and attractiveness of steam power, the demand for it in the vicinity of our large cities and other places, wherein the population is more dense, and circumstances tend to make it the cheaper motor, [and] have drawn the attention of our scientific men, and even of the more scientific of our experimenters, away from the force of water." The reviewer cited as exceptions to the rule only "the Report of the Franklin Institute Committee on Water Power, a work yet incomplete, and the as yet unpublished experiments of Mr. Zebulon Parker, the inventor of what may with justice be called the American Turbine."[27]

Of Rumsey's reaction wheel, patented in 1791, we know very little apart from the short article in the *Transactions of the American Philosophical Society* in 1792 and the briefest of notices given it in Evans's *Young Mill-wright and Miller's Guide.* The wheel of Calvin Wing, with its widely imitated success, was described by a twentieth-century authority as an "outward flow turbine without guides."[28] About the same time, two country millwrights of eastern Ohio, Zebulon and Austin Parker, plagued by the problem of backwater during flood periods, did some experimenting with reaction wheels, with mixed results. Erecting a gristmill in 1827, they installed a reaction wheel, reportedly of the Rumsey type but improved by them in some details of construction. The water entered the wheel from below through a central opening and passed out horizontally through channels formed by iron plates. The modified wheel ran with indifferent results until by chance a plank falling into the forebay came to rest in a tilted position that gave the water entering the penstock a whirling motion corresponding to the motion of the wheel runner. The result was astonishing: the speed of the wheel at once doubled, as did the power it developed. Thus was born the "helical sluice," which, by guiding the entering water so that its direction corresponded to that of the wheel's motion, caused the wheel to perform much more work with a smaller consumption of water. According to Zebulon Parker's own account of the wheel's origin, the speed of their reaction wheel, when the spiral guide was adapted to it, rose from 80 to 280 rpm, "and with an aggregate orifice of 250 square

[27] *JFI* 55 (1853): 215–16.

[28] Robert E. Horton, in Safford and Hamilton, "The American Mixed-Flow Turbine," pp. 1304, 1338.

Z. & A. Parker,

Water Wheel,

Patented Oct. 19. 1829.

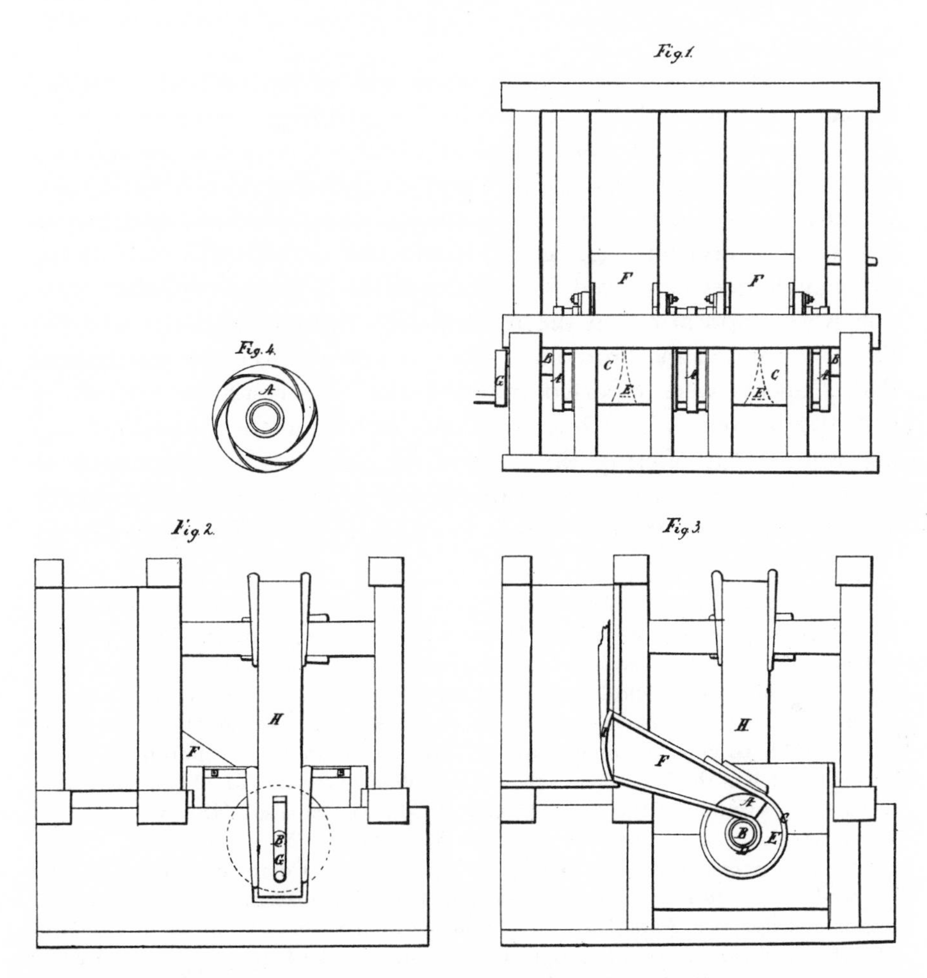

Fig. 67. Z. and A. Parker waterwheel, patented Oct. 19, 1829. (Patent Office specification.) *Fig. 1,* front. *Fig. 2,* end. *Fig. 3,* transverse section. *Fig. 4,* section through buckets. *A,* the wheel. *B,* shaft. *C,* outer cylinder. *D,* inner cylinder. *E,* partition dividing cylinders. *F,* spout. *G,* crank. *H,* hanging post. *I,* gate.

inches" admitting the water, "it sawed 3,000 feet of lumber in the same time that a flutter [undershot] wheel, at the same fall, with a gate of 450 square inches, sawed 2,000 feet."[29] The Parker brothers took out their first patent October 19, 1829. A second, covering certain improvements, was taken out June 27, 1840, six years after the death of Austin Parker, and was later followed by an extension of the original patent to 1850.

The effectiveness of the Parker wheel was evidenced both by widespread adoption and frequent imitation. Although eventually built with a capacity of 100 horsepower and over, its principal use appears to have been in grist- and sawmills and small industrial establishments. Under a practice common at the time the wheel was built in different places under licensing arrangements and cost perhaps $75 to $125 for the smaller sizes employed in common mills.[30] The many other reaction wheels appearing on the market with striking similarities to the Parker wheel led the brothers to exact payments from the purchasers. Parker agents were reported roaming the countryside in regions as widely separated as Ohio and Vermont, spying out the alleged infringing wheels and offering the owners the unpleasant alternatives of paying the royalty fee or accepting a suit and having their property attached. One report alleged that in 1851 alone, $2,000 was collected

[29] Account of the wheel's origin by Zebulon Parker in *Scientific American* 6 (3 May 1851): 264. What in time became known as the "scroll case" was in one form or another adopted widely by turbine designers and builders, including Uriah A. Boyden at Lowell. Continued typically by the wheel guides proper, which were used in all forms of turbines, the helical sluice or scroll case eliminated the sharp changes in direction of flow injurious to wheel efficiency. See Safford and Hamilton, "The American Mixed-Flow Turbine," pp. 1273 ff. See a similar account by Zebulon Parker, "Sketch of the Invention of the Parkers' Water Wheel," *JFI* 52 (1851): 49, and an earlier description of this wheel ibid. 9 (1830): 33. See also "History of Turbine Water-Wheels," *Scientific American*, n.s. 6 (15 Mar. 1862): 165.

[30] An advertisement of the Warden, Nicholson, and Company foundry in the *Pittsburgh Weekly Advocate and Emporium*, 8 Jan. 1840, described the Parker Patent Percussion and Reaction Water Wheel as "of one entire piece of *cast iron* and from 27¼ to 42 inches in diameter" (italics supplied). The sawmill wheels weighed 1,000 to 2,600 pounds; the gristmill wheels from 600 to 1,200 pounds. Prices were 5 cents per pound of metal plus $10 to $20 for "shop work" and $25 to $40 for the right to use. These figures would give a range in cost at Pittsburgh from $85 to $130 for sawmill wheels, $65 to $120 for gristmill wheels. This foundry had rights to sell these wheels in western Pennsylvania, Ohio, and Kentucky. John Martineau, inventor of "an improved water wheel" of which some 300 were reported in use, declared, "Where the cost of applying power by overshot wheels will average $1,000 to each run of stones, by my wheel may be reduced to $125–150 each run" (*Journal of the American Institute* 3 [1837]:167).

from millowners in a single Vermont county and that a similar fate awaited New Hampshire millowners in the season ahead. Parker's agents were said to be collecting royalty fees for the use of more than forty different reaction wheels, most of them patented, on the ground of infringement.[31]

These measures and the attendant litigation aroused considerable public feeling against Parker, but the outcome of court action in most cases was in his favor. Communications on the subject in the *Scientific American* during 1850 became so numerous—with the greater number opposing the claims of Parker and the extension of his patent—that the editors at length refused to publish more.[32] Some judges went much beyond a mere confirmation of the validity of the Parker patents. In the case of *Parker* v. *Brant et al.*, involving some two hundred sawmills and gristmills, a circuit court judge in Philadelphia, in March 1850, declared that the patent in question was not only valid "but of the greatest importance to the country." The judge described the defendants as "a combination of two hundred wealthy mill owners" who had best settle with Parker or face the possibility of heavy damages. In a case the following year contesting the originality of the Parker wheel, the judge told the jury the value of the improvement was such as to place Zebulon Parker in the same class of inventors as Eli Whitney and Oliver Evans.[33]

To the helical sluice, which, in the words of Parker, caused the water to enter the wheel "with a lively circular motion," was added in the patent of 1840 a so-called draft box, or tube, a water- and airtight casing extending from the wheel outlet into the tailrace. This device originated with Austin Parker in 1833 and was further developed following his death in 1834. It permitted the elevation of the wheel from its customary submerged position without affecting the efficiency of operation, permitting ready access to the wheel at all times for examination, maintenance, and repair and simplified connection with the millwork. The advantages of this arrangement were so obvious that the draft tube eventually became a permanent and widely adopted feature of turbines.[34] The French inventor Feu Jonval is also credited

[31] *Scientific American* 7 (20 Mar. 1852); see also ibid. 5 (6 Oct. 1849); 5 (22 June 1850).

[32] Ibid., 5 (6 Apr., 4 May, 10 Aug. 1850).

[33] Ibid. 5 (4 May 1850):261; 6 (5 July 1851):336.

[34] *JFI* 52 (1851): 49. See also Tyler, "Evolution of the American Type of Water Wheel," pp. 879 ff.; Safford and Hamilton, "The American Mixed-Flow Turbine," p. 1284 ff. Frizell later declared (without reference to its originator) that "the adoption of the draft-tube marks a distinct era in the development of appliances of water-power" (Joseph P. Frizell, *Water-Power*, 2d ed. [New York, 1901], p. 308).

with the invention of the draft tube, which was commonly employed with Jonval wheels and eventually was adopted in most turbine installations.[35]

Competent independent authority gave testimony to the merit of the Parker wheel. After examination and tests at several installations, the Committee on Science and the Arts of the Franklin Institute in 1846 made a highly favorable report, describing this wheel as a "true wheel of pressure, or turbine," whose sluice served the same function as the curved guides of the Fourneyron turbine. In recommending the Parker wheel to millowners, the committee listed the familiar virtues of the turbine: freedom from interference from backwater or ice, ability to run on high or low falls, and the reduced need of gearing. As to useful effect, "it ranks with the overshot water wheel and the turbine. . . . It is simple in its construction, and of durable materials, and is for these reasons not expensive, and not liable to get out of order." The efficiency as determined by committee experiments upon Parker wheels at three different mills ranged from 62 to 68 percent. According to the *Scientific American*, the Parker wheel's efficiency was raised to 70 to 75 percent by 1848.[36]

With whatever modification in design, this wheel by 1850 was being built in a range of sizes capable of developing as much as 140 horsepower under a fall of thirty-one feet.[37] In common with the Wing wheel it was built for use with either a vertical or a horizontal shaft,

[35] J. Allen, "Hydraulic Engineering," in Charles Singer et al., eds., *A History of Technology,* 5 vols. (Oxford, 1958), 5:531.

[36] Committee on Science and the Arts, "Report on Parker's Water Wheel," *JFI* 42 (1846):35 ff. According to Zebulon Parker, the efficiency of the common reaction wheel in Ohio increased gradually from 25 percent to 40 to 45 percent in 1828 (*Scientific American* 3 [19 Aug. 1848]: 381).

[37] See communication to *JFI,* 3d ser. 48 (1849): 401–3. A modification of the Parker wheel by Andrews and Kallbach of Bernville, Pa., ranked among the three top wheels tested in the Fairmount Waterworks turbine competition, 1859–60 (*Scientific American,* n.s. 3 [7 July 1860]:22). See also laudatory reference to the Parker wheel in "History of Turbine Water Wheels," *Scientific American* 6 (3 May 1862): 278–79. In an as yet unpublished paper based on extended research in court records and other unused materials, Professor Edwin T. Layton, University of Minnesota, has corrected and added substantially to what has hitherto been written of the work of the Parker brothers, thereby assuring them a position of distinction among Americans contributing to the evolution of the reaction wheel. He shows that they developed a theoretical framework for guidance in their long-continued experimental works based upon the writings of Newton and the extended discussion of mechanics by Oliver Evans in his *Guide.* They built their own wheel-testing flume, eventually with glass walls for the better observation of the action of the water; and on the basis of a published description of the Prony dynamometer built and used one like it.

permitting in the latter case a direct connection with factory shafting without use of gearing. It possessed operating characteristics and a useful effect approaching those of the first Fourneyron turbines installed in this country, as described below, in 1842–43. It was brought into wide use, first at the level of the country mill and, following development through painstaking experimentation and observations, in a capacity suited to the needs of medium-scale mills and factories. The wheel held its own in the vigorous competition arising among turbine builders in the 1850s, making up for a somewhat lower efficiency in water use by the simplicity and ruggedness of its construction.

Samuel Howd and the Inward-Flow Wheel

Still another rural mechanic to rise well above the common level of achievement in waterwheels was Samuel B. Howd of Ontario County, New York. Howd reversed the usual practice in reaction wheel design—as, for example, in the wheels of Wing and the Parkers (and also in Fourneyron's turbine)—of admitting water at the center of the wheel and discharging it at the circumference. By placing the guides outside of the wheel runner and making an appropriate rearrangement of water supply, he introduced in practical use the first inward-flow wheel, a change with important consequences, both practical and theoretical, in the evolution of waterwheels.[38] As adapted and improved by James B. Francis and made the subject of a celebrated series of waterwheel experiments, the Howd-Francis wheel became the prototype of what in the 1860s and 1870s was developed into the justly celebrated "American mixed-flow turbine." Under the common practice of selling territorial rights to build a patented device—which resolved at a stroke the problems of distribution and sales under conditions of difficult communications—the Howd wheel (patented 1838) was introduced in considerable numbers throughout the country during the 1840s and 1850s. It sometimes bore the name "United States Wheel," from the United States Water Wheel Company, which held rights to build and sell the Howd wheel over an extensive area.[39]

[38] Howd later succumbed to the conventional wisdom in "Howd's latest improved water-wheel," patented in 1842. Howd's definition of this 1842 wheel is quoted at length in Safford and Hamilton, "The American Mixed-Flow Turbine," p. 1353. Whether Howd's reversal of his own position weakens the claim made for him by Horton and others for a larger share of credit for the invention of the inward-flow turbine raises a nice point. According to Crozet-Fourneyron, Benoît Fourneyron (his great uncle) had from the beginning of his work contemplated inward- and axial-flow as well as outward-flow wheels (*Invention de la turbine,* p. 36).

Among the purchasers of licenses to build the wheel in Massachusetts was the Locks and Canals company of Lowell, James B. Francis, superintendent and chief engineer. The company paid some $1,200 for rights to build and sell the Howd wheel in Middlesex County.[40] The lightly tinted drawings accompanying the 1838 patent specifications of the Howd wheel, submitted by the inventor, antedate the first published descriptions of the Fourneyron and Jonval turbines in the United States.[41] They reveal the essential character of the Howd wheel as a turbine, that is, a waterwheel acting not by weight, impact, or impulse but by pressure and reaction. It was a wheel that operated with its passages continuously filled with water, directed by fixed guides and gliding over the curved surfaces of what in the trade were termed "buckets" (which in overshot and breast wheels actually carried the water) placed at intervals around the wheel's circumference. That the designation of the Howd wheel as a turbine is proper may be seen by the definition of a contemporary authority, Professor W. P. Trowbridge of Columbia College. "The turbine wheel," he declared, "consists essentially of two rims or crowns, firmly attached to an axis, numerous curved vanes or *buckets* being fixed between the crowns, and the main stream of water which passes through the wheel is divided by guide blades, not attached to the axis, into numerous jets or smaller streams, which impinge upon these buckets simultaneously at all points of the circumference, and produce motion in the wheel."[42] To this description further details are provided by Samuel Webber, a New England hydraulic engineer. Webber stated that in Howd's wheel "the action of the Fourneyron wheel is reversed, and the converging guides, which were straight, were placed outside the wheel, which had curved buckets, revolving inside the guides, and was, in fact, only one form of the old 'tub wheel'. . . . the regulating gates were placed outside the guides or 'chutes.' The buckets were cast iron, fastened by bolts to wooden top and base plates, and the discharge was central."[43]

[39] Francis, *Lowell Hydraulic Experiments,* p. 61, and pl. 9, figs. 3–4.

[40] Sworn statement of J. T. Morse, 20 Jan. 1847, LCP, A-17, no. 98.

[41] Allen, "Hydraulic Engineering," pp. 525–26, 530. James Thomson's "inward flow vortex turbine" was patented in January 1851 and described in a paper before the British Association in 1852; for an illustrated description, see William Fairbairn, *Treatise on Mills and Millwork,* 2 pts. (London, 1861–63), pt. 1, pp. 163–72. Fairbairn described Thomson, of Belfast, Ireland, as a former pupil.

[42] William P. Trowbridge, *Turbine Wheels* (New York, 1879), pp. 10–11.

[43] Samuel Webber, "Water-Power—Its Generation and Transmission," *TASME* 17 (1895–96): 47–48.

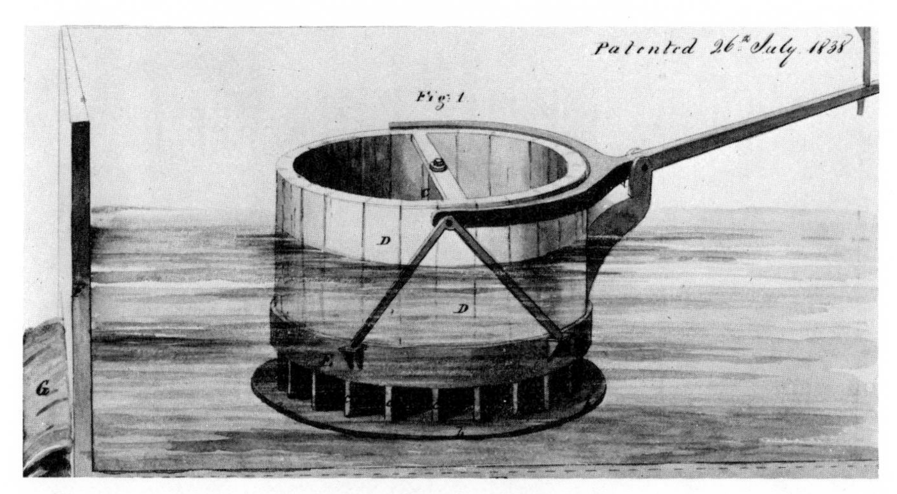

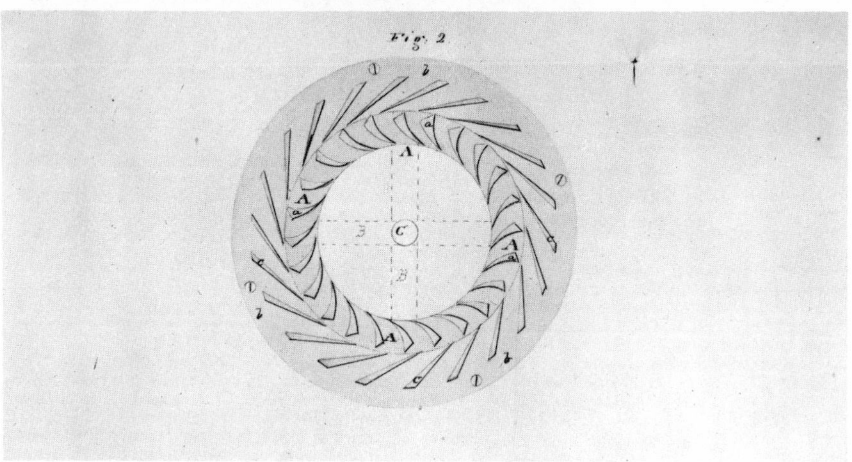

Fig. 68. Howd's Improved Waterwheel, patented July 26, 1838. (Photo by Barry Donahue. LCP, vol. A-19, file 98, Baker Library, Harvard University.)

AAA (Fig. 2) —Lower rim of wheel runner (upper rim cutaway to show "buckets": *AAA*
aaa (Fig. 2) —Vertical dividers between lower and upper rims of runner forming passages termed "buckets," curved as shown

ccc (Figs. 1 & 2) —Long, narrow wedges forming fixed guides directing water *inward* through buckets to and through center vent to tailrace (not shown)

DD (Fig. 1) —Circular curb resting on outer edge upper rim of fixed guide ring

EE (Fig. 1) —Cylindrical gate, raised or lowered to admit or shut off water to runner

BB (Fig. 2) —Arms attaching wheel runner to wheel shaft *C*

FF (Fig. 1) —Forebay or flume in which entire wheel assembly is immersed (water: shaded portion)

HOWD'S IMPROVED WATER WHEEL.

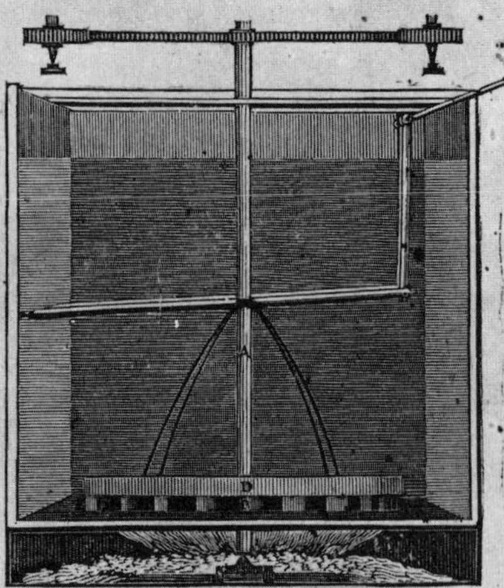

The subscribers having obtained the exclusive right of making, vending and useing the above improved water wheel, in the counties of Grafton, Coos, and Carroll in the state of New Hampshire, and also the right of making, vending, and using the same in the state of Vermont, will dispose of individual rights, or of portions of territory, by towns or otherwise, on liberal terms; and will attend to building wheels, and putting them into operation in any part of the country, on short notice and on terms to suit purchasers.

This wheel, being constructed of cast iron, is of superior durability, and from the manner of its being enclosed is not, like other wheels, affected by ice or back water. It is peculiarly adapted to streams where but small head of water can be obtained, and will propel machinery, with from two to ten feet head, with greater force and less water than any other kind of horrizontal wheel now in use.

These wheels are in successful operation in many places in New England, New York and other sections of the United States. The operation of them may be seen by examining wheels now in use in this vicinity and in Vermont, at the following places: At the Scythe Works in Littleton, at the mill of Mr. Charles Bellows in Northumberland, at the Grist mill in Lyman and in Lebanon, New Hampshire—at Montpelier village and at the Woolen Factory in Waterville, Vermont. Any persons wishing for information in regard to this improvement, will please address the subscribers (postage paid,) and their communications shall be immediately atttended to.

PETER PADDLEFORD,
PHILIP H. PADDLEFORD.

Littleton, N. H., May 6th, 1844.

CERTIFICATES.

This may certify that there has recently been put in operation, in the Scythe Factory in Littleton, N. H. one of S. B Howd's Cast Iron Water Wheels, in place of an under shot Wheel 8 feet in diameter, length of Buckets 5 feet. This wheel is used to carry 2 grindstones, 6 feet in diameter, a Straw Tinster and Polishing wheel and requires about half the power with 7 feet head; while the one displaced was insufficient, with full power, under same head, another wheel of the same kind 4 feet in diameter is about being put into the same Factory, for driving a fan Bellows, in place of a Tub wheel 8 feet in diameter.

ELY & REDINGTON.

Littleton N. H. April, 19, 1844.

I have recently put in operation in my wire factory one of S. B. Howd's Direct Action cast iron Water Wheels, 6 feet in diameter, in the place of a 12 feet breast wheel, 12 feet length of buckets. I verily believe that this wheel does more work with the same water than one displaced. I believe it is better adapted to the purpose I use it for than any other wheel, as it gives a very uniform motion.

A. DENSLOW.

Coun, Windsor, Ct., Jan. 6th, 1842.

We hereby certify, that during the last season we have put into our Custom Mill one of Howd's improved Water wheels, under a 4 1-2 feet head, with which we drive two runs of mill stones, with smut mill screen and bolting geer attached, to our satisfaction, and we do not hesitate to say, in regard to the cheapness of construction and economy of power, it is the best wheel with which we are acquainted.

WM. H. IMLAY & Co.

Hartford, Jan. 7, 1842.

Wm. H. Imlay & Co. have purchased the right for the town of Hartford, giving over $400 for the same.

This may certify that we have used one of Howd's Patent Water Wheels since December last, by the side of a re-action wheel and we think that Howd's will do nearly double the business that the re-action wheel will do. We have had but three feet head, and can grind with that eight bushels per hour. We are subject to back water. This wheel will do as good business under back water as the re-action, and we recommend it to the attention and patronage of the public.

SIMON BURTT,
MILES S. LEACH.

Lyons, Sept. 3, 1838.

L. J. M'INDOE, PRINTER, NEWBURY, VT

Fig. 69. Advertising broadside for Howd's waterwheel. (Photo by Barry Donahue. Baker Library, Harvard University.)

The business done in Howd wheels under the license obtained by the Locks and Canals through the associated Lowell Machine Shop was not impressive. However, Howd's unique and wholly original inward-flow principle was adopted by Francis, who combined it with what he had learned from the carefully designed Fourneyron turbines built at Lowell by Uriah A. Boyden and produced the skillfully engineered wheel that in its generalized form became known throughout the world as the Francis turbine. In his influential *Lowell Hydraulic Experiments* Francis explained that he had "so modified the form and arrangement of the whole as to produce a wheel essentially different from the Howd wheel" although possibly "technically covered" by the Howd patent.[44] The Francis wheel was not at this time or in the years immediately following placed in production, a few only being built for experimental and corporation purposes, as described in the *Lowell Hydraulic Experiments.* By the standards set by Boyden and Francis in their wheels—in which considerations of cost were disregarded in the determination to secure excellence—the Howd wheel was indeed a rude and roughly executed machine, not inaccurately, if somewhat disparagingly, described by Francis as "constructed in a very simple and cheap manner, in order to meet the demands of a numerous class of millers and manufacturers, who must have cheap wheels if they have any."[45] The Howd wheel, like the Parker wheel during the same period, met the conditions and served the requirements of an early industrial age.

The influence of the Howd turbine extended well beyond the range of its practical introduction in mills and small factories. Safford and Hamilton in their 1922 paper traced a direct line of descent from Howd through the experimental inward-flow turbine of the *Lowell Hydraulic Experiments* to the American "mixed-flow" turbines dominating the post–Civil War period. Since these formative years, Safford and Hamilton concluded, "almost every wheel has been a combination of the

[44] Francis, *Lowell Hydraulic Experiments,* pp. 61–62, and pl. 9, figs. 3, 4. This is the testimony of an interested party, of course. The words of fellow engineer Samuel Webber may be added: "In 1849 Mr. Francis took this matter up, and built, for the Boott Mills in Lowell, an inward discharge [Howd] wheel in which he employed the carefully designed curves of the 'Boyden' wheels, and which gave excellent results, nearly equal to those of the outward discharge" ("Water-Power," p. 47). For correspondence respecting the Howd wheel and purchase of rights for its use, see LCP, A-19, 98. For later views on Howd's role, see Safford and Hamilton, "The American Mixed-Flow Turbine," pp. 1242, 1352 ff., and especially the letter of Robert E. Horton, ibid., pp. 1302–6.

[45] Francis, *Lowell Hydraulic Experiments,* p. 2.

tub wheel in its many forms, the center-vent Howd-Francis, and the axial-flow Jonval."[46] In the discussion of Safford and Hamilton's paper on the emergence of the American type of turbine in the *Transactions of the American Society of Civil Engineers* the view was strongly expressed by several contributors that Howd had received much too little recognition for his seminal role in its development.

It is clear that American mechanics and inventors early showed an awareness of the need of industry for new waterwheels more suitable to their requirements and in their own empirical fashion set about providing them. They seem to have worked quite independently of Old World developments and in most if not all instances in ignorance of them. Taking the common tub wheel with its simplicity, compactness, and high rotating speeds as the point of departure and, after Rumsey, the reaction wheel with the added ability to run submerged and in either a horizontal or vertical position, such leading mechanic-inventors as Wing, the Parkers, and Howd did much to meet needs not served by traditional wheels, as Francis testified. The Howd-Francis inward-flow turbine may be taken to represent the convergence of the differing methods of waterwheel improvement, American empiricism, and the more fundamental scientific-engineering approach followed in Europe.

Birth of the True Turbine

By 1840 word was reaching this country of the successful results of a different method of waterwheel development undertaken on the Continent, principally in France. Published accounts of the new and greatly improved waterwheels, given the name "turbines" from the Latin stem denoting the motion of spinning or whirling, stressed their superior efficiency and operating characteristics. Within several years efforts to design and build wheels of the new class were meeting practical success; Lowell and Philadelphia were the cities in which this activity chiefly centered. Within less than ten years French turbines had been adapted to American conditions and needs, had been in

[46] Safford and Hamilton, "The American Mixed-Flow Turbine," pp. 1255–56. Reviewing the early years of turbine equipment in the United States, the authors remark: "The result of this period of scientific design has been the production of one thoroughly American wheel, the Howd wheel, as developed by Francis. This wheel was to be the forerunner of all modern reaction wheels, and the next step [in this paper] will be to trace its development into the mixed-flow turbine" (ibid., p. 1255).

some respects significantly improved, and had been successfully built in the large horsepower capacities required to meet the needs of some of the largest industrial establishments of the time, the great textile mills of New England.

European developments followed a very different course and employed different methods from those appropriate and congenial to a young and undeveloped nation. The editor of the *Journal of the Franklin Institute*, reviewing in 1853 the American translation of d'Aubuisson's *Treatise on Hydraulics for the Use of Engineers*, remarked upon the slight contributions of Americans to practical hydraulics despite their great dependence on waterpower.[47] The *Lowell Hydraulic Experiments,* which were to establish the engineering reputation of Francis and carry his name throughout the engineering world, lay two years into the future. The elaborate and carefully planned and executed experiments of the Franklin Institute on full-scale waterwheels, intended to update the mid-eighteenth-century experiments Smeaton conducted with small models, were dragged out over the decade 1830–40 and left dangling in midair for want of an overall summary and interpretation of findings to meet the needs of millwrights and wheel builders (see Appendix 8). In France, as the *Journal's* editor remarked, "the large body of able engineers which her schools of practical science have so long and so continually given her, has led to the fullest and most scientific investigation of the subject, and hence it is to France principally that we are to look for the most ample and most reliable data for the establishment of hydraulic works."[48] There, behind the millwright and practical mechanic, were ranged engineers, scientists, and mathematicians. Such men, with the encouragement of government and learned societies, sought in systematic ways to develop new and more efficient water motors, which were given special meaning in many regions by the limited fuel resources of wood and coal. The experimenters combined an understanding of scientific method and the use of mathematical tools of investigation with a concern for results of practical value to the manufacturer and millowner. "We owe the development of turbines chiefly to Continental mathematicians," declared William Fairbairn.

[47] *JFI* 55 (1853): 215–16. Signed "F." The review was evidently written by the editor, John F. Frazer. See also *Appleton's Dictionary of Machines,* s.v. "Water-Wheels"; George Rennie, "Report on the Progress and Present State of Our Knowledge of Hydraulics as a Branch of Engineering," *JFI* 19 (1835): 56 ff., 125 ff., presented initially before the British Association for the Advancement of Science.

[48] Review of d'Aubuisson, *Treatise on Hydraulics,* in *JFI,* 55:215–16.

European efforts to improve conventional waterwheels understandably began much earlier than American efforts but were directed to much the same deficiencies. Such eighteenth-century investigators as Segner, the Eulers, Valernod, Belidor, and Borda were followed in the early nineteenth century by Navier, Burdin, Poncelet, and others. In 1823 the Société d'encouragement pour l'industrie nationale offered a substantial prize for development of an efficient horizontal wheel of large capacity for industrial use. The deficiencies of conventional wheels as enumerated by the Société were virtually the same as those that American inventive efforts were attempting to overcome: the inability to operate when submerged; large bulk and weight in the case of overshot wheels and the gearing required to obtain operating speeds; the low efficiency of horizontal wheels discouraging their use except where water was abundant. Much attention had already been given to vertical wheels. One developed by Poncelet, which has since borne his name, was awarded a prize in 1825, but the common horizontal wheel, widely used in various forms in the gristmills of southern France, was the principal object of improvement.[49]

The long-continued efforts brought success in the wheel of Benoît Fourneyron, a pupil of the French engineer Claude Burdin, whose own wheel proved a failure in everything but the name he gave it, the turbine. The Fourneyron turbine, first given practical application in 1827, demonstrated such high efficiency and excellent operating characteristics as to effect a revolution in hydraulic motors. Several years passed, however, before Fourneyron adapted his wheel to the power requirements of ironworks and textile mills. According to Crozet-Fourneyron, he prepared a paper outlining a complete theory of the new wheel, for which he obtained a patent in 1832. In 1833, having admirably fulfilled the conditions of the competition sponsored by the *Société*, Fourneyron was awarded the grand prize.[50] Careful experiments by the distinguished engineer Arthur Morin showed that in actual industrial use the large Fourneyron turbines operated with an efficiency ranging from 70 to 78 percent. Word of the astonishing new waterwheel spread rapidly through Europe, and by 1839–40 descriptive accounts by European authorities were appearing in the *Journal of the Franklin Institute.*[51]

[49] D'Aubuisson, *Treatise on Hydraulics,* pp. 361 ff., 406 ff.; Jean Victor Poncelet, *Mémoire sur les roues hydrauliques à aubes courbes, mues par dessous* (Theil, 1827).

[50] Crozet-Fourneyron, *Invention de la turbine;* d'Aubuisson, *Treatise on Hydraulics,* pp. 418 ff.; *La Grande Encyclopédie* (Paris, 1886–1902), 17:922.

[51] M. Savary, "Reports to the Academy of Sciences of Paris of Experiments . . . on

The new turbine was a relatively simple mechanism with three principal components: a central fixed disk on which were mounted a number of iron guides that curved downward and outward, forming spiral passages by which the water passed from the penstock to the wheel proper; a horizontal wheel, or runner, mounted on a vertical shaft and having two outer rims, separated by vertical metal strips dividing the space between them into a number of curved passages, or buckets, through which the water received from the fixed guides moved outward; and a gate mechanism by which the admission of water from the penstock to the wheel was regulated. In sharp contrast with the buckets of the overshot and breast wheels, which held and carried the falling water, the turbine's buckets simply presented curved surfaces against which the water exerted force by pressure and reaction in passing through the wheel.[52] In the conventional vertical wheels—overshot or undershot—the water operated, by weight or impulse, on only a portion of the buckets, or floats, at a time. In the turbine all the working surfaces were simultaneously subject to the pressure of the column of water passing through the wheel. Like the small horizontal wheels preceding and culminating in it, the turbine was fundamentally a reaction wheel, its power being derived principally from the reactive pressure of the water upon the surfaces of the buckets or passages from which it issued. It went far toward meeting the conditions laid down by Jean Charles de Borda in 1767 for the ideal waterwheel: the water must enter the wheel without shock and leave it without velocity.[53]

The Fourneyron turbine was the first of several major types, each of which in time underwent innumerable variations in proportions,

---

the Turbines of M. Fourneyron," trans. J. Griscom, *JFI* 27 (1839):133 ff.; George Rennie, "Description of the Turbine, or French and German Water Wheels," reprinted ibid. 29 (1840):403–8; Arthur Morin, "Experiments on Water Wheels, Having a Vertical Axis, Called Turbines," trans. Ellwood Morris, ibid. 36 (1843): 234–52, 289–97, 301–2, 370–77.

[52] The use of the terms *pressure, reaction,* and *impulse* by different writers on waterwheels is a subject of no little confusion. That this problem arose early is suggested by a statement published in 1852: "The effects of these different varieties of wheels arise from three sources—weight, impulse, and reaction. But in stating these as the primary and simple elements of hydraulic power, it is to be remarked that we very rarely find the effect reducible to a single mode of action; more commonly we find two, and sometimes even the three acting simultaneously, and not unfrequently in nearly equal degrees" (*Appleton's Dictionary of Machines,* s.v. "Water-Wheels," p. 786).

[53] Crozet-Fourneyron cites Borda's *Mémoire* presented to the Academie des sciences, noting that "l'eau devait entrer sans choc et sortir sans vitesse" (*Invention de la turbine,* p. 18). See also Rennie, "Description of the Turbine," pp. 403 ff.

arrangement, and construction of the principal components and supporting equipment. The effectiveness and efficiency of each wheel depended largely on the skill and care with which the forms and dimensions of the several parts, especially the guides and wheel buckets, were designed and fabricated. Turbines of the Fourneyron type, called "outward flow," were distinguished by the fact that the water, as with most reaction wheels, was admitted at the center, and after passing downward through the curved guides and outward through the buckets, was discharged at the periphery of the wheel. The Jonval turbine employed a downward, or axial, flow; the water entering at the top passed down through the guides to and through the wheel and discharged into the tailrace below. The inward-flow turbine first devised by Howd and perfected by Francis admitted water at all points of the circumference of the wheel; after passing through the buckets, the water was discharged downward at the center, reversing the position of guides and buckets employed in the outward-flow turbine. In each type the curves of the guides and the buckets were different. Each type possessed certain distinctive operating characteristics and was adapted to operate within a more or less favorable range of efficiency under different conditions of head and streamflow. Eventually a turbine combining the characteristics of the inward- and the downward-flow types, known as the mixed-flow turbine, was developed and brought into very wide use in the United States and to a less extent elsewhere.

The Turbine Reaches America

The introduction of the European turbine was aided by the activities of several interested Americans. Ellwood Morris of Philadelphia combined an enthusiastic advocacy of the Fourneyron wheel in articles published in the *Journal of the Franklin Institute* with the design and installation of a number of small turbines in factories in his area. Uriah A. Boyden at Lowell designed and built Fourneyron-type turbines of such size, precision, and efficiency as to secure the early adoption of the new wheel in the great cotton mills of the textile centers of New England. Andrew Robeson, Jr., and George Kilburn of the Fall River textile industry independently introduced the Fourneyron turbine in New England and engaged in its commercial production in the small to medium powers adapted to the needs of smaller scale manufacturing. These pioneering events all fell within the brief span

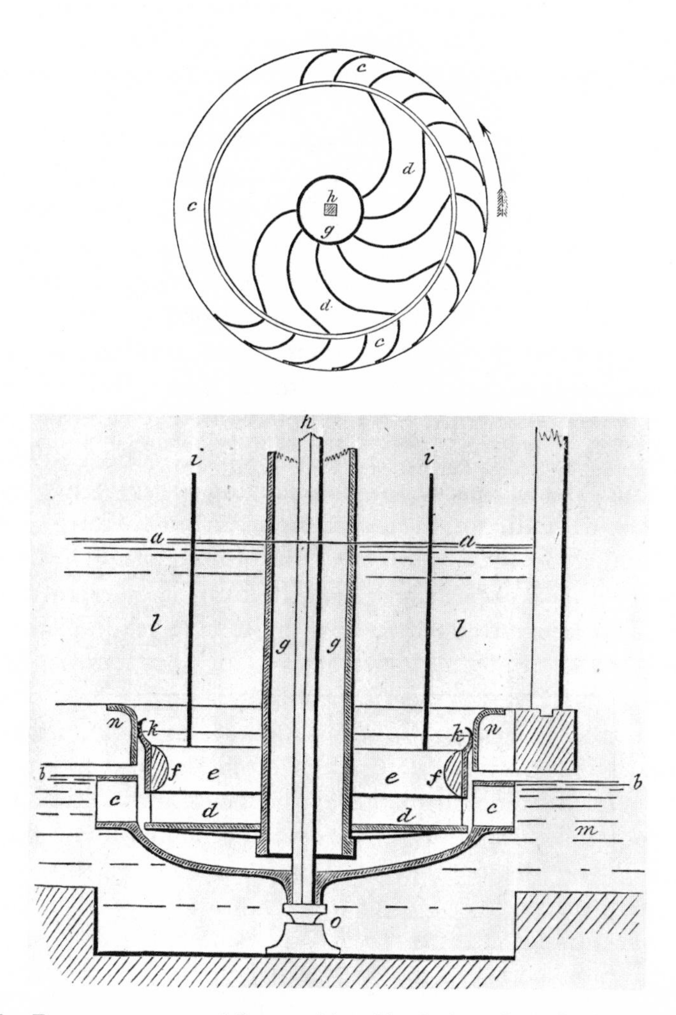

Fig. 70. The Forneyron outward-flow turbine. Vertical section of a turbine showing: "*a a*, the surface of water in the upper level, or forebay; *b b*, the surface of water in the lower level, or tail race; *c c, the wheel* with buckets curved in plan; *d d, the fixed disk* and *curved guides . . . ; e e, the annular sluice gate* with its wooden internal cushions *f f . . . ; g g, the shaft pipe; h, the shaft* upon which the *wheel c c* is firmly fixed at its lower part . . . ; this shaft runs upon a suitable step, or pivot, at *o; i i* two vertical rods . . . attached to the annular sluice gate at equal distances apart . . . [that] raise or depress the annular gate with perfect regularity; *k k, a leather collar* . . . which . . . being pressed outwards by the water, against the concave surface of the concentric *fixed cylinder n n*, effectively secures this joint in a manner that prevents leakage; *ll, the forebay*; . . . and finally, *m* is *the tail race* through which the water escapes, after having actuated the turbine." (Ellwood Morris, "Remarks on Reaction Wheels," *JFI* 34 [1842].)

of years 1842–46. By 1850 the Jonval, or axial-flow, turbine had been introduced from France by Emile Geyelin of Philadelphia, who was for many years identified with its successful and extensive adoption here.

Ellwood Morris, an engineer of some distinction, first brought the importance of the European turbine forcefully to the attention of American engineers and millowners. In 1842 and 1843 Morris carefully reviewed in the *Journal of the Franklin Institute* the various types of waterwheels, old and new, then in use. As background for his discussion of the Fourneyron turbine, he gave particular attention to reaction wheels, then at the height of their vogue, finding the advantages of their operating submerged often more than offset by their disappointingly low efficiency.[54] He supplemented a careful description of the new wheel, illustrated with drawings, with a translation of Capt. Arthur Morin's paper on the exacting experiments confirming the large claims made for the Fourneyron turbine. Morris went much further. Guided by little more than the accounts and general drawings appearing in European technical literature, he designed and himself built—and later arranged to have built by a leading Philadelphia machine shop—a number of Fourneyron turbines. Morris conducted experiments under varied working conditions on these turbines, ranging from 4 to 30 horsepower and installed at various manufactories in the region. The tabular results, published in the *Journal of the Franklin Institute*, on the turbines installed at the Rockland Cotton Mills and the Du Pont Powder Works, confirmed the claims made for the turbine abroad and demonstrated the practicability of this waterwheel under American conditions of construction and use. The experiments showed that a useful effect of fully 70 percent of the energy present in the falling water could safely be counted upon.[55] By December 1843 seven Morris-designed turbines were in use.

In southern New England a similar role was played by two men of Fall River, Massachusetts. Andrew Robeson, Jr., operator of a cotton print works, saw a Fourneyron wheel in operation in France. He arranged for the master mechanic of his firm, George Kilburn, to build a turbine, which was placed in operation in 1844, reportedly with complete success. Two years later the Fall River firm of E. C. Kilburn and Company, textile-machinery manufacturers, began the commercial production of this turbine, at first for local use but soon

[54] Morris, "Remarks on Reaction Wheels," pp. 217–27.

[55] Ellwood Morris, "Experiments on the Useful Effect of Turbines in the United States," JFI 36 (1843): 377–84.

for a widening market. Although the Kilburn-built wheels were of simple construction, relatively low capacity, and probably of but moderately good efficiency, they were superior in many respects to conventional waterwheels for establishments with modest power needs. They signified the beginning of continuous commercial production of standard turbines in a range of sizes—from 21 to 108 inches in diameter—adapted to the requirements of all but the largest classes of industrial establishments.[56] The prices cited for the Kilburn wheels indicate that they must have been built with only a small part of the care given the Boyden-Fourneyron wheels at Lowell.[57]

For reasons possibly associated more with a failure aggressively to exploit the Fourneyron turbine than with its capabilities, the pioneering work of Morris and of Robeson-Kilburn had, by present evidence, but a modest influence in the Americanization of the French-born wheel. A new and larger phase of the turbine's development in this country occurred in the middle and late 1840s with the work of Uriah A. Boyden and James B. Francis in the adaptation of the Fourneyron wheel to the requirements of large textile mills and in the aggressive and effective campaign of Emile Geyelin to introduce the Jonval wheel. Jonval's axial-flow turbine, with guides and wheel blades arranged so that the water followed a course generally parallel to the shaft, or axis, in its downward passage through the wheel, was marked by simplicity in design and efficiency in water use. Moreover, the manner of construction favored its use with a draft tube, a castiron, airtight cylinder extending from the wheel into the tailrace. As in the Parker wheel, this arrangement placed the wheel proper well above the water level, where it was readily accessible for cleaning and repair, a feature understandably endearing it to millowners and millwrights.

Upon purchasing the Jonval patents, the well-known machinery

[56] Jonathan T. Lincoln, "Material for a History of American Textile Machinery: The Kilburn-Lincoln Papers," *JEBH* 4 (1931–32):265 ff.; William R. Bagnall, *Sketches of Manufacturing Establishments in New York City, and of Textile Establishments in the Eastern States,* ed. Victor S. Clark, 4 vols. (Washington, D.C., 1908), 3:1846. Bagnall makes much of a few months' priority of installation of the first Kilburn turbine over the first Morris-designed one; he also emphasizes a reported visit of Uriah A. Boyden to see the Kilburn wheel in operation, inferring that Boyden learned much of benefit in designing his first turbines at Lowell. Such tales, true or false, are indispensable to the folklore of technology.

[57] Lincoln cites sizes and prices from an 1857 letter of the Kilburn firm ("American Textile Machinery," p. 272). See also the Kilburn, Lincoln Machine Company Papers, 1846–1869, Baker Library, Harvard University, Cambridge. The extensive literature on the American turbine contains scant reference to the Kilburn wheel.

building firm A. Koechlin and Company of Mulhouse, France, carried out various improvements and by numerous experiments devised formulas for calculating the dimensions of wheels appropriate to any given situation. By 1850 more than 300 of its wheels were in successful operation in Europe.[58] In 1849 Emile Geyelin, formerly associated with the Mulhouse company, came to the United States, bringing with him the rights to the manufacture and sale of the Jonval turbine in this country. He settled in Philadelphia, establishing his office in the Hall of the Franklin Institute, and embarked on a very effective campaign of promotion for the Jonval-Koechlin wheel. He arranged for the production of the wheel by a number of the leading machine shops in the East, including the I. P. Morris firm of Philadelphia and the West Point Foundry near New York City. He followed the installation of his wheels by tests of their power and efficiency, accounts of which were published in the *Scientific American* and the *Journal of the Franklin Institute* and in promotional pamphlets, along with testimonials, drawings, and comparisons with the performance of the overshot wheels the Jonval turbines so often displaced. Guarantees of efficiency ranged from 75 percent on falls of twelve to thirty feet down to 60 percent on falls of four to six feet.

Within two years of his arrival at Philadelphia, Geyelin witnessed the installation of one of his wheels at that city's Fairmount Waterworks, the leading establishment of its kind in the country, with a long record of engineering innovations and achievements. The performance of this wheel, declared the annual waterworks report, "has exceeded the most sanguine expectations." Owing to the peculiar situation of the Fairmount Waterworks on a tidal river, the breast wheels, hitherto the only source of power in pumping, were unable to operate during six hours of the twenty-four, but the Jonval wheel ran contin-

[58] Much of the information which follows is presented in a ten-page pamphlet with four plates of drawings, evidently published by Emile Geyelin, *Description of Jonval's Turbine Built by E. Geyelin, Hydraulic Engineer* (Philadelphia, ca. 1851). See also communications by Geyelin in *JFI* 52 (1851): 418–20, 58 (1854): 332–35. Other articles about the Jonval-Geyelin turbine appear in *JFI* 50 (1850): 189–91, 66 (1858): 12–13, 164–65. See also *Scientific American* 7 (10 July 1852): 340, 12 (11 Apr. 1857): 246. Geyelin explained the conditions under which an overshot wheel would give more satisfactory service than a turbine, ibid. 3 (18 Aug. 1860): 115. An illustrated description of the Geyelin-Jonval wheel appears in *Appleton's Dictionary of Machines*, 2:835 ff. As introduced in France, the Jonval turbine employed the draft tube; by one account, Geyelin built the Jonval wheels in the United States under a license from Zebulon Parker, presumably because the draft tube was covered in the Parker patent of 1840 ("Early Development of the Turbine Waterwheel in France," *Scientific American*, n.s. 6 [5 Apr. 1862]: 212).

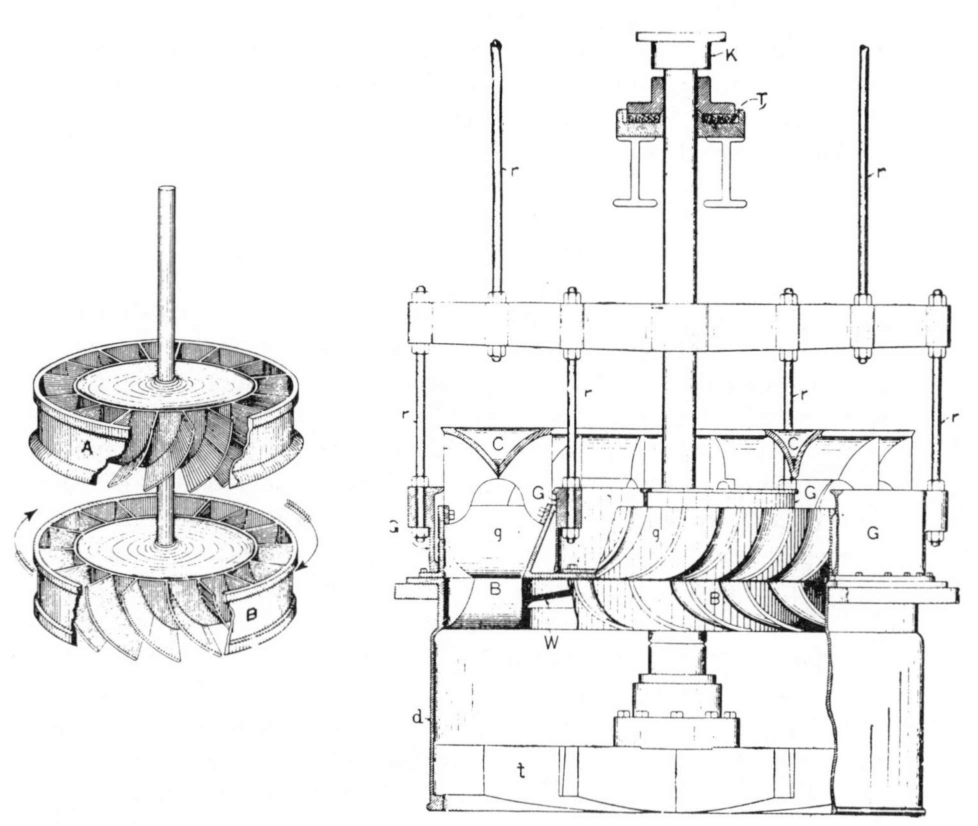

Fig. 71. Geyelin-Jonval axial-flow turbine as built in the United States. (Daniel W. Mead, *Water Power Engineering* [New York, 1908].)

uously and increased the daily capacity more than a half-million gallons. Moreover, it opened up the grateful prospect of raising the total capacity of the waterworks from 4 million to 6 million gallons daily without erection of new buildings or resort to steam power at far greater expense.[59] The wisdom of the installation was amply confirmed by a series of freshets occurring in 1852 that seriously reduced the capacity of the breast wheels for many days. For a period of four hours these wheels were entirely out of operation, while the Jonval wheel continued in a dramatic demonstration of its value.[60] The

[59] Philadelphia, Watering Committee, *Annual Report . . . for the Year 1851 to the Select and Common Councils* (Philadelphia, 1852).

[60] Philadelphia, Watering Committee, *Annual Report . . . for the Year 1852 to the Select and Common Councils* (Philadelphia, 1853).

prestige of the Jonval turbine was raised even higher in the competition conducted by the Fairmount Waterworks in 1859–60 to select a turbine for the further expansion of pumping capacity. Of the nineteen scale models of turbines participating, the three Jonval wheels led the field. The Jonval wheel by Geyelin was awarded the contract, primarily on the basis of past experience since it did not demonstrate the highest useful effect.[61]

Contributions of Boyden and Francis

While the Jonval turbine was obtaining a clear lead in the Middle Atlantic states, in New England the Fourneyron wheel as modified by Boyden was making waterwheel history in an even more striking way. In scale and mechanical perfection of installations, in its percentage of useful effect, and in its general adoption throughout the large-scale textile industry of the region, the Boyden-Fourneyron turbine ruled the hydraulic field for a generation. Lowell was the focalpoint of this development. The majority of the early Boyden wheels were designed for its mills and were built by the Lowell Machine Shop. The Locks and Canals company, as owner, developer, and lessor of Lowell's waterpower, encouraged and aided the introduction of a wheel that promised to make its limited water resources go further. With the available flow at Lowell in the mid-1840s used nearly to capacity and the Locks and Canals about to invest heavily in supply reservoirs in New Hampshire, the replacement of existing breast wheels, giving typically no more than 60 percent useful effect, by wheels with efficiency factors of 75 percent or better was obviously desirable. Since the breast wheels in use at Lowell developed in some instances as much as 350 horsepower each, wheels of 50 to 100 horsepower as developed by Ellwood Morris had no practical meaning for Lowell's mills.[62]

[61] Henry P. M. Birkinbine, *Report on the Experiments with Turbine Wheels Made in 1859–60, at Fairmount Works, Philadelphia* (Philadelphia, 1861). See also extracts from the Birkinbine report, *Scientific American*, n.s. 3 (7 July 1860): 22. An account of the installation of the new Jonval wheels at Fairmount appears ibid., n.s. 6 (14 June 1862): 386. For references to the Jonval turbine at later dates, see *Appleton's Cyclopaedia of Applied Mechanics*, 2:838 ff., 918 ff.; Clemens Herschel, "Niagara Mill Sites, Water Connections and Turbines," *Cassier's Magazine* 8 (1895):245–46.

[62] For example, at one period each of the four mills of the Merrimack Manufacturing Company was driven by two breast wheels, each 30 feet in diameter and 12 feet wide across the face, developing some 380 horsepower (Francis, *Lowell Hydraulic Experiments*, p. i). A British ordnance commission visiting Lowell in 1854 reported in a cotton mill under

It remained for a Massachusetts civil and hydraulic engineer, Uriah A. Boyden (1804–1879), to adapt the Fourneyron turbine to meet the exacting requirements of the great textile mills. Boyden not only solved the problems of designing and building wheels of unprecedented size—among the largest were wheels nearly nine feet in diameter and rated at 700 horsepower under the thirty-three-foot fall at the Merrimack Company's mills—but he attained an efficiency, as well as a capacity, in excess of those attained by Fourneyron himself.

With only a common school education, Boyden combined unusual mechanical skills with great patience and methodical procedures in undertaking the careful experiments and laborious calculations needed to determine the form and proportions of his wheels. His early experience included surveying, work on railroads and dry docks, building of mills as at Lowell, and the design of the hydraulic facilities by which the waterpower at Manchester, New Hampshire, was developed. At the age of twenty-nine he established himself as a consulting engineer at Boston; much of his time from 1844 was devoted to the design of hydraulic turbines.[63] He was said to have been seriously handicapped in making the elaborate calculations necessary in designing turbines by ignorance of calculus. "Mr. Boyden made his calculations by arithmetic approximations, but as Mr. Francis told me, Professor Peirce said that the results were correct, but showed that the work of months by Mr. Boyden, with his usual checks by different calculators, could have been resolved in minutes by the use of calculus."[64]

Although the Boyden turbine in its principal features was modeled on that of Fourneyron, Boyden's own contributions were substantial, including important modifications in the general design and significant additions to the wheel's equipment, which he patented. These improvements included a cone-shaped flume, to give a spiral course to the entering water; an improved gate mechanism for regulating the flow of water to the wheel; a mode of wheel suspension from above

construction two large turbine wheels of 750 horsepower each (extracts from Great Britain, House of Commons, *Sessional Papers*, vol. 50, "Report of the Committee on the Machinery of the United States of America," p. 21). See also "On Breast Water-Wheels," *JFI* 39 (1845):8 ff.; Nathan Appleton, *Introduction of the Power Loom, and Origin of Lowell* (Lowell, 1858), p. 36.

[63] *DAB*; Francis, *Lowell Hydraulic Experiments*, chap. 1. We know fewer details of Boyden's professional career than of most other leading figures in the development of waterwheels. Later efforts to persuade Boyden to write "a historical and scientific account" of the development of the turbine were unsuccessful. See letters of Francis et al. in LCP, DE-9, 20, 29, 133.

[64] Robert Allison, "The Old and the New," *TASME* 16 (1894–95):749.

Fig. 72. Guides and buckets of the Boyden-Fourneyron turbine. (Daniel W. Mead, *Water Power Engineering* [New York, 1908].)

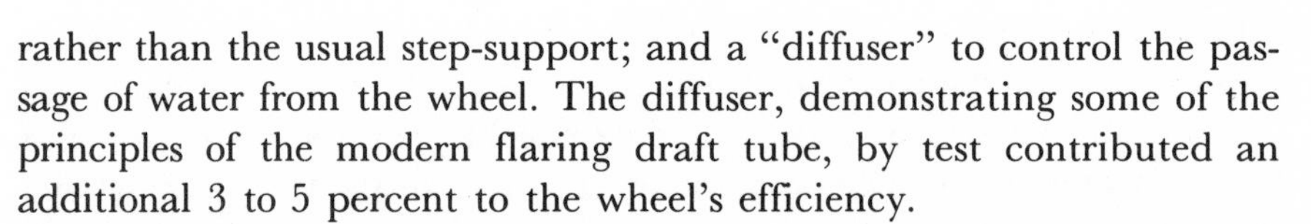

rather than the usual step-support; and a "diffuser" to control the passage of water from the wheel. The diffuser, demonstrating some of the principles of the modern flaring draft tube, by test contributed an additional 3 to 5 percent to the wheel's efficiency.

The impact of the Boyden-Fourneyron turbine was prompt and positive. Four years after its first installation at the Appleton Mills of Lowell in 1845, Francis, as chief engineer of the Locks and Canals company, drew up a balance sheet on the turbine versus the breast wheel. In favor of the turbine: greater power from a given quantity of water; reduction of the harmful variations of fall due to backwater or low water; more perfect regularity in machine speeds; and greater compactness with lower space requirements. On the debit side: the somewhat greater difficulty and expense of maintenance and repair and a greater risk of damaging accidents. Nonetheless he summed up

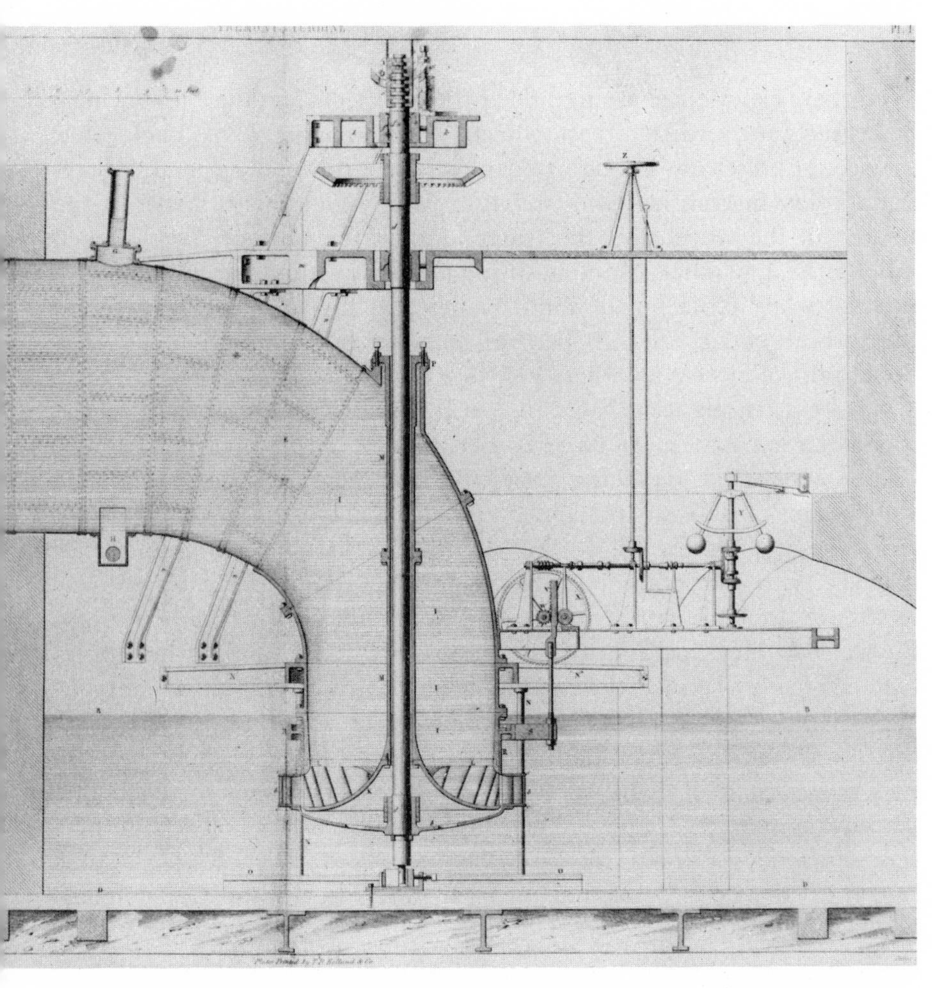

Fig. 73. Tremont (Boyden-Fourneyron) turbine in cross section showing the governing mechanism on the right. (James B. Francis, *Lowell Hydraulic Experiments* [Boston, 1855].)

the experience with the turbine to date as satisfactory; the assumption was fairly warranted "that no new breast wheels will be put in the [Lowell] mills." A fair statement: yet, in practice, turbines appear to have been installed only when the old breast wheels wore out or new capacity was required. A survey of forty-seven New England cotton mills in 1866 gave the ratio of breast and overshot wheels to turbines as 88 to 119. One of Lowell's massive breast wheels was still in operation in the 1890s.[65] For a comparison of an early Boyden turbine with the celebrated Burden overshot wheel see Appendix 9.

[65] JBF, "Memo on Mr. Boyden's Patents," LCP, A-19, no. 96. By 1857, according to a British engineer, David Stevenson, six of the largest Lowell textile mills had com-

Further experience strengthened the respect for the new wheel. Since turbines, unlike breast wheels, were unaffected by backwater, they could be set low enough to capture the entire fall, instead of leaving a safety margin of some twelve to twenty-four inches between the bottom of the wheel and the water level of the tailrace. In a Lowell mill of the Lawrence Manufacturing Company this gain, added to a superior wheel efficiency of about 13 percent, gave a total advantage to the turbine of virtually 40 percent, enough power to drive an additional mill of 30,000 spindles.[66]

Boyden's wheels were built to special designs adapting each to the special circumstances of its use. The best materials were employed, bronze eventually replacing iron in the guides and buckets of the wheel. The parts were machined and fitted together with precision, "as carefully," declared R. H. Thurston, "as the chronometer is made." With corporate customers of large resources first cost was secondary to high performance and long life; they obtained both. Boyden had built his first turbine in 1844 for use in one of the Appleton Company's mills; developing some 75 horsepower, it demonstrated under test conditions an efficiency of 78 percent.[67] His contract with the same company to build three more turbines of 190 horsepower each in 1846 called for additional payment for each one percent of added efficiency. When the new wheels gave on test 88 percent efficiency, Boyden received the base guarantee of $1,200 plus $4,000 for the extra performance. In 1848 the Locks and Canals com-

---

pletely changed over to turbines. In the remaining five mills there were as many turbines as breast wheels (*Sketch of the Civil Engineering of North America*, 2d ed. [London, 1859], pp. 212–13). The present writer was informed by George E. Whitaker, engineer, Locks and Canals, December 1956, that the last breast wheel at Lowell was removed from the Appleton Mills in the 1890s. The figures for 1866 are from New England Cotton Manufacturers' Association, *Statistics of Cotton Manufactures in New England, 1866* (Boston, 1866), no. 1. See also Francis, *Lowell Hydraulic Experiments*, pp. 2–14 and pls. 1, 2, 3.

[66] JBF to Henry Hall, 21 Feb. 1853, LCP, A–18, no. 89; to William Gray, 21 Jan. 1854, DA–4, 72. In 1848 Boyden reported that the replacement of breast wheels by turbines in the Appleton Company mills made it possible with "about the same quantity of water" to operate 17,920 spindles "with the other necessary machinery" where only 11,776 spindles had been operated before, an increase in capacity of more than 50 percent (U. A. Boyden, "Article on Turbines Written for the 'American Cabinet,' " ms., 11 Apr. 1848, Baker Library, Harvard University, Cambridge, Mass.).

[67] Thurston, "The Mechanical Engineer," pp. 81–83. My account largely follows Francis, *Lowell Hydraulic Experiments*, chaps. 1, 2. See also JBF, "Memo on Mr. Boyden's Patents," 1 Dec. 1848, in LCP, A–19, no. 96. Francis referred to some of the Boyden turbines as built "almost regardless of cost" ("Address . . . as President of the ASCE," *TASCE* 10 [1881]:192).

pany purchased from Boyden the right to build the Boyden turbines for use in Lowell—for which $18,000 was asked and $16,000 paid.[68] Thereafter Francis undertook the detailed calculations necessary to adapt the turbines to specific requirements of the Lowell mills. In 1858 fifty-six of these turbines were in use at Lowell, developing more than 12,000 horsepower. They ranged in capacity from 35 to 650 horsepower (average, 225) and averaged in cost $100 per horsepower, completely installed. The pair of 650 horsepower wheels placed in the mills of the Merrimack Manufacturing Company in 1855 cost about $100,000, an average of about $77 per horsepower, including all costs of equipment and installation "from the Canal to the River, and the delivery of the power to the main pullies of the Mills."[69] According to a later comment the Boyden turbines were "as expensive as steam engines of the same power."[70] But, it must be added, with operating expense that was but a small fraction of the steam engine's.

What distinguished the work in turbine development at Lowell was the emphasis at the outset on the thorough testing of the results in operation of the wheels when installed and the often expensive investigation attending innovations in design and construction. The difficulties met in the early years were not readily overcome. Boyden remarked in 1848: "Unlike most other machines, their operations, that is, the acting of the water in the wheels, is done out of sight, and there seems to be no means of comprehending the cause of a failure, but by such abstruse investigations as caused such men as Sir Isaac Newton and Leonhard Euler, to fall into great absurdities." By this date, however, according to Boyden, turbines were being made very accurately and successfully at Manchester, Lowell, Lawrence, and Chelmsford—adjacent to Lowell and the site of a leading machine shop, Gay, Silver and Company. Rules and tables were being made for the guidance of builders, and machinery was devised for their construction. With developing experience Boyden anticipated substantial reductions in their heavy cost, all "without understanding their principles any better than we understand astronomy."[71] *Apple-*

[68] JFB, "Memo on Mr. Boyden's Patents," 1 Dec. 1848, and "Memo of an Agreement between U. A. Boyden and PL & C," 22 Dec. 1848, LCP, A-19, no. 96.

[69] JBF to H. M. Birkinbine, 18 Dec. 1858, ibid., DA-5, 70-71. See also JBF to L. Child, 18 July 1856, A-19, no. 96; to John Weale, 13 Oct. 1851, A-18, no. 90.

[70] "American Engineering as Illustrated at the Paris Exposition of 1878," *TASCE* 7 (1878): 371-72.

[71] Boyden, "Article on Turbines," p. 5. Fourneyron understandably was not eager to disclose the results of his own experiments and investigations. An interesting but brief account of the efforts of an Irish millwright to obtain information for his own

ton's Dictionary remarked in 1852: "The construction of turbines suggests the most complicated problems of hydraulics, and theory has not yet afforded the means of solving them *a priori*. Practice alone gives any solution at present. The greatest difficulties in the turbine are in the details of execution. The water, to produce the maximum effect, must enter without shock, and leave without velocity. M. Fourneyron has constructed several turbines, but he has not made known the proportions which he gives to them."[72] Boyden was confident that these difficulties would be overcome and made much progress. His work was continued by Francis, who formulated a number of empirical rules by a comparison of the proportions of the wheels designed by Boyden and added to these by experiments on a turbine built under their guidance.[73]

For some years, before turning to other interests, Boyden continued to design large turbines for numerous manufacturing plants in New England. Under special licensing arrangements, the Boyden turbine also came to be made in a range of smaller sizes more suited to general needs.[74] When the Boyden-designed wheels were new and in good order, they yielded 85 percent efficiency and occasionally more; but after a few years of use and under ordinary working conditions 70 to 75 percent was the usual level of performance. A loss of about 10 percent in useful effect resulted from injury to the fine edges and curves of the guides and buckets from leaves and other debris in the water.[75]

guidance by visits to France, during which he obtained an unfruitful interview with Fourneyron and met with other rebuffs, is found in William Cullen, *A Practical Treatise on the Construction of the Turbine, or Horizontal Water-Wheel* (London, 1860), pp. 5 ff. From what he gleaned on his visits to wheel installations and workshops and later correspondence with French engineers, Cullen laid down his own rules by which the proportions of the various components and the more critical dimensions could be obtained (ibid., pp. 17–18). I am indebted to Edwin Layton for calling my attention to the brief but interesting treatise by Cullen.

[72] See the report on hydraulic machines at the Exposition de l'industrie française, Paris, by M. Burat, quoted in *Appleton's Dictionary of Machines*, s.v. "Turbines." It was remarked that although Fourneyron had provided much general information in the *Mémoir* on his wheel, he had been silent on matters of "constructive details," evidently as a builder of turbines to protect his monopoly. Repeated failures to rival his mode of construction made it doubtful that in this country a machine demanding such nicety of technical detail and knowledge of hydraulic principles would be successful (ibid., 2:828).

[73] Francis, *Lowell Hydraulic Experiments*, 2d ed. (New York, 1868), p. 44.

[74] See the arrangement with the Ames Manufacturing Company of Chicopee, Massachusetts, in Orra L. Stone, *History of Massachusetts Industries*, 4 vols. (Boston, 1930), 1:631–32.

[75] JBF to T. Jefferson Coolidge, 1860, LCP, DB-5, 280; to T. C. Keefer, 3 July 1868,

Other shortcomings of the Boyden-Fourneyron wheel were its high cost and a marked decrease in efficiency when operated at less than full capacity.[76] In 1868 Francis commented on the latter drawback: "It is the great objection to the turbine, that to use the water economically it must be run to its full power or nearly so. To do this the economical use of the water power must control the amount of machinery run, which usually does not suit the manufacturer."[77] In time the Boyden-Fourneyron turbine gave way to more compact and rugged wheels of much lower first cost, equal or better efficiency at full gate, and much better performance when operating at less than full capacity.[78]

Francis, too, played an important role in the development and introduction of the large-capacity, precision-built turbine, but one in the first instance clearly subordinate to that of Boyden. As for his independent achievement, it is difficult to distinguish between his contributions as chief engineer and superintendent of the Lowell hydraulic system as a whole, as engineer-designer responsible for adapting Boyden wheels for specific mill installations, as experimenter with turbines and the flow of water over weirs widely publicized in the several editions of *Lowell Hydraulic Experiments* (1855), and as "inventor" of the inward-flow, or center-vent, turbine—an innovation which, despite engineering voices of dissent and protest, has gone down in the history of engineering as "the Francis turbine." As the official responsible for the development and improvement of the Lowell power system as a whole and its management in the interests of the textile corporations, Francis was deeply concerned with the improvement of wheel design, construction, and efficiency. He was concerned partly because of the bearing wheel efficiency had on the capacity of the Lowell system as a whole, but even more directly because his consulting duties to the corporations required him to

---

DB–25, 266. "The Fourneyron turbine is used here in preference to the Jonval as we think it the more perfect motor; that is, we think we can get a higher coefficient of useful effect. The best wheels are now with bronze guides and buckets."

[76] "Memo on Mr. Boyden's Patents," 1 Dec. 1848, LCP, A–19, no. 96; JBF to T. J. Coolidge, 22 July 1868, A–18, no. 89. Also Francis, *Lowell Hydraulic Experiments*, 2d ed., pp. 6, 230–32. A later opinion of the Fourneyron-Boyden wheel declared that it "was expensive to build, it clogged easily, it was not efficient at part-gate, and it ran too slowly" (Tyler, "Evolution of the American Type of Water Wheel," p. 881). Francis made these points, except the last, in his "Memo on Mr. Boyden's Patents."

[77] Francis to T. J. Coolidge, 22 July 1868, LCP, A–18, no. 89.

[78] The discussion that follows is based largely on Francis, *Lowell Hydraulic Experiments*, pt. 1, "Experiments on Hydraulic Motors."

CENTRE VENT WHE
Fig. 3.
Scale for Figs 3 & 4
Scale for Fig 1

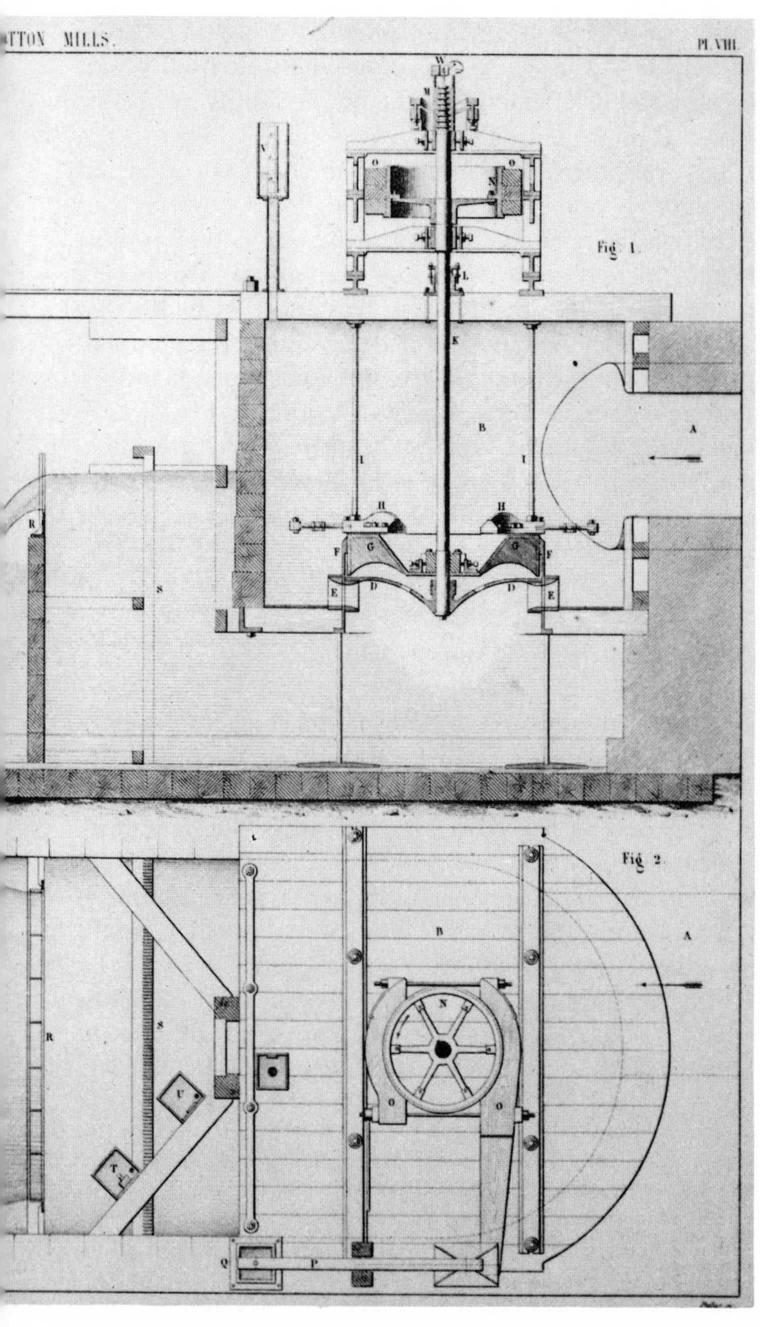

Fig. 74. The Howd-Francis inward-flow, or center vent, turbine. (James B. Francis, *Lowell Hydraulic Experiments* [Boston, 1868].) In the drawing at the left the portions of the guides and the wheel blades, or floats, used in the Fourneyron and Boyden-Francis outward-flow wheels are reversed. The outer circle shows the fixed guides, the inner the blades of the wheel runner.

assist in the selection of wheels and advise on their design, construction, and installation. He did much to encourage the shift from breast wheels to turbines as a means of extending the power supply at his disposal.

With methods and equipment of his own design, Francis conducted the long series of rigorous trials for gauging the flow of water over weirs (see above, chapter 6), essential for measuring the volume of flow and the consumption of water by mills and individual wheels. He undertook these measurements in the first instance with Boyden wheels, testing their efficiency as a means of determining the compensation due Boyden under his contract with the Locks and Canals.[79] Boyden, as Francis noted, gave him copies of many of his turbine designs, together with the results of experiments, but "very little theoretical information," leaving Francis to be guided, as he put it, "principally, by a comparison of the most successful designs, and such light as he could obtain from other writers on this intricate subject."[80]

Who Invented the "Francis" Turbine?

The later debated issue of the Francis contribution to turbine development centered in the assignment of credit for the wheel which down to our own day has generally been termed the Francis turbine. Known from its mode of operation as the center-vent (Francis's designation), or inward-flow, turbine, this wheel has been called at different times the Howd, Howd-Francis, and, by a leading hydraulic engineer of the post-Francis generation, "the so-called Francis type, which might properly be called the Francis-Swain-McCormick-Leffel type."[81] By his own account Francis designed, built, and tested three inward-flow turbines at Lowell: a small wheel in 1847 and two full-scale wheels, 250 horsepower each, installed at the Boott Mills in Lowell in 1849.

[79] According to Bagnall, Francis's careful and precise experiments reduced the possible error in gauging water from 10 to 2 percent or less (*Sketches of Manufacturing Establishments*, 3:2127–28). See also JBF to Gustav Ritter von Wex, 14 Apr. 1888, LCP, DF-1, 1885–92. For the circumstances leading to the experiments, see Francis, *Lowell Hydraulic Experiments*, pp. 71 ff.; O. Chanute et al., "Engineering Progress in the United States," *TASCE* 9 (1880):222–23; "American Engineering as Illustrated at the Paris Exposition of 1878," pp. 371–72. See also George R. Bodmer, *Hydraulic Motors: Turbines and Pressure Engines* (London, 1889), pp. 308 ff., 320 ff., and chaps. 11, 12.

[80] See Francis, *Lowell Hydraulic Experiments*, pp. 55 ff., 61 ff. The results of these experiments are presented in elaborate tabular form.

[81] Freeman, "General Review of Current Water Power Practice," pp. 335 ff.

The three wheels were essentially experimental and appear to have had no immediate successors. Francis here undertook, as noted earlier, to adapt the Howd wheel, hitherto uniformly built "in a very simple and cheap manner" to the larger and more exacting requirements of industrial use. The result was "a wheel essentially different from the Howd wheel." As with wheels of the Boyden type, he applied not only the design data and procedures of Boyden but certain distinctive features covered by Boyden's patents, now at the disposal of the Locks and Canals, and rules for proportioning turbines based on comparisons of Boyden wheels. From careful tests Francis found the completed center-vent wheels on the whole successful, despite certain design defects. He was encouraged to believe that when this wheel "has received the same degree of attention as the turbine [*sic*] it will not be much behind that celebrated motor, in its economical use of water."[82] The mean efficiency of the small Howd-Francis wheel of 1847 was but 69 percent, despite the great care given to its design and construction. The 79 percent efficiency of the full-scale Boott wheel fell well below the best performance of the Boyden wheels. Its efficiency, especially at part gate, also fell below that of the first Francis-engineered Boyden wheels built for the Tremont mills, installed and tested in 1851.[83] Here possibly was the cause of the general indifference to the Howd-Francis wheel, which, as Safford and Hamilton later remarked, "in its original form was little used in the United States and soon disappeared."[84]

The most persuasive evidence, perhaps, of the at best minor role of Francis in the development of the inward-flow turbine, which increasingly dominated the American scene from the 1850s, is the judgment of his fellow engineers, expressed by implication. In two obituary "memoirs" reviewing his career and achievements, prepared in each instance by committees of three leading engineers and presenting in highly laudatory terms the career of the man generally regarded as America's leading hydraulic engineer, only incidental and slight

[82] Francis, *Lowell Hydraulic Experiments*, p. 69. Francis was careful to reserve the use of the term "turbine" for wheels of the Boyden-Fourneyron type and to use for his improvements on the Howd wheel the designation "center-vent water-wheels."

[83] See ibid., for Tremont turbine, pp. 32–35; for model of center-vent wheel, p. 58; for the Boott center-vent wheel, pp. 66–67.

[84] Safford and Hamilton, "The American Mixed-Flow Turbine," p. 1255. Francis's description does not support the explanation of Safford and Hamilton in some respects. In the analysis presented here I have relied chiefly on the evidence of the performance of the Howd-Francis and the Boyden-Francis wheels in respect to such matters as quantity of water, extent of gate opening, and ratio of useful effect to power

attention was given to the turbine that was to bear his name.[85] Yet with the coming of hydroelectricity the adaptability of the inward-flow wheel to an almost indefinite increase in capacity requirements brought a gradual if partial reversion of the mixed-flow turbine to the simpler prototype, first in Europe, where it had obtained but limited acceptance, and increasingly in large installations in the United States, beginning with Niagara Power House Number 2. Understandably the name of Francis, which was identified with this wheel from the drawings, from carefully conducted and recorded experiments, and from explanations appearing in the several editions of the widely disseminated and consulted *Lowell Hydraulic Experiments*, became permanently applied to the type.[86] It was not, of course, that Francis

---

expended for each of the several wheels in question, together with the observations of Francis himself. Many years later in correspondence respecting the experiments on the Boott Mills center-vent wheel, Francis began his explanation of his approach to the design of this wheel with a consideration of the path of the water entering, passing through, and leaving the wheel. This, he added, was the first wheel made of that form "and the success attending it indicates that the mode of arriving at its form is something better than 'cut and try.' I have no doubt if the construction of such wheels had been continued [note this!], still better results would have been obtained by making small changes on the 'cut and try' principle and noting the results" (JBF to D. H. Morrison, 19 Apr. 1880, LCP, DB-10, 237).

[85] See Desmond FitzGerald et al., "James Bicheno Francis, A Memoir," *JAES* 13 (1894):1–9; "Memoirs of Deceased Members: James Bicheno Francis, Past President and Honorary Member," *PASCE* 19 (1893):74–77, followed by "Index to Professional Papers of James B. Francis, on File at Lowell, Mass.," pp. 78–88. The memoir was prepared by a committee of three past presidents of the ASCE, including Francis's old friend and onetime professional associate, William E. Worthen. In a paper delivered before the local historical society, Worthen described briefly the work of Francis with the turbine ("Life and Works of James B. Francis, Read February 20, 1893," *Contributions of the Old Residents Historical Association* (Lowell) 5 [1892–93]: 227–42).

[86] The views of several engineers, post–1900, on the role of Francis in turbine design are of interest here. Arnold Pfau, in a paper, "Wheels of the Pressure Type," declared that the "radical inward discharge wheel" was invented by an American engineer, J. B. Francis, as early as 1849" (*Transactions of the International Engineering Congress, 1915*, 12 vols. in 13 [San Francisco, 1916], 7:448). A. Streiff of the Swiss Society of Engineers, in discussion of the Pfau paper, declared that the "classic investigations" by Francis for his center-vent wheel "are the foundations of the modern Francis turbine" (ibid. p. 498). Robert L. Daugherty, professor of mechanical and hydraulic engineering at Rensselaer Polytechnic Institute, stated: "The first inward flow turbines were built in the United States in 1838 but they were very crude. The turbine designed by J. B. Francis in 1849 was such a vast improvement upon any other that he is rightly credited with being the originator of the modern inward flow turbine" (*Hydraulic Turbines*, 2d ed. [New York, 1914], p. 28). Daugherty later modified without greatly changing his opinion: "Because of the publicity given to these wheels [built by Francis in 1849] due to the

anticipated the future and its needs; it was not that the credit was fully his. The future presented conditions which, in part through chance and circumstance, were matched in fundamental respects by the wheel type first conceived and applied successfully to practical use in this country by Howd. The type was greatly refined by Francis with the aid of the design methodology and models developed by Boyden and was to be exploited with great commercial success, as we shall see in the chapter ahead, by a whole generation of wheel builders bent on practical results and indifferent alike to hydraulic principles and conventional engineering procedures.

In the United States during his lifetime Francis was identified with neither wheel design nor wheel building but with the far larger and more significant achievement, the development from an early stage of the pioneer and most impressive example of direct-drive waterpower as a base for a large industrial city. The attention Francis gave to the perfecting and introduction of the turbine was incidental to his concern with stretching the power of the Merrimack River to its limits to meet the ever increasing power requirements of the mill corporations whose dedicated servant he was. He enlarged and perfected the impressive system of waterpower supply and management and brought it close to the practicable limits of capacity. By the stimulus he gave both to the adoption of the turbine and to the replacement of the original equipment of Boyden wheels designed by himself with the superior mixed-flow wheels to come, Francis increased the capacity of the Lowell power system by one-fourth, as a fellow engineer testified. It remained for another Lowell resident, no engineer but, like Howd—and Wing and the Parkers before him—an ingenious and inventive mechanic, to rescue the inward-flow turbine from the desue-

very precise tests which he conducted on them his name became attached to them and today modern inward-flow turbines are known as Francis turbines, even though they differ considerably from his original design" (*Hydraulics*, 3d ed. [New York, 1925], p. 239). Substantially the same conclusion was reached more recently by Allen: "Francis was largely responsible for attracting attention to the inward-flow turbine and he formulated rules for the design of the runner. As a result his name has been attached generically to the broad class of reaction wheels having inward flow characteristics" ("Hydraulic Engineering," p. 529). An important recent contribution, yet to be published, is Edwin Layton, "Scientific Technology, 1845–1900: The Hydraulic Turbine and the Origins of American Industrial Research," prepared for the Bicentennial Meeting of the Society for the History of Technology, 17 Oct. 1975, Smithsonian Institution, Washington, D.C. Professor Layton reviews the evolution of the turbine in Europe and the United States within the context of hydraulic theory as understood at the time. He gives primary attention to the interrelations of science and technology and to the contributions of Boyden and Francis to an emerging synthesis.

tude to which Francis, with his larger concerns, had consigned it. By giving the Howd-Francis wheel literally a new twist, Asa Methajer Swain in the late 1850s set it upon a new course which, as the succeeding chapter will disclose, resulted in the justly celebrated American mixed-flow turbine.

8
A Generation of Declining Use and Advancing Technology

THE MOST STRIKING feature of the power scene in the generation following the Civil War was the accelerating decline of waterpower's role in the national economy. Presumably still in the lead as late as 1860, waterpower in 1870 accounted for slightly less than one-half the total power used in manufacturing, as reported in the first federal census to enumerate power by kind and amount. Steam power, which about 1840 was of marginal importance in manufacturing, moved sharply ahead during the postbellum years. Its share, vis-à-vis hydraulic power, rose from 52 percent of the total in 1869 to 64 and 79 percent in 1879 and 1889, respectively. The share of direct-drive waterpower, despite a marked recovery in absolute terms in the 1890s, continued its decline from 21 percent in 1889 to 15 percent in 1899 and 11 percent in 1909.[1] The nadir was reached during the 1870s and 1880s, when steam power doubled each decade and waterpower's increase fell to 8 and then to less than 3 percent. The recovery to some 16 percent increase during the 1890s is a matter for conjecture: did the upward surge of hydroelectricity give direct-drive waterpower a needed stimulus?

Although the growth of waterpower declined drastically in the postbellum period, the generation was one of marked technological progress. These years witnessed the rise of a large and flourishing industry engaged in the development of a distinctive and distinguished class of waterwheels, departing significantly in design and greatly superior in practical utility to the hydroturbines introduced from western Europe in the 1840s. At the scientific and engineering levels, these were admittedly the doldrum years. The experimental and theoretical work of Boyden and Francis did not prove contagious. The *Lowell Hydraulic Experiments* of Francis was brought out in new editions without prompting new investigations. Empiricism carried the day, with profit more precious than understanding and with operating results, as will be seen, that were impressive.

The ferment of innovation in waterwheels during the 1840s and

[1] Carroll R. Daugherty, "The Development of Horsepower Equipment in the United States," in *Power Capacity and Production in the United States*, U.S. Geological Survey, Water-Supply Paper no. 579 (Washington, D.C., 1928), p. 71, table 23.

Fig. 75. John B. McCormick in his shop-office, J. & W. Jolly Co., Holyoke, Mass. (c. 1890). McCormick was among the leading contributors to the development of the American mixed-flow turbine. (Courtesy of the Smithsonian Institution.)

1850s was but a prelude to the more feverish tempo and more effective improvements of the years that followed. There was, the *Scientific American* declared in 1863, "a constant and laudable struggle between inventors to see who can produce the cheapest and most efficient motive power . . . [producing] forms and varieties which are almost endless in their details and general construction."[2] By the mid-fifties turbines were reportedly manufactured in nearly every state;[3] as we have seen, more than twenty wheel builders entered turbines in the Fairmount Waterworks competition of 1859–60. Early in 1861 the *Scientific American* reported that while "the undershot breast wheel is the kind almost universally used in the cotton and woolen manufactories of New England . . . turbines are now being quite extensively

[2] *Scientific American*, n.s. 8 (7 Feb. 1863): 88.

[3] Ibid. 11 (10 May 1856): 279; see also ibid. 4 (23 June 1849): 317.

introduced." James Emerson, operator of the pioneer commercial turbine-testing flume at Holyoke, Massachusetts, reported forty makes of wheels in production in 1873 and twice this number five years later.[4] One historian of the time declared that "the iron turbine has now almost superseded the great wooden wheel of our forefathers."[5] If we except the rank and file of small gristmills and sawmills, still numbered in the tens of thousands throughout the country, this generalization may not have been wide of the mark. In that early stronghold of waterpower, Massachusetts, according to the state census of 1875 turbines comprised 82 percent of all waterwheels.[6]

The rapid introduction of the hydraulic turbine from the 1850s proceeded during the very years when steam power was overtaking and then moving ahead of waterpower in manufacturing industry, accounting by 1879 for nearly 64 percent of all installed horsepower equipment.[7] It was occurring, too, in the face of the gradual leveling off, after 1870, of the actual increase of waterpower in manufacturing. No doubt the new types of waterwheels with their superior efficiency and operating characteristics did something to retard industry's conversion to steam power. But since the greater part of industrial growth was concentrated in urban communities, few of which possessed waterpower in appreciable amount, this influence was evidently marginal. It may be surmised that in many, possibly most, instances the turbine was introduced as the simplest and often the only available means of meeting the growth needs for power at a given mill privilege. In many situations the available power could be increased from 25 to 40 percent by installation of a well-designed turbine, and at less cost and with less dislocation of arrangements than the introduction of auxiliary steam power would occasion. In such manner, too, that curiously dread day could be postponed when a steam engine, unfamil-

[4] James Emerson, *Fourth Annual Report of Turbine Tests* (Lowell, Mass., 1873), pp. 11–43; idem, *Treatise Relative to the Testing of Water-Wheels and Machinery*, 2d ed. (Springfield, Mass., 1878), pp. 105–7. Some fifty American turbines were exhibited at the Philadelphia Centennial Exhibition (Great Britain, Executive Commission, Philadelphia Exhibition, *Report on the Philadelphia International Exhibition of 1876*, 3 vols. [London, 1877], 1:207–8).

[5] Albert S. Bolles, *Industrial History of the United States* (Norwich, Conn., 1878), pp. 303 ff.

[6] Massachusetts, Bureau of Labor Statistics, *Census of Masschusetts, 1875*, 3 vols. (Boston, 1876–77), 2:325. This calculation excludes 144 wheels of a type "not designated." Of the total number of 2,950 wheels, 2,317, or 79 percent, were turbines. Horsepower data are wanting. Overshot and breast wheels comprised 12 percent and flutter, float, undershot, and tub wheels the remainder.

[7] Daugherty, "Development of Horsepower Equipment," pp. 49–50, table 5.

iar and with a not readily shaken reputation for high cost, must be installed.

Apart from establishments with a need for additional power, the replacement market itself was sufficiently large to keep many manufactories of turbines busy. Old-style waterwheels of wooden construction had a useful life in most instances of no more than ten or twelve years; the best built iron turbines lasted three times as long.[8] The number of waterwheels reported in use in manufacturing throughout the United States rose from 51,000 in 1870 to 55,000 in 1880. The wheels averaged some 22 horsepower each. The aggregate number of wheels was estimated to have reached some 70,000 in 1892, of which no more than 2,000 wheels of the larger capacities were found at the larger and well-known waterpowers in New England, New York, and the Middle West.[9] With direct-drive waterpower increasing 61 percent between 1870 and 1900, there was a large new, as well as replacement, business to be done by the wheel-building industry.

There were other branches of industry to be served. The waterpower capacity of mines and quarries, reported separately, rose from 75,000 to 250,000 horsepower from 1869 to 1889, thereafter falling precipitously.[10] From the 1880s the rapidly multiplying hydroelectric installations brought a new and steadily accelerating demand for wheels of ever larger capacity and higher efficiency. Any slack in the wheel-building business attending the gradual leveling off and eventual decline of direct-drive waterpower was amply offset by the new and more exacting demands of hydroelectricity. The expansion of the wheel-building industry was most pronounced in the middle regions of the country, where the use of waterpower had previously lagged. Robert H. Thurston, one of the most competent of observers as well as an authority on steam power, remarked in 1873 that "the opening of the immense fields of the West and the growth of manufactures have caused the utilization of thousands of the best located mill sites along the many rivers and streams which intersect that part of the country."[11] Some of the largest and most successful manufacturers of turbines in the country were located in southern Ohio: for example, the

[8] A. W. Hunking, "Notes on Water Power Equipment," *JAES* 13 (1894): 209.

[9] George F. Swain, "Statistics of Water Power Employed in Manufacturing in the United States," *PASA*, n.s. 1 (1888): 35; Hunking, "Notes on Water Power Equipment," p. 198.

[10] Daugherty, "Development of Horsepower Equipment," p. 49, table 5.

[11] Robert H. Thurston, "Report on Manufactures and Machinery," in *Reports of the Commissioners to the International Exhibition Held at Vienna, 1873*, ed. Robert H. Thurston, 4 vols. (Washington, D.C., 1875), 3:175 ff. On the replacement of old-style wheels

Stillwell and Bierce and the Stout, Mills, and Temple companies in Dayton, and James Leffel and Company in Springfield.

The Stock-Pattern Turbine Industry

The great increase in the turbine-building industry, often simply termed the wheel-building industry, was something more than an expansion of a branch of enterprise tracing its origins back to the 1830s and 1840s and the numerous small-scale and local builders of reaction wheels.[12] The native wheels, though authentically turbines, as with those of Howd and Parker, were for the most part confined in service to the rank and file of industry. It remained for the wheels of foreign origin, the Fourneyron and Jonval turbines as improved and adapted in this country during the 1840s and 1850s to meet the larger and more exacting requirements of what for that day was "big industry." But by 1860 a significant thing happened: empiricism resumed its sway. The work of Boyden, Francis, and the European savants was largely brushed aside and for most practical purposes ignored. The tinkers, backwoods inventors, smalltown millwrights, and foundry and machine-shop mechanics resumed their once dominant role from an age before the preeminence of the Lowell Machine Shop and West Point Foundry. Now, however, they worked on a more sophisticated level, with far more advanced metal working equipment, larger financial resources, and often with a broader background of production experience. The all-purpose machinery-building firm in these years, too, was giving way increasingly to the specialized shop producing machine tools, millwork, textile machinery, or steam engines. From the evidence of their advertisements as well as from other sources we find at least the larger and more successful turbine builders concentrating primarily if not wholly upon waterwheels.

Instead of making up a wheel by casting parts from the old, well-worn patterns drawn from the stockroom each time a turbine order

---

by turbines see Robert F. Fries, *Empire in Pine: The Story of Lumbering in Wisconsin, 1830–1900* (Madison, 1951), pp. 62 ff.; Jacob A. Swisher, *Iowa: Land of Many Mills* (Iowa City, 1940), pp. 182–84.

[12] William E. Worthen, "Address of the President," *TASCE* 17 (1887):5 ff.; John R. Freeman, "General Review of Current Water Power Practice in America," *Transactions of the First World Power Conference, 1924*, 5 vols. (London, [1925]), 2:374. Both works are retrospective reviews by hydraulic engineers representing different generations of experience.

came in, the new wheel builders were continually introducing this change or that in the shape, dimensions, or proportions, seeking improved performance, greater efficiency, or simpler or better methods or materials of production. Within a short span of years a new and distinctive class of wheels gradually emerged, each make with its own characteristics and peculiarities but reflecting the often common results of a widely shared experience, each builder influencing, and, outside the uncertain limits of patent infringement, copying others. What eventually became known as "the American," or "mixed-flow," turbine gradually took form and rose to ever higher levels of achievement.

This class of turbines was distinguished typically by novelty and unconventionality in design; simplicity and durability in mechanical arrangements and equipment; quantity or small lot production in standardized sizes to stock rather than to order; avoidance of refinements in detail and materials; and, at every stage, low cost. Within a relatively few years the leading makes of turbines were performing with an efficiency which closely approached and then excelled that attained by the individually designed, precision-built, and costly wheels of Boyden and Francis. The response of European engineers, as toward some American innovations in steam engines, was not always favorable. One not unsympathetic critic, remarking upon a rather savage attack by a continental authority, conceded that many of the American "wheels of obviously irrational design" nonetheless had reported efficiencies "equalled only by very good European wheels."[13]

The commercial exploitation of the new class of waterwheels fell squarely within the emerging tradition of a free, competitive, and wholly pragmatic business enterprise, to which, in their own fashion, the turbine builders made some contribution. If the small- to medium-scale firms that continued to rely on water as a source of power found themselves somewhat outside the mainstream of industrial development, their needs for waterwheels to replace and enlarge older equipment provided a promising market to wheel builders who could see and meet their requirements. These were, as Robert Thurston pointed out, "primarily cheapness of first cost and repairs, and secondly, economy in the use of water."[14] To this task the designers and builders of turbines applied themselves with great competitive zeal. They were aided by the accumulating body of experience in the quantity produc-

[13] George R. Bodmer, *Hydraulic Motors: Turbines and Pressure Engines* (London, 1889), pp. 189 ff.

[14] Thurston, "Report on Machinery," p. 176.

tion of durable goods, which in varying degrees they sought to apply to their own field. For the most part they directed their attention chiefly to wheels ranging in diameter from 18 to 54 inches that under heads of 15 to 20 feet developed some 25 to 75 horsepower.[15] With scores of firms trying their hand at the business, the number and variety of wheels placed on the market was indeed formidable.

In many instances the distinctive features of the wheels appeared to rest on nothing more substantial than the hunches and hopes of their builders. Many were crudely designed and wretchedly built and performed badly as well as inefficiently. As one builder confessed in respect to the early years: "There is no use denying that our object was to make money. We had seen these parties build an ordinary casting weighing about 300 lb. and worth when finished $30, and charge for it $231. Such profits as that were well worth working for, so we made the experiment." James Emerson reported that nearly all of the early wheels sent to him for testing required alterations and repairs before they could be run in tests.[16] To sell their wheels, builders often depended upon heavy advertising and exaggerated claims of performance. As Emerson was to point out, "The poorest wheel builders circulate the most astonishing certificates." Another commentator declared that "every free-born American citizen considers it among his inalienable rights and privileges to invent a patent medicine and a water wheel, and he usually does both with usual ignorance of and indifference to the laws of both hygiene and hydraulics."[17]

[15] See "Reports of Turbines Tested at the Holyoke Flume," in Emerson, *Fourth Annual Report*, and idem, *Treatise Relative of the Testing of Water-Wheels.* The testing flume at Holyoke could test wheels under no greater head than eighteen feet (Joseph P. Frizell, *Water-Power: An Outline of the Development and Application of the Energy of Flowing Water*, 3d ed. [New York, 1903], p. 630).

[16] Turbine builder quoted in Arthur T. Safford and Edward P. Hamilton, "The American Mixed-Flow Turbine and Its Setting," *TASCE* 85 (1922): 1255; Emerson, *Treatise Relative to the Testing of Water-Wheels*, p. 27.

[17] Emerson, *Fourth Annual Report*, p. 7; Samuel Webber, "Ancient and Modern Water-Wheels," *Engineering Magazine* 1 (1891): 324 ff. "There is a good deal of poetry in the general talk about water wheels, and to bring any of them down to hard and solid fact is very trying" (JBF to James Leffel and Co., 7 Oct. 1868, LCP, DB-5, 317). Many years earlier Frederick Overman had remarked on the problem resulting from a great many wheels on the market, with each builder claiming superiority for his product while "the public are in the dark as to the merits of the respective claimants. We cannot too often recommend the use of the friction-brake in cases of doubt . . . the amount of water is easily ascertained, the yield of the wheel is readily found" (*Mechanics for the Millwright, Machinist, Engineer, Civil Engineer, Architect & Student* [Philadelphia, 1858], p. 322). Drawings and use of the friction-brake were given ibid., chap. 8, "The Measurement of Moving Power."

James Emerson and the Holyoke Testing Flume

Into this chaotic situation a measure of order was at length introduced. Time brought its own remedy as incompetence was revealed in demonstrably poor performance, but a less costly form of relief came from an innovation unique in the business experience of the time. This was the introduction in 1869 of the commercial testing of turbines. Turbine testing not only provided the customer with a measure of protection but gave the designer and builder clear evidence of performance and the results of changes in design and construction. The objective evaluation of performance by commercial testing became a major factor in turbine improvement, since most turbine builders lacked both the engineering skills and the elaborate equipment required for measuring efficiency. The early turbine designers—Morris, Geyelin, Boyden, and Francis—had all tested the performance of their wheels by means of dynamometers and flow measurements. The competitive turbine trials at the Fairmount Waterworks in Philadelphia, 1859–60, despite some controversy over results and the adequacy of the methods employed, had indicated the value of certification of performance by an independent agency.

The pioneer commercial testing flume, providing service to anyone for a fee, was established at Lowell in 1869 by James Emerson, an eccentric former seaman. The inventor of a cheap and simple dynamometer for measuring power, Emerson had been engaged by Asa M. Swain to assist in testing Swain-designed wheels in a flume built for the purpose. Possibly as a result of a competition held at this flume, Emerson embarked on turbine testing as a business. By improving equipment and methods, he reduced testing costs from about $2,500 per wheel to no more than 2 or 3 percent of this figure, placing the service within the reach of most turbine builders.[18]

Emerson transferred his operations to Holyoke, Massachusetts, in 1872, prompted by the interest shown there in water use and measurement. Ten years later the flume was taken over by the Holyoke Water Power Company. Enlarged and improved, the flume served for many years as a national center for testing the efficiency and performance of

[18] See Emerson's works on testing and Robert H. Thurston's comprehensive "The Systematic Testing of Turbine Water-Wheels," *TASME* 8 (1886–87):365 ff. The cost of a test at the Holyoke flume was 10 percent on the list price of the wheel, with a minimum charge of $30. In 1883 the number of wheels tested was 185 (William C. Unwin, *On the Development and Transmission of Power from Central Stations* [London, 1894], pp. 87–89).

hydraulic turbines.[19] With turbines whose water discharge rates were ascertained by impartial testing, the otherwise complicated and expensive methods of measuring the waterpower used by customers were vastly simplified. A turbine of known capacity could be used as its own water meter. Moreover, by encouraging or requiring the use of the more efficient turbines, as determined by test, the waterpower company could promote the most effective use of the limited water supply.

For some years Emerson published the results of all wheel tests at his flume in annual reports available to anyone on request. Not content to let the figures speak for themselves, he frequently added pungent comments on the quality of the wheels and their performance. "The comparative merits of any wheel," Emerson declared, "may be determined at a testing flume as perfectly as groceries can be weighed."[20] Whether published or, as later, given in confidential reports to the wheel builder, test results provided a check on the extravagant claims of builders. Systematic testing gradually forced from the market wheels that fell much below the general level of achievement. Manufacturers who avoided tests eventually found themselves unable to sell turbines. Emerson, in the ebullient prose that marked his reports, summed up in 1873 the success to date: "Five years of hard, persistent labor have done much to bring order out of this confusion, for many of the loudest mouthed, have been silenced and their wheels consigned to the 'Old Junk Shop' . . . every wheel of any general reputation has been subjected to thorough and decisive trial and its merit determined . . . the time has indeed passed for running Turbine Water Wheels by wind . . . no honorable turbine builder will sell wheels without first ascertaining their value as motors."[21]

Increasingly it became common practice to base the sale contract upon guaranteed performance of the wheel as determined in the test flume. Assertions respecting the efficiency of new turbines, declared Joseph P. Frizell, "command no attention whatever, unless backed by a responsibly certified test." As an official of the Holyoke Water Power

[19] Thurston, "Systematic Testing of Turbine Water-Wheels," pp. 382 ff. The scale of the company's interest is suggested by the fact that by 1885 its lessees used some 15,000 horsepower by day and 8,000 by night. This water drove 139 turbines in some seventy factories (ibid., pp. 382–83).

[20] James Emerson, "Report of Water-Wheel Tests at Lowell and Other Places," *JFI* 93 (1872):177.

[21] Emerson, *Fourth Annual Report*, p. 3; see also idem, *Treatise Relative to the Testing of Water-Wheels*, p. 31.

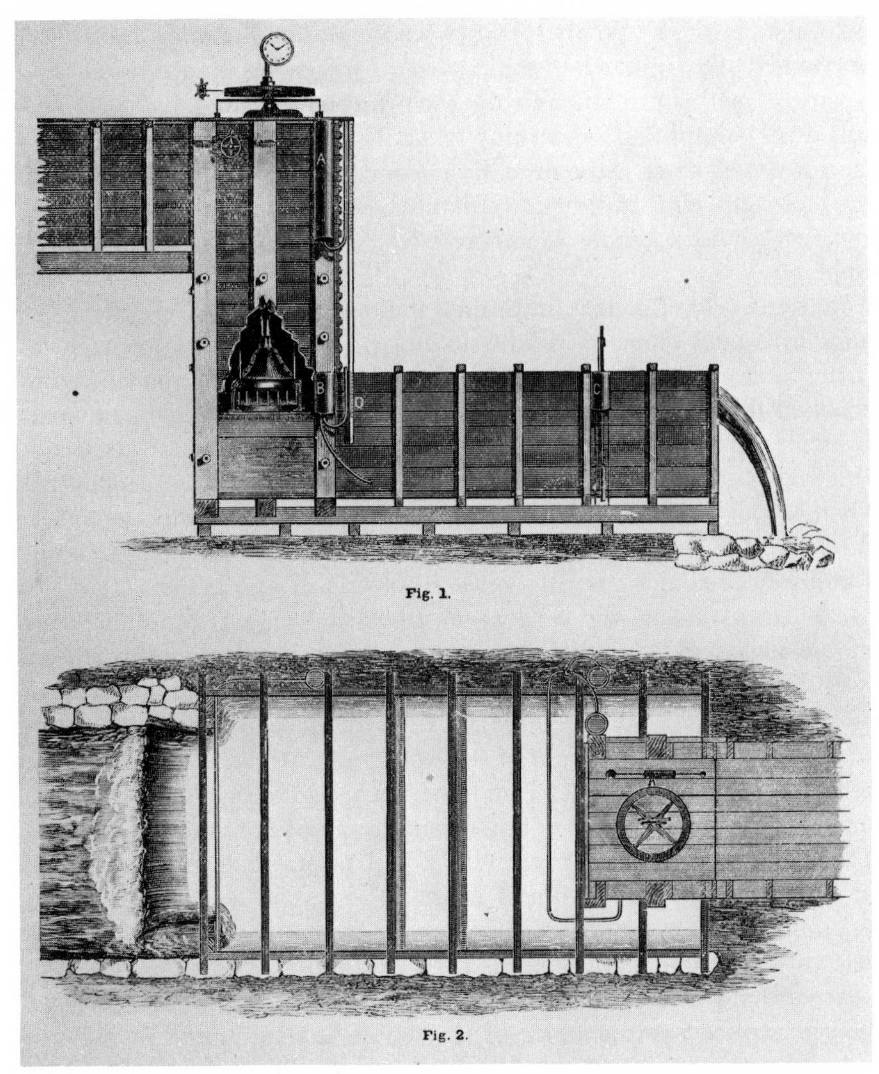

Fig. 76. Emerson's testing flume at Lowell. (James Emerson, "Report on Water-wheel Tests at Lowell," *JFI* 63 [1872].)

Company said, "The testing of turbines is the only way to perfect them."[22] Eventually the larger turbine builders set up their own testing flumes to supply the continuous checking on performance essential in developing improvements.[23] The results were reflected in the high general level of performance reached by the ten or twelve makes of turbines that by 1890 had come to dominate the field. As early as 1872 the Stillwell and Bierce Company of Dayton was reported as testing in its own flume each one of its Eclipse turbines before delivery.[24] Similarly the firm of T. H. Risdon and Company, of Mount Holly, New Jersey, was reported to have tested sixty different forms of wheels in its own flume, making nearly 800 different experiments, and for purposes of comparison testing seventy-five different wheels of other makers.[25] James Leffel devised a miniature flume with glass sides permitting him to observe the behavior of wheel models ten inches in diameter, which were modified until the desired results were obtained. An account of the Leffel turbine as manufactured by Poole and Hunt of Baltimore reported: "Mr. Leffel constructed and experimented with over one hundred different forms of water wheels, among which were the original outward discharge of Fourneyron, the vertical discharge of Jonval, the inward discharge, etc. Each class underwent in his hands numerous modifications in form."[26]

To come into general industrial use, turbines had to be cheap as well as reliable and efficient. They had to compete not only with old-style wheels but with steam power. Although turbine builders did no more than retard the shift from direct-use waterpower to steam power, they scored a striking success in supplying establishments depending on waterpower with efficient wheels at low cost. In important part this success was accomplished by major innovations in design, discussed below, which reduced sharply the size and weight of wheels of a given capacity. Costs were further reduced by keeping to a minimum constructional features calling for close dimensional accuracy, machining, and hand fitting. American turbines, reported Thurston, "are

[22] Frizell, *Water-Power*, pp. 538 ff.; Thurston, "Systematic Testing of Turbine Water-Wheels," p. 367. See also Daniel W. Mead, *Water Power Engineering: The Theory, Investigation, and Development of Water Power* (New York, 1908), chap. 15.

[23] Thurston, "Systematic Testing of Turbine Water-Wheels," p. 414; also, *Manufacturer and Builder* 5 (1873): 2–3.

[24] *Scientific American* 27 (14 Sept. 1872):162.

[25] *Manufacturer and Builder*, 5:2–3.

[26] W. W. Tyler, "The Evolution of the American Type of Water Wheel," *JWSE* 3 (1898):888–89; *Manufacturer and Builder* 11 (1879):198. See also Frizell on Hunking, "Notes on Water Power Equipment, p. 223.

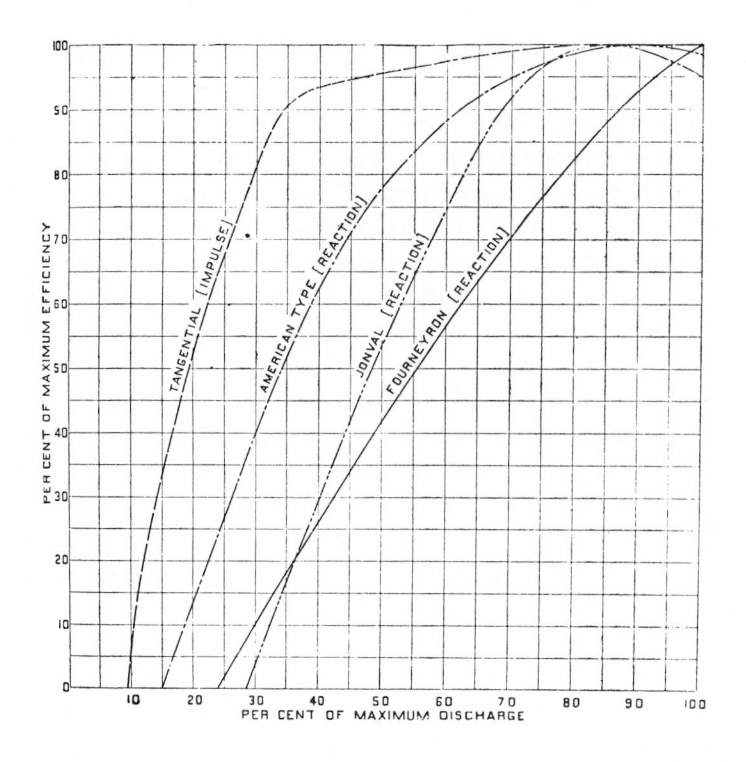

Fig. 77. *Above,* comparative efficiencies of various types of turbines; *below*, performance of turbines at different gate openings. (Daniel W. Meade, *Water Power Engineering* [New York, 1908].)

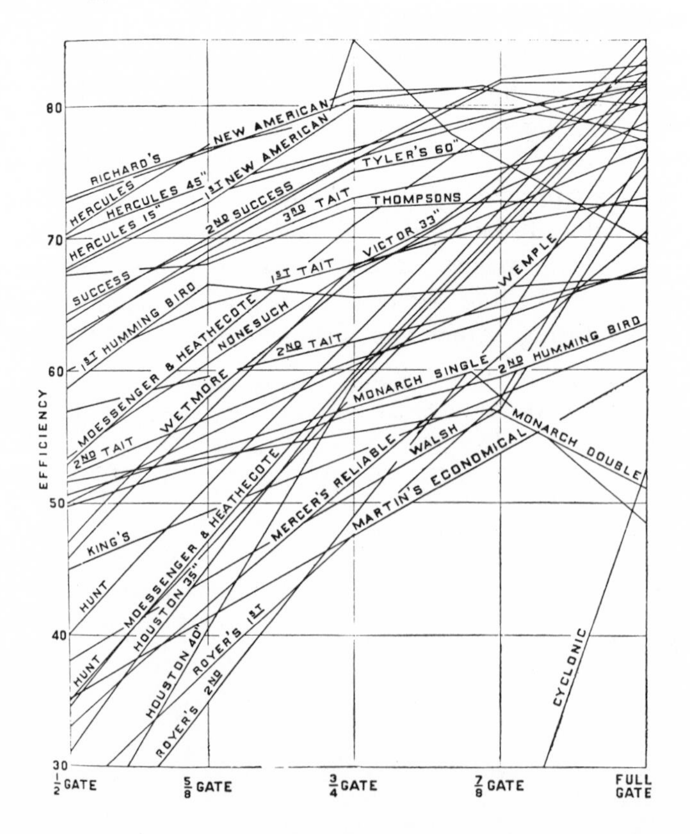

usually cast with the buckets in place, and are set at work very nearly as they come from the foundry. A sharp competition in the market usually forbids the expenditure of much time and labor in finish, or in machine shop work. In a few cases . . . the buckets are cast separately and bolted in; but even in this case, the surfaces or the guides and buckets are left with the skin upon them, just as they leave the foundry."

Believing that the high finish given the wheel surfaces of the Boyden turbines was responsible for an added 10 to 15 percent efficiency, judging by the reduced efficiency of worn wheels, Francis subsequently was puzzled by the high efficiency obtained by later wheels lacking such finish. Thurston noted that the results of experimental trials appeared to show that, contrary to previous belief, "friction in a well-formed wheel, becomes partly a means of transfer of energy from water to wheel."[27]

Equally important for cost reduction was the "American system" of quantity production on the basis of standard dimensions and interchangeable parts, combined in varying degree when quantities justified with special machine-tool setups and gauges. By the 1860s this system was spreading from such light articles as firearms, clocks, locks, and sewing machines into heavy goods, such as agricultural machinery, steam engines, and even locomotives.[28] Quantity production of turbines was favored by the large number of manufactories using relatively modest amounts of power and by the relatively uniform hydraulic conditions throughout the northeastern United States, where the bulk of industrial production was concentrated. With turbines, as with steam engines, the application of "the American system" led inescapably to production of a range of standard sizes to open stock rather than to specification on individual order. Typically, American

[27] Thurston, "Systematic Testing of Turbine Water-Wheels," pp. 361–62; JBF to James Leffel and Co., 7 Oct. 1868, and to T. H. Risdon and Co., 25 Sept. 1872, LCP, DB–5, 317, and DB–6, 466; Robert H. Thurston, "The Mechanical Engineer—His Work and His Policy," *TASME* 4 (1882–83):81–83. In 1877 the Ames Manufacturing Company, of Springfield, Mass., builders of Boyden wheels and equipped with their own testing flume, offered turbines in three grades with guides and buckets (1) of cast iron; (2) with wrought-iron buckets grooved and mortised in; (3) with brass guides and bronze buckets grooved and mortised in. The three classes were priced as follows for wheels 16 to 84 inches diameter: (1) $260–$1,550; (2) $320–$2,150; (3) $375–$2,750 (*Catalog: Ames Manufacturing Company, Ames-Boyden Turbine Water-Wheels* [Springfield, Mass., 1877], pp. 5 ff., 52). By 1877 several thousand wheel tests had been made in the testing flume and 125 wheel forms tried.

[28] Joseph Whitworth, *The New York Industrial Exhibition* (London, 1854); *American Artisan*, 5 June and 10 July 1867, 18 Nov. 1868; *Manufacturer and Builder* 1 (1869): 300.

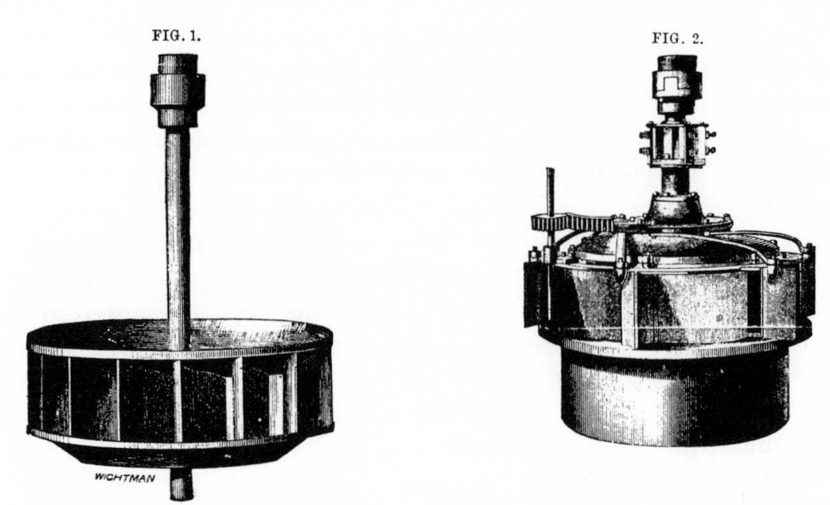

Fig. 78. American turbine. (Globe Iron Works, *The American Turbine* [Dayton, Ohio, 1875].)

wheels were made up in lots for the market; their builders, declared Chief Engineer Herschel of the Holyoke Water Power Company, ordinarily have stocks of wheels on hand "and turn them out as they would shelf-hardware."[29]

Interestingly enough, two companies in southern Ohio led the field in this development. The "American" wheel built by Stout, Mills, and Temple in Dayton and widely distributed by 1870, especially for use in small gristmills and sawmills, has been described as "the first stock wheel" and the beginning of quantity production of standardized wheels.[30] By September 14, 1872, the *Scientific American* reported of another Dayton wheel, the "Eclipse" built by Stillwell and Bierce, that "every part of the wheel and case is fitted up by machinery to standard gauges, so that all parts can be readily duplicated." The company founded by James Leffel in 1862 in the not distant town of Springfield early produced standard-size wheels to stock and by the mid-seventies was advertising the ready availability of spare parts for

[29] Clemens Herschel, "Niagara Mill Sites, Water Connections and Turbines," *Cassier's Magazine* 8 (1895): 242–43; Thurston, "Report on Manufactures and Machinery," pp. 176–77.

[30] Safford and Hamilton, "The American Mixed-Flow Turbine," p. 1261.

repairs.[31] Associated with these developments was the eventual trend toward larger and fewer turbine manufacturers; by 1906, according to Robert E. Horton, "a large majority of the turbines used in this country are built in half a dozen factories."[32]

European development of turbines followed a quite different course, possibly in some part owing to differences in industrial philosophy but more largely reflecting differences in hydraulic conditions and in the nature of the market. Continental waterpowers presented a far wider range of conditions, especially with reference to fall; high heads were common in the mountainous regions. Built-to-order turbines, which became the exception in the United States, were "all but the invariable rule in Europe."[33] Each European turbine, stated one report, "is designed for the special conditions under which it is to operate," with turbine builders relying more largely on theoretical analysis and the application of mathematical tools than in the United States.[34]

In the United States improved design combined with quantity production brought about a striking decline in turbine costs and prices. Cost per horsepower fell from $100 to $200 for the costly, finely built Boyden wheels as used on the usual fall at Lowell in the 1840s to about $33 for a Boyden wheel tested at Holyoke in 1880.[35] But at the latter date such "American type" mixed-flow turbines as the Victor and the Hercules ranged in cost from $4.50 to $8.00 per horsepower.[36] Thurs-

[31] *Manufacturer and Builder* 9 (1877): 225. In time the larger American turbine builders introduced modifications of their basic wheel designs in order to meet important differences in conditions and requirements, producing two or more series of wheels in a range of sizes (Mead, *Water-Power Engineering*, pp. 286–87).

[32] Robert E. Horton, *Turbine Water-Wheel Tests and Power Tables*, U.S. Geological Survey, Water-Supply and Irrigation Paper no. 180 (Washington, D.C., 1906), p. 15.

[33] Herschel, "Niagara Mill Sites," pp. 242–43.

[34] Horton, *Turbine Water-Wheel Tests*, p. 15; also Mead, *Water Power Engineering*, pp. 277 ff.

[35] The Locks and Canals of Lowell was not the only company authorized to build the Boyden wheel. For example, in 1859 the Ames Manufacturing Company of Springfield, Mass., referred to above, obtained this right. "A large number of [Boyden] turbines, varying from one hundred to five hundred horse-power, have been in use from ten to twenty-five years in the large cotton and other mills in New England and New York" (J. D. Van Slyck, *Representatives of New England Manufacturers*, 2 vols. [Boston, 1879], 1:27).

[36] These data are from a number of sources, especially reports of turbines tested in Holyoke Water Power Company, *Holyoke Hydrodynamic Experiments* (Holyoke, Mass., 1880); Safford and Hamilton, "The American Mixed-Flow Turbine," p. 1252; Emerson, *Treatise Relative to the Testing of Water-Wheels and Machinery*, 3d ed. (Springfield, Mass., 1881), pp. 155 ff.; *Scientific American* 6 (5 and 12 Oct. 1850): 19, 29. See also Thurston, "Report on Manufactures and Machinery," pp. 176–77.

ton reported that prices of Gwynne and Company's Girard turbine, built in London, were nearly 50 percent higher than for American wheels of the same capacity, notwithstanding the lower cost of labor and material in Britain.[37] In 1894 A. W. Hunking declared that the cost of all "the iron work pertaining to a complete water power equipment" was less than one-fourth that of the old-time equipment used with Boyden wheels.[38]

Even more impressive was the improvement in operating characteristics and in efficiency. Size in relation to capacity was steadily reduced in an outstanding demonstration of engineering skill. Efficiency was raised until the best stock wheels produced in quantity equaled and then moved ahead of the onetime brilliant performance of the Boyden and Francis wheels. These results were obtained in what is usually thought to be a characteristically American manner, not by scientific analysis supported by mathematical calculations and tested by experiment but, it was the engineer's boast, by almost wholly empirical methods in the hands of practical men. "All the scientific teachings of Boyden and Francis were thrown to the winds, and the great god, 'Cut and Try,' came into his own," explained Safford and Hamilton: "If a wheel did not come up to expectations, its buckets were chipped back, up, or down, or its blades pounded, until it gave something better." Turbine improvements, declared Thurston, "had been *felt out* by the makers, working often in the dark—for few builders claim to understand the principles of their art, and no two ever, ever agree in their statements of the principles underlying their practice."[39] On the basis of the practice of several leading wheel builders, another engineer summarized the American manner of turbine development: "An experienced millwright is given the opportunity of making patterns for buckets, etc., of a wheel, which is then built and shipped to a testing flume where its efficiency, whatever it proves to be, is determined. Little alterations then suggest themselves to the patternmaker's mind, and he proceeds to repeat the process, perhaps several times, until finally the 80 percent mark is passed, when he pronounces the wheel as good as can be made. All through it he is indifferent to fine-spun theory, being guided apparently by instinct alone."[40]

[37] Thurston, "Report on Manufactures and Machinery," p. 178.

[38] Hunking, "Notes on Water Power Equipment," p. 209.

[39] Safford and Hamilton, "The American Mixed-Flow Turbine," pp. 1266, 1256; Thurston, "The Mechanical Engineer," pp. 81–83. See also Thurston, "Systematic Testing of Turbine Water-Wheels," pp. 361–62.

[40] G. I. Rockwood, in discussion of Robert Allison, "The Old and the New," *TASME* 16 (1894–95): 753–54; see also *Appleton's Dictionary of Machines, Mechanics, Engine-Work & Engineering*, s.v. "Water-Wheels."

James Leffel, a highly successful builder and an important contributor to the turbine's advance, has been described as "an ingenious mechanic of limited education, who owned a small machine shop, and [began] . . . the manufacture of water wheels. His theories were wrong . . . but ignorance . . . did not prevent his success." There was John B. McCormick, inventor of the astonishing Hercules wheel, which at a stroke trebled the output of wheels of its diameter. Setting himself to the task of providing designs for the many wheel sizes desired, "he had no drawings or method of precedure, and the 'cut and try' process was the system followed." No wonder that in his comment on waterwheel progress the compiler of a handbook for millwrights exclaimed with pontifical assurance: "And may I ask to whom we are indebted for this valuable light? To the man of scientific knowledge, or to the practical mechanic? We say, to the latter. . . . Learned theoretical investigations have never accomplished much for our advantage in the improvements of the mechanic arts of the country."[41]

The Mixed-Flow Turbine

The end product of a quarter century of such "cutting and trying" was the distinctively American mixed-flow turbine. This wheel was the result of changes prompted by certain design and operating limitations of the Boyden-Fourneyron, Howd-Francis, and, in less degree, Jonval-Geyelin turbines: slow speed, bulk, proneness to obstruction by trash, unsatisfactory gate mechanism and inefficiency at part gate, that is, when operating at less than full capacity with gate partially closed. To handle the large volumes of water necessary at low heads to develop needed power, the wheels were large in size, weight, and cost, with a proportionately high cost of installation. Since rotating speed was in inverse ratio to diameter, they ran slowly. Speed was increased by reducing the wheel diameter. Greater capacity was obtained by increasing the depth and later the width and openings of the buckets and extending them toward the wheel center—for the inward-flow principle was early adopted. This last change increased the volume of flow and the wheel's capacity, but left little room to accommodate the

[41] Tyler, "Evolution of the American Type of Water Wheel," pp. 888–89; A. C. Rice, "Notes on the History of Turbine Development in America," *Engineering News* 48 (1902): 208–10; William C. Hughes, *The American Miller and Millwright's Assistant*, rev. ed. (Philadelphia, 1867), p. 103. See also Thurston, "Systematic Testing of Turbine Water-Wheels," pp. 361–62.

Fig. 79. Hercules turbine, designed by John B. McCormick. (James Emerson, *Treatise Relative to the Testing of Water-wheels and Machinery,* 6th ed. [Williamsett, Mass., 1894].)

increased discharge at the center. It was necessary, therefore, to turn the buckets downward and then outward to dispose of the water. The end result was the movement of water through the wheel in a continuously and smoothly curving spiral path. Then it was found that reducing the number of buckets and widening the distance between them both increased the wheel's capacity and minimized its clogging from trash.

In the aggregate these changes, introduced bit by bit over many years, altered the wheel almost beyond recognition. The intricate play of hydraulic forces was neatly summarized many years later by Professor of Engineering George E. Russell:

Under the low heads which were first to be developed in this country, it was necessary to meet the demands for increased speed and power by decreasing the

Table 13. Characteristics of American waterwheels, 1847-1900, on basis of 30-inch wheel

Date.	Type.	1-Ft. Head.			Efficiency.			N_s (full power.)	Remarks.
		Revolutions per minute at maximum efficiency.	Horse-power, maximum.	Horse-power at maximum efficiency.	Full power.	Half power.	Maximum.		
1851	Boyden	38.3	0.47	....	79.4	46	79.4	26	Tremont turbine, inside diameter of 30 in.
1847	First Francis Model	44	0.19	....	....	....	71.6	19	2 13/16-in. bucket height; no gate.
1849	Boott-Francis	39	0.20	....	79.7	(55)	79.7	17	First Francis wheel installed.
1869	Swain	44	0.59	0.59	82.2	67.7	82.2	34	H. F. Mills, Lowell, 42-in. wheel, 5.35-in. buckets.
1870	Swain	45	0.69	0.69	81.9	....	81.9	37	do. Same model. Buckets. 7.4 in; lower band lowered; all else (except gate) the same.
1874	Swain	46	0.76	0.73	83.5	72.1	83.85	40	James B. Francis. Test on present Boott No. 2 on wheel shaft; no draft-tube.
1872	"American"	38.5	0.41	0.33	76.8	72	80.4	25	Emerson at Holyoke.
1871	Leffel	43	0.48	....	74.3	(60)	74.3	30	Emerson at Lowell.
1873	Tyler Scroll	44	0.42	....	81.6	57	81.6	29	" " Holyoke
1873	Houston	44	0.52	....	88	54	88	32	" " "
1873	Risdon	49	0.51	....	90.5	73	90.5	35	" " " , with diffuser.
1873	Risdon	50	0.49	....	88.8	....	88.8	35	" " " , without diffuser.
1876	"Hercules"	40	1.44	1.29	83.8	73	89.2	48	" " "
1878	"Victor"	42	1.82	....	86.7	65	86.7	48	" " "
1883	"Hercules."	40	1.44	1.28	85.14	....	86.94	48	In present Holyoke Flume.
1897	"Samson"	51	1.85	1.71	82.03	75.6	84.30	69	Holyoke Test No. 979
1897	Special "Hercules"	40	1.79	....	84.0	70.5	84.0	55	" " " 1051
1897	"Victor"	42	1.80	1.72	80.59	68.8	81.07	56	" " " 1061
1899	Smith-McCormick	53	1.65	....	84.55	69.3	84.55	68	" " " 1201
1876	"Green Mountain" Propeller	126	0.98	....	....	....	(60)	125	Calculated from maker's tables; 10-ft. head readings taken for 28 and 32-in. wheels used and averaged; water and speed assumed to be correct; wheel efficiency assumed at 60% and horse-power calculated from this.

Source: Arthur T. Safford and Edward P. Hamilton, "The American Mixed-Flow Turbine and Its Setting," *TASCE* 85 (1922): 1272.

diameter and deepening the wheel axially. As the decrease in diameter prevented the free escape of water from the radial inward-flow wheel of Francis, it became necessary to change the shape of the buckets in order to turn the water in them to an axial direction. Thus the Francis wheel was gradually metamorphosed into the vortex type. This wheel has been developed to a high degree under low heads by experimental testing at Holyoke and elsewhere. Very little, if any, of this class of work has been done under high heads, and there are enough reliable data to show that the general [mixed-flow] type of wheel so far produced in this country is not as efficient as designs which are based on the radial flow type.[42]

Suggestion of the magnitude of these changes in wheel design is offered by Figure 80, showing outlines of the wheel buckets with wheel runners of approximately the same diameter. Table 13 summarizes the results of these changes in terms of efficiency and power for wheels of the same size.[43] From the Howd-Francis inward-flow wheel of 1847 to the Hercules wheel of 1876, capacity increased sevenfold for the same diameter of wheel, and construction costs were reduced in about the same proportion. The result of this evolution was a wheel that combined the inward flow of the Howd-Francis wheel, the axial, or

[42] In discussion of C. M. Allen, "The Testing of Water Wheels after Installation," *TASME* 32 (1910): 302–3.

[43] Safford and Hamilton, "The American Mixed-Flow Turbine," pp. 1271–72.

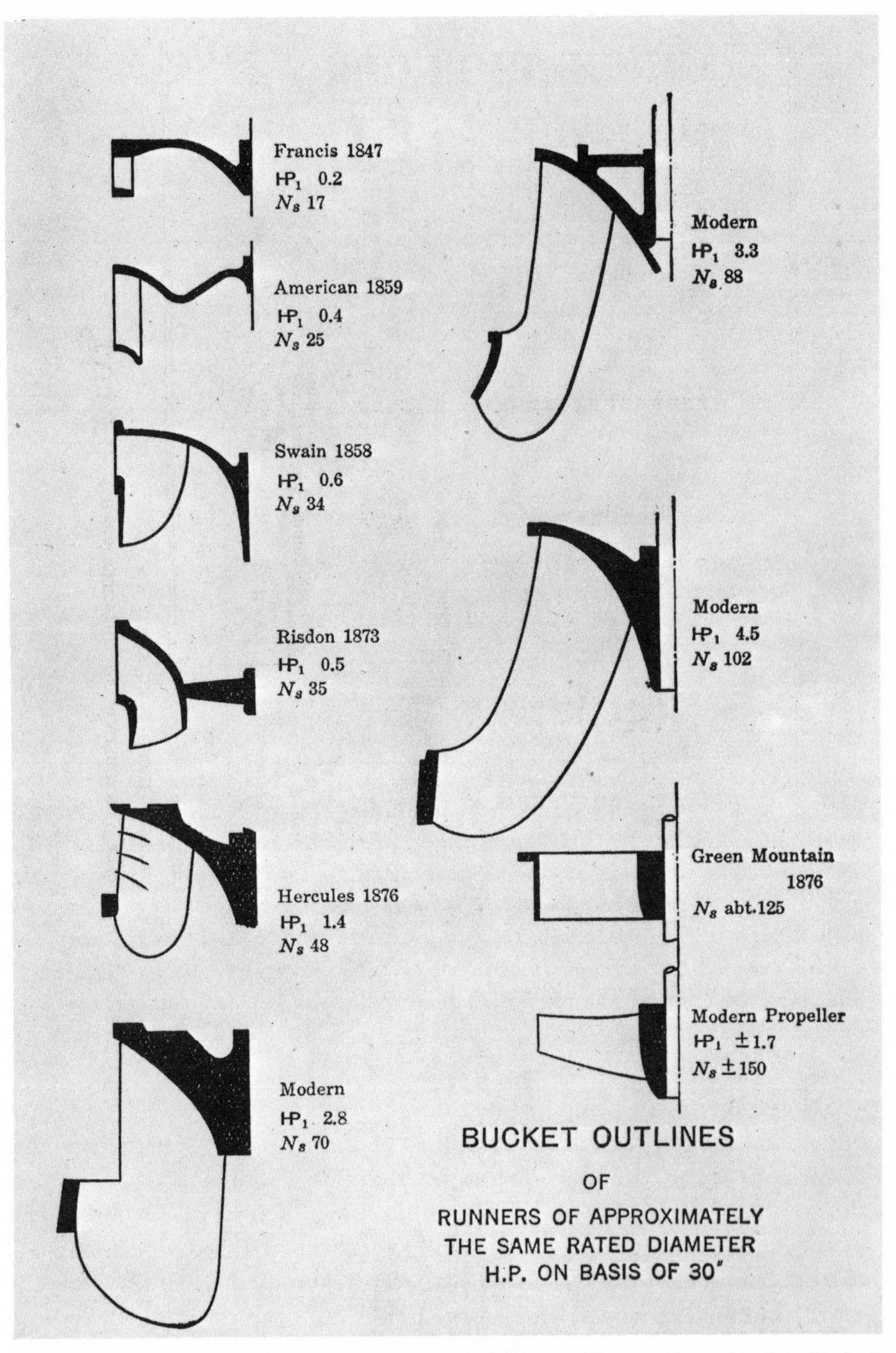

Fig. 80. Evolution of bucket outlines in American mixed-flow turbines. (Arthur T. Safford and Edward P. Hamilton, "The American Mixed-Flow Turbine and Its Setting," *TASCE* 85 [1922].)

Fig. 81. Swain runner. The buckets were shaped to direct the discharge of water inward and downward, which resulted in a marked increase in efficiency. (Photograph courtesy of the Smithsonian Institution.)

parallel, flow of the Geyelin-Jonval wheel, and the outward flow of the Boyden-Fourneyron wheel to form the American mixed-flow turbine. With innumerable variations in details of design, proportions, construction, and such accessory equipment as guides, gates, and settings devised by the many turbine builders, the mixed-flow wheel came into general use in industry, almost to the exclusion of the older types in this country, although it was slow to find favor abroad. Not until advances in hydroelectric technology after 1900 led to single wheels with a capacity of the entire power system of a Lowell or Holyoke was there a gradual reversion to the simpler lines of the earlier years.

Designer-Builders of the Mixed-Flow Turbine

The first turbine designer-builder to combine the inward-flow Howd-Francis principle with the axial-flow principle of the Geyelin-Jonval wheel was a Lowell patternmaker, Asa Methajer Swain (1830–1908),

Table 14. Tests of discharge and efficiency of the Swain turbine

72-INCH WHEEL AT BOOTH COTTON-MILLS, LOWELL, AUGUST, 1874.

Number of Experiment.	Height of Speed-gate.	Head Acting on the Wheel.	Discharge in Cubic Feet per Second.	Ratio of the Velocity of the Exterior Circumference to the Velocity due the Head.	Ratio of the Useful Effect to the Power Expended.
	In.	Feet.			
1	3.25	13.96	70.86	0.706	0.595
4		13.93	72.20	0.600	0.609
5		13.93	72.49	0.523	0.580
6		13.93	69.54	0.772	0.589
7	6.50	13.36	108.11	0.852	0.715
8		13.28	111.55	0.779	0.757
10		13.27	113.45	0.734	0.774
14		13.32	116.13	0.677	0.779
15		13.25	116.60	0.555	0.721
23	2.00	14.31	47.48	0.902	0.232
26		14.28	49.21	0.792	0.380
31		14.19	51.67	0.510	0.463
32	3.00	13.98	68.75	0.512	0.541
40		14.05	63.10	0.903	0.453
41	4.00	13.90	72.76	0.967	0.423
45		13.76	81.49	0.766	0.646
49		13.70	83.89	0.648	0.666
50	5.00	13.44	96.86	0.695	0.717
53		13.44	97.82	0.517	0.657
57		13.55	94.86	0.761	0.706
60		13.66	86.85	0.931	0.551
69	6.00	13.28	110.09	0.669	0.761
72	7.00	13.39	103.59	0.981	0.543
75		13.16	116.71	0.783	0.766
80		13.09	121.56	0.657	0.787
90	8.00	12.97	130.25	0.698	0.806
103	9.00	12.84	138.84	0.710	0.829
108	10.00	12.91	128.57	0.969	0.620
111		12.74	143.26	0.762	0.831
113		12.68	144.77	0.708	0.836
122	12.00	12.77	143.44	0.950	0.690
123		12.52	156.70	0.749	0.838
127		12.49	158.46	0.600	0.775
128	13.08	13.10	120.39	1.174	0.206
129	Full	12.88	137.87	1.068	0.499
130	Ht.	12.603	149.13	0.968	0.690
131		12.480	158.88	0.866	0.803
133		12.37	162.54	0.770	0.836
136		12.37	164.24	0.727	0.830
142		12.40	165.03	0.570	0.746
143		12.40	162.85	0.438	0.627
144		13.17	117.84	1.193	0.000

36-INCH WHEEL AT HOLYOKE TESTING-FLUME, JANUARY, 1897.

Number of the Experiment.	Height of Speed-gate, Full Height being 1.000.	Head Acting on the Wheel.	Discharge in Cubic Feet per Second.	Ratio of the Velocity of the Exterior Circumference to the Velocity due the Head.	Ratio of the Useful Effect to the Power Expended.	Percentage of Full Discharge.
	In.	Feet.				
1	0.067	16.49	12.80	0.739	......	0.161
2	0.125	16.11	18.73	0.778	0.283	0.239
3		16.18	19.24	0.712	0.377	0.245
6		16.17	19.62	0.653	0.430	0.249
7		16.07	19.76	0.619	0.449	0.252
8		16.01	19.89	0.582	0.460	0.254
9		16.03	20.06	0.545	0.464	0.257
10	0.250	15.79	27.64	0.885	0.440	0.356
12		15.70	29.26	0.808	0.572	0.378
14		15.61	30.59	0.725	0.623	0.396
16		15.61	31.46	0.661	0.639	0.407
17		15.57	31.83	0.628	0.642	0.412
18		15.54	32.11	0.596	0.645	0.416
19		15.53	32.40	0.564	0.640	0.420
27	0.375	15.30	38.57	0.837	0.682	0.504
21		15.21	40.01	0.765	0.724	0.524
24		15.21	41.58	0.670	0.746	0.545
25		15.19	42.06	0.641	0.748	0.552
26		15.21	42.50	0.604	0.738	0.557
28	0.500	15.96	49.62	0.801	0.773	0.635
30		15.91	50.87	0.738	0.795	0.652
33		15.78	52.54	0.647	0.805	0.676
34		15.74	52.98	0.608	0.791	0.683
35	0.625	15.65	55.32	0.829	0.776	0.715
39		15.43	58.95	0.709	0.845	0.767
40		15.38	59.45	0.673	0.836	0.775
42	0.750	15.20	62.06	0.818	0.831	0.814
44		15.62	64.09	0.780	0.850	0.829
45		15.54	64.64	0.744	0.854	0.838
46		15.70	65.41	0.722	0.852	0.844
53	0.875	15.28	70.01	0.705	0.845	0.916
54		15.15	70.42	0.683	0.845	0.925
56	1.000	15.33	70.58	0.807	0.798	0.922
57		15.44	71.76	0.787	0.809	0.934
58		15.47	72.73	0.767	0.821	0.945
59		15.48	73.45	0.750	0.830	0.954
60		15.43	74.19	0.730	0.836	0.966
61		15.42	74.74	0.718	0.843	0.973
62		15.40	75.52	0.700	0.846	0.984
63		15.25	75.98	0.677	0.848	0.995
64		15.16	76.49	0.656	0.848	1.004

SOURCE: Joseph P. Frizell, *Water-Power* (New York, 1900), p. 282.

whose work on turbines began when he was a mechanic building Boyden and Francis wheels in the Lowell Machine Shop. Swain built his first six-inch wheel model in 1858. The following year he produced a full-scale turbine, then launched into the wheel-building business. This early Swain wheel, Safford and Hamilton noted, "was much like the Howd-Francis, except that its buckets were deeper, many in number and curved outward from the inner discharge edge, so that the discharge was inward and downward." Still large for its capacity and with rather shallow buckets, "it was a really good wheel and was the direct predecessor of the modern low-speed reaction wheel."[44] The textile and hydraulic engineer Samuel Webber in 1895 declared that Swain "conceived an idea which produced the prototype and exemplar of all the modern American turbines. He combined the inward and downward flow wheels, curving the buckets both laterally and vertically, and discharging the water mainly downward, where a reversed curve in the base on which the wheel rested threw it outward again, so that the path of the water was a semi-circle."[45] Some twenty-five years later Forrest Nagler, a turbine designer with Allis-Chalmers of Milwaukee, contributing to the discussion of the Safford-Hamilton interpretation of the development of the American turbine, took a similar positon, holding that Swain's innovation "permitted the greatest single increase in efficiency . . . in the history of turbine development."[46] While differing with Nagler on some points, Safford and Hamilton cited the excellent characteristics and performance of the Swain wheel and termed it "the father of the modern wheel, just as the Howd-Francis was the grandfather."[47] The Swain turbine came into general favor at the Lowell mills, where, when new wheels were needed, it was widely adopted in preference to the Boyden wheel, owing evidently to the favorable appraisal and actual tests by Francis himself (see Table 14).[48]

[44] Ibid., p. 1256 and fig. 12. Swain's own account of the manner in which he became interested in the improvement of turbines, together with a description of his apprenticeship in woodworking and carpentry and work as a mechanic in and about the Lowell mills and machine shops is found in his testimony in the case of *Swain Turbine and Manufacturing Company* v. *Ladd*, transcript of record, Supreme Court of the United States, no. 299 (filed Oct. 15, 1877), pp. 129–68. This reference is courtesy of Professor Edwin T. Layton.

[45] Samuel Webber, "Water-Power," *TASME* 17 (1895–96): 48; Tyler, "Evolution of the American Type of Water Wheel," pp. 879 ff.; and Rice, "Turbine Development in America," pp. 208–10.

[46] Safford and Hamilton, "The American Mixed-Flow Turbine," pp. 1310–11.

[47] Ibid., pp. 1260–61.

[48] See letters of JBF, 1872–78, LCP, DB-6, 385; DB-8, 30, 61, 149, 187, 373.

The Swain wheel was followed shortly by the "American Turbine," of Stout, Mills, and Temple, also employing the inward and downward mixed-flow principle. It too underwent a long and successful development, chiefly for use in small capacities, and was said to be the first turbine built to stock in standard sizes. A few years later, in 1862, James Leffel formed a company in Springfield, Ohio, to manufacture a unique wheel of his own design. The Leffel turbine combined two wheels in a single casting: the upper one was given conventional inward-flow buckets, the lower one buckets that curved inward and then downward (Fig. 82). This quite "unorthodox" double wheel worked very well, although no one was quite sure why.[49] Leffel himself claimed 92 to 95 percent efficiency for his "double-turbine wheel." In the early years of manufacture, most Leffel turbines fell within the thirty- to forty-inch diameter class, selling at $350 to $500. Output rose from 47 wheels in 1862 to 153 wheels in 1864, and by 1870 may have reached 400 to 500 wheels with an aggregate value of $250,000, produced by a labor force of sixty-five hands. Within a decade more than 6,000 Leffel wheels were reported to be in use; by 1880 the number was placed at more than 8,000.[50] The performance of this wheel is further discussed below.

Many other wheels followed, for the most part with only minor variations on existing patterns. An occasional new wheel presented greater novelty without, in most instances, bringing appreciable improvement in performance. One exception was the Risdon wheel of 1873, which achieved 90 percent efficiency by using a draft tube and diffuser; the Risdon Company was the first wheel builder to apply this device since Boyden. Another notable exception was the celebrated Hercules wheel, designed in the 1870s by John B. McCormick, a Pennsylvania mechanic. With its inward-downward-outward buckets of unusual depth and very wide openings, it not only operated with high efficiency but astonished turbine builders and testers alike by developing three times as much power as other wheels of the same diameter. Further developed by the Holyoke Machine Company and produced by a number of manufacturers, the McCormick wheel had a significant impact upon the development of the mixed-flow turbine.[51]

[49] Safford and Hamilton, "The American Mixed-Flow Turbine," pp. 1261 ff., 1259, and fig. 14.

[50] Carl Becker, "James Leffel, Double-Turbine Water Wheel Inventor," *Ohio History* 75 (1966): 257–70; *Manufacturer and Builder* 4 (1872): 244; ibid., advertisement, end of volume for 1881.

[51] Safford and Hamilton, "The American Mixed-Flow Turbine," pp. 1262, 1265 ff.;

Fig. 82. Leffel's double turbine waterwheel. The upper section of this wheel is of the conventional inward-flow type. In the lower section the buckets are of mixed-flow design, which channels the water inward and then downward. (James Leffel Co., *Leffel's American Double Turbine Water-wheel* [Springfield, Ohio, 1869].)

Important as were the contributions of such men as Swain, Leffel, and McCormick, the mixed-flow turbine, as Safford and Hamilton contended, "was not a sudden invention": "it was merely the crystalization and modification of principles toward which all had been working. Many an old, self-trained mechanic contributed his mite to the development of this new type. This was a real American production, the result of evolution during a changing period in American history. The need arose, made itself felt, and eventually was met, not by the work of one great scientist, but by the multitudinous efforts of an army of old Yankee millwrights and machinists, many of the names of whom are either unknown or forgotten."[52]

Many years later a Swiss engineer, while expressing his admiration of Francis's "classic investigations" on the center-vent, inward-flow wheel in 1850, believed it hardly regrettable that "the Francis turbine was not reared in the same scientific atmosphere in which it was born, since this might have stifled the innumerable original creations of the inventors who followed."[53] To other engineers abroad the rude empirical methods of Americans produced wheel designs that were strange if not grotesque; it would seem, declared George R. Bodmer, "as if the designers of such motors had aimed at giving the water the most tortuous course possible, with what object it is difficult to imagine." Such design methods, it was later to be protested, defied mathematical analysis. Bodmer noted the "energetic measures . . . taken to introduce the mixed-flow turbine abroad by claiming in many cases, extraordinary efficiency." Yet Bodmer was inclined to accept the efficiencies as determined in the Holyoke flume. In a comparative analysis of the mixed-flow wheel and the Jonval turbine, generally accepted in Europe, he concluded that the American turbine possessed both distinct theoretical and great practical advantages over the Jonval wheel.[54]

The progress of the American type of turbine is illustrated by the results of two comparative trials, the first at Philadelphia's Fairmount Waterworks in 1859–60 and the second at the Philadelphia Centennial Exhibition in 1876. As we have seen, the Jonval wheel led the field

---

Rice, "Turbine Development in America," pp. 208–10; Tyler, "Evolution of the American Type of Water Wheel," p. 893; Mead, *Water Power Engineering*, pp. 266–67.

[52] Safford and Hamilton, "The American Mixed-Flow Turbine," p. 1266.

[53] A. Streiff, paper before International Engineering Congress, 1915, quoted ibid., p. 1269.

[54] Bodmer, *Hydraulic Motors*, pp. 439–41; Bodmer's chap. 12, "American Turbines," is devoted largely to the mixed-flow turbine. Charles Singer et al., eds., *A History of*

in efficiency at Fairmount; the seven turbines representing the beginnings of the American type averaged more than 10 percent lower in efficiency than the Jonval. At the Centennial Exposition seventeen years later, fifteen of the sixteen wheels in competition were variant forms of the American type. The sixteenth wheel was a Jonval turbine built by the same makers who won the competition at Fairmount; it displayed admirable workmanship and employed the finest of materials. The best of the American-type wheels, the Risdon, although a plain iron casting, gave 5 to 10 percent greater efficiency at full load and 33 percent greater efficiency at part gate than the Jonval wheel.[55] "Manufacturers," declared Frizell some years later, "cannot afford to use wheels of low efficiency at any price." He pointed out that in an installation of 1,000 horsepower using water at a moderate annual cost of $20 per horsepower, a 5 percent difference in efficiency would mean a difference in aggregate cost of $500 a year, warranting an added capital expenditure of as much as $10,000 to obtain better wheels.[56] As demonstration of this principle, the Swain turbine, with its lower cost, smaller space needs, and superior efficiency at part gate, was in use in the Lowell mills by 1871, quickly establishing superiority over the older Boyden wheels.[57]

Next in importance to efficiency—to many doubtless of first importance—was the size of a wheel for a given capacity, since cost for wheels

---

Technology, 5 vols. (Oxford, 1958), 5:530–31, attributed an alleged low efficiency of the mixed-flow turbine at part gate to the indifference of wheel builders owing to the great abundance of water. Quite the contrary: concern with wheel efficiency as a means of extending water supplies, already limited at many large centers before 1850, was present well before the mixed-flow turbine had its beginnings. As table 1 in Safford and Hamilton indicates, mixed-flow wheels by the early 1870s were attaining more than 70 percent efficiency at half power ("The American Mixed-Flow Turbine," p. 1272). Careful flume tests on a Swain wheel by the builders in 1869 gave the mean maximum efficiency of a forty-two-inch wheel built for a customer with results as follows: at full, three-quarters, and one-half gate, respectively, 81.3, 77.3, and 69.3 percent (Hiram F. Mills, "Experiments upon a Central Discharge Water-wheel," *JFI* 89 [1870]: 190–91). See also Thurston, "Report on Manufactures and Machinery," p. 177.

[55] Henry P. M. Birkinbine, *Report on the Experiments with Turbine Wheels Made in 1859–60 at Fairmount Works, Philadelphia* (Philadelphia, 1861); Tyler, "Evolution of the American Type of Water Wheel," pp. 879 ff.

[56] Frizell, *Water-Power*, pp. 558 ff.

[57] See JBF to Oliver Ames and Sons, 16 Feb. 1870, to J. Converse, 20 Apr. 1872, to T. J. Younglove, 7 Nov. 1871, to Hurlbert Paper Co., 14 Mar. 1872, LCP, DB-6, 10, 72, 338, 385. According to Tyler, by 1876 the American type of wheel almost completely controlled the New England textile trade ("Evolution of the American Type of Water Wheel," p. 881).

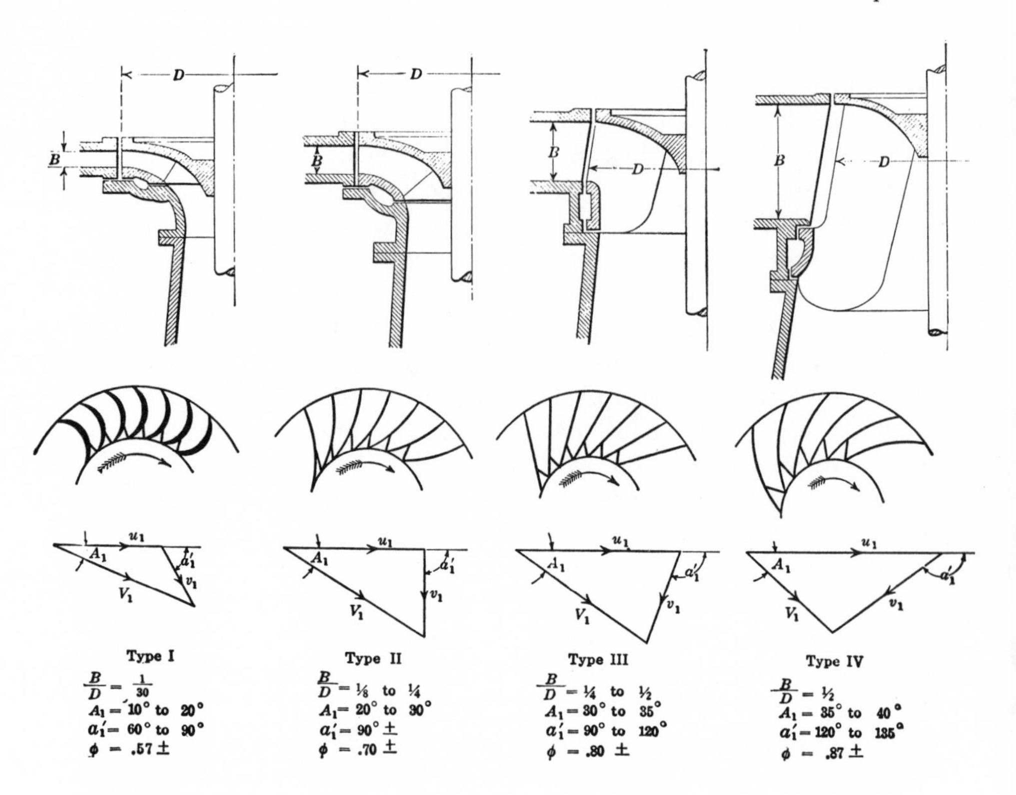

Fig. 83. Four styles of "Francis," or "American," type runners. All differ somewhat from J. B. Francis's original radial inward-flow design. Higher speed and power were obtained from wheels of the same diameter by increasing the depth of the runner (dimension *B*, above), using fewer vanes, and extending the vanes further toward the center of the wheel. As a result, more of the water had to be discharged axially, or downward, hence the inward and downward, or "mixed-flow," design. Type IV (above) illustrates the extreme high-speed, high-capacity mixed-flow runner. A tabular summary showing increases in speed and power resulting from changes in runner design can be seen below. (Robert L. Daugherty, *Hydraulic Turbines*, 2d ed. [New York, 1914].)

TABLE 1.—COMPARISON OF 12-IN. WHEELS UNDER 30-FT. HEAD

Type	Discharge, cu. ft. per minute	H.p.	R.p.m.
Tangential water wheel	7.9	0.37	380
Reaction turbines:			
Type I	99.0	4.3	460
Type II	329.0	14.9	554
Type III	741.0	33.4	600
Type IV	1209.0	55.5	730

Table 15. Capacity, speed, and power of 48-inch "American" turbines under a 16-foot head

	Year brought out	Discharge in cu. ft.	rpm	Horse-power
American	1859	3,271	102	79.1
Standard New American	1884	5,864	102	141.8
New American	1894	9,679	107	234.0
Special New American	1900	11,061	107	267.0
Improved New American	1903	13,234	139	325.0

SOURCE: Daniel W. Mead, *Water Power Engineering: The Theory, Investigation, and Development of Water Power* (New York, 1908), Table 23.

Table 16. Increase in speed of "American" turbines for same power under a 16-foot head

	Size of wheel (inches)	Horse-power	rpm
American	48.0	79.1	102
New American	36.0	81.5	136
Special New American	27.5	87.3	186
Improved New American	25.0	87.5	267

SOURCE: Mead, *Water Power Engineering*, Table 24.

of the same class were approximately proportionate to size, as were also space requirements and foundation and installation costs. In 1890 six high-speed wheels were installed at the Tremont and Suffolk mills in Lowell in the three masonry pits previously occupied by three 160-inch low-speed Boyden wheels. They nearly doubled the power developed within the same space and increased the wheel shaft velocity 35 percent.[58]

The reduction in wheel size and cost relative to capacity—that is, power—was pursued with a success suggested in the accompanying tables and illustrations covering the period 1860–1900. Table 15 shows the striking advances in wheel capacity for a given size under a given head in successive models of the "American" turbine, 1859–1903. The general form of the original 1859 wheel runner is shown in Figure 78.

[58] Hunking, "Notes on Water Power Equipment," p. 204.

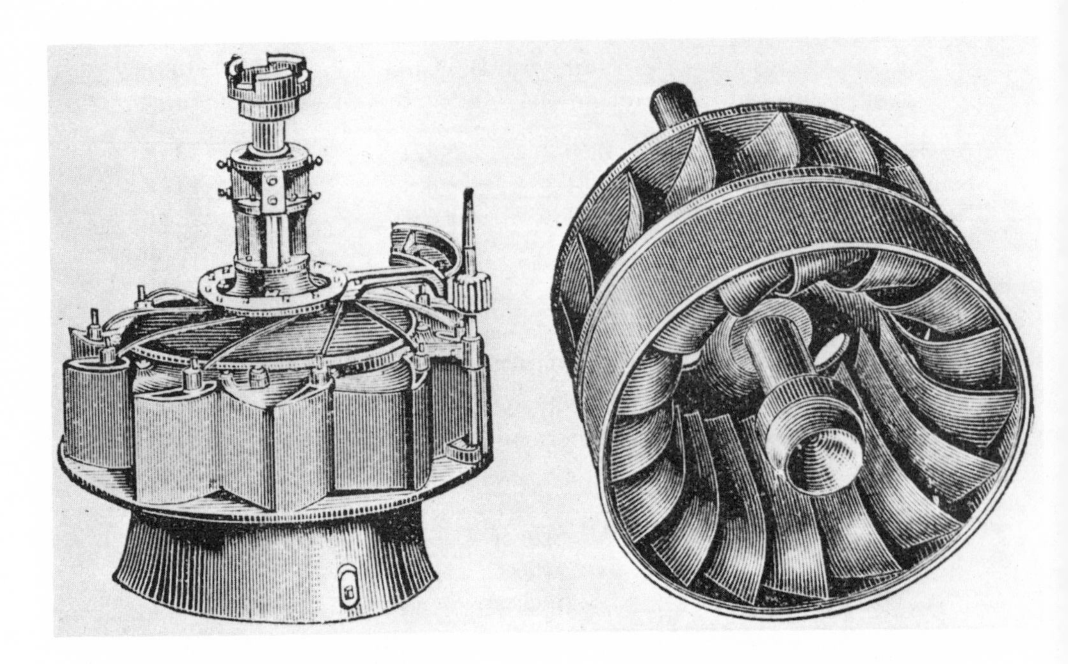

Fig. 84. New American turbine, casing and runner. (James Emerson, *Treatise Relative to the Testing of Water-wheels and Machinery,* 6th ed. [Williamsett, Mass., 1894].)

The flow of water was inward through the bucket openings around the periphery of the wheel, discharging at the central opening. In the model of 1884, the Standard New American, the buckets after a short passage inward are lengthened and turned downward, thereby increasing somewhat the bucket area subject to the pressure and reaction of the water (Fig. 84). The power of the new wheel, which had about the same diameter and a greatly increased volume of dis charge, was nearly doubled, from 79 to 142 horsepower. The further lengthening and deepening of the buckets can be observed in succeeding models. By 1900 in the Special New American model capacity had been raised to 267 horsepower with little change in speed; the Improved New American wheel of 1903, with its speed increased about one-third, was rated at 325 horsepower. Table 16 shows that in the succession of American models of similar capacity but with rotating speed increased more than 2.5 times, the size of the wheels was reduced nearly one-half.

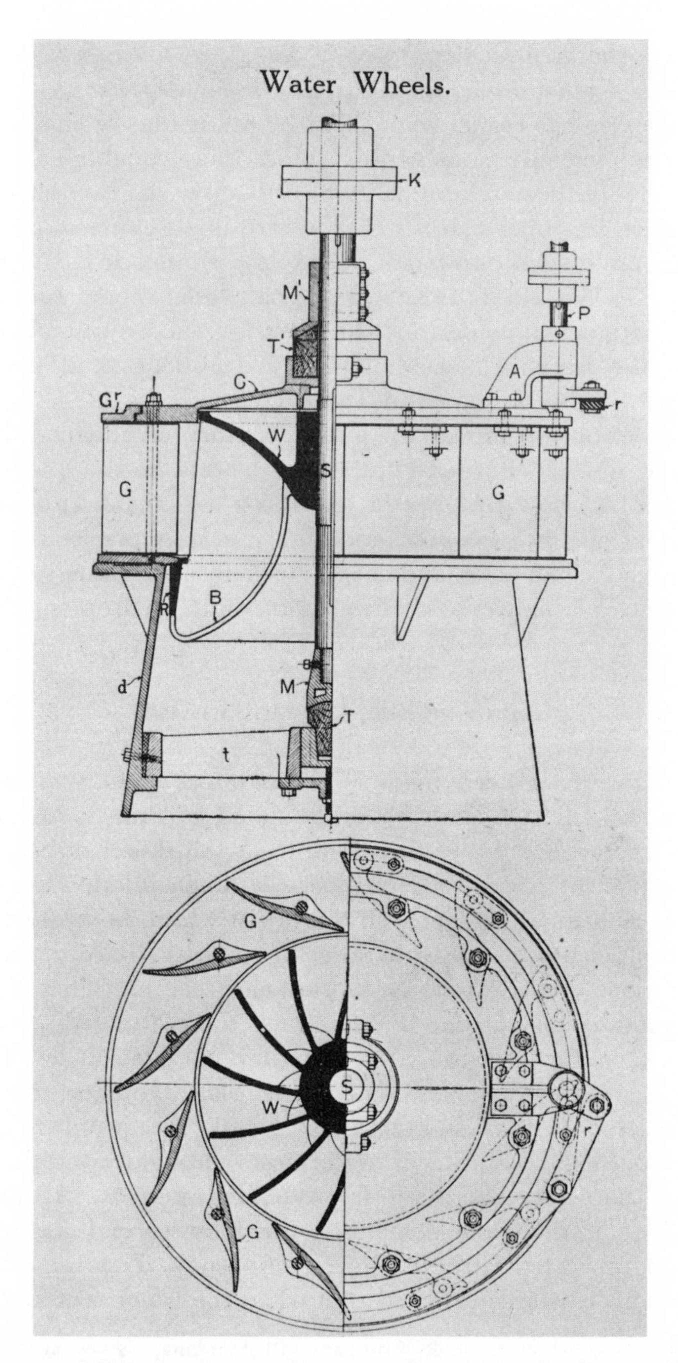

Fig. 85. Improved New American turbine, manufactured by the Dayton Globe Iron Works Co. *W*: the crown and hub of wheel; *S*: shaft on which wheel turns; *B*: the buckets; *T*: step-bearing on which wheel shaft turns; *K*: coupling to drive shaft; *GG*: wicket gates controlling admission of water, operated by ring *G* through an eccentric and rod, *r*, connected with governor through the shaft, *P*. (Daniel W. Mead, *Water Power Engineering* [New York, 1908].)

"Perhaps the greatest departure of American inventors," according to Professor Daniel Mead in his *Water Power Engineering*, "from the lines of the original Francis type of turbine was that of James Leffel." Among commercially successful turbines this "double" wheel was possibly the farthest removed from orthodoxy, as we have noted, combining in its upper half a radial inward-flow runner of the Francis type with an inward- and downward-flow runner in its lower half (Figs. 86, 87). As late as 1922 the original Leffel model continued to sell well. With only a modest increase in speed, successive models of this wheel in the forty-inch size under a sixteen-foot head more than trebled in capacity, as shown in Table 17.

Samuel Webber summed up a half century of advance from the remarkable pioneer Boyden-Fourneyron wheels of the 1840s to the perfected mixed-flow turbines of the 1890s: equal power in one-half the space, at one-fifth the cost, and with much simpler construction.[59] The later wheels had fewer and deeper buckets, with runners—the wheel proper—cast in a single piece instead of built up from many parts.

Improving Efficiency at Part Gate

Much progress had been made in overcoming what was for many years a major defect in the turbine: as stated by Francis in 1868, "To use water economically it must be run to its full power or nearly so."[60] In this important respect the turbine was much inferior to the overshot wheel and in lesser degree to the breast wheel, both of which ran with high efficiency over a wide range of capacity. Since the early turbine's efficiency fell off rapidly at part gate, the efficiency figures so much publicized by turbine builders were those attained at full gate. But most manufactories much of that time operated at less than the capacity of their waterwheels. When water was abundant, this did not greatly matter; but as shortages developed, the problem assumed larger proportions. Insufficient water from drought conditions often compelled part-gate operation for extended periods. At just such times, when limited flow demanded the most careful use of water, the turbine operated with the lowest efficiency. Tests of early Boyden and Francis wheels revealed that when the water required for full

[59] Webber, "Water-Power," p. 48. Compare with Hunking, "Notes on Water Power Equipment," p. 199.

[60] To T. J. Coolidge, 22 July 1868, to S. S. Fosher, 15 Mar. 1875, LCP, A-19, no. 89, DB-8, 91. See also *Manufacturer and Builder* 5 (1873): 123.

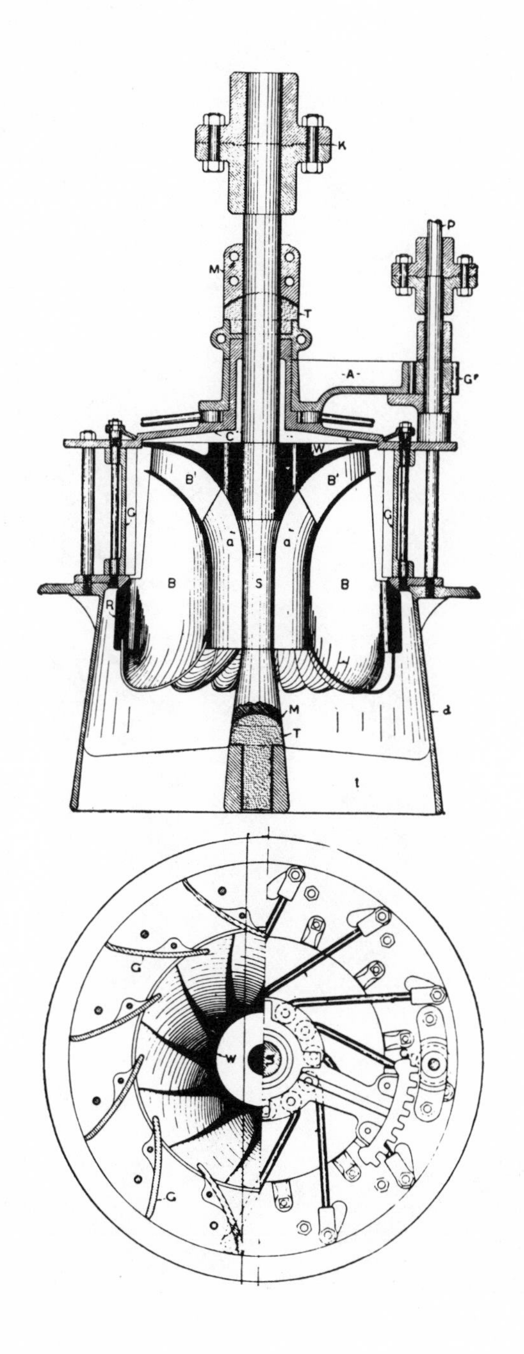

Fig. 86. Section and plan of a Leffel Samson turbine, manufactured by James Leffel & Co. (Daniel W. Mead, *Water Power Engineering* [New York, 1908].)

Fig. 87. Runner of a Samson turbine. (Photo courtesy of James Leffel & Co., Springfield, Ohio.)

Table 17. Capacity, power, and speed of 40-inch Leffel wheels under a 16-foot head

	Year brought out	Discharge	rpm	Horse-power
Standard	1860	2,547	138	64.5
Special	1870	3,672	138	93.0
Samson	1890	6,551	158	155.0
Improved Samson	1897	8,446	163	207.0

SOURCE: Mead, *Water Power Engineering,* p. 260.

operation was reduced about one-half, their efficiency fell from 76 to 79 percent to 33 to 35 percent. Other early turbines did no better if as well. What the manufacturer almost invariably needs, wrote James Emerson in 1878, is a wheel "that will economize water at any stage of gate opening." The real value of a turbine, another engineer held, was determined by its efficiency when operating between half and full gate; the fairest basis of comparison was three-quarters gate, the average opening at which wheels operated.[61]

Turbine builders recognized the problem and tackled it with their customary empiricism. By 1868 Swain was obtaining nearly 75 percent efficiency from his wheel at full gate and 66 and 50 percent, respectively, at three-quarters and half-gate operation. The leading turbines participating in the 1876 Centennial Exposition at Philadelphia showed substantially higher part-gate efficiency than the early Boyden and Francis wheels. At a competitive trial at the Holyoke testing flume in 1879, several turbines reached their highest efficiency at seven-eighths gate and were nearly as efficient at three-fourths gate as at full gate. Progress continued, and by 1900 many makes of wheels had as high or even higher efficiency at three-fourths gate as at full gate and declined in efficiency less than 10 percent from full gate to half gate.[62]

[61] James B. Francis, *Lowell Hydraulic Experiments* (Boston, 1855), pp. 31–34; Holyoke Water Power Company, *Holyoke Hydrodynamic Experiments*, p 143.

[62] Emerson, *Fourth Annual Report*, p. 11; Thurston, "Systematic Testing of Turbine Water-Wheels," p. 370; Emerson, *Treatise Relative to the Testing of Water-Wheels*, 3d. ed., pp. 114–47. See also Mead, *Water Power Engineering*, pp. 362–63, and App. D, pp. 703 ff.; Samuel Webber, "Efficiency of Turbines as Affected by Form of Gate," *TASME* 3 (1882): 78–80. On the limitations of flume testing, see Allen, "Testing of Water Wheels after Installation," pp. 300 ff. "One of the greatest stumbling blocks in the path of turbine development in this country has been the lack of systematic testing after installation" (G. E. Russell in discussion, ibid.).

Advances in Setting, Governing, and Positioning

In the early and typically small turbine installations only incidental attention was given to the wheel setting, the builders doing little more than providing structural support and the means of getting the water to and from the wheel casing. The wheels were often simply placed in an opening in the floor of the wooden flume or trunk supplying them. In time wheel builders were to discover that provision for the smooth and turbulence-free movement of the water into and from the wheel was second in importance only to its movement past and through the guides and buckets of the wheel itself. As earlier noted, the Parker brothers had by chance discovered the marked gains in performance resulting from what they termed a "helical sluice." Similar arrangements were devised by others in various forms of what were usually called "scroll cases," whose curving and gradually narrowing form conveyed the water in a spiral path to all portions of the wheel's circumference. In his adaptation of the Fourneyron turbine, Boyden introduced a cone-shaped penstock for the same purpose. With the general shift from wood to iron construction, the difficulty of obtaining the desired smoothly curving spiral course with scroll cases made of sheet iron, adopted because of its cheapness, led to the sacrifice of hydraulic to structural and cost considerations. For many years, in the phrase of Safford and Hamilton, "the boiler-maker ruled." A return to sound hydraulic design awaited the coming of large power plants, especially for hydroelectric installations, where every means of improving wheel efficiency was exploited. Hydraulic principles were carefully studied and given priority, aided by improved materials and methods of construction, especially reinforced concrete. The draft tube, first introduced by the Parkers in 1840 and eventually adopted generally, was simply a means of raising the wheel out of the water to a height convenient for access in maintenance and repair. Its form and proportions received attention as means of avoiding turbulence and obtaining the velocity of flow desirable for improving wheel output and efficiency. Gate design and mechanism became similar objects of attention and improvement.[63]

[63] The best brief discussion of the development of wheel settings is Safford and Hamilton, "The American Mixed-Flow Turbine," pp. 1273–84. See also Tyler, "Evolution of the American Type of Water Wheel," pp. 885–88, 891–93, and for a more detailed treatment, Mead, *Water Power Engineering*, chaps. 13–14; Robert L. Daugherty, *Hydraulic Turbines*, 2d ed. rev. (New York, 1914), chap. 5. "The utmost care is taken in design to prepare smooth passages and easy curves for the water passing through and

Two further improvements were necessary to adapt the turbine to the industrial needs of the late nineteenth century: an adequate governor and the introduction of the horizontal-shaft turbine. Waterwheels were long used without any automatic device comparable to the steam-engine governor for maintaining a uniform speed. There was little need for such regulation in most industrial operations to which waterpower was chiefly applied—gristmilling, sawmilling, fulling, furnace blowing, forging, and the like. Although in Britain governors of the fly-ball type common on mill engines from Watt's day were adapted for use with waterwheels, we find almost no reference to their use in this country before the introduction of the turbine.[64] To a varying degree bucket wheels of the overshot and breast types adjusted themselves automatically to changes in load; a speeding up or slowing down of their operation caused the buckets to receive less or more water in consequence, countering to some extent the change in speed.[65] With significant changes in the load carried by the waterwheel, the adjustment of the gate opening by hand was of course necessary. Where such changes were frequent, as with sawmill or forge hammer, the machine operator would alter the gate opening by a lever placed at his convenience. In the case of the celebrated sixty-foot waterwheel at the Burden Iron Works in Troy, New York, "a man [was] seated on an elevated platform, in front of it, having his hand on a lever, by which he increases or diminishes the volume of water," thereby adjusting its speed to the power requirements of the moment, depending upon the number and kinds of machines in operation.[66] In the continuous-process method of papermaking, tub wheels were reported in 1845 to be frequently employed, despite their waste of water, on account of the regularity of their motion; to the same end it was "almost invariable practice . . . to employ an independent wheel to

---

a spiral approach for lessening eddy loss is now universal" (John R. Freeman, "The Fundamental Problems of Hydroelectric Development," *TASME* 45 [1923]: 534–36).

[64] The article "Governors" in *Appleton's Dictionary of Machines*, has the briefest of references to the possibility of adapting a steam-engine governor for waterwheel use. The same is true of *Appletons' Cyclopaedia of Applied Mechanics*. In England in the 1830s, at the textile mills of the Messrs. Strutt at Belper, a self-acting governor attached to each waterwheel was reported as "never in a state of repose, but . . . incessantly tightening or slackening the reins of the mill-geering, so to speak, according to the number of machines moving within, and the force of the stream acting without" (Andrew Ure, *The Philosophy of Manufactures* [London, 1835], pp. 343–44).

[65] See discussion between Emile Geyelin and editor, *Scientific American*, n.s. 3 (18 Aug. 1860): 115; also ibid., n.s. 4 (30 Mar. 1861): 206; n.s. 7 (6 Sept. 1862): 155.

[66] Margaret E. Proudfit, *Henry Burden* (Troy, N.Y., 1904), p. 64.

each machine," to prevent outside disturbance of the wheel and machine speed.[67]

As late as 1840 a War Department report considering a site and power for armory use stated that waterwheels, unlike steam engines, were not susceptible to regulation.[68] This was not the case, yet waterpower, with or without governors, continued nearly to 1900 to be less satisfactory than steam power in respect to uniformity of motion. The size and weight of the penstock gate, controlling admission of water to the wheel, made necessary a special train of gears to convey from the wheel shaft the power required for its operation.[69] The governors devised for the purpose were slow and cumbrous in operation, providing but indifferent regulation and in many situations giving preference to steam power over waterpower. The operational needs of an advancing industrialism gave increasing importance to uniformity of motion, as affecting machine operation, productivity, and frequently quality of product. These factors and the widespread introduction of the turbine stimulated the improvement of the waterwheel governor. The practice of incorporating the gate mechanism in the turbine assembly made possible a more direct and responsive means of speed regulation; and the more exacting demands of manufacturing made its use desirable.

From the 1860s a growing number of new waterwheel governors appeared on the market; the manner of their operation and improvement received increasing attention from engineers.[70] Superintendent Francis, advising an independent manufactory in Lowell concerning a turbine installation in 1865, remarked that if a governor was wanted, the type used by the Lawrence Manufacturing Company "could be adapted to this wheel. Some of their wheels have no governor, but have an apparatus for moving the gate by hand."[71] Thurston in 1875

[67] *JFI* 39 (1845): 123.

[68] U.S., Congress, "Site for a Western Armory," House Ex. Doc. 133, 27th Cong., 3d sess. (Serial 421, pt. 4), p. 21.

[69] See Frizell, *Water-Power*, pp. 373 ff.; Mead, *Water Power Engineering*, chap. 14. See also Francis, *Lowell Hydraulic Experiments*, Pl. 1, which shows a fly-ball governor patented by N. Scholfield, Norwich, Conn., 17 May 1836. In 1855 Scholfield assigned the Locks and Canals company at Lowell full and exclusive rights to making, using, and vending his regulator (LCP, A-18, no. 92). References to waterwheel governors begin to appear in the technical journals shortly before 1850. See *Scientific American* 3 (23 Oct. 1847): 40; 3 (1 Apr. 1848): 220; 4 (10 Feb. 1849): 164; 7 (6 Mar. 1852): 196.

[70] *American Artisan*, n.s. 1 (25 Oct. 1865): 398; Thurston, "Report on Manufactures and Machinery." See also *American Artisan*, n.s. 4 (14 Nov. 1866): 17; Emerson, *Treatise Relative to the Testing of Water-Wheels*, 3d ed., p. 61.

[71] JBF, memorandum for H. W. Butler and Company, 6 July 1865, LCP, DB-3, 654.

called attention to the "very large number of regulating devices and forms of gate" patented in the United States, remarking that it was "in this direction that some of the best work in the improvement of the turbine has been done." Yet as late as 1888–89 Thurston and C. T. Main agreed that governing was superior with steam power and that a governor was yet to be found that was capable of maintaining turbine speed in a truly satisfactory manner.[72] The final push toward success came from the use of waterpower to generate electricity and the exacting demands of generators for closely regulated speed. The result was the development of often elaborate mechanisms combining some form of rotating ball or spring governor usually with hydraulic or electrical transmission of power to operate the gate mechanism and providing close and sensitive response to changing power requirements.[73]

To place the turbine on its side to obtain the advantages of a horizontal wheel shaft—the direct drive of line shafting or machinery and the elimination of the costly and heavy bevel gearing that was standard turbine equipment—was to break with a millennia-old tradition. Evolving as the early turbines did from the horizontal gristwheel and operating in a submerged position, the upright position of the shaft was both natural and necessary. Such early developers of reaction wheels as Wing and the Parkers made much of the ability of their encased wheels to operate in either a vertical or a horizontal position. The draft tube introduced by the Parkers enabled the wheel to operate when elevated some distance above the water, making a horizontal position of the wheel shaft both feasible and advantageous. Yet tradition and mechanical difficulties discouraged the practice.[74] Such

[72] Thurston, "Report on Manufactures and Machinery," p. 177, and comments of Thurston and Main in Charles T. Main, "On the Use of Compound Engines for Manufacturing Purposes," *TASME* 10 (1888–89): 79, 87; also Charles H. Manning, "Comparative Cost of Steam and Water Power," *TASME* 10 (1888–89): 505. For illustrated description of various types of waterwheel governors, see Mead, *Water Power Engineering*, chap. 14. For detailed discussion, see "The Lombard Water Wheel Governor," *Electrical Engineer* 19 (1897): 312 ff.; Mark A. Replogle, "Some Stepping Stones in the Development of a Modern Water-Wheel Governor," *TASME* 27 (1906): 642 ff.; idem, "Speed Government in Water-Power Plants," *JFI* 145 (1898): 81 ff.; John Sturgess, "Speed Regulation of Water-Power Plants," *TASME* 27 (1906): 682 ff.

[73] Webber, "Water-Power," pp. 50–51; "Electricity," *Engineering Magazine* 6 (1893–94): 526; and Edward H. Sanborn and Thomas C. Martin, *Power Employed in Manufactures*, Twelfth Census, 1900, Bulletin no. 247 (Washington, D.C., 1902), pp. 49–50.

[74] See William C. Hughes, *The American Miller, and Millwright's Assistant*, rev. ed. (Philadelphia, 1859), pp. 46 ff.; and Overman, *Mechanics for the Millwright*. In the case of the Parker wheel, the draft tube invented by the Parker brothers, Austin and Zebulon,

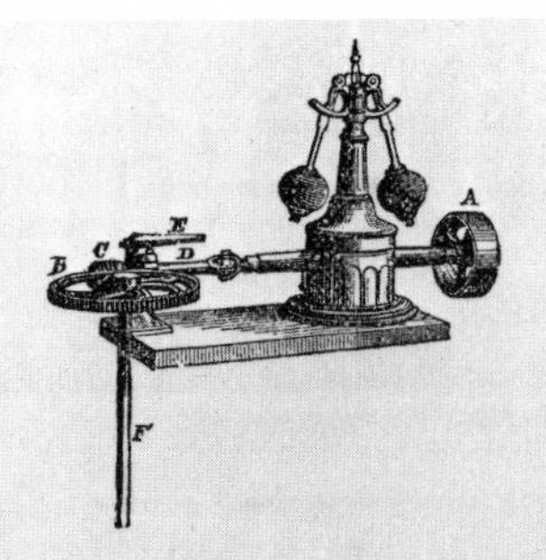

Governor or Regulator.

The above cut is designed to show the construction of the Governor or Regulator which we now manufacture and are prepared to furnish. It also shows the manner in which it is attached to the water wheel. A is the driving pulley, receiving a belt from any convenient shaft and pulley in the mill. D is a short shaft carrying the screw C, which works into the worm-wheel B. The shaft D is connected by a knuckle-joint to the shaft within the casing. This joint permits a lateral motion of the shaft D by the handle E, thus disconnecting the Governor from the worm-wheel B whenever occasion may require. F is the gate-rod extending down to water wheel. To give the proper motion to the Governor, the pulley A should make 92 revolutions per minute.

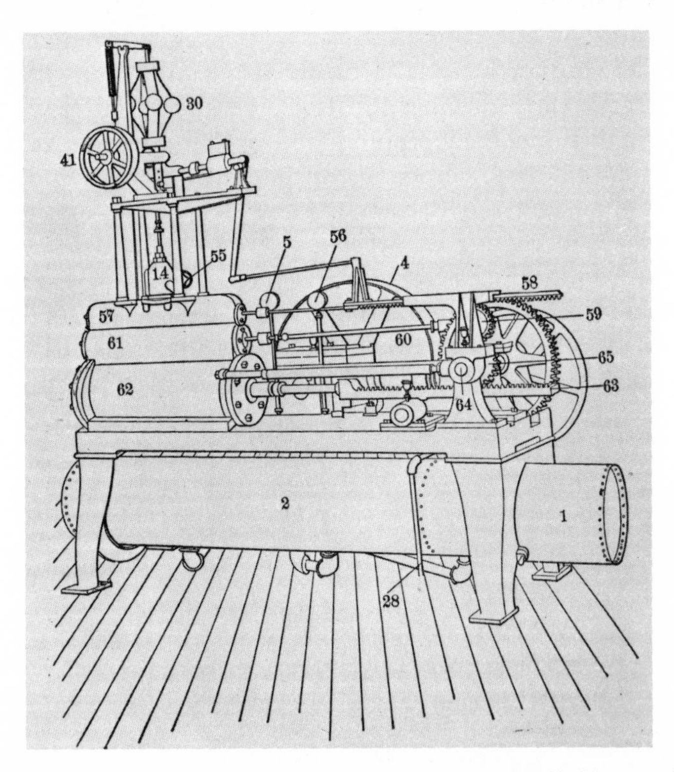

Figs. 88–89. Governors. *Above,* a simple mechanical governor supplied by a major turbine manufacturer in the 1860s. (James Leffel Co., *Leffel's American Double Turbine Water-wheel* [Springfield, Ohio, 1869].) *Below,* a complex Lombard hydraulic waterwheel governor. (Joseph P. Frizell, *Water-Power* [New York, 1903].)

prominent turbine builders as Geyelin in the 1850s and, twenty years later, Swain, followed by Leffel, occasionally supplied horizontal wheels for particular situations.[75] There was a manifest advantage of a wheel shaft position permitting a direct or belted connection with a horizontal mainline-shaft or with machinery. In addition to eliminating the weight and expense of the bevel crown gear above the turbine, it also did away with gear-caused friction losses, ranging from 10 to 20 percent of the power output of the wheel—not to mention the noise, nuisance of lubrication, and possibility of gear wheel breakage and shutdowns.[76] The weight of a turbine installed at Lowell in 1857 was 7,131 pounds; that of its crown gear, 4,145 pounds—with costs somewhat in the same proportion. For instance, bids received from three builders by the Philadelphia Waterworks for three turbine wheels and the matching gearing for each were: $6,900 and $16,500; $16,000 and $14,800; and $12,390 and $13,080.[77]

The chief compulsion to change in this as in other matters relating to prime movers from the 1880s came from the new electric lighting and power industry. The dynamos of this period were almost invariably mounted on horizontal shafts. In order to supply the now generally recognized advantages of a direct connection with the prime mover, builders of turbines for hydroelectric stations necessarily conformed to the horizontal shafting, adapting wheel settings and draft tubes to the horizontal position of the wheel shaft. This not only eliminated the bevel-gearing power takeoff of the horizontal wheel but also the often troublesome step bearing on which it rested and turned. The bearing was usually of tallow-soaked hardwood. On large wheels the tallow was supplemented by a piped stream of water

---

and patented in 1840, was devised as a means of permitting the use of a horizontal shaft required for the operation of an up-and-down sawmill without resort to gearing (Edwin M. Layton, "Millwrights and Engineers . . .," unpub. article, 1977).

[75] Emile Geyelin as early as 1854 provided a horizontal-shaft wheel for a Mexican installation. See Webber, "Water-Power," pp. 50, 54; Tyler, "Evolution of the American Type of Water Wheel," pp. 893 ff.; Rice, "Turbine Development in America," pp. 208–10; Sanborn and Martin, *Power in Manufactures*, pp. 49–50; Safford and Hamilton, "The American Mixed-Flow Turbine," pp. 1280 ff.

[76] W. M. Barr, "Hydraulic Generation and Application of Motive Power," *Engineering Magazine* 29 (1905): 865 ff.

[77] "Memorandum," LCP, DB-1, 46; Jean F. d'Aubuisson, *A Treatise on Hydraulics for the Use of Engineers*, trans. J. Bennett (Boston, 1852), pp. 431 ff.; "Extracts from the Report of Chief Engineer Birkinbine," *Scientific American*, n.s. 3 (7 July 1860): 22; and on friction losses, Thurston, "Systematic Testing of Turbine Water-Power," pp. 375–76.

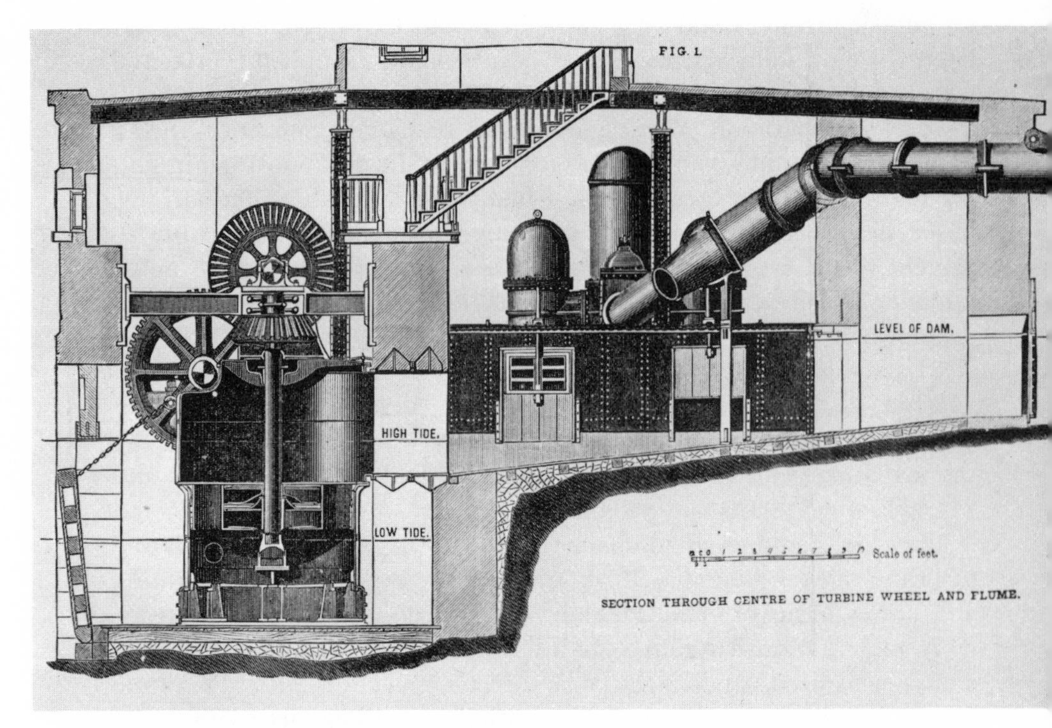

Fig. 90. Massive crown gearing required with vertical shaft turbines. Philadelphia Water Works, 1868. (*JFI* 87 [1869].)

coolant to the bearing to prevent damage from overheating. The combined requirements of large power and high speed led to the mounting of from two to six small-diameter wheels on the same shaft (Fig. 91).[78] Experience in time revealed flaws in what was hailed as a happy solution. The practice, for economy, of using a single draft tube to serve a

[78] See discussion of the shift from vertical to horizontal shafts in Tyler, "Evolution of the American Type of Water Wheel," pp. 879 ff., esp. pp. 893 ff.; Rice, "Turbine Development in America," pp. 208–10. See also Richard A. Hale, discussion of Hunking, "Notes on Water Power Equipment," pp. 221–22. Safford and Hamilton explain: "By about 1900, and for a considerable time thereafter, a majority of the plants had horizontal shaft units, which either drove generators or were direct-connected to machinery. Electrification called for a fairly high speed, and this was accomplished by using batteries of runners of small diameter. Until the high-speed wheel was developed further, no other alternative was possible, as long as high speed was required" ("The American Mixed-Flow Turbine," p. 1230). See also Lewis F. Moody, "The Present Trend of Turbine Development," *TASME* 3 (1921): 1113–14; J. T. Fanning, "Progress in Hydraulic Power Development," *Engineering Record* 47 (3 Jan. 1903): 24–25; and Mead, *Water Power Engineering*, chap. 20.

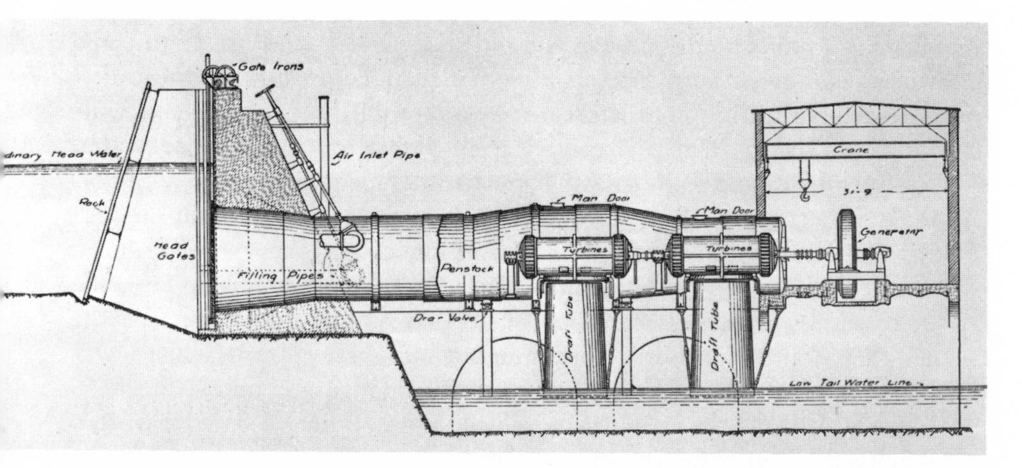

Fig. 91. Four turbines mounted in tandem on a horizontal shaft, Winnepeg Electric Railway Co. (Daniel W. Mead, *Water Power Engineering,* 2d ed. [New York, 1915].)

pair of wheels was shown on test to cause a loss of several percent in wheel efficiency because of interference between the two discharge streams. The mounting of multiple wheels on the same shaft proved to be not only a cumbersome, space-demanding arrangement but a source of lower wheel efficiency and higher unit costs. Before many years wheel builders and station operators were to be rescued from this situation by the shift to the vertical-shaft "umbrella" type of generator, which permitted a seemingly indefinite expansion in size and capacity at speeds satisfactory to both the moving and moved machines.[79]

The American Stock-Pattern Turbine

The development of the American mixed-flow, brand-name turbine was, in sum, a notable technological and industrial achievement. It was an achievement effected largely outside the framework of hydraulic science and the engineering professions. Over a period of some twenty-five years varied forms of this wheel were devised, improved, and brought into wide use. They developed an efficiency in water use

[79] Freeman, "Fundamental Problems," pp. 534–36. At times the multiwheel arrangement continued to prove preferable (idem, "General Review of Current Water Power Practice," pp. 370–71).

superior to all but the best of the old-style wheels and the engineered wheels of Boyden and Francis alike. In their compactness, durability, and such operating characteristics as adaptability to use over a wide range of heads, relative freedom from interference from backwater, trash, and ice, and high rotating speeds, they demonstrated a superiority that eventually drove the old-style wheels largely from the field.[80] As they improved steadily in performance, turbines were reduced drastically in operating cost and price. In the commercial phase, the more successful wheels came to be turned out in stock sizes. To the simplicity and durability that marked the better built wheels was added a ruggedness that seemed to defy rough usage and neglect. In his account of the new waterwheel from France, Fairbairn had remarked upon "the defects of this class of machines, requiring a nicety of design and execution, and being susceptible to injury from small bodies carried into it by the water." The problem of obstruction and damage from trash, shared by the finely built turbines of Boyden and Francis, gradually ceased to trouble significantly the more successful of the mixed-flow wheels, made with fewer buckets and

[80] Freeman, "General Review of Current Water Power Practice," pp. 382–83. Of a selected group of manufacturing industries employing over 5,000 water horsepower nationally, as reported by the census of 1880, the average capacity per waterwheel was under 25 horsepower in five industries and between 25 and 50 horsepower in four others. Only in the case of cotton and worsted textiles did the average capacity exceed 100 horsepower: 112 and 121 horsepower, respectively. The national average per wheel for all manufacturing was 22 horsepower, as compared with 39 horsepower for steam engines. For the leading northeastern waterpower states at this date, the average capacity for all waterwheels in manufactures was: Maine, 27, Vermont, 24, New Hampshire, 34, Massachusetts, 45, Rhode Island and Connecticut, 34. For New York, New Jersey, and Pennsylvania the comparable figures were 22, 23, and 16 horsepower, respectively (U.S., Census Office, Tenth Census, 1880, vol. 2, *Manufactures* [Washington, D.C., 1883], pp. 483 ff., 501, 502). Undoubtedly these low average capacities were owing in large part to the persistence of small, old-style waterwheels in grist- and sawmilling. Though efficiency occupied a prominent position in technical literature and was seized upon as a point of competitive advantage by builders, its practical importance among rank-and-file mills and factories must have been relatively small. Probably not one wheel user in a hundred was in a position to check or have checked the accuracy of the wheel builders' claims to efficiency. The prevailing practice evidently, encouraged understandably by wheel builders, was to have ample capacity to avoid interruption in power supply on this score. Of some 400 turbines in use on the Black River in New York in 1893, a fourth were operated at no more than 25 to 75 percent of capacity (Horton, *Turbine Water-Wheel Tests*, pp. 15, 85–88). As power shortages increased with the expansion of production, improvement in wheel efficiency became a matter of importance to mills and factories finding themselves in a marginal position with respect to power supply.

enlarged flow passages for increased capacity and improved efficiency.

The chief shortcomings of the stock-pattern turbines built to standard sizes seemed to arise from the manner of their selection and installation and with the need of wheel builders to enlarge their markets. Millowners and millwrights, whose experience and traditional lore were variously outmoded by the requirements of the new wheels, were often at the mercy of the wheel builder's cupidity and the frequently extravagant and misleading assertions in advertisements and catalogs. Most purchasers were dependent in wheel selection upon catalog data with respect to wheel size and performance under different conditions of head, streamflow, and operating speed, fortified in the wheel builder's catalog with bland assurances and well-screened user testimonials. The catalogs usually supplied, too, directions for gauging streamflow and estimating capacity. In the absence of streamflow knowledge extending over some years, there was a serious measure of uncertainty that the wheel builder was hardly inclined to exaggerate.

As we shall discuss elsewhere, purchasers of steam engines were faced with much the same problems but with an important difference. Specifications for engines and boilers were matched without great difficulty, however excessive at times the margin of capacity in relation to actual need, supplied or prescribed as they usually were by the same builder. Also, control over pressure and volume of steam through the rate of firing provided the means for generous variations in power and speed.

The problems of selecting a wheel were greatly to be magnified with the coming of hydroelectricity, since what in the ordinary mill or factory was a minor item among production expenses became here the product itself, with output and unit costs determining the viability of the enterprise. In 1906 a hydroelectric engineer recalled the experience of the earlier years, when wheels of 200 to 300 horsepower were considered large. The purchaser of a wheel seldom required the aid of a consulting engineer. "The sales department [of the wheel builder] consisted of a catalogue and scores of testimonials. Efficiency was of little consequence, and thanks to the foresight of the designing engineer, the usual testimony was: 'Your wheel gives more power than required.' Connection between the turbine and the driven shaft was by means of a belt, rope drive or gearing, making one design available for many different speeds and a few designs for all speeds. Regulation was sometimes attempted and the results as often ignored."[81]

[81] E. F. Cassel, "Design and Manufacture of Hydroelectric Installations as a Whole,"

Ten years later a paper on hydraulic turbines delivered before an international audience warned against two influences dominant in wheel selection in the earlier days: the existence of designs and patterns previously used and the customary practice of a country. However desirable from the commercial viewpoint, especially with the "non-technical financier," the first was the dominant factor during the years "when the stock trade turbines were in vogue." Real engineering always suffered in that day, as witness the many failures, owing to advice "from parties either incapable of making a proper analysis, or short of the necessary foresight to comprehend the final results."[82]

Influence of Hydroelectricity

In 1887 William E. Worthen, successful textile mill superintendent and onetime aide to Boyden in testing the early Fourneyron-Boyden wheels, reflected the general acceptance and approval of the stock-pattern turbines that had widely displaced the Boyden wheel in the New England textile mills. He remarked, "Water wheel construction has passed out of the hands of engineers into those of the manufacturers, who furnish wheels ready made, simple, efficient and strong, guaranteed to give 80 percent effective motive power."[83] Yet conditions and requirements were changing rapidly. The 1880s witnessed the rise and extraordinary growth of the electric lighting industry and the practical transmission of power by electricity. By 1890 hydroelectricity was a demonstrated success in Europe and spreading rapidly in the Far West; the foundations of the electric power industry were being laid. Production of hydroelectricity for transmission and sale provided powerful incentives for raising wheel efficiency. The cut-and-try methods of practical wheel builders came under the fire of criticism. In 1894 Joseph P. Frizell, a younger-generation professional hydraulic engineer having much the same background as Worthen, made this strong indictment:

Transactions of the National Electric Light Association, 29th Annual Convention, 1906 (New York, 1906), 1:648 ff.

[82] Arnold Pfau, "Wheels of the Pressure Type," *Transactions of the International Engineering Congress, 1915*, 12 vols. in 13 (San Francisco, 1916), 7:471–72.

[83] William E. Worthen, "Address of the President," p. 5. Worthen (1819–1897) began his career as assistant to Loammi Baldwin in the Boston area and later worked under Francis at Lowell.

At present the business of making turbines is in the hands of mere machinists. Good machinists undoubtedly and men of good judgment and business capacity, but with very little knowledge of the mechanical principles involved in the action of the turbine wheel. I have never yet seen the circular of a wheel maker which evinced any rational conception of the principles of hydrodynamics or any intelligent comprehension of the action of the water upon his wheel. The forms now in the market are the result of experiment uncontrolled by accurate knowledge. They are, it is fully conceded, durable and convenient in design, ready of application, excellent in workmanship and of reasonable efficiency; but I desire to say emphatically that this efficiency could be improved by more thorough knowledge and more intelligent application of mathematical principles. It is customary to say that turbines go by water but it cannot be denied that the sale of water wheels goes largely by wind, and the success of wheel makers depends to a much greater extent upon their skill in the use of the latter agency than of the former.[84]

As will be seen in a later volume, the Niagara project, in 1894 approaching successful completion, revealed in a striking way the wide gap between the methods and achievements of the practical wheel builders of this country and those of the engineering firms that led in the design and building of turbines in Europe. The larger and more progressive of the American wheel-building companies in time responded to the opportunities presented by the new fields of application, especially hydroelectricity, and made appropriate adjustments. Aware of the limitations of their stock in trade for all but the smaller classes of hydroelectric installations, some companies extended the scope of their operations to include wheels of large capacity designed to meet the conditions of specific sites. Methods of wheel design, construction, and selection for direct-drive applications rarely exceeding several hundred horsepower proved quite unacceptable in hydroelectric stations of any scale and importance. There was no significant basis of comparison between such wheels operating under heads of some twenty to forty feet and driving mills and factories directly and

[84] Discussion, Hunking, "Notes on Water Power Equipment," pp. 223–24. Frizell (1832–1910), following textile-mill experience in Manchester, New Hampshire, entered the office of the city engineer at Lowell in 1854 and later became assistant to Francis at the Locks and Canals. A hydraulic engineer of similar but later experience, John R. Freeman, gave high praise to the stock-pattern turbine for its remarkably low cost and durability. "It was by refining [the details of these wheels] that the magnificent turbines of today have been produced and standardized" ("General Review of Current Water Power Practice," pp. 371).

the 5,000-horsepower generating units under 150 feet of head of the initial Niagara powerhouse.

Stock-pattern wheels enlarged manyfold were simply out of the question for large hydroelectric installations. Other considerations apart, the performance of such standard wheels within the narrow limits demanded by large capacity alternators was unpredictable. In the absence of soundly based and tested theory, American turbine builders relying on traditional methods simply could not design turbines to precise specifications of performance. Cut-and-try methods had brought them to a dead end. "The shape of the American runner passage," declared Professor Russell, "is so intricate as to defy mathematical analysis of the water's action."[85] Central station operators became unwilling to hazard large investments on costly equipment that could be brought within performance limits prescribed by generator requirements only by costly and uncertain modifications after installation. The issue was raised inescapably and settled decisively by the course of action adopted in the Niagara enterprise. Preliminary investigations convinced the leaders of this project that in so unprecedented an undertaking there was no alternative to reliance on the very different methods of turbine design and construction practiced in Europe. An engineer prominent in the "new school" of American turbine design described the European experience:

> In Europe, the turbine developed along different lines. There it was the outcome of mathematical investigation and not experiment, and its evolution had been based largely on theory. In Europe one rarely found two turbines alike, for there it had been customary to design each machine for the actual conditions of power, head, and speed governing its installation. All questions directly or indirectly related to the theoretical laws which govern the passage of water through the turbine, had been most thoroughly studied in Switzerland and in Germany. For years machines had been built to order in these countries. Each of the various parts which go to make up the complete turbine had been made the object of the most careful investigation—the wheel casing, the guide vanes, the runner, and the draft tube; with the idea of increasing the efficiency of the unit as a whole by perfecting the various parts of which it is constructed.[86]

[85] Russell, in discussion of Allen, "Testing of Water Wheels after Installation," pp. 302–3.

[86] Henry Birchard Taylor, "The Development of the Hydraulic Reaction Turbine in America," *Engineering Magazine* 38 (1909–10): 842.

Turbine Design after Niagara

The lessons learned from the Niagara experience soon enabled American engineers to break away from dependence upon European engineering practice and models in wheel design. By 1900 the Twelfth Census reported that it had become general practice to build turbines for the particular conditions of each installation's intended use, instead of to standard sizes.[87] By 1915 the period of tutelage was over; and before many years had elapsed leadership in the design of hydraulic turbines, in the view of some American engineers, had passed from Europe to America.[88] A noteworthy feature of these years was the return to prominence of what had come to be known as the Francis turbine. This prototype of the inward-radial-flow, central-discharge turbine had been thrust into obscurity by the popularity of the low-cost, standard mixed-flow wheels that had evolved from it during the 1860s and 1870s. As we have seen, the mixed-flow wheel had obtained the desired results of increased speed and power by reducing the diameter and deepening the wheel runner axially. The goal of a low-cost wheel was attained but at the expense of comprehension and future development. The distortion of the simple flow lines of the Francis wheel complicated design beyond theoretical comprehension and mathematical analysis, although the product met admirably the power needs of most direct-drive users.

While the Francis turbine in its original radial inward-flow design

[87] Edward H. Sanborn, "Motive-Power Appliances," in U.S., Census Bureau, Twelfth Census, 1900, vol. 10, *Manufactures,* Pt. 4, *Special Reports on Selected Industries* (Washington, D.C., 1902), pp. 399–400.

[88] See esp. Taylor, "The Hydraulic Reaction Turbine in America," pp. 841 ff.; idem, *Fifteen Years: The Golden Age of the Francis Turbine in America* (New York, 1945); and Pfau, "Wheels of the Pressure Type," pp. 443 ff., esp. pp. 471 ff. Before coming to the United States, Pfau was, about 1897, in the employ of the leading Swiss engineering firm, Escher-Wyss and Company. In a commentary on the Pfau paper, A. Streiff, the Swiss engineer, declared that "the author might have pointed out more forcibly the prominent part taken by American engineers in the development of the turbine. . . . It is a matter of fact that every marked improvement in the design of pressure-type wheels originated in America." Streiff discussed the nature of these contributions, beginning with Francis's first inward-flow turbine and drawing his most recent illustrations from the Swiss National Exhibition at Berne in 1914 (ibid., pp. 498–500). On the leadership of European turbine designers, 1890–1905, see comments of H. A. Hageman in Safford and Hamilton, "The American Mixed-Flow Turbine," pp. 1298–1302.

was introduced in Europe during the 1870s, it was not until the 1890s that it was taken up by leading European builders, who were attracted by its high part-gate efficiency, superior mechanical design, and low maintenance expense.[89] From Europe it was brought home to America, attracting wide attention from its adoption in a Swiss design as original equipment in Niagara Powerhouse Number 2 in 1902–4; it was to replace the Fourneyron turbines in Powerhouse Number 1 several years later. In the hands of the new school of American turbine designers the Francis turbine underwent refinements in every major component, including especially its setting, long largely ignored in direct-drive installations. The emphasis was upon simplification of parts and arrangement and the attainment of capacities scaled to hydroelectric installations of ever increasing size. By 1915 the Francis turbine had replaced all other wheels of the pressure-and-reaction type in important new installations. For many years to come, in company with the tangential impulse wheel for use with the highest heads, it led the field both in this country and abroad.[90] The Francis turbine, declared Eugene Wittick many years later, is "really a universal machine in that it may be adapted to any head of water from the lowest to the highest." It was used at the Keokuk, Iowa, Mississippi River project with only 13 feet of head and at the Boulder Dam Colorado River project with a head of 510 feet.

The leading problems of turbine design for large hydroelectric plants were those of improving wheel efficiency, obtaining compatibility between wheel and generator, resolving the mechanical and hydraulic problems attending the rapid and continuing increases in wheel capacity, and adapting the turbine for use with heads measured in hundreds of feet. Rarely before Niagara were heads more than forty to fifty feet.[91] Despite the promotional emphasis of stock-pattern wheel builders upon efficiency, this feature was probably not of

[89] See esp. articles by Taylor, Pfau, and Safford and Hamilton just cited; Allen, "Testing of Water Wheels after Installation," pp. 274 ff.; Barr, "Hydraulic Generation," pp. 872–74; Edward D. Adams, *Niagara Power: History of the Niagara Falls Power Company, 1886–1918,* 2 vols. (Niagara Falls, 1927), 2:125 ff., 455.

[90] Taylor, "The Hydraulic Reaction Turbine in America," pp. 843–49; Taylor, *Fifteen Years,* pp. 11–15; Pfau, "Wheels of the Pressure Type," pp. 443 ff.; Eugene Wittick, *The Development of Power* (Chicago, 1939), pp. 19–20.

[91] Horton states, "Stock patterns of the turbines are seldom applied under heads exceeding 60 feet" (*Turbine Water-Wheel Tests,* p. 82). For heads over 60 feet, he noted, "American builders commonly resort to the use of bronze buckets and 'special wheels,' not designed along theoretical lines, as in Europe, but representing modifications of standard patterns" (p. 15).

pressing importance in most direct-drive installations. Even when given careful attention as a means of stretching a limited water supply to meet expanding power requirements, as at the Merrimack River textile centers, efficiency of wheel performance, once the inevitable auxiliary steam plant had been installed, was probably a matter more of professional pride of the engineering staff than of operational or financial importance. Although individual stock-pattern wheels on test gave higher and much publicized results, the efficiencies of the best wheels as determined at the Holyoke flume as late as 1900 gave efficiencies within one percent, plus or minus, of 80 percent.[92] Average working efficiency probably did not exceed 70 percent.

In commercial hydroelectric operations wheel efficiency was of prime importance. The return on the heavy investment represented by the entire complex of dam, reservoir capacity, and power plant was dependent upon the output of wheel installations accounting perhaps for no more than 10 to 15 percent of the aggregate first cost. C. M. Norris in 1924 stated that the permanent structures, as distinguished from the electrical and mechanical equipment, accounted for some 60 to 85 percent of the total cost of a hydroelectrical project. With maximum wheel output as a major engineering objective, wheel efficiency was gradually raised until by 1925 the record figure of 93.8 percent, subject to relatively small impairment with use, was reached, a level only slightly improved upon in the years ahead.[93]

The difficulties of obtaining compatibility between hydraulic turbine and generator centered in the adjustment of rotating speeds and the adaptation of wheel and generator shaft positions required by a direct connection between them. As machines there was a basic similarity between hydraulic and steam turbines, but their operating speeds were far apart. As media for the development, control, and use

[92] Ibid., pp. 20–21. See also Safford and Hamilton, "The American Mixed-Flow Turbine," p. 1271, tables 6–9; Allen, "Testing of Water-Wheels After Installation," p. 276.

[93] C. M. Norrie, "Factors of Efficiency in Hydroelectric Development," *Transactions of the First World Power Conference, 1924,* 5 vols. (London, 1925), 2:227. The maximum efficiency of turbines during the period 1910–25 rose from 88.0 to 93.8 percent, thereafter remaining unchanged until about 1945, when 94.5 percent was reached on an 85,000 horsepower turbine at the Shipshaw development in Canada (Taylor, *Fifteen Years,* p. 11). In 1915 Streiff reported that design advances by American engineers during the previous five years had "jumped the average efficiency of high speed runners five percent or more" (discussion of Pfau, "Wheels of the Pressure Type," p. 499). *Recent Economic Changes in the United States,* 2 vols. (New York, 1929), 1:135, gives figures for test model wheels, 1910–28, ranging from 91.0 to 93.7 percent.

of high pressures, steam and water were very different.[94] Waterwheels had always presented greater difficulties in speed regulation than steam engines, and these were accentuated greatly by the demanding requirements for close governing in central station operations.[95] Yet under the pressure of necessity—and the prospect of gain—these difficulties were overcome by the development of increasingly sensitive and complex regulating devices, such as the Replogle, Lombard, and Woodward governors, with mechanical, hydraulic, or electrical relay systems for operating the gate mechanisms.[96] The 1900 census report on prime movers declared it was at last possible to obtain regulation "practically as sensitive and efficient as the government of the best steam engines," in part owing to improvements in the hydraulic turbine itself.[97]

The problem of matching engine speed and shaft position was for some years resolved by accepting the long established horizontality of the generator and placing the hitherto upright hydroturbine on its side in the manner described earlier. This awkward and inefficient arrangement was by 1920 being reversed in favor of the Niagara solution: the generator was upended in the so-called umbrella dynamo with its vertical shaft, and the hydroturbine resumed its traditional upright stance.[98]

[94] Cassell, "Design and Manufacture of Hydroelectric Installations as a Whole," pp. 648 ff.

[95] Replogle, "Speed Government in Water-Power Plants," pp. 81 ff. There was "a decided tendency on the part of engineers and investors to study the proposition more from a scientific view, laying bare, if possible, the principles that underlie it. These principles then may be used for foundations from which to reason out a solution to the problem." At an early hydroelectric station in the State of Washington, the only satisfactory means of governing required a workman to raise or lower the penstock gate, guided by voltmeters and ammeters and a bank of lights (A. McL. Hawks, discussion of M. S. Parker, "Governing of Water Power under Variable Loads," *TASCE* 37 [1897]: 24).

[96] See B. C. Washington, Jr., "Water-Power Electrical Plants," *JFI* 148 (1899):164 ff.; Barr, "Hydraulic Generation," pp. 876–77; and articles by Replogle, Henry, and Buvinger in *TASME* 27 (1906).

[97] Sanborn, "Motive-Power Appliances," p. 400; Pfau remarks upon the increasing obsolescence of mechanical governors and the great variety of oil-pressure "hydraulic governors" on the market. Some concerns built governors exclusively ("Wheels of the Pressure Type," pp. 479 ff.).

[98] See ibid., pp. 460 ff. and the very detailed and well-illustrated discussion in Mead, *Water-Power Engineering,* chap. 20, pp. 500 ff. See also Sanborn and Martin, *Power Used in Manufactures,* pp. 49–50; and Fanning, "Progress in Hydraulic Power Development," pp. 24–25. Vertical shaft turbines, declared Pfau, are indispensable where large runners are involved ("Wheels of the Pressure Type," pp. 460 ff.). The

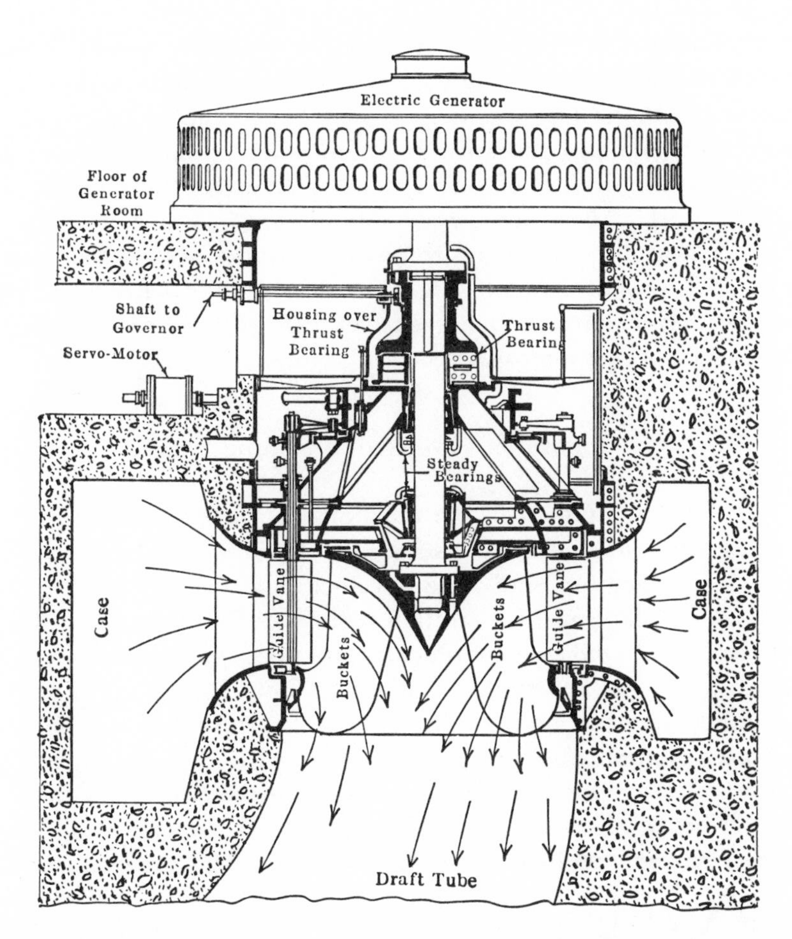

Fig. 92. Umbrella generator direct-connected to vertical shaft turbine with overhead suspension by thrust bearing. (Robert L. Daugherty, *Hydraulics*, 3d ed. [New York, 1925].)

In capacity of generating units the largest hydraulic turbines were not long in approaching the capacity of those driven by steam. By 1905, within ten years of the first Niagara installation, with its unprecedented 5,000-horsepower units, the first 10,000-horsepower unit, prototype of the "modern superturbine," was placed in operation at a Canadian Niagara plant. Less than twenty years later, in 1923, a "large capacity" hydraulic turbine was defined as one of not less than 25,000-horsepower capacity and having a revolving weight of not less than 200 tons.[99] By this date the 50,000-horsepower mark had been passed and the ceiling was not yet in view. In the exultant prose of a prominent engineer, the great hydraulic turbine had become "one of the most marvelous machines ever fabricated by the hand of man," extracting 94 percent "of a great falling current of water in the short space of 3 to 5 feet of travel through its runner within less than one-tenth of a second of time!" And it could do this twenty-four hours of the day, seven days in the week, year after year "with marvelously small cost for care and oversight."[100]

Innovations in Western Mining States

While the development of the reaction, or pressure, turbine was proceeding with such effective results east of the Mississippi, advances in power technology of similar importance and even greater originality were taking place in the mountain regions of the West. Here the basic meteorological and topographical conditions of the eastern United States were just reversed. In place of the abundant rainfall and streamflow and the modest falls of the industrial Northeast, there were low precipitation and aridity combined with the high heads characteristic of the lofty mountain regions from the Rockies to the Pacific coast. The value of the available waterpower was magnified by the general scarcity of fuel resources that made the adoption of steam

modern hydraulic turbine is almost invariably a vertical machine, reported Taylor in 1925 (*Fifteen Years,* p. 10).

[99] George F. Wittig, "Technical Developments," in U.S., Bureau of the Census, *Census of Electrical Industries, 1927: Central Electric Light and Power Stations* (Washington, D.C., 1930), pp. 85–87; also H. G. Acres, "Modern Hydraulic Turbines of Large Capacity," *TASME* 45 (1923): 41 ff. Acres calls attention to the essential auxiliary equipment for the largest class of hydraulic turbines: the Johnson automatic plunger valve and differential surge tanks and the Kingsbury thrust bearing as employed with the 55,000-horsepower generating units at the Queenston plant in Ontario.

[100] Freeman, "Fundamental Problems," p. 534.

power a matter of last resort. The economy of California, oldest region of the Far West in terms of settlement, rested almost wholly on agriculture and stock raising when at mid-century the discovery of gold occasioned the extraordinarily rapid and extensive growth of the mining industries. Next to labor, water was the most critical supply requirement in the mining and processing of the precious metals, whether conducted by individual prospectors working the gold-bearing sand and gravel beds with pan, rocker, or long tom or by corporate enterprise exploiting the quartz lodes by conventional underground methods. In a manner suggestive of the onetime argument from design, the most extensive and rewarding of the gold fields of California occupied the lower western slopes of the great Sierra Nevada ranges, where the water resources were most abundant.[101] In contrast to the meager rainfall in the agricultural lowlands of the extensive central valley of California, a heavy precipitation of snow on these western slopes, deposited by the moisture-bearing winds from the Pacific, formed a power reservoir of great value. A flow of snowmelt from winter through early summer supplied more than a score of tributary streams of the Sacramento and San Joaquin rivers of central California.

If the volume of streamflow was unimpressive by eastern standards, it was accompanied in many districts by gradients of from 100 to 250 feet per mile, in sharp contrast with slopes of five to ten feet per mile characteristic of the best waterpower streams of New England.[102] Where in the East heads of as much as 40 to 50 feet were uncommon, here and in somewhat lesser degree in many other mountain states from Colorado to Nevada and Idaho, heads were typically measured in hundreds of feet, not infrequently passing the thousand-foot mark. While subject typically to wide seasonal variations of flow, here was a great power resource especially welcome in fuel-scarce regions.

The waterpowers of the gold regions were exploited with great skill and imagination, as we shall note in a later volume, by a mining population recruited in no small part from the eastern states, where waterpower had long been used to advantage. Water was employed not only in the conventional manner by means of waterwheels, but its kinetic energy was applied directly by means of sluices and pipes to raise water, elevate sand and gravel, operate powerful cranes for dis-

[101] See esp. Titus F. Cronise, *The Natural Wealth of California* (San Francisco, 1868), pp. 596–97; Hamilton Smith, Jr., "Water Power with High Pressures and Wrought-Iron Water Pipe," *TASCE* 13 (1884): 15 ff.; Augustus J. Bowie, *A Practical Treatise on Hydraulic Mining in California,* 8th ed. (New York, 1898), chaps. 3, 8.

[102] Bowie, *Hydraulic Mining in California,* pp. 62 ff.

posing of boulders, and through the techniques of ground sluicing and booming to mechanize the prospector's manual methods of washing out gold by pan and rocker. The climax in the direct application of the power of water under high head to the handling of gold-bearing gravels was reached in hydraulic mining, in which hills were leveled and mountain sides demolished. As the tide of frontier improvisation receded and the boom moved hopefully to new regions and as placer mining gave way to hard-rock operations underground, two innovations survived as a permanent addition to hydraulic technology and prime movers: the rude but effective hurdy-gurdy waterwheel and the ditch system of high-head water supply.

What was variously termed the tangential, impulse, or, in common usage, Pelton waterwheel was developed from the hurdy-gurdy wheel during the 1860s and 1870s. It was a simple and efficient wheel of large capacity which took its place along with the established pressure, or reaction, turbine as a leading wheel type.[103] The tangential wheel was a response to the distinctive combination of circumstances found in the mining regions of California and neighboring states as far eastward as Colorado. The primary factor in its development was the gradual but cumulatively decisive shift from the near-surface operation of placer mining to the accumulated beds and bodies of auriferous gravels and then to the underground mining of hard-rock ores, with its growing demands for power in the materials-handling operations of hoisting and drainage. The minute amounts of metal present even in the richest of ores made necessary their reduction to a fine powder, a process accomplished in so-called stamp mills located at or near the mine mouth. Further processing by chemical or mechanical means to extract the fine metal often required as much or even more power than the operation of the mine itself. Other major factors in the wheel's development were the scarcity and cost of fuel in most mining districts, the difficulties and expense of importing and operating steam engines and boilers, and the rapid decline of hydraulic mining in the 1880s, owing to the stringent legal restrictions placed upon its practice. This last factor made available for deep mining and stamp-milling operations the water resources of the ditch companies hitherto largely used in hydraulic mining.[104]

[103] Perhaps the earliest competent professional account of the tangential wheel is Smith, "Water Power with High Pressures," pp. 15 ff.

[104] Although the stamp mill was an indispensable adjunct of the mining of precious metals and some base ones, particularly copper, it was more akin to the factory than the mine and is given only incidental attention here. See *Appleton's Cyclopaedia of*

The widespread availability of high-head waterpower in combination with the scarcity and high cost of fuel for steam power made these two factors largely controlling in the development of the tangential waterwheel. In a number of important mining regions of the Far West, including the ranking Comstock Lode in Nevada, power accounted for 30 to 50 percent of total mining and milling costs. Coal deposits in most of this vast region were small, scattered, and mostly unsuited to development; petroleum fields along the Pacific coast were not brought into production until the 1890s and were largely out of reach. Wood varied in amount from district to district. Even when initially abundant, the timber under the heavy joint demands for fuel and pit props was rapidly depleted within the immediate environs. Costs rose rapidly under conditions of wagon haulage over poor roads through rough country. Cordwood prices during the 1860s and 1870s ranged several times as high as in the industrial Northeast, where in these years railroads were making the shift to coal as locomotive fuel as rapidly as feasible. On the Comstock Lode in 1866 wood consumption for all purposes was over 200,000 cords at an average cost to mine operators of $10 a cord, with retail prices for local consumption ranging, according to the season, between $15 and $30.[105]

Even in an area where forest-covered hills made wood cheap in the early 1880s and a round price had to be paid for ditch water, the much lower cost of waterpower warranted a changeover from steam. Twenty years later A. P. Brayton reported that if a mining company lacked a

---

Applied Mechanics, s.v. "Stamps, Ore," and *Appleton's Modern Mechanism,* s.v. "Ore Crushing Machines." For early quartz mining and milling processes in stamp mills in California, see *Mining Magazine* 1 (1853): 145–48. See also James Douglas, "American Improvements and Inventions in Ore Crushing and Concentration," *TAIME* 22 (1893): 321 ff.; T. A. Rickard, "Limitations of the Gold Stamp Mill," ibid. 23 (1894): 136 ff. For early stamp-milling methods in the copper region of Lake Superior, see John F. Blandy, ibid. 2 (1873–74): 208 ff., and Charles H. Rolker, ibid. 5 (1876–77): 584 ff. An account of the ditch system as it developed in the mining regions of the Far West as a source of water and power supply in mining appears in a later volume.

[105] A. P. Brayton, "Water Power in Mining," *Cassier's Magazine* 22 (1902): 429 ff.; Eliot Lord, *Comstock Mining and Miners* (1883; reprint ed., Berkeley, Calif., 1959), pp. 116, 202–4, 256–57, 348–51 (tabular data). When a short local railroad was built in 1869 to ease the waggoning problem between mines and mills on the Comstock Lode, cordwood prices at Virginia City fell from $15.00 to $11.50 (ibid., pp. 249 ff.). One probably not wholly representative example was the itemized cost of pumping in the course of shaft-sinking at the Ophir Mine, Comstock Lode, January 1872. Of an aggregate monthly cost of $2,483, covering wages, fuel, lubricants, and interest on cost of pumping machinery installed at 1 percent, 141 cords of wood at $10 each cost $1,410 (*Report of the Commissioners on the Sutro Tunnel* [Washington, D.C., 1872], p. 809).

fuel supply in its own timbered land and had to purchase fuel, the annual cost per horsepower might run as high as $250, several times the figure paid in fuel-scarce New England for steam power.[106] Transport difficulties and expense often discouraged if they did not prevent bringing in steam engines and boilers, whose smallest components weighed hundreds or thousands of pounds. Freight rates ranged from one to six cents per pound about 1880.[107]

From the Hurdy-Gurdy to the Tangential Waterwheel

While the states of the Far West led the nation in the abundance of waterpower resources, the low volume and high heads characteristic of the region were poorly adapted to the old-style waterwheels and the reaction, or pressure, turbines employed in the eastern United States.[108] Except for limited purposes and small powers, undershot and overshot wheels were ill suited to use with high heads. Turbines developed in response to conditions prevailing in the East initially gave quite unsatisfactory service, although eventually design changes permitted their use with medium-high heads. Part of the difficulty was owing to the necessity of gearing down to the lower speeds necessary for direct-drive use in hoisting, pumping, and stamp milling. A more immediate and serious problem arose from the sand and silt in the water supplied by mountain streams in flood, causing heavy wear, declining efficiency, and a short working life. This wear, declared Hamilton Smith in 1884, was so objectionable that of nearly 800 wheels in operation in California in 1884, there was not, to his knowledge, a single turbine at work.[109] Together with high maintenance and repair

[106] Smith, "Water Power with High Pressures," pp. 25–27; Brayton, "Water Power in Mining," pp. 429 ff.

[107] U.S., Census Office, Tenth Census, vol. 13, *Statistics and Technology of the Precious Metals* (Washington, D.C., 1885), p. ix. Freight costs were especially high in the early development of a mining district. See Rossiter W. Raymond, *Mineral Resources West of the Rocky Mountains* (Washington, D.C., 1869), pp. 141–42, citing 37 cents a pound for milling machinery freighted across the plains to Montana Territory; for freight to the Comstock Lode from California, see Lord, *Comstock Mining and Miners,* 190 ff.

[108] Except for traditional applications in grist- and sawmilling, overshot and undershot wheels appear to have found relatively little use in mining regions. Two important exceptions widely employed in riverbed mining were current and dip wheels. For their construction and use, see California, State Bureau of Mines, *Ninth Annual Report of the State Mineralogist* (Sacramento, 1890), pp. 265 ff.

[109] Discussion following Smith, "Water Power with High Pressures," p. 38. A fellow engineer in California added, "I do not know of a single one."

costs these circumstances gave turbines a bad name in the mining regions.

The tangential wheel as developed from the hurdy-gurdy wheel of the placer-mining frontier proved almost ideally suited both to the conditions of high-head water supply and to the requirements of mining and milling the precious metals. The small volume of flow that made possible the conveyance of high-potential water for miles across rugged mountain and piedmont terrain by ditching, fluming, and wooden or iron piping also made possible the delivery of water directly at the waterwheels without resort to dams, canals and related structures. The ditch system as developed extensively in California and other mining states was the western analog of the commercial system of power distribution and supply under low heads employed so advantageously throughout the Northeast. The course and extent of its development are reserved for a chapter in a later volume.[110]

Except for efficiency in the use of water—the best built wheels of the type reportedly developed no more than 40 percent—the hurdy-gurdy wheel possessed most of the basic characteristics needed in a prime mover. So called from resemblance to the musical instrument by which the hardships of life in mining communities were in some part relieved, the hurdy-gurdy wheel was initially a narrow vertical wheel of perhaps ten feet in diameter and less than a foot in width. Arranged about the circumference at close intervals were numerous "buckets," or cavities, in profile resembling straight-cut sawteeth, measuring but a few inches in depth and width and enclosed on either side by wooden shrouding. The wheel was placed in operation when a stream or jet of water under considerable head and pressure was directed tangentially against its periphery, striking the buckets in rapid succession. In its original form this wheel was simply a crude but effective impact wheel, deriving no advantage from the force of reaction. The early hurdy-gurdies employed heads of no more than forty or fifty feet; water was supplied through square wooden pipes, bolted or clamped together. As more power was required, heads were increased, wooden pipes were replaced by metal ones, and wooden by metal buckets. Nozzles were improved and additional power secured either by adding nozzles directed at different points on the periphery of the wheel or by mounting a second wheel on the same shaft.

As improved upon by wheel designers and builders in west coast foundries and machine shops, the hurdy-gurdy evolved into a new

[110] A comprehensive quantitative review of hydraulic mining in 1880 is Tenth Census, *Statistics and Technology of the Precious Metals,* pp. 187 ff.

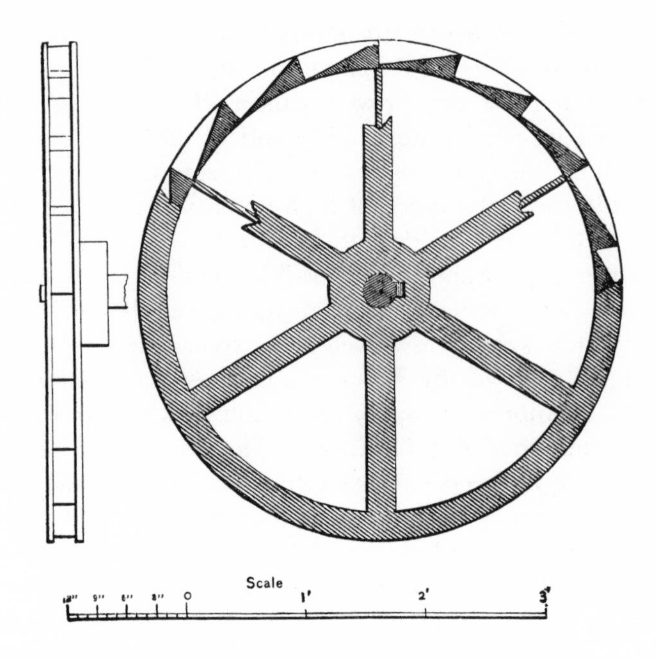

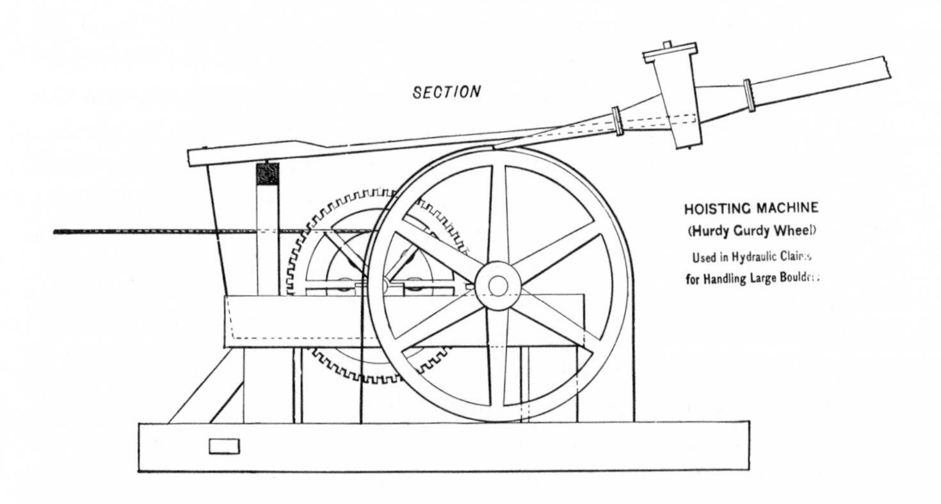

Fig. 93. Hurdy-gurdy wheel in cross section and in use with derrick hoist. (Augustus J. Bowie, Jr., *A Practical Treatise on Hydraulic Mining in California* [New York, 1885].)

and important hydraulic motor, the tangential wheel. Simplicity in construction, installation, and maintenance, flexibility in location and operating characteristics, ease in transportation and handling, and a low delivered cost accounted for the popularity and widespread introduction of the new wheel. Tangential wheels required neither masonry foundations, wheelpits, close-fitting casings nor ironwork settings and draft tubes.[111] A simple, inexpensive timber structure on which to mount the wheel was usually adequate; a rough wooden casing was sufficient to confine the splashing water. All parts were immediately accessible for inspection, repair, and replacement; maintenance of alignment presented little difficulty. Rotating speed was largely determined by the wheel's diameter; even with very high heads rpm could be kept relatively low by increasing the size of the wheel. The customary horizontal position of the wheel shaft simplified the mounting of the wheel, avoided problems with step bearings, and eliminated the need for the costly, power-wasting bevel gearing long required by turbines. In wheels of large size and weight running at high speeds, a steadying flywheel effect was obtained. Since the tangential wheel's efficiency was largely independent of its size, the mounting of several wheels on the same shaft made possible the desired relationships of rpm and power output. The relatively small size of supply pipes gave much freedom in the location of the wheel within the plant. Buckets were bolted on and could be readily replaced when worn. There was, finally, in the words of Hamilton Smith, Jr., "almost absolute immunity from accidents, the wear and tear being practically *nil*."[112]

The development of the tangential wheel followed much the same course as that of the American mixed-flow turbine east of the Mississippi during about the same years, the mid-fifties to the early eighties. The power requirements of precious-metal mining increased tremendously with the shift from placer to what was variously termed hard-rock, quartz-lode, or deep mining. Hoisting and drainage requirements, which reached massive proportions in the largest and

[111] Smith, "Water Power with High Pressures," pp. 16–17. The only situation in which the turbine could operate to better advantage than the tangential wheel, in Smith's view, was when the wheel was apt to be submerged by backwater. Although the turbine could be used with high heads, the small size and accompanying high pressures gave rotating speeds too high for most uses except by belting or gearing down to ordinary operating speeds (U.S., Census Office, Eleventh Census, 1890, vol. 7, *Mineral Industries in the United States* [Washington, D.C., 1892], p. 148). The fully developed tangential wheels operated with an efficiency a little lower than the turbine.

[112] Smith, "Water Power with High Pressures," p. 17.

deepest mines, were commonly much exceeded by the reduction of the ores preliminary to extraction of the metal. Stamp mills multiplied and spread rapidly through the mining regions; the thunder of batteries of cam-operated iron stamps, each weighing several hundred pounds, contrasted sharply with the mild cacophony of picks, shovels, and running water of placer operations. A correspondent of the eastern-based *Mining Magazine* in 1856 remarked upon the rapid rise of this highly mechanized version of an ancient device: "Where now our [quartz] mills are counted by the dozens, they will soon number hundreds. These quartz mills are to California what the cotton and woolen mills are to New England."[113]

The demand for greatly increased supplies of power was accompanied by a concern for its cost and the efficiency of its generation and use. The improvised on-site, rough-hewn, and often jerry-built hurdy-gurdy wheels gradually gave way to the engineered, factory-built, and field-assembled tangential wheel, possessing the same essential characteristics but greatly improved in respect to efficiency, capacity, and durability. The necessity for precision in design, strength in construction, and durability in use led to the replacement of wood by iron, first in the buckets and eventually in all components. While ideas for wheel improvement often originated in the field, ironworks in San Francisco and the larger mining-supply towns came to play the leading role in resolving problems of design and construction.[114] The design of the nozzle was of critical importance for the control and direction of a water jet under high pressure, but principal attention early centered on the improvement of the wheel buckets.

The rude, angular—sawtooth—wooden buckets of the hurdy-gurdy wheel were early recognized as a source of power loss attending the

[113] *Mining Magazine* 7 (1856): 414–15. Within two years of the beginning of the mining rush to the Comstock district, roads from California were clogged with wagon loads of engines and mill machinery bound for stamp mills springing up along the watercourses. The first stamp mill was completed in the summer of 1860. By the end of 1861 seventy-six had been erected and twenty more were building or planned (Lord, *Comstock Mining and Miners,* pp. 113–15).

[114] This development is carefully traced by W. A. Doble, "The Tangential Water-Wheel," *TAIME* 29 (1899): 857 ff. See also John Richards, "Notes on a Problem in Water Power," *TASME* 13 (1891–92): 331 ff.; F. F. Thomas, "Water-Wheels," in California, State Bureau of Mines, *Eighth Annual Report of the California State Mineralogist* (Sacramento, 1888), pp. 785 ff. Hamilton Smith, Jr.,'s article, "Water Power with High Pressures," covers supply-pipe design and construction and intraplant transmission as well as early wheel developments. Smith was a distinguished hydraulic engineer in California for some years from 1868. See DAB.

interference between the entering and escaping water. The eventual goal of wheel designers was to secure the greatest advantage both from impact by the entering stream and from reactive force in leaving the wheel bucket, with minimal interference in the process.[115] The first recorded incident in this development was the building at a San Francisco ironworks in 1866 of a "cast-iron wheel" for a sixteen-stamp (medium-scale) mill at the Swin mine in nearby Calaveras County, which with adjacent Amador County served as a kind of proving ground for the emerging tangential wheel. The marked superiority in performance of this wheel over the hurdy-gurdy, in the opinion of W. A. Doble, a wheel designer and builder, "proved to be the turning point in the building of this class of wheels."[116]

For the next twenty-five years wheel modification via the familiar "cut-and-try" route brought continued improvement in wheel performance. More than a score of tangential wheels were brought to the point where the names of their builders got into the record. The more successful ones, in approximate order of introduction, were the Collins (1866), Moore (1874), Knight (1875), Pelton (1880), Donnelly (1882), and Dodd (1889). Developed over some five years and patented in 1880, the Pelton wheel gave its name to the type; whether strictly on the basis of superior merit in efficiency and performance, claimed priority of achievement, or as a result of the production capabilities and commercial drive of its inventor is not entirely clear. In the absence of a commercial testing flume in the area, Pelton constructed a Prony brake and weir to measure power and water volume, with assistance from Francis's *Lowell Hydraulic Experiments*, and tested a succession of experimental wheel models. "I crossed the plains from Ohio in 1850," he wrote in 1897,

> and engaged in mining almost continuously until 1864, when I took up millwrighting, in connection with mining, at Camptonville, Yuca county, and other places north of that town, in which business I was employed until 1878; and during this period I constructed a number of water wheels, of the type commonly known as hurdy-gurdy wheels, having an efficiency of 40 percent and upwards, according to the style of buckets used. Here I conceived, was a chance for improvement; and early in 1878 I procured the necessary appliances for testing the efficiency of buckets for pressure- or jet-wheels, and devoted most of the time for two years following to designing a bucket which

[115] See Smith, "Water Power with High Pressures," p. 37.

[116] Doble, "The Tangential Water-Wheel," p. 858. see also J. D. Galloway, "Hydraulic Power Development and Use," *Transactions of the International Engineering Congress, 1915,* 12 vols. in 13 (San Francisco, 1916), 7:369.

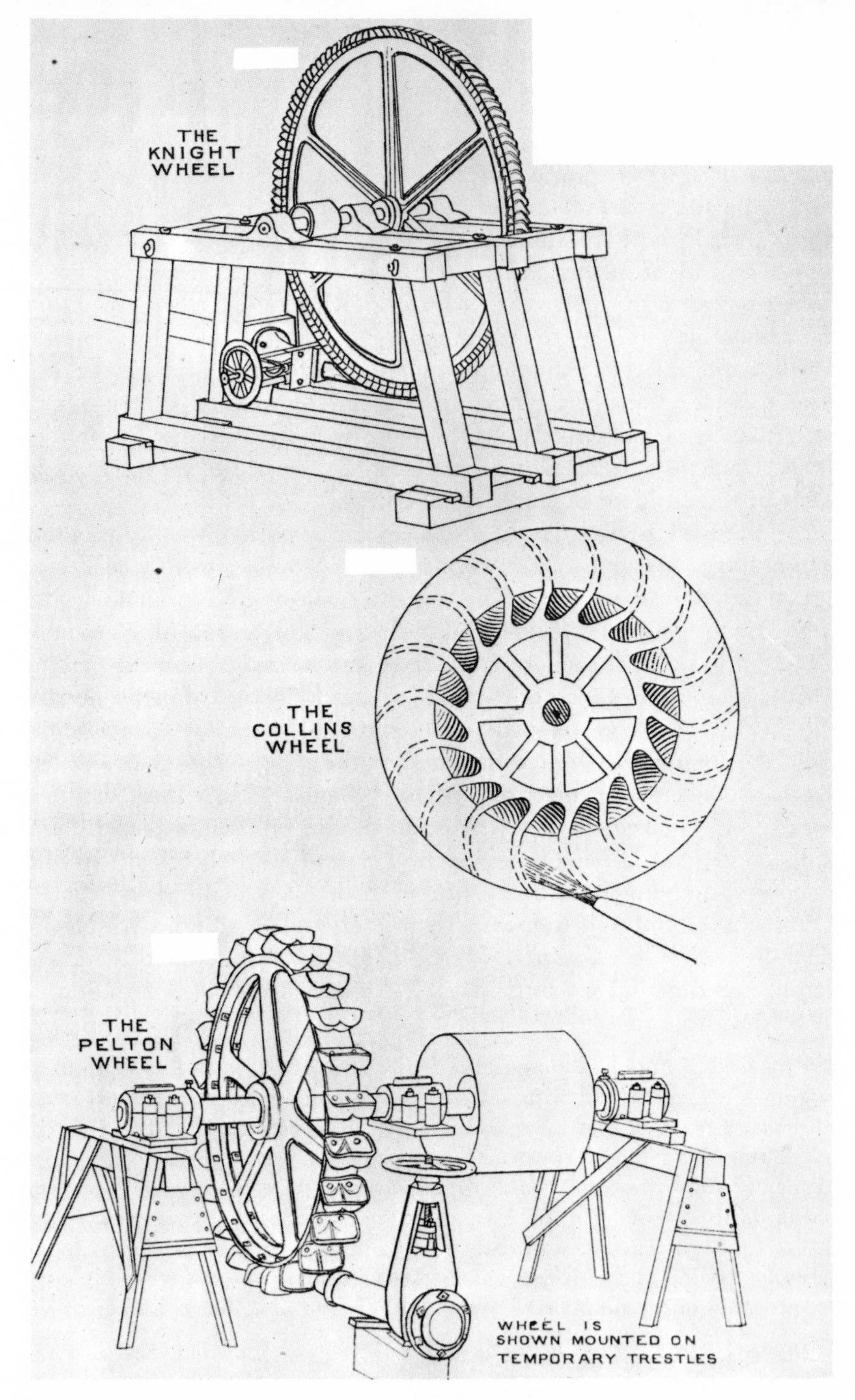

Fig. 94. Three types of tangential wheels. (Hamilton Smith, Jr., "Waterpower with High Pressures and Wrought Iron Water Pipe," *TASCE* 13 [1884].)

would give a higher efficiency. I tested between thirty and forty different shapes of buckets, and finally noticed that a curved bucket having a jet strike on the side . . . instead of in its center . . . gave a marked increase in the efficiency of the wheel, but caused an end thrust against one bearing. To avoid this I experimented with placing the buckets alternately . . . when it was but a step to combining the two curved buckets [on opposite sides of the wheel rim] and splitting the stream. . . . This bucket, when tested, gave such astonishing results that I immediately took steps to secure my invention . . . obtaining a patent in October, 1880.[117]

The critical element in the most successful wheels proved to be a central knife-edged wedge, or ridge, termed the "splitter." This divided the entering jet into two streams, which in following the contours of the bucket were gradually reversed in direction and discharged on opposite sides of the wheel, their energy largely spent through impulse followed by reaction (Fig. 95). The issue of priority of invention tended to center upon the adoption of the jet-splitting ridge. The available evidence indicates that in two and possibly three other wheels its introduction preceded that in the Pelton wheel. The relative merits of the different "splitters" in buckets of varying forms seem never to have been subjected to objective professional test, allowing each claimant the comfort of self-assurance. The evidence assembled in the 1890s by W. A. Doble gives weight to his conclusion, which finds support in many areas of technological change, that a great many Californians—mechanics, millwrights, owners and superintendents of ironworks, and the like, including even professors—participated in the long development and contributed variously to the end result: a waterwheel that came nearly as close as the reaction, or pressure, turbine to realizing the long-proclaimed goal of waterwheel design, a wheel in which the water would enter without shock and leave without velocity.[118]

[117] Quoted in Doble, "The Tangential Water-Wheel," p. 863. As early as 1853 an American, Jearum Atkins, applied for a patent, not issued to him owing to an intervening illness until 1873. This wheel incorporated what has been termed "the true principle" underlying the operation of the tangential wheel: a combination of impulse and reaction (ibid., pp. 853–54). Doble in 1915 attributed the "Pelton" designation of this type of wheel to the circumstance that Pelton "developed" the characteristic dividing wedge and "was the first to develop the wheel from the commercial standpoint" (W. A. Doble, Chief Engineer [note this], Pelton Water Wheel Co., "Water Wheels of the Impulse Type," *Transactions of the International Engineering Congress, 1915,* 12 vols. in 13 [San Francisco, 1916], 7:503).

[118] Doble, "The Tangential Water-Wheel," pp. 865, 890; see also Smith, "Water Power with High Pressure," pp. 16–22. Smith regarded the Pelton wheel as superior in point of efficiency to the various competing wheels (ibid., pp. 20–21). An extended

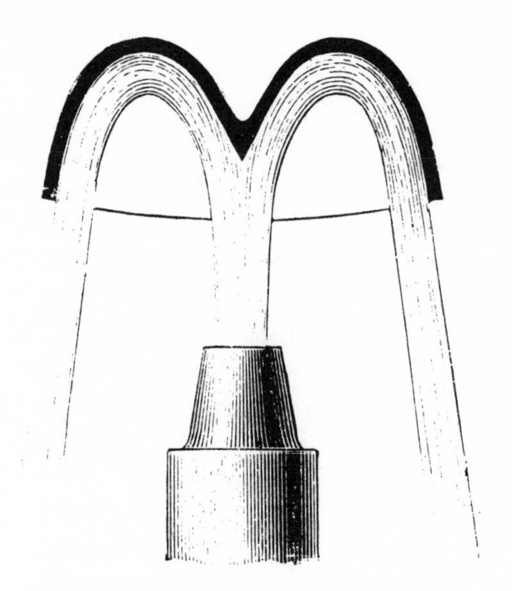

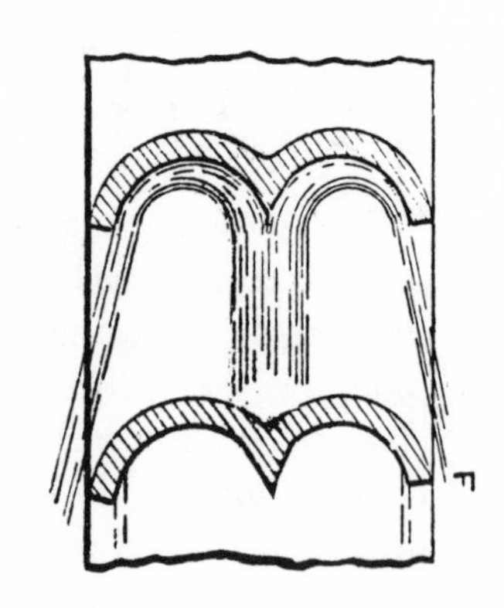

Fig. 95. Action of water in bucket with jet-splitting ridge. ("The Pelton Water Wheel," *JFI* 139 [1895].)

The efficiency of the improved wheels rose from the 40 percent attributed to the better hurdy-gurdies of the 1860s to 80 percent and higher. The *Report of the California State Mineralogist* in 1888 claimed 90 percent efficiency under special conditions and 85 percent "as an everyday running proposition."[119] Ten years later Doble was content with the more modest claim of "over 80 percent," a figure supported by eastern laboratory experiments on stock wheels.[120] The basic simplicity of the wheel itself—horizontal shaft, moderate diameter, buckets of small dimension, supporting framework with two bearings and loose-fitting housing—underwent no essential change. In a manner reminiscent of the engineering world's belated recognition of the Cornish pumping engine in Britain more than half a century earlier, the tangential wheel was slow to arouse interest in the eastern United

discussion of the design and role of the "splitter" and other elements in bucket design is found in Doble's account; see also Galloway, "Hydraulic Power Development and Use," p. 369.

[119] *Eighth Annual Report of the California State Mineralogist,* pp. 790 ff.

[120] Doble, "The Tangential Water-Wheel," pp. 790 ff. See also Committee on Science and the Arts, "The Pelton Wheel," *JFI* 140 (1895): 173, and tests at the U.S. Naval Academy cited ibid., pp. 196–98.

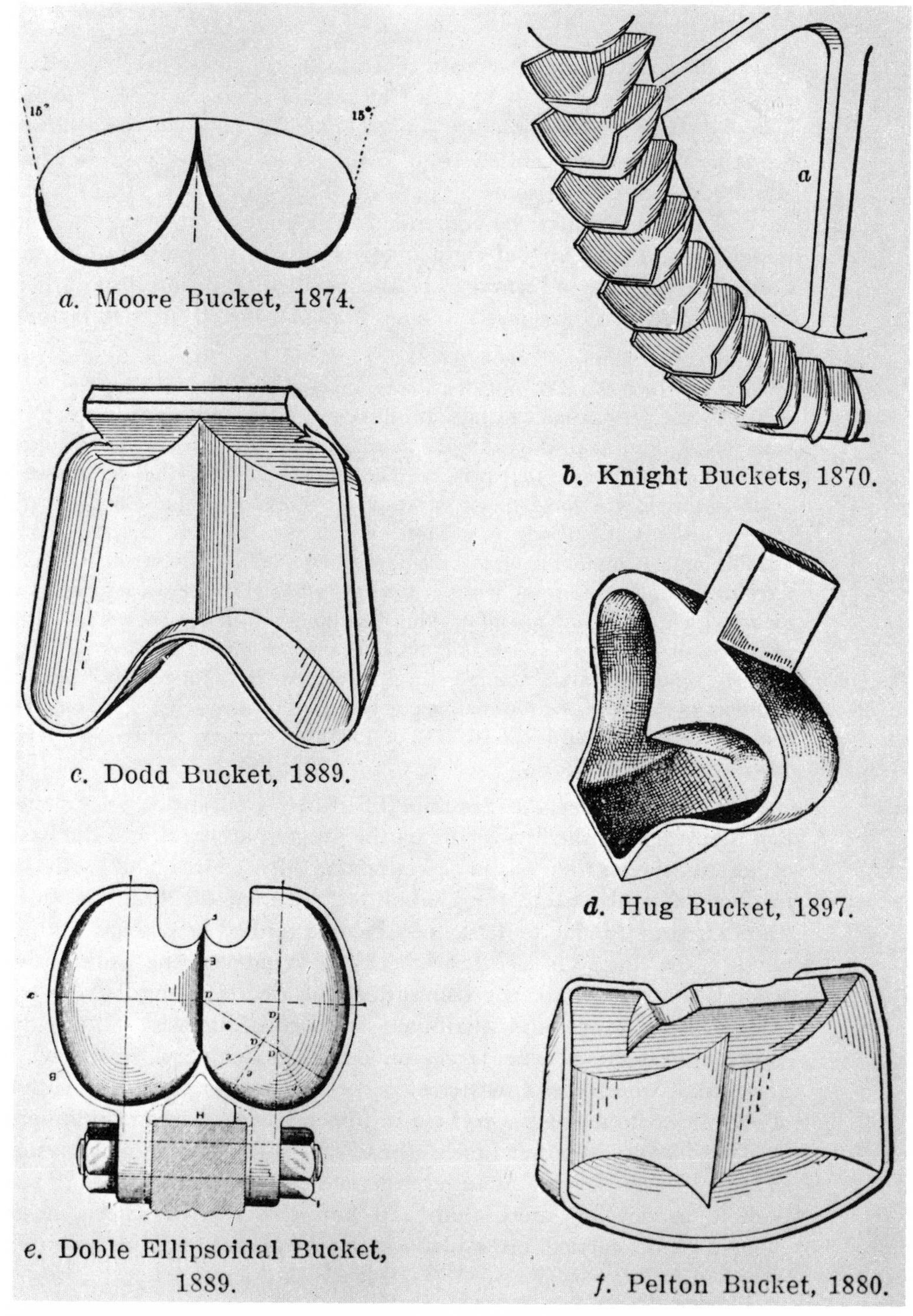

Fig. 96. Buckets used on different makes of tangential wheels. (Daniel W. Mead, *Water Power Engineering* [New York, 1908].)

States. In a brief paper before a meeting of the associated "mechanicals" at San Francisco in 1892, a Californian engineer chided his fellows for their conventionality in recognizing only the traditional forms of turbines descended from Fourneyron, Jonval, and Boyden, admittedly comprising some 80 percent of all waterwheels in use. The force of spouting water, he contended in introducing the case for the impulse wheel, was virtually equal to its gravity or pressure. Only in California had the advantages of the impulse, or tangential, wheel been grasped and exploited. These hardly suffered in itemization:

Figuratively speaking, when a wheel is changed from the pressure to the impulse system it is taken out of its case, mounted in the open air, in plain sight. All the various inlet fittings are dispensed with and are replaced by a plain nozzle and stop valve. Its diameter is made to produce the required rotative speed, whatever that may be. The shaft and its bearings are divested of all strains except those of gravity and the stress of propulsion when the water is applied at one side only. Most important of all there are no running metallic joints to maintain against the escape of water, no friction and no leaks; there are, indeed, no running joints or bearings whatever, except the journals of the wheel shaft . . . . there are no working conditions which involve risk or which call for skill. If a vane [bucket] is broken, another one is applied in a few minutes' time. If a large or small wheel is wanted, the change is inexpensive and does not disturb the foundations or connections. Capacity is at complete control; the wheels can be of 10, 100, or 1,000 horsepower, without involving expensive special patterns.[121]

Several years later the Franklin Institute's Committee on Science and the Arts published its report on the tangential wheel. On the basis of unpublished criteria and investigatory procedures and with no mention of rival wheels, the Committee awarded the Franklin Institute's Cresson Medal to Lester A. Pelton, termed somewhat ambiguously "the inventor of this wheel."[122] Whatever the substantive grounds for the award, the committee was clearly enthralled by the wheel's performance and attributed large consequences to its widespread introduction. The large amount of machinery driven by a quite small wheel was a matter of perpetual wonder. The application of several jets to a single wheel ten to fifteen feet in diameter produced 3,000 to 5,000 horsepower under a head of 150 feet. In the not unusual situation where several hundred feet of head were available, it was possible to develop more than 100 horsepower with wheels small enough to be carried on muleback, an important consideration in

[121] Richards, "Notes on a Problem," pp. 331–36.

[122] Committee on Science and the Arts, "The Pelton Water Wheel," p. 184.

mining country. The committee cited two striking examples. Six tangential wheels were installed at one of the Comstock mines in Nevada. Each developed 125 horsepower at 900 rpm under a head of 1,680 feet; yet each wheel was but 40 inches in diameter and weighed 220 pounds. At a light and power station in Aspen, Colorado, eight Pelton wheels, each 24 inches in diameter and weighing 90 pounds, developed 175 horsepower each at 1,000 rpm under 820 feet of head, or about two horsepower for each pound of weight. In 1890 the price of a six-foot wheel, fully equipped for installation and capable of developing 24 to 755 horsepower under heads of 50 to 500 feet, ranged from $400 to $550.[123] "Scores of the largest producing and most profitable mines on the Pacific Coast," declared the committee's report, "could not be worked today but for this [wheel]." Further evidence of "the change wrought by the introduction of these wheels, is the high price that low-grade mines are commanding" when a few years ago these mines "could hardly be given away." Pelton Company literature reported 840 companies using its wheels, many running ten to fifteen wheels under heads varying from 18 to 2,100 feet.[124]

The coastal drainage basins of central California were the principal scene of these pioneer developments in high-head waterwheels. The area was favored by the extensive ditch systems, an important innovation originating in the requirements of hydraulic mining. In fact, the distribution of mining ditch systems is suggestive of the distribution of the tangential wheels they chiefly served. Of a national total of ditch lines reported in a Twelfth Census survey—175 in number and 3,049 in miles—California and Oregon together accounted for 121 ditch systems and 2,333 miles of ditching. Idaho, Montana, Colorado, and the Dakotas contained the remaining systems, and in the East the state of Georgia accounted for most of the 9 ditch lines and 141 miles of ditches.[125] Little known outside its areas of use and suffer-

[123] Ibid., pp. 164 ff.; *Trade Catalog of the Pelton Water Wheel Company, The Pelton Water Wheel,* 3d ed. (San Francisco, 1890); and *Trade Catalog of the Oakland Iron Works, The Tuthill Patent Water Wheel* (Oakland, Calif., ca. 1900), pp. 79 ff.

[124] Committee on Science and the Arts, "The Pelton Water Wheel," pp. 168–69; Eleventh Census, *Mineral Industries,* pp. 106–7; Doble, "The Tangential Water-Wheel," pp. 890–91. The 1890 Pelton trade catalog reported more than 900 Pelton wheels in operation, representing every variety of service. The builders guaranteed 85 percent efficiency under all heads.

[125] See Tenth Census, *Precious Metals,* p. 209 and tables; also Introduction by Clarence King, pp. vi ff. See also lists of numerous high-, medium-, and low-head installations in the United States and Europe, with data on size, capacity, and so on, in Galloway, "Hydraulic Power Development and Use," pp. 402–13.

Fig. 97. Various forms and arrangements of Pelton wheels. ("The Pelton Water Wheel," *JFI* 140 [1895].)

ing perhaps in general estimation from the somewhat derogatory implications of the the term, the ditch system was an indispensable feature in the effective use of high-head waterpower. Its creation was an impressive engineering and practical achievement that bears comparison with other constructional feats on the mining frontiers of the West.

Eventually the use of the tangential wheel spread to Europe. There the concept of the impulse wheel had long been familiar in hydraulic engineering, but previous to the complicated Girard wheel had not led to a practical wheel of wide utility.[126] Not until the development of

[126] On the Pelton wheel as a replacement for the Girard impulse wheel in Europe in

hydroelectric power in this country did the tangential wheel begin to realize its full potential. The Pacific Coast led the way in this new field, having most of the hydroelectric installations made before 1900. Under conditions of central station work that were far more demanding than at mines or mills, generating units increased rapidly in heads and capacity. By 1915 two Pelton-Doble generating units, each developing 20,000 hp at 360 rpm under a 1,330-foot head, were in operation at the Drum central station on the south Yuba River in California. The waterwheel with its casing occupied but a small fraction of the space of the generator driven by it.[127] At another California central station two generating units were installed in 1919, each rated at 30,000 horsepower under an effective head of 1,008 feet. A wheel barely 13 feet in diameter operating at 171 rpm was driven by a jet of water eleven inches in diameter.[128] The efficiency of such tangential wheels was only a few percentage points below that of pressure turbines of comparable capacity. There was, in short, an important carryover from the pioneer age of development dominated by the requirements and conditions of the mining industries to the far more diversified and sophisticated requirements of the hydroelectric age ahead.

In Brief Summation

Although the second half of the nineteenth century witnessed the gradually accelerating shift from waterpower to steam power in industry, technologically the period was a creative one. There were both opportunity and stimulus for introducing substantial improvements in traditional waterpower facilities and equipment, chiefly in prime movers. The early American turbines, built as singles to meet the needs of large textile mills and without due regard for costs, were the work of such engineers as Boyden and Francis, proceeding with great care and disciplined thought, employing precise measurements and calculations, checked by experiments. Their wheels were cum-

1915, see H. Zoelly, "Developments in Modern Water Turbine Practice," *Transactions of the International Engineering Congress, 1915,* 12 vols. in 13 (San Francisco, 1916), 7:433. See also J. Allen, "Hydraulic Engineering," in C. Singer et al., eds., *History of Technology,* 5 vols. (Oxford, 1958), 5:531–38.

[127] For the tangential wheel as applied in American hydroelectric stations, see Doble, "Water Wheels of the Impulse Type," pp. 503–39 and illustrations.

[128] F. Nagler, "The Cross-Flow Impulse Turbine," *TASME* 45 (1923): fig. 1.

Fig. 98. Runner for a Pelton turbine developing 25,000 horsepower, built for the Hetch Hetchy plant, City of San Francisco, 1925. (Photograph courtesy of the Smithsonian Institution.)

brous, costly, and slow moving, ill suited to general industrial use. During the 1850s the engineers were superseded by the wheel builders, practical men whose goals were high performance and low costs with slight concern for the reasons why. Aided by a spreading railway network and a rapidly widening market economy, these manufacturers developed by wholly empirical methods turbines that progressively superseded the traditional waterwheels. The American mixed-flow turbine—superior in durability, compactness, and cost; better adapted to conditions of streamflow; more economical in water use; and turned out in quantity and standard sizes—in its varied forms met admirably the needs of small- to medium-scale manufacturers. To such creative mechanic-businessmen as Swain, Risdon, Leffel, and McCormick—successors to the Wings, Parkers, and Howds of an earlier generation—must be added the name of the caustic-tongued former sailor James Emerson of Lowell and Holyoke, whose pioneer turbine-testing flume and reports on wheel performance—emulated eventually by other leading builders—contributed greatly to the perfecting of the mixed-flow turbine and the success of the standard-pattern wheel-building industry.

During these same years in the western mountain states the same practical gifts of technological adaptation and improvisation were operative. The urgencies of survival and getting rich quickly led to the development of the direct use of hydraulic energy in the varied forms of placer mining and to the revival of ancient wheel forms rarely seen in the eastern states: the Persian wheel and the common current wheel. The climax of adaptation and innovation was reached in what began as the jerry-built but very effective hurdy-gurdy wheel and culminated in the tangential impulse waterwheel, a technological advance of the first order, carried out ith minimal indebtedness to hydraulic science and engineering. The tangential, or Pelton, wheel provided an often indispensable means in fuel-scarce mining districts for the conduct of hard-rock mining of precious metals, meeting alike the needs of hoisting and pumping and the on-site milling of ores. The apprenticeship of a new and basic wheel type with a large future in the hydroelectric field was served in the mining districts of the Far West.

9
The Transmission of Power through Millwork

THE ROLE OF motive power in industry was early recognized as falling into three major divisions: the generation of power, the transmission of power, and the application of power to the purpose or purposes at hand. The generation or production of power was readily defined as the conversion of energy in one form—falling water or steam under pressure—to rotary motion on the wheel shaft or the reciprocating action of the push and pull of the piston. Application nearly always meant the driving of the machines whose endless proliferation has been a central feature of industrialization since 1750.[1] So vast is this last field that it virtually defies description, let alone analysis, save for listing of the broadest categories of mechanization; typically it has been dealt with selectively by way of occasional reference or illustration. Such will be the practice in this study.

The role of transmission seems simple enough. The prime mover, which produces power, and the machines, which in their operation consume it, are usually separated by more or less space. To bridge such distance by appropriate mechanical arrangements, to join the driver and the driven, has been the principal function of transmission. Millwork, the machinery of transmission, consists of varied combinations of shafts or shafting, pulleys, bands or belts, rope or rods, and assorted toothed wheelwork or gears and accessory equipment for support and servicing to place in motion the machinery and equipment of mill, mine, and factory.[2]

[1] Charles Babbage, *On the Economy of Machinery and Manufactures* (Philadelphia, 1832), p. 15; Andrew Ure, *The Philosophy of Manufactures* (London, 1835), pp. 26–27.

[2] The term *millwork* appears in the titles of two leading nineteenth-century English treatises on transmission and related matters: Robertson Buchanan's *Practical Essays on Mill Work and Other Machinery,* 3d ed. rev. (London, 1841), and William Fairbairn's *Treatise on Mills and Millwork,* 2 pts., 2d ed. (London, 1864–65). See also "Mill-Work," in Abraham Rees, ed., *The Cyclopaedia,* 1st Am. ed. (Philadelphia, 1810–24); John Nicholson, *The Operative Mechanic, and British Machinist*, 2d Am. from 3d London ed., 2 vols. (Philadelphia, 1831), 1:vi. The older term, *geering*, although becoming obsolete, was still in use in *Appleton's Dictionary of Machines, Mechanics, Engine-Work & Engineering*, 1852: "Geering is the general term employed to denote a combination of mechanical organs, interposed between the prime mover and the working parts of the machinery" (1:784).

Histories of industry and technology have given but scant attention to power transmission in earlier ages and in most instances fail adequately to describe and explain its role during the nineteenth century.[3] This indifference to and neglect of transmission evidently arises from two circumstances. Most important, probably, were the conditions that during the preindustrial centuries obscured the role of transmission. Before the late eighteenth century the most numerous and important industrial establishments were water mills whose most distinctive mechanical feature, perhaps, was the incorporation of the prime mover, a waterwheel, in the machinery of the mill. Indeed the typical water mill may in its entirety properly be termed a machine of which the waterwheel was a principal component. For many centuries, too, the vast majority of occasions for the use of machinery and power were found in water mills for grinding the bread grains supplying man's principal food. Here, as noted earlier, the waterwheel comprising the prime mover and the millstone assembly comprising the machine were so directly and immediately joined as to form integral components of a single machine. Even in the larger and more elaborate preindustrial mills—such as forges, rolling and slitting mills, and mines with their hoisting and pumping equipment—transmission in the conventional sense of conveying power some distance was largely lacking. Not until the growth of industrial establishments was accompanied by the multiplication of machines did transmission in the full sense of the term become an important feature of industrial operations.

A second circumstance undoubtedly tending to subordinate the role of power transmission has been the far more dramatic role of prime movers. The machine that replaces the directing mind and hand and the prime mover that takes over the burden from straining muscles have understandably been the chief and often the exclusive object of the historian's attention. The emancipation of man from his ancient burden of labor and the replacement of his manipulative skills by devices that bring vast increases in productivity quite naturally enthrall the imagination and fix the attention upon the literally superhuman means by which these results are obtained. By contrast the dispersed and often obscure arrangements of line shafts, belts, pulleys, and gearing used to distribute power throughout an industrial establishment presented in such portions as were apparent a confusion

[3] For example, the first four volumes of the Oxford *History of Technology,* edited by Charles Singer et al., contain only an occasional reference to transmission and gear wheels.

of sights and sounds. Nor did the meaning of such apparatus become more manifest when, in the fanciful language of Andrew Ure, it was termed "the grand nerves and arteries which transmit vitality and volition, so to speak, with due steadiness, delicacy and speed to the automatic organs. . . . if they be ill made or ill-distributed, nothing can do well."[4]

The function of millwork was not simply to distribute power throughout the mill and factory where needed but to subdivide the aggregate power into such parcels of energy as were required to place each machine, large or small, in motion at the rotating speeds required for the operation or operations performed.[5] The only alternative to power distribution and subdivision within the factory was to provide each machine, each industrial operation or process, with its own source of power; that is to say, with its own waterwheel or steam engine. The early steam engines were much too large and cumbrous for such practice, save in occasional instances; but waterwheels were not infrequently employed in this manner when power requirements of individual machines were small and physical arrangements suitable. As late as the 1830s the Ordnance Department's armories for the manufacture of small arms at Springfield, Massachusetts, and Harpers Ferry, Virginia, relied largely on numerous small waterwheels to drive individual machines.

Yet in the early industrial years in the United States as in England the practice of a motor for each machine proved quite impractical in most instances. Not only were small prime movers much less efficient than larger ones, but inconvenience and expense in respect to space requirements and such matters as the supply and removal of water or of fuel, ashes, and combustion fumes quickly reached the impasse of a reductio ad absurdum. English millowners and millwrights in meeting the requirements of ever larger production units soon moved to the adoption of the central power plant with supporting arrange-

[4] Ure, *Philosophy of Manufactures,* pp. 32–35. "It is a noteworthy historical fact," declared Coleman Sellers, a prominent mechanical engineer associated with the manufacture of millwork, "that economy in the generation of power in the motor, and economy in the utilization of the power in the machine, have been in most countries far in advance of the economical transmission of power from one to the other" ("On the Transmission of Motion," *JFI* 94 [1872]:233).

[5] "That power is cheapest and most beneficial which can be subdivided among the largest number of users, with a decided advantage to each over his former method of obtaining power; or which gives to new users power which they otherwise could not obtain" (Allen R. Foote, *Economic Value of Electric Light and Power* [Cincinnati, 1889], pp. 16 ff.).

ments of millwork for the distribution and parceling out of power as required.[6] With the increase in the scale of production it became necessary to provide ever larger buildings with more floors and a mounting acreage of floor space to accommodate the operations and machinery. The problems of transmission became correspondingly larger and more difficult. In the larger establishments, more commonly than not the record, if it were available, would probably show that planning for, providing, and maintaining the millwork imposed a much greater burden upon millwrights than the generation of the power itself. The burden in cost of installation and operation, power loss in friction apart, was in many instances at least of comparable magnitude.[7]

The elements of millwork—with certain exceptions, as for example the operation of pumping equipment in mines, mills, and waterworks, which used a reciprocating action—usually provided for the conveyance of power through the rotation of shafts and wheelwork.[8] The basic unit was the shaft, a round or polygonal bar of wood or iron, some inches in diameter. The bars were coupled together in lengths that were convenient to manufacture and support, usually six to fifteen feet. Assemblages of such rotating members, known as line shafts,

[6] See Fairbairn, *Treatise on Mills and Millwork,* pt. 2, sec. 5, chap. 1.

[7] In reports on capital expenditures millwork is typically lumped with machinery. However, in the estimated cost of a proposed steam-powered cotton factory at Portsmouth, New Hampshire, in the 1830s two 60-horsepower engines with boilers and engine house were given a cost of $12,000; belting, main shafting, and drums were to cost the same figure (*Journal of the American Institute,* 1 [1836]:469–70). Estimating the cost of establishing a cotton mill of 4,000 spindles in the late 1840s, the superintendent of a cotton mill in New York gave the cost of the prime mover as $1,300 for a waterwheel (no mention of supporting facilities) or $6,000 for a steam engine fully installed. Gearing for the mill, including shafting, pulleys, hangers, and so forth, was $2,000; belting of the same, $1,040. The horsepower allowed was 68; that actually consumed, 52 (William Montgomery, communication in *Scientific American* 3 [5 Aug. 1848]:365).

[8] Fairbairn, *Treatise on Mills and Millwork,* pt. 2, sec. 4, chaps. 1–3, provides a concise, well-illustrated description. A briefer, more informal account of American equipment and practice by a comparable authority is Sellers, "Transmission of Motion," pp. 233–43, 307–19. A supplement to Sellers is the series by John Richards, "The Principles of Shop Manipulation for Engineering Apprentices," *JFI* 96–98 (1873–75), esp. "Machinery for Transmitting and Distributing Power," 96:394 ff. See also *Encyclopaedia Britannica,* 11th ed., s.v. "Power Transmission," which dismisses "mechanical transmission" quickly to reach hydraulic, pneumatic, and electrical transmission. Conventional millwork continued for some time to have its place in engineering handbooks. See the treatment of shafting, pulleys, belting, gearing, and the like in William Kent, *The Mechanical Engineers' Pocket-Book,* 9th ed. (New York, 1916), pp. 1130–80.

were carried by means of hangers or brackets just below ceilings or along walls at a level convenient for connection with the machines installed on the floor. Power was carried from the line shafts by means of smooth-faced pulleys and belts of leather or other material directly to the drive pulleys of the machines or, if demanded by the size of the room or number of machines, by way of secondary or counterlines or short sections of countershafts often provided for each machine. In the power takeoff from the engine or waterwheel shaft, as well as in the main line shafts carrying power to the upper floors, it was for many years the practice to use, not pulleys and belting, but toothed wheelwork in the form of spur and bevel gear wheels. Such geared drives were favored, and in England long continued in use, for their positive action (no slippage as with belting), more rugged character, and the precise manner in which changes in direction of motion and in rotating speeds could be effected.

This system of millwork effected power subdivision, within the limits of the capacity of the prime mover and the main line shafts, by regulating the rotating speeds of the shafting. As power in the form of motion was communicated from drive shafts to line shafts, and, according to need, from these to counterlines, countershafts, and machine shafts, the regulation was effected by varying the pulley (or gear wheel) diameters on the driving and driven shafts and the width of the belts employed. Further variation of machine speeds within the control of the operator was often provided by means of a pair of matching cone or step pulleys on the machine and on the countershaft above.

The primary components of main and line shafts and countershafts, geared wheelwork, pulleys, and belting were supplemented by such auxiliary equipment as hangers, brackets, and couplings; various kinds of clutches and fast-and-loose pulleys for engaging and disengaging whole lines of shafting or individual machines; and lubricating cups at journals or bearing boxes on which the shafting rested and turned. With line shafts of unusual length—eventually as much as 1,000 feet in the largest establishments—shaft flywheels were often introduced to counter the "whipping" effect attending major changes in machine loads.[9] Determination of the design, dimensions, running speeds, and operating characteristics of millwork components and

[9] See Fairbairn, *Treatise on Mills and Millwork,* pt. 2, pp. 50 ff., 60 ff., 107; also Sellers, "Transmission of Motion," p. 310. In Worcester, Massachusetts, a three-story building 1,400 feet long had line shafts extending the length of the structure (*American Artisan,* n.s. 9 [14 July 1869]:23). On balance wheels see below, n. 66.

the management and maintenance of the system as a whole were major responsibilities of millwrights and mill engineers.

This system of power transmission was a product of the first Industrial Revolution and an indispensable partner of the direct-drive waterwheel and steam engine. During the century and a half in which it was the chief reliance of industry, it underwent little change in its essential character. Yet, as will be reviewed below, there were important changes in the materials of fabrication and many and marked improvements in design, construction, and operating characteristics and efficiency. There were no major innovations comparable to those marking the evolution of prime movers during the eighteenth and nineteenth centuries, a circumstance possibly accounting for the obscurity of this subject in the history of technology, although not in the literature of engineering.

There was no occasion for power transmission in the sense of the term used here in the water-milling practice of the preindustrial age, since the waterwheel was in most instances an integral part of the mill. In the simple tub or grist mill the upper, or runner, millstone was mounted on the upper end of the tub wheel's vertical shaft. The larger gristmills and flouring mills employed a considerable amount of toothed gearing to change the direction and increase the speed of motion, but this gearing functioned in much the same manner as the gear wheels of a watch or clock. They comprised distinctive elements of the machine itself and were essential to its functioning. So, also, with other types of water mills used for sawing, fulling, forging, blowing, and the like. In most instances the power of the waterwheel was communicated, directly or through a connecting rod, crank or tappet cam, from the wheel shaft or its extension. It may well be that occasionally in the larger manufacturing establishments of the preindustrial period the operation of several simple machines called for some form of power transmission. The use of ropes and rods for conveying power was to some extent practiced in mining operations from the sixteenth century in Europe if not earlier.

From Gristmill to Flouring Mill

Occasional exceptions apart, the first significant use of power transmission in this country was evidently that employed in the larger flouring and grist mills using the ingenious—and in time generally adopted—arrangements developed by Oliver Evans toward the close of the eighteenth century. In contrast to the traditional country grist-

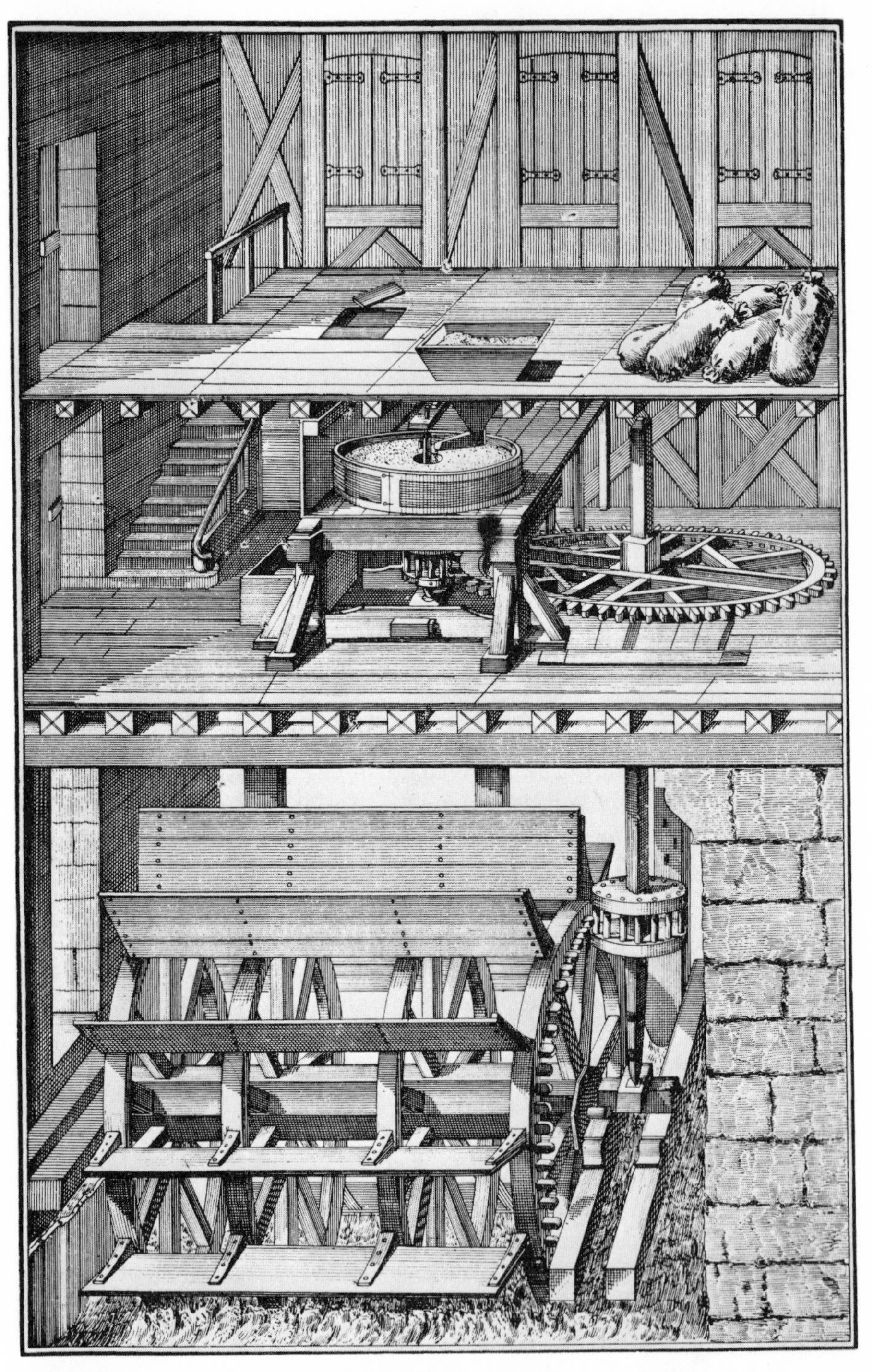

Fig. 99. Eighteenth-century European gristmill equipped with traditional wooden gearing. (Drawing based on Diderot. Photograph courtesy of the Smithsonian Institution.)

mill in which the waterwheel was an integral component, the Evans mill consisted of a group of machines arranged in different parts of and at different levels in the same building. These machines performed successive operations in the conversion of grain to refined flour: cleaning, grinding, cooling, bolting, and packing.[10] Beginning with its removal from vessel or wagon, the grain underwent a series of treatments, starting with the action of the rubbing stones, which removed dirt, and passing to a rolling screen, which eliminated loose dirt, other foreign matter, and broken and immature grain. The grain then passed to the millstones, which reduced it to meal. The meal was elevated to an upper floor, where it underwent a spreading, cooling, and gathering process by means of a revolving rake known as the "hopper boy." Finally it passed through the bolting reels, which separated the fine flour from the bran and middlings, and in large commercial operations was packed, with power assistance, in barrels for shipping. In the course of this series of operations the grain, meal, or flour was conveyed from floor to floor and from place to place automatically by several types of devices, termed by Evans the elevator, the conveyor, the drill, and the descender. In the common gristmill there was only one process—grinding grain to meal between the millstones—providing only one use of the power supplied by the mill wheel and no occasion for power distribution. In the Evans system the grain underwent three preliminary cleaning processes, followed by grinding, cooling, and bolting; and it was conveyed to and from each machine as required. Each process and operation was performed automatically at different speeds and with power.

Since these various operations and processes were distributed over several stories of a substantial building, the power supplied by the mill wheel had to be carried throughout the structure to the points of use (Fig. 100). Transmission was effected by a number of shafts, gear wheels, and belts taking their motion from the wheel shaft. Most of the machinery on the upper floors was driven by a vertical shaft driven in turn by gearing hidden beneath the main floor of the mill. Mounted on the vertical shaft at different points were crown wheels, the wooden predecessors of iron bevel gears, from which several horizontal shafts and conveyors took their motion. These horizontal shafts in turn

[10] The account that follows is based on the illustrated description in the first (1795) and, with only slight changes, later editions of Oliver Evans's *The Young Mill-wright and Miller's Guide,* esp. pt. 3. See also Greville and Dorothy Bathe, *Oliver Evans: A Chronicle of Early American Engineering* (Philadelphia, 1935) on the development of Evans's system of milling.

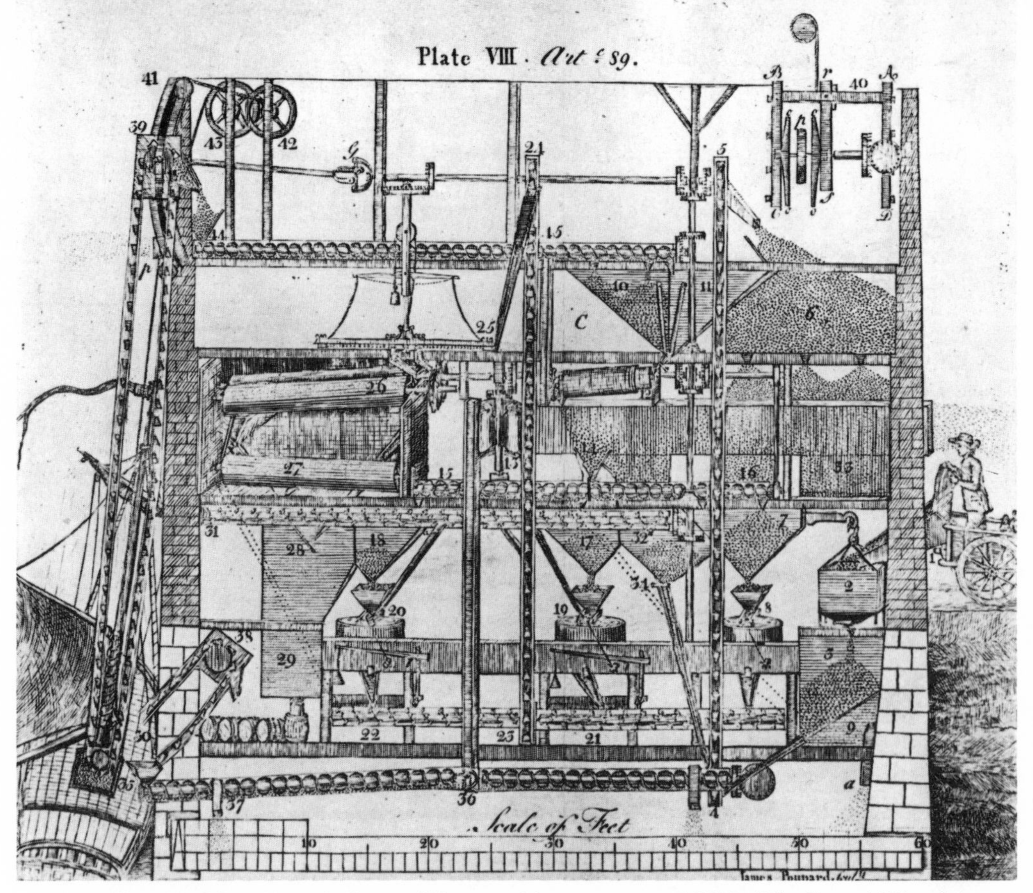

Fig. 100. Oliver Evans's flour-milling machinery, patented 1791. (*The Young Mill-wright and Miller's Guide,* 1795. Legend from Greville and Dorothy Bathe, *Oliver Evans* [Philadelphia, 1925].) For a complete description of the process of milling, reference may be made to Article 89–90 and 91 of *The Young Mill-wright and Miller's Guide.*

1—The wagoner emptying grain into scale pan.
2—Scale pan, to weigh the grain.
3—Small garner and wind chest for cleaning wheat.
4–5—Elevator, to top floor of mill.
6—Main store for wheat.
7—Garner, feeding the shelling or rubbing stones.
8—The rubbing stones.
9—Grain is again elevated and deposited in garners 10–11.
12—Rolling screen.
13—Fan for cleaning the grain.
15–16—The conveyer to garners 7–17 and 18.
18–19–20—Millstones.
20–21—The conveyer, collecting meal after it is ground.
23–24—Elevator to hopper-boy 25.
25—Hopper-boy, which spreads and cools the meal and supplies the bolting chest.
26–27—Bolting reels.
28—The chest containing the super-fine flour.
29—Spout for filling barrels.
35–39—Elevator for unloading ships. This rises and falls in the curved slots and is driven by the universal coupling at G.
40—One view of the mechanism 42–43 for hoisting the elevator clear of the ship.
38—A temporary elevator for short lifts.

drove, directly or through connecting vertical countershafts, the elevators, bolting reels, revolving screen, and other apparatus. For the most part the distances were relatively short and except for the millstones the power required in the several operations quite small; rotating speeds in most instances were quite low. The total length of the main shafting in its vertical and horizontal components did not greatly exceed 100 feet. With possibly one or two exceptions a fractional horsepower was probably sufficient to operate any of the various conveyors, elevators, cleaning equipment, and bolting reels.

Characteristic of the machinery of the day, the millwork—shafts, gear wheels, pulleys—was fabricated of wood, except for the small amounts of iron used in bearings and fastenings and the leather or canvas used in the endless-belt elevators and descenders. The vertical drive shaft, taking its power from the master cogwheel on the waterwheel shaft (in the basement of the mill), was a scantling six or seven inches in diameter. The horizontal shafts by which power was distributed on each floor were about five inches in diameter. All the shafting was polygonal in cross section. In the words of the millwright they were six-, eight-, or twelve-square to permit the ready attachment of cogwheels and pulleys with "eyes" (center openings) of the same cross section by means of wedging. The belt, which was to have so large a place in the years ahead, had an important role in the Evans system. Typically referred to in the preindustrial years as "straps" or "bands," belts were simply endless strips of leather or canvas with small scoops of wood (later metal) attached at regular intervals. They moved continuously around upper and lower pulleys, the upper or driven pulley taking its motion from a horizontal shaft. The elevator band, besides raising the grain as required in milling operations, also served as a transmission belt conveying power to drive two of the horizontal endless screw conveyors by which meal was moved where needed on the same level. The drill and descender were also endless-belt conveyors. Although the descender conveying meal some distance downward might be operated by the weight of the meal it carried, it worked more reliably, as Evans noted, when power driven.[11]

Such, then, was the transmission equipment, or millwork, of what has justifiably been termed America's first fully mechanized factory, the Evans flouring mill. This mill brought together a number of machines in a compact and coordinated mechanical system, taking and distributing power from a single source, a waterwheel. None of

[11] See Evans, *Miller's Guide,* chap. 9, Art. 88; chap. 11, Arts. 95–101; chap. 12, Art. 102. On "straps" in conveying power see 1850 ed., pp. 230–32, 237–39, 247 ff., 326–27.

Fig. 101. Ground plan of the Evans flour mill. (*The Young Mill-wright and Miller's Guide,* 1850 ed.)

The Ground Plan of a Mill

Fig. 1 and 8—bolting chests and reels, top view
2 and 4—cog-wheels that turn the reels
3—cog-wheel on the lower end of a short upright shaft
5 and 7—places for the bran to fall into
6, 6, 6—three garners on the lower floor for bran
9 and 10—posts to support the girders
11—the lower door to load wagons, horses, &c., at
12—the step-ladder, from the lower floor to the husk
13—the place where the hoisting casks stand when filling
14 and 15—the two meal-troughs and meal spouts
16—meal-shaking sieve for Indian and buckwheat
17—a box for the bran to fall into from the sieve
18 and 19—the head-block and long spur-block, for the big shaft
20—four posts in front of the husks, called bray posts
21—the water and cog-wheel shaft
22—the little cog-wheel and shaft, for the lower stones
23—the trundle for the burr stones
24—the wallower for do
25—the spur-wheel that turns the bolts
26—the cog-wheel
27—the trundle, head wallower, and bridge-tree, for country stones
28—the four back posts of the husk
29—the two posts that support the cross-girder
30—the two posts that bear up the penstock at one side
31—the water-wheel, 18 feet diameter
32—the two posts that bear up the other side of the penstock
33—the head-blocks and spur-blocks, at water end
34—a sill to keep up the outer ends
35—the water-house door
36—a hole in the wall for the trunk to go through
37—the four windows of the lower story

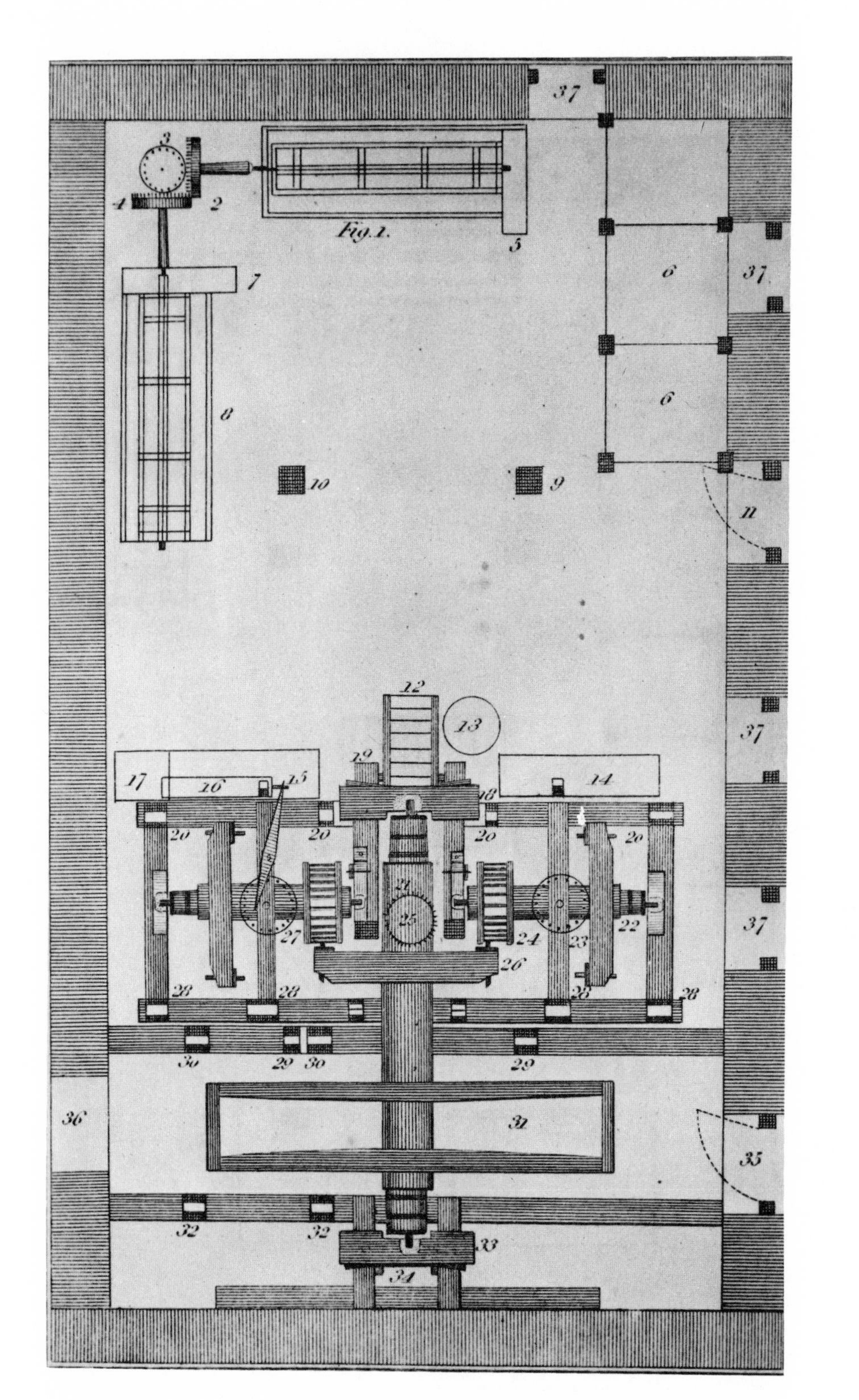

37
3
4
2
Fig. 1.
5
6
37
7
6
8
10
9
11
12
13
37
17
16
15
19
18
14
20
20
20
20
21
25
27
24
23
22
37
26
28
28
28
28
30
29
30
29
36
31
35
32
32
33
34

Fig. 102. Second floor of Evans flour mill. (*The Young Mill-wright and Miller's Guide,* 1850 ed.)

Second Floor

Fig. 1 and 9—a top view of the bolting chests and reels
2 and 10—places for the bran to fall into
3 and 8—the shafts that turn the reels
4 and 7—wheels that turn the reels
5—a wheel on the long shafts between the uprights
6—a wheel on the upper end of the upright shaft
11 and 12—two posts that bear up the girders of the third floor
13—the long shaft between two uprights
14—five garners to hold toll, &c
15—a door in the upper side of the mill-house
16—a step ladder from 2d to 3d floor
17—the running burr mill-stone laid off to be dressed
18—the hatch way
19—stair way
20—the running country stone turned up to be dressed
21—a small step-ladder from the husk to the 2d floor
22—the places where the cranes stand
24—the pulley-wheel that turns the rolling screen
25 and 26—the shaft and wheel which turn the rolling screen and fan
27—the wheel on the horizontal shaft to turn to bolting reels
28—the wheel on the upper end of the first upright shaft
29—a large pulley that turns the fan
30—the pulley at the end of the rolling screen
31—the fan
32—the rolling screen
33—a step ladder from the husk to the floor over the water-house
34 and 35—two posts that support the girders of the third floor
36—a small room for the tailings of the rolling screen
37—a room for the fannings
38—do. for the screenings
39—a small room for the dust
40—the penstock of water
41—a room for the miller to keep his books in

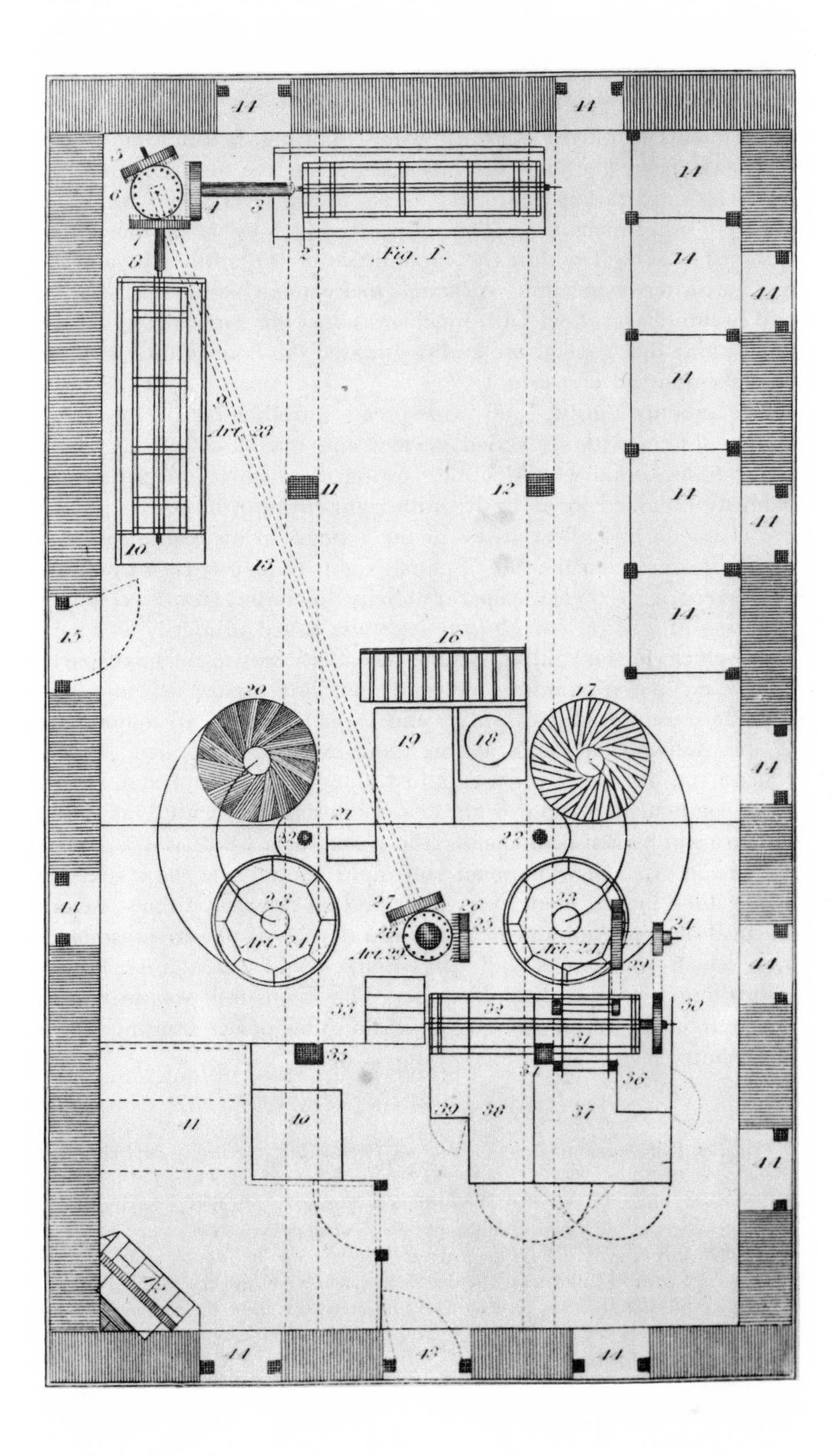
Fig. I.
Art. 23
Art. 24
Art. 29.
Art. 24
Art. 24

the elements employed in transmission—shafting, belting, and wheelwork—was new. The most original feature was the use of belting in a dual capacity: raising materials to an upper level while conveying power to other machinery. The millwork, like most of the machinery it served, was well within the capabilities of the millwright and the mill carpenter of the time. Although to Evans's chagrin this mechanized milling system met with much resistance for some years, it eventually came into general use and dominated the flour-milling industry for three-quarters of a century.[12]

The practical utility and widespread introduction of the Evans milling system with its varied devices and overall arrangements for power transmission were doubtless owing as much to the widespread publicity obtained for it by its millwright inventor as to the innovative character and effectiveness of the system. In his efforts to secure royalty income from the sale of patent rights to its use, Evans not only used every form of newspaper publicity, including controversy, but his *Young Mill-wright and Miller's Guide* was issued primarily as a publicity vehicle for his milling system. The *Guide* was in no small part a primer on power transmission and handling materials, supplying numerous and detailed drawings and explanations of all major parts of the millwork as well as the flour-milling equipment. Precise dimensional data were often supplied along with bills of material for major components. Whatever the degree of originality with Evans of the various transmission components, the system as a whole was a demonstrated success. The widespread and rapid adoption of the system, by licit or illicit means, made every flour mill on this plan a showcase and practical demonstration of methods and devices of power transmission from which millowners and millwrights could through experience absorb much of general applicability. The Evans mill was in truth a milestone in the history of mechanization and power transmission in this country.

[12] Charles B. Kuhlmann, *Development of the Flour-Milling Industry in the United States* (New York, 1929), pp. 96–101. See Evans's summary of the advantages of his system, *Miller's Guide,* chap. 11, Art. 101. The common European practice of having a separate waterwheel for each pair of millstones does not appear to have gained much of a foothold in this country, where the ordinary practice was to "double gear" a waterwheel to two pairs of millstones. For the European method, see fig. 247 in Fairbairn, *Treatise on Mills and Millwork,* pt. 2, p. 111. I observed this to be the common arrangement in gristmills in the less developed parts of southern and eastern Europe whenever vertical wheels were employed.

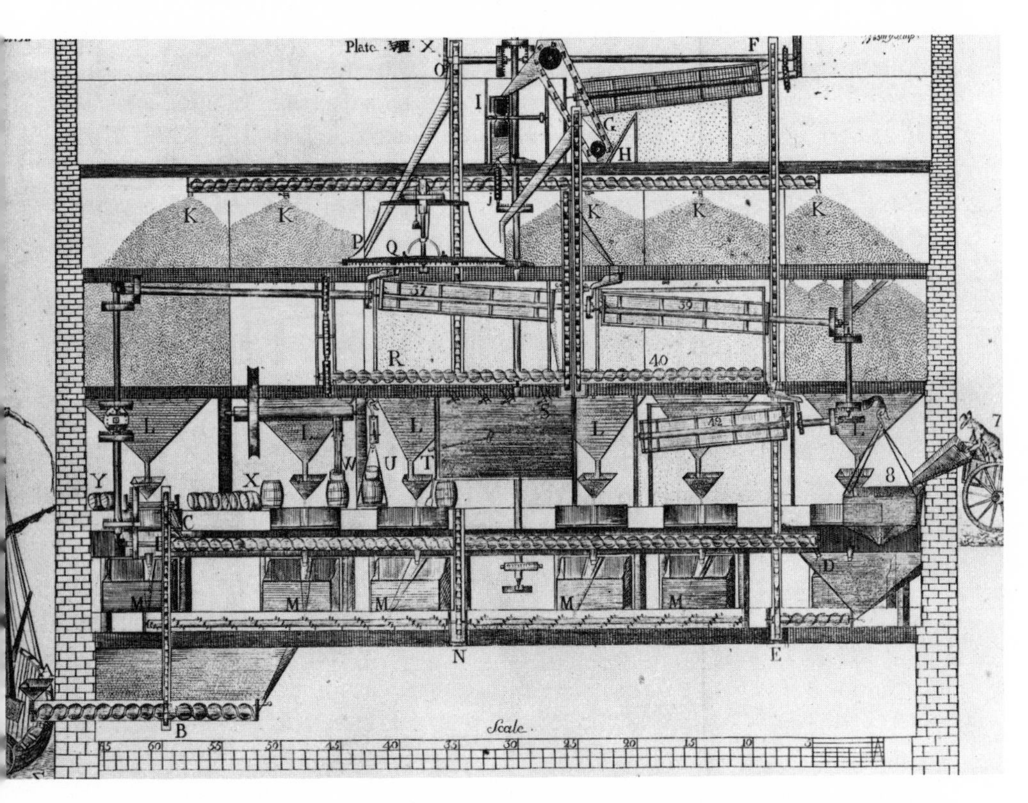

Fig. 103. Ellicott-built flour mill on the Occoquan River, Va., using three waterwheels to power six pair of stones. (Oliver Evans, *The Young Mill-wright and Miller's Guide,* 1807 ed.)

Variations in Transmission Requirements

The shafting, wheelwork, belts, and pulleys employed so effectively in large merchant flour mills served as the chief means of power distribution in the manufacturing industries generally until the transmission revolution (to be treated in a later volume) gathered momentum in the closing decades of the nineteenth century. Each branch of industry had its own needs and problems of power distribution; rarely even in the same industry did two plants present quite the same conditions. Means and methods of distribution reflected the character of the major fabricating operations, the nature of the machinery employed, and space requirements—as well as the amount of concentration of power in specific areas, rotating speeds of equipment, and regularity of motion. Although at present the evolution of the system in the United States can be presented only in general outlines, the record is strewn with illuminating fragments and with the recollections of old-timers looking back with wonder at the changes within their own lifetimes. We learn, for example, from Benjamin H. Latrobe's correspondence from Pittsburgh in 1814 that the horse-powered sweep mill of the riverbank shop set up to build machinery for one of the Fulton group's steamboats drove "a perpendicular shaft, wheels and drums to turn the lathes and drills." The equipment of a large white-lead works in New York State, reported the federal census of 1820, included a 12-horsepower steam engine, four pairs each of paint stones and cast-iron rollers, and two each of cast-iron air pumps and forcing pumps, together with several lifting pumps, lead stirrers, and bolting screens. All these were "connected by an extensive machinery so as to put the whole in motion at the same time or such part as may be required." A hundred years later a hydraulic and mechanical engineer, reviewing the development of water mills in New England before 1815, reported that "in order to simplify and shorten the distribution of power by shafting and belts, a separate water wheel in each factory was the common practice and not infrequently there was a separate water wheel for driving each principal machine."[13]

Power Distribution in Textile Mills

Illustrations may be drawn from key industries in the early advance of industrialization: the textile mill, the ironworks, and the shaft mine. The central power-distribution problems in each of these three instances were those attending the degree of dispersion, concentration, and remoteness in the application of motive power. The cotton mill has particular interest because, in this country as in Britain, it was in important respects the first true factory. It did much to establish the trend toward ever greater production employing large assemblages of power-driven machinery and workers. There has come down to us

[13] Benjamin H. Latrobe to Robert Fulton, 21 May 1814, with drawing, Latrobe Letterbooks, Maryland Historical Society, Baltimore; see also letters of 24 May and 16 June 1814. U.S., Fourth Census, 1820, Schedules of Manufactures, New York, National Archives. John R. Freeman, "General Review of Current Practice in Water Power Production in America," *Transactions of the First World Power Conference, 1924,* 5 vols. (London, [1925]), 2:382–83.

through the recollection of a nonagenarian a vivid account of the excitement aroused by the opening in 1794 of the first woolen mill of Newburyport, Massachusetts. The mill, we are told,

created a great sensation throughout the whole region. People visited it from near and far. Ten cents was charged as an admission fee. . . . Never shall I forget the awe with which I entered what then appeared the vast and imposing edifice. The large drums that carried the bands on the lower floor, coupled with the novel noise and hum, increased this awe, but when I reached the second floor, where picking, carding, spinning and weaving were in progress, my amazement became complete. The machinery, with the exception of the looms, was driven by water, the weaving was by hand.[14]

In cotton mills the distribution of power was controlled by the dispersal of manufacturing operations over an extensive area, owing to the bulky character of the spinning, weaving, and preparatory machinery and to the small power requirements of the individual mechanical units, especially spindles and looms. The power required for driving a single spindle was slight; one horsepower was sufficient to place in motion 100 to 300 spindles, depending upon the type, together with the necessary preparatory machinery. A single horsepower, too, would drive ten to a dozen looms.[15]

The basic elements of power distribution in cotton mills were much the same as in the large commercial flouring mill, but they were multiplied many times and required far greater attention in planning, installation, and management. The millwright had to deal with scores and in time hundreds of machines, falling for the most part into several main categories. This consideration virtually compelled the ranking of machines in rows and the segregation of operations by floor, hardly less for convenience in attendance and maintenance of the

[14] Royal C. Taft, *Some Notes upon the Introduction of the Woolen Manufacture into the United States* (Providence, 1882), pp. 7–8, quoting Joshua Coffin, *A Sketch of the History of Newbury, Newburyport, West Newbury* (1845).

[15] Zachariah Allen reported in 1829 that at Lowell 44 horsepower were required to operate 4,000 "dead" spindles (equivalent to 4,400 throstle spindles and including all the preparatory machinery for spinning "no. 30" yarn) and that 12 horsepower were required to drive 144 looms (*The Science of Mechanics* [Providence, 1829], pp. 148–49). See also ibid., pp. 344–45. Experiments at Lowell reported by James Montgomery about 1840 showed that with yarn no. 14 one horsepower moved 77 throstle spindles "with preparation" and 105 without, 8.5 power looms "with dressing" and 12 looms without (*A Practical Detail of the Cotton Manufacture of the United States of America* [Glasgow, 1840], p. 108). Andrew Ure reported in 1854 that at one of the largest and most modern cotton mills in Britain 1 horsepower would drive 500 mule spindles, 300 self-actor spindles, or 180 throstle spindles, including the preparation processes (*Dictionary of Arts, Manufactures, and Mines,* s.v. "Cotton Factory").

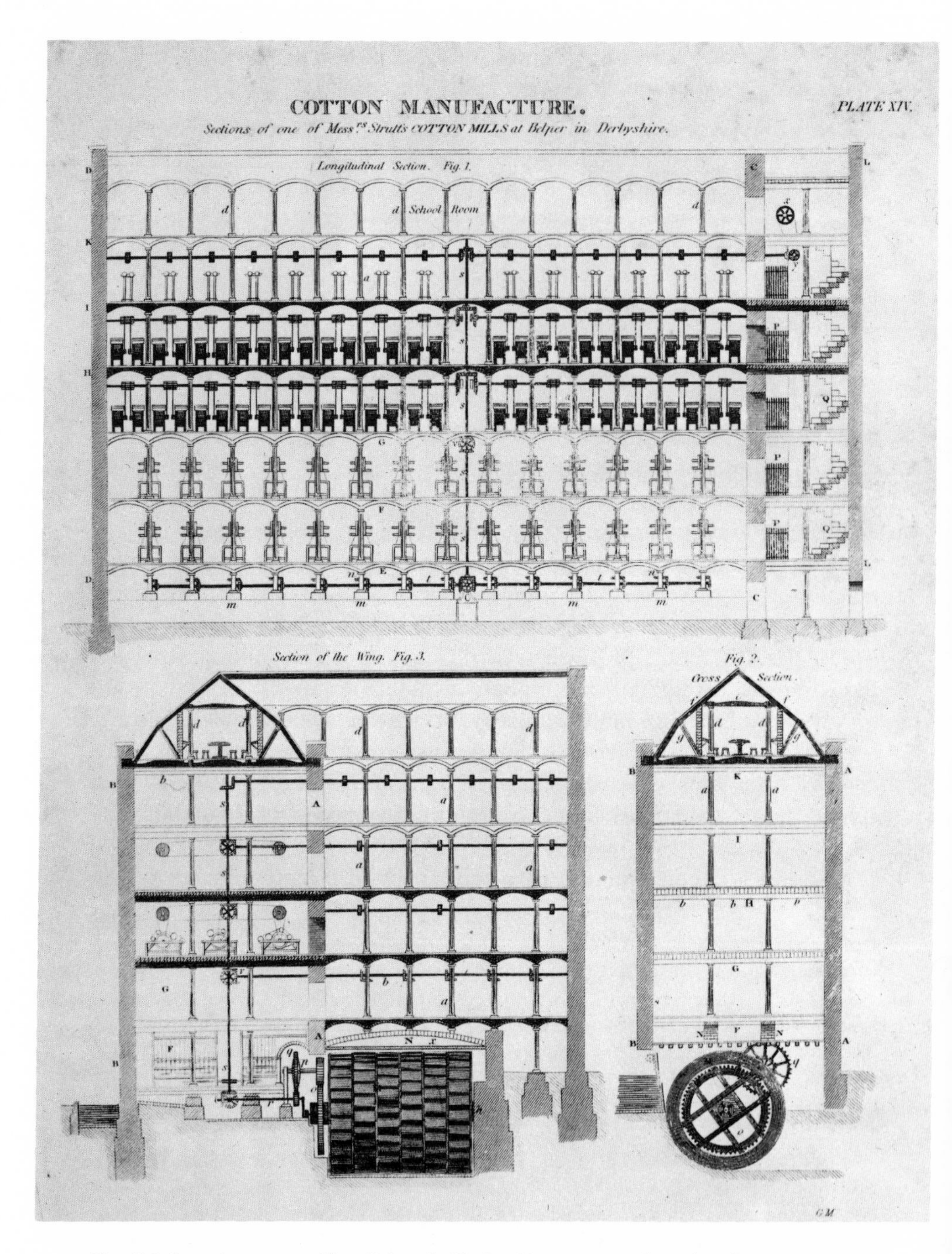

Fig. 104. Strutt's cotton mill at Belper in Derbyshire, cross sections showing power transmission by metal gearing. (Abraham Rees, ed., *The Cyclopaedia,* 1st Am. ed., plates, vol. 4.)

machinery than for economy and efficiency in the arrangement, installation, and maintenance of the millwork. By the 1830s a typical medium-scale cotton mill occupied a building 150 to 200 feet long, 40 to 50 feet wide, and four or five stories high above the basement, where usually the waterwheel or steam engine was located. Machinery occupied the greater portion of each floor, and lines of shafting with pulleys, belting, and countershafting much of the space between machines and ceiling. Power had to be conveyed from the basement over fifty feet vertically to the upper floors and for as much as several hundred feet on each floor for the line shafts on either side, countershafting apart. With an aggregate length of as much as 1,500 feet and an average diameter of three inches, iron shafting with couplings, pulleys, and hangers would have a total weight of twenty or more tons. The power for whose distribution this ponderous equipment was required ranged between 50 to 75 horsepower.[16]

More typically, perhaps, mills grew rather than were planned in neat rectangular proportions. One of the earliest steam-powered cotton mills in Philadelphia, established in 1808, was successively enlarged by additions, climaxed in 1840 by the erection of a five-story structure. Its power was obtained from a large steam engine, with a fifteen-inch cylinder bore and a five-foot stroke. We are told that

> there was a connection by a square 2½-inch iron shaft from engine to large mitre [bevel gear] wheels in the center of the main basement, and from thence to rear buildings. Power was transmitted to the second floor of the main mill by massive spur cogwheels of eight and three feet diameter with five-inch faces, one of the large wheels being on the main shaft of the engine projected through the end wall; and when working this gearing made a loud and unpleasant clanging. A wooden drum on the north wall of the rear buildings, furnished, by belting, most of the power to that portion of the mills.[17]

In 1843–44 a large number of power looms were installed in the two upper stories of the main mill, "driven from long wooden drums of 12

[16] See description of the Hamilton Mills in Ithamar A. Beard, "Practical Observations on the Power Expended in Driving the Machinery of a Cotton Manufactory at Lowell," *JFI* 15 (1833):6 ff. This mill contained 4,288 spindles and 144 looms, both with preparatory equipment. By dynamometer tests in 1830 the power used in the several departments of the mill was as follows: carding room, 18.63 hp; spinning room, 4,288 spindles, 9 warping and 12 dressing frames, 48.95 hp; weaving room, 144 looms, 15.19 hp; dressing room, 9 warping and 12 dressing frames, 6.13 hp; main driving gear, 3.24 hp. Total hp: 92.13. See also R. H. Baird, *The American Cotton Spinner and Managers' and Carders' Guide* (Boston, 1854), pp. 30 ff.

[17] Samuel H. Needles, "The Governor's Mill and Globe Mills, Philadelphia," *Pennsylvania Magazine of History and Biography* 8 (1884):387–88.

inches diameter fixed on rough iron shafts." In the lower story a two-and-one-half-inch iron shaft ran through the center of the building carrying pulleys entirely of wood.

By mid-century the largest cotton mills were of impressive size and capacity. One of the largest and best designed cotton mills in England, the Orrell Mill at Stockport, contained nearly 46,000 spindles and some 1,100 looms. In 1858 the five largest textile corporations at Lowell, Massachusetts, averaged some 60,000 spindles and 1,800 looms, divided among several mills served by waterwheels or hydraulic turbines. By the mid-fifties the Saltaire woolen mills in Yorkshire, England, employed shafting nearly two miles in aggregate length and over sixty tons in weight to join steam engines of 1,250 horsepower with the vast array of machinery that produced 30,000 yards of alpaca cloth daily. The mill proper was 550 by 50 feet and six stories high, and the loom shed measured 300 by 200 feet. The floor area was equivalent to twelve acres.[18]

American textile engineers were not long in reaching and passing this achievement. Harmony No. 3 Mill on the Mohawk River at Cohoes, New York, was built by sections, 1868–72. Its main structure was nearly 1,100 feet long and housed 70,000 spindles and 2,700 looms. With virtually the same horsepower as the Saltaire works, Harmony No. 3's power plant consisted of three Boyden-type turbines mounted on the same massive shaft, along with six twelve-foot pulleys belted to main line shafts serving different levels of the five-story structure. The three wheels placed in motion over two miles of shafting, 1,400 main-shaft pulleys, and more than ten miles of belting of varying widths between the main drive pulleys and the driven machines.[19]

Ironworks stood well apart from most other industrial establishments in respect to power transmission and at the opposite end of the spectrum from the textile industry. Machines were few in number, heavy in power demands, and were commonly located close to their source of power. Except where required to service auxiliary equip-

[18] Ure, *Dictionary of Arts, Manufactures, and Mining,* 1:504–5; table in David Stevenson, *Sketch of the Civil Engineering of North America* (London, 1838), pp. 212–13; William Pole, *The Life of Sir William Fairbairn* (London, 1877), pp. 327–28. Fairbairn gave the number of wheels (evidently gear wheels and pulleys) at Saltaire as 600, and declared that "in mills of my own construction there have been on the average not less than 450 wheels and 7,000 feet of shafting in motion" (*Treatise on Mills and Millwork,* pt. 2, pp. 98–100).

[19] Diana S. Waite, "Number 3 ('Mastodon') Mill 1868 and 1872," in Robert Vogel, ed., *A Report of the Mohawk-Hudson Area Survey,* Smithsonian Studies in History and Technology no. 26 (Washington, D.C., 1973), pp. 98 ff.

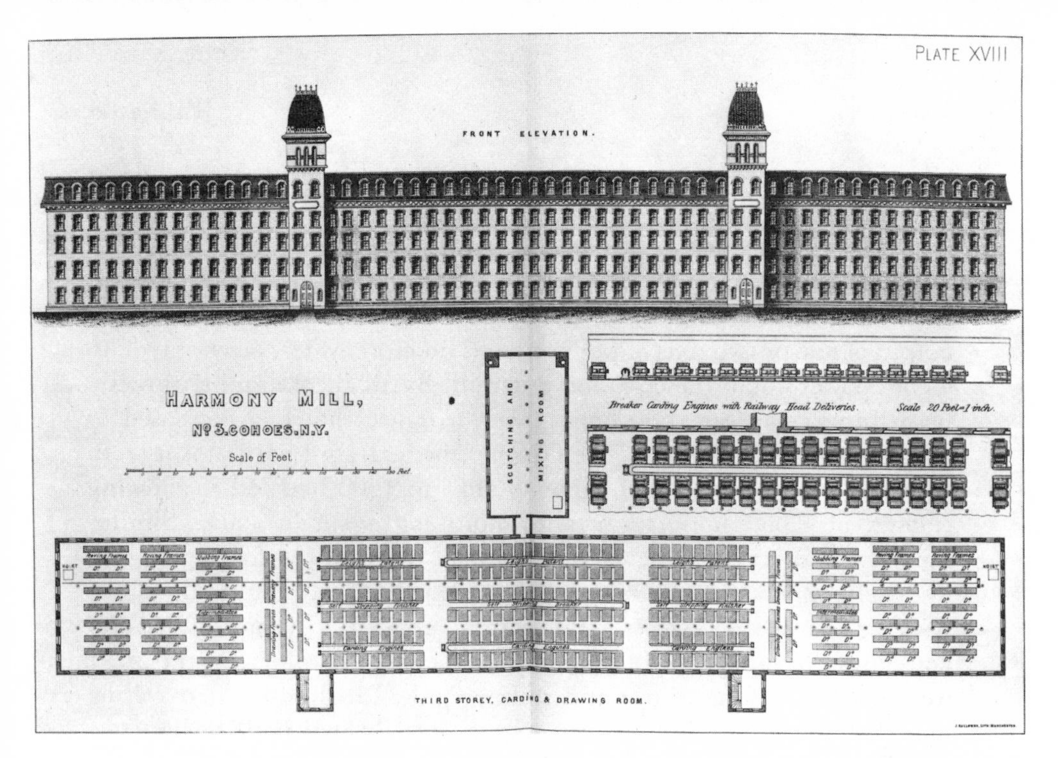

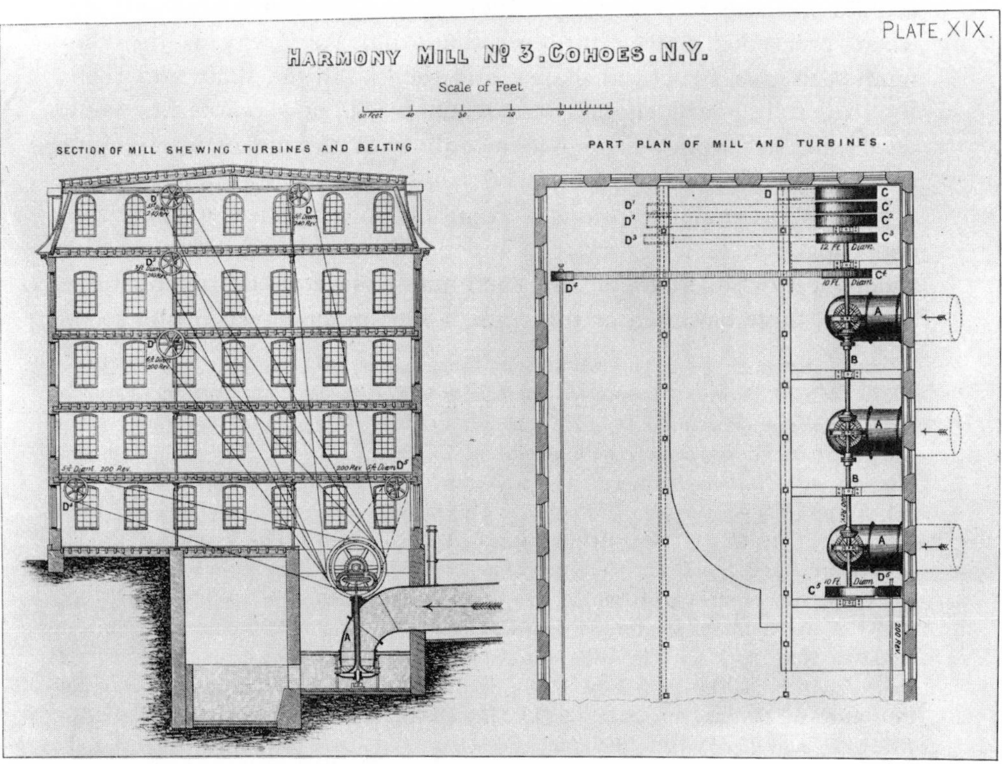

Fig. 105. Harmony Mill No. 3, Cohoes, N.Y., showing power transmission by belting. (Evan Leigh, *The Science of Modern Cotton Spinning*, 2d ed. [Manchester, 1882].)

ment, such millwork as shafting and belting was little used; direct connection was the rule. Ironworks ranged widely in size and power needs. From the opening decades of the nineteenth century the trend in the scale of operations and power needs was progressively upward. The small, rural charcoal blast furnace of the early 1800s with its daily output of one or two tons of pig iron used no more waterpower—say, 2 to 4 horsepower—than the common gristmill. By 1850 coke and anthracite blast furnaces averaged 50 horsepower, a figure that had increased to 200 horsepower by 1880. Then came the extraordinary upsurge in coke blast furnace size and capacity that in 1900 had led to blowing engines of 2,000 horsepower and more for a single stack. Similar though less spectacular developments took place in the forging of iron by trip-hammers and roll trains. At one extreme were the small water-powered rural bloomeries and forges, requiring typically several horsepower to supply the blast and drive the forge hammer. At the other were the heavy forge shops and rolling mills. Pittsburgh's first rolling mill, 1812, had a 70-horsepower engine; one erected in 1818 had two engines of 120 horsepower each. Increase in engine size here and elsewhere proceeded slowly until merchant bar gave way as the chief product to rails, structural shapes, and the like in the 1850s and 1860s. By 1870 rolling-mill engines had reached 600 horsepower but could do the heavier work assigned them only with the aid of massive and costly gearing.[20]

With occasional exceptions a common rule applied to all classes and capacities of ironworks: an independent power source, whether waterwheel or steam engine, for each major piece of equipment: blowing tubs, forge hammer, or roll train.[21] This practice was owing to the

[20] S. W. Roberts, "Observations on Blast Furnaces for Iron Smelting," *JFI* 3 (1842): 29–30; John B. Pearse, *A Concise History of the Iron Manufacture of the American Colonies up to the Revolution* (Philadelphia, 1876), p. 98; Frederick Overman, *The Manufacture of Iron in All Its Various Branches* (Philadelphia, 1850), pp. 334–38, 460; U.S., Congress, "Report on a National Armory," House Report 43, 37th Cong., 3d sess. (Serial 1161, vol. 5), pp. 138–39; William P. Shinn, "The Genesis of the Edgar Thomson Blast-Furnaces," *TAIME* 19 (1890–91):674 ff.; T. W. Robinson, "The Economic Production of Iron and Steel," *Cassier's Magazine* 22 (1903):383; *Pittsburgh Gazette,* 5 Mar. 1818. See also Frederick Overman, *The Manufacture of Steel* (Philadelphia, 1873), pp. 85–88. A trip-hammer of unusual size with head weighing 13,000 pounds at an ironworks in New York City in 1840 was driven by a 30-horsepower engine (*Pittsburgh Daily Advertiser and Gazette,* 5 Nov. 1840). See descriptions of rolling mills in Ohio and Pittsburgh in *American Artisan*, n.s. 6 (13 May 1868):282; *Scientific American* 27 (31 Aug. 1872):132, and 20 (23 Jan. 1869):50.

[21] Overman, *Manufacture of Iron,* pp. 334 ff., 462 ff.; Fairbairn, *Treatise on Mills and Millwork,* pt. 2, chap. 10; and Thomas Egleston, "The American Bloomery Process for Making Iron Direct from the Ore," *TAIME* 8 (1879–80): 536–44.

Fig. 106. Water-powered tilt hammer: *a*, cast-iron hammer head (50 to 400 lbs. according to work done); *b*, wooden hammer helve (handle) vibrating on fulcrum near tap-ring end; *c*, cast-iron anvil resting on iron "stock" *d; e,* vertical log base of anvil, 3–4 ft. diameter and 6–8 ft. long; *h,* wooden recoil beam for tap-ring end of helve; *i,* iron tap ring on helve end receiving blows from cams; *l,* cams on cam wheel, *M,* mounted on wooden waterwheel shaft. (Frederick Overman, *The Manufacture of Iron* [Philadelphia, 1850].)

crucial role of heat in all operations and the importance of giving full and instant control of power and speed to the operator, who, especially in forging and rolling, was in effect a blacksmith manipulating the incandescent but rapidly cooling slab of bar iron with the powerful tilt hammer or roll train responding immediately to hand or foot controls regulating water or steam supply. Reliance on a single prime

Fig. 107. *Above,* model of the interior of the Lukens Iron Co. rolling mill at Coatsville, Pa., c. 1830–40. An overshot waterwheel equipped with a 12–15-ft. iron flywheel drove the roll train and operated two shears with power conveyed by means of heavy cast-iron cogwheels. *Below,* another view of the same diorama shows the close proximity of furnaces to rolling-mill operations. (Diorama, Hagley Museum, Wilmington, Del.)

mover or central power plant as practiced increasingly in cotton and other factories never found favor and was rarely employed in forging operations, in part because of the difficulties of power transmission and subdivision to meet the heavy concentration of power required, and also because of the difficulty of avoiding harmful speed variations when pieces of heavy equipment were thrown on and off the power line, as with a roll train idling at one moment and a few moments later operating at full capacity. Flywheels weighing as much as fifty tons were used to obtain regular, continuous motion from the early, slow-moving, long-stroke engines. Even with massive flywheels and responsive prime movers, uniformity of motion was hardly feasible, and speed variations to meet individual machine operations were not possible. Different stages of forging required variations in the force and rapidity of blows; different roll trains required different speed ratios and, in the manner of forge hammers, variations in speed and power. Such requirements could be met only with an independent power source. As rolling mills and forge shops increased in size—and blast-furnace plants similarly—the number of prime movers multiplied with the large roll trains and forge hammers. Smaller equipment might be driven alternately by engine or waterwheel of appropriate capacity, as also such auxiliary equipment as shears, saw, or pump rig.

For such reasons conventional millwork of shafting, wheelwork, and belting from the beginning played no more than an incidental or auxiliary role in the manufacture of iron. In the small rural establishments of the early years, bellows or blowing tubs and forge hammers took their motion directly from the waterwheel shaft by cam tappets or by cranks. In the largest works, especially rolling mills, the waterwheel and, in time, the steam engine were placed beside, and were connected with, the roll train either directly or more commonly through gearing to obtain the necessary speed. By means of simple clutches the prime mover might be connected alternately to roll trains of different size or to a forge hammer. The gearing used with slow-moving waterwheels of large diameter was often of great size and weight. Even with steam power, direct-drive dispensing with gearing was slow to become practicable because of the high roll speeds required. To obtain the 50 to 100 rpm demanded by roll trains called for gear wheels of large diameter and massive construction. We read of spur gear wheels twenty feet in diameter with two-foot breadth of face. According to a leading British engineering journal at least one-half the whole power developed was expended in keeping this gearing in motion and its first cost represented one-half the capital investment.[22]

[22] *Engineering,* quoted in *Van Nostrand's Engineering Magazine* 1 (1869):399–400. See

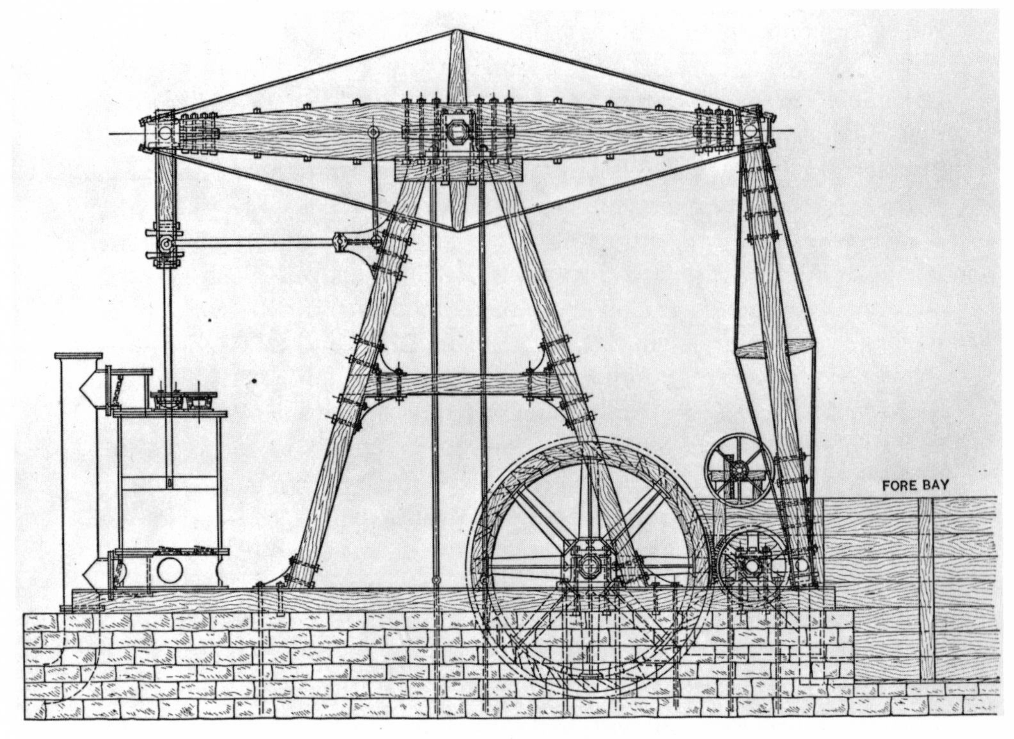

Fig. 108. Waterpower blowing engine, Crane Iron Works, Catasauqua, Pa. Erected 1839–40. Vertical longitudinal section. Scale: ⅛ in. = 1 ft. (Samuel Thomas, "Reminiscences of the Early Anthracite-Iron Industry," *TAIME* 29 [1899].)

The conditions in the primary iron industry responsible for the virtual elimination of transmission in the usual sense of the term were present in lesser degree in that branch of secondary manufactures engaged in making wrought-iron goods, such as edge tools and hardware. Here developed what was in effect the extensive mechanization of activities long conducted in most rural communities in blacksmith shops. The basic shaping of such ironware was carried on with light water-powered trip-hammers; there was the same need as in heavy forging operations, although on a lesser scale, for operator control of the force and speed of the hammer. In establishments of any size a separate motor for each trip-hammer presented manifest difficulties in getting the water to and from the hammer wheel. One of the earliest and largest of such establishments was the Springfield Armory, which, until joined by the armory at Harpers Ferry, was the Ordnance

the complaint of such gearing in rolling mills about 1850 by John Fritz, "Early Days of the Iron Manufacture," *TAIME* 24 (1894):600 ff.

Department's principal source of small arms, chiefly muskets. The experience of these two armories, dependent almost solely upon waterpower before the 1840s, provides a useful primer on the difficulties and methods of distributing and subdividing power in the early industrial years.

By 1824 the five workshops of the Springfield Armory were producing with a work force of some 250 hands 12,000 to 15,000 muskets annually.[23] Three of the workshops were provided with waterpower. Twenty-seven waterwheels (twenty-two in active use), for the most part undershot and tub wheels of several horsepower or less, drove the dozen or more trip-hammers used in the basic forging work and powered such varied operations as milling, drilling, boring, slitting, grinding, and polishing. Waterpower was employed in ninety-six operations; ninety-seven operations were manual, most of them hand-forging and filing. Power was supplied by the Mill River, a modest enough stream about twenty-five feet wide and some two feet deep, but more than adequate to develop power ranging between 50 to 70 horsepower. The initial subdivision of power was arranged by nature in the three privileges serving the upper, middle, and lower water shops, together described at the time as "the greatest assemblage of mills and other waterworks to be found in the State." The limitations of power transmission at the time are suggested by the use of the twenty-seven wheels, the majority developing less than several horsepower. A board of the Ordnance Department engaged at this time in a survey of possible sites for a western armory reviewed the choice of power in relation to its distribution:

> For all purposes that require or will admit of a concentration of power, a single engine might be fully competent, and employed at Pittsburgh, would be cheaper than almost any power but the power employed at an armory is only employed as auxiliary to the labor of a large number of mechanics; and the nature and necessary distribution of the work to be performed by them render a considerable division of the power indispensable to the convenience and economy of such an establishment. [It is ascertained that three engines or wheels will be indispensable] and to make the power, as it ought to be subservient to the most advantageous arrangement of the work [will require at least four engines or water-wheels].[24]

[23] *American State Papers, Military Affairs,* 2:538–39, 729–91. See also correspondence, esp. from and to Col. Roswell Lee, 1823 et seq., in Springfield Armory Records, National Archives.

[24] *American State Papers, Military Affairs,* 4:492–94 and 2:538, 745.

The situation at Springfield and at Harpers Ferry alike underwent little change for many years to come. In the various plans for rearranging or for expanding operations at both armories, the reduction in the number of prime movers received much attention. In correspondence with Col. Roswell Lee, superintendent of the Springfield Armory, Eli Whitney in 1824 declared the armory's power with its many wheels was too much divided. Although the armory used in a number of instances a single waterwheel to drive several machines engaged in milling, drilling, and the like, in the critical operation of forging gun barrels this practice was held not to be feasible. Whitney on the basis of his own experience proposed that the many undershot and tub wheels of low efficiency be replaced with several breast wheels of 15 to 20 horsepower supplying the requirements of the entire armory. He believed "that a hammer of 500 to 1000 lbs may in most situations be driven by a belt with great advantage," and argued that "a number of hammers, say 8 or 10, may be operated by one wheel with great convenience."[25]

At the Springfield Armory at least, Whitney's proposals for the concentration and distribution of power were eventually adopted in part. A description of the armory in 1852 shows that in welding of gun barrels and the later grinding and polishing operations, transmission by line shafts, pulleys, and belts was employed on a striking scale. The welding operation was performed in one of the middle water shops, where "a range of tilt-hammers extend up and down the rooms, with forges in the center of the room, one opposite to each hammer, for heating the iron. The tilt-hammers are driven by immense water-wheels, placed beneath the building—there being an arrangement of machinery by which each hammer may be connected with its moving power, or disconnected from it, at any moment, at the pleasure of the workman." With eleven heats required to finish the drawing of each barrel, the sight and sound of eighteen trip-hammers was to the visitor a startling experience. At operating speeds of 600 blows to the minute the shop resounded "with the incessant and intolerable clangor and din."[26]

[25] Whitney to Lee, Aug. 22, 1824, Springfield Armory Records. This point was later paraphrased by W. C. Johnson in U.S., Congress, "National Foundry," 12 Jan. 1819, House Report 229, 27th Cong., 3d sess. (Serial 427).

[26] Jacob Abbott, "The Armory at Springfield," *Harper's New Monthly Magazine* 5 (July 1852):148. A drawing shows the massiveness of the structures in which the trip-hammers were mounted. However, neither drawing nor description makes clear the precise means of the transmission between "the immense water-wheels" and the tilt hammers. Probably it was a modification of the heavy line shaft with correspondingly heavy pulleys and belting shown in the drawing of the grinding equipment, ibid., p. 152. For the operation of what were variously termed cams, tappets, or wipers,

The plans for establishing a western armory reported in the 1820s and at later dates were not accepted primarily for political reasons, but none was so bold as to recommend less than three or four prime movers, whether engines or waterwheels. As late as 1861 a committee of manufacturers urging the location of the western armory at Pittsburgh declared that the great advantage of steam power, apart from its lower cost, over waterpower lay in the ease and convenience with which it could be distributed "over a vast amount of surface and a great variety of speeds and of powers," simply by "multiplying the engines, which can be located at pleasure."[27] The use of waterpower, the manufacturers argued, would involve "an enormous expense" in waterwheels and supporting equipment; "or if one large water wheel is used, and all the power taken from it, then the shafting, geer-wheels, pulleys, belts, &c., necessary for the proper distribution and regulation of this power, will cost several times as much as the first cost of the engines, and the expenses of keeping up this machinery, many times more than the cost of the fuel for the engines."[28]

Heavy forging equipment such as that at Springfield in 1852 was quite unsuited to general factory use. Even light tilt hammers could not take their motion from general purpose line shafts owing to the disturbing and damaging effects upon shafting, couplings, and gearing of the sudden and varied resistances attending their use.[29] The rapidly growing use of iron and other metalwork in a wide range of manufacturing activities and the need to eliminate slow and costly hand-hammering led in the 1850s and 1860s to the development of new types of light power hammers.[30] A number of these came on the

---

devices for converting rotary to reciprocating action in various forms of trip- or tilt hammers, ore stamps, fulling mills, and the like, see brief accounts, with drawings, in Frederick Overman, *Mechanics for the Millwright, Machinist, Engineer, Civil Engineer and Student* (Philadelphia, 1851), pp. 258 ff., and *Appleton's Dictionary of Machines*, 2:722–23.

[27] Letter of William Wilkins, Dec. 1861, in House Report 43, 37th Cong., 3d sess., p. 73.

[28] Ibid., pp. 83–84.

[29] The best discussion to come to my attention of the forging process, the character of forge hammers, and the problems of driving trip-hammers is J. Richards, "The Principles of Shop Manipulation for Engineering Apprentices," esp. "Forging," 98:197 ff. On the difficulty of driving old-style trip-hammers through line shafts, see also John G. Winton, *Modern Steam Practice* (Philadelphia, 1883), vol. 2, supplement, pp. 47–48.

[30] One of the early light power hammers in England is described in James Kitson, "Description of an Improved Friction Hammer," *PIME,* 1854, pp. 133 ff. The Palmer spring hammer made in Manchester, New Hampshire, had "the advantage of being operated by the same power as is used to run the works" (*Manufacturer and Builder* 5 [1873]:270).

market in the United States in the 1870s. Their common characteristic was the ability to take their motion without disturbing effects from the line shafts serving the factory as a whole. Operating on various principles and employing varying forms of spring and cushioning action with arrangements for varying the force and number of strokes, these hammers, within the limited range of their capacity, overcame at last the problems that in respect to power transmission had placed the forging of iron in a class by itself.[31]

Arrangements for the distribution of power were as varied as the character of the industrial enterprise and the requirements of the production processes and the machinery used. The so-called engineering works occupied a central role in any advanced industrial economy because here was produced the greater part of the machinery and equipment on which industry itself was dependent. While the engineering works itself took on varied forms according to the nature of the machinery produced, typically it comprised certain basic components or divisions. These were the foundry with its cupolas, pattern shop, and mold room; the machine shop with its assortment of standard and special machine tools; the forge or smith shop with its forge fires and power hammers; and the associated fitting or assembly shops. These departments and the supporting facilities, including power plant and transmission equipment, are shown in a drawing from a mechanical engineer's graduating essay published in *Scientific American*, 15 September 1860 (Fig. 109). Various considerations influenced the arrangement of the several shops, or departments, shown in a ground plot measuring 250 by 300 feet, the size of a small city block: space requirements; convenience of communication between related shops; segregation of noise, vibration, dirt; and, with the power plant employing steam, the explosion and fire hazard of the boiler house. Of the several shops only the storeroom and the foundry proper, having no machinery at this period except the provision of the blast for the

[31] See comment on power hammers, described as "peculiarly American in design," in Great Britain, Executive Commission, Philadelphia Exhibition, *Reports on the Philadelphia International Exhibition,* 3 vols. (London, 1877), 1:229–30. See also Charles H. Fitch, "Report on the Manufacture of Engines and Boilers," in Tenth Census, 1880, vol. 22, *Power and Machinery in Manufactures* (Washington, D.C., 1888), pp. 18–19. Descriptions appear in *Manufacturer and Builder* 5 (1873):145, 172 and 6 (1874):129; *American Artisan,* n.s. 1 (25 Oct. 1865):391, n.s. 4 (24 Apr. 1867):395 and (10 July 1867).

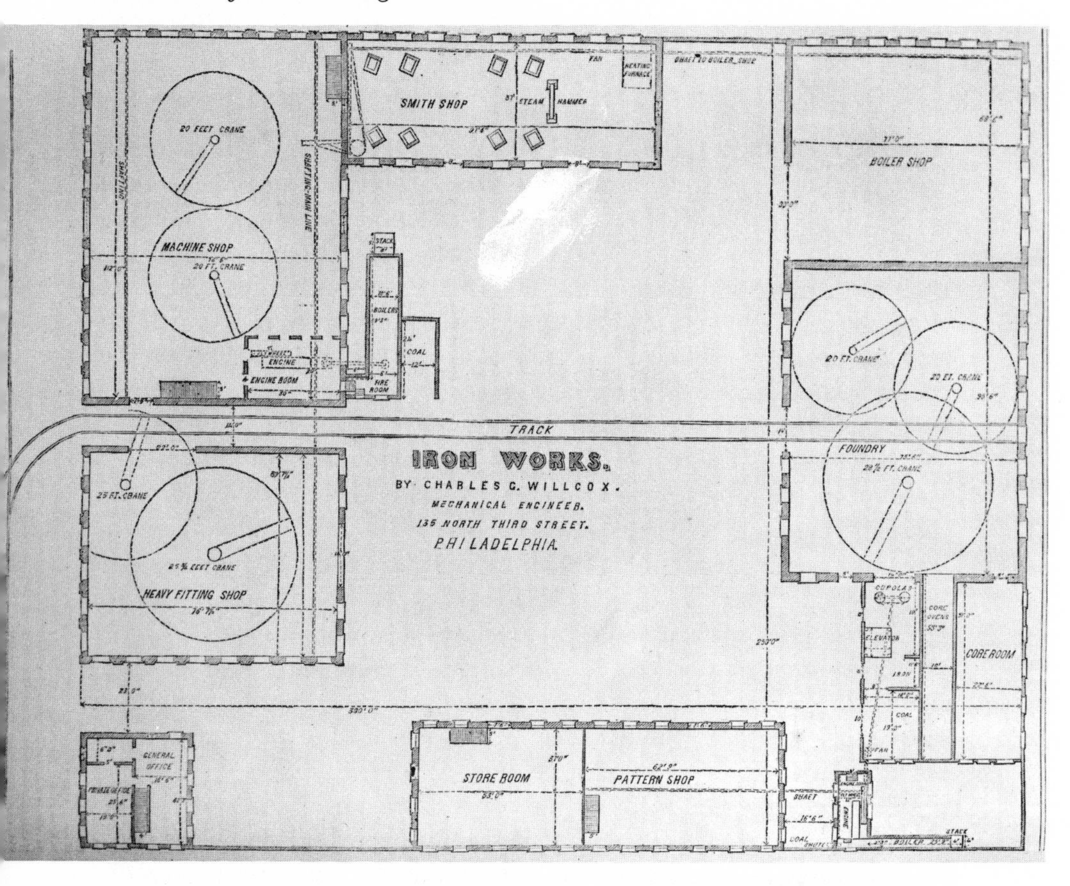

Fig. 109. An ironworks "for the purpose of manufacturing iron into boats, buildings, engines, machinery, and tools." Two engines power the works by means of shafting. The engine in the upper left powered the heavy fitting shop, the machine shop, smith shop, and boiler shop. The engine in the lower right of the drawing powered the pattern shop and the fan for the cupola blast. (*Scientific American,* Sept. 15, 1860.)

melting cupola, required no shafting for power distribution. The student designer remarked:

It is essential, in locating engines, to have them near their heaviest work, and so placed that power can be transmitted from them in all directions with the least possible expense of shafting. It is difficult, in extended works, to make one engine do all the work. It is generally necessary to divide it, and place two or three engines at points where they are most needed.

The boilers should be placed in buildings entirely independent of the main ones. Experience has shown that in explosions the large buildings suffer comparatively little if the boilers are somewhat removed from them.

Power Transmission in Mining

In mining operations (to which fuller consideration will be given in a subsequent volume) the problems and conditions of transmission were very different from those in surface establishments. The horizontal was replaced by the vertical as the predominant plane of motion, rotary by reciprocating motion, and the conventional millwork of shafting, wheelwork, and belting by what may be termed the "mine work" of rope and rod. During the greater part of the nineteenth century the use of mechanical power was largely confined to shaft or deep mines and to handling materials rather than to operations upon the materials, that is, the rocks and minerals themselves. This handling fell for the most part into two major divisions: the removal of minerals and waste from the mines by winding (hoisting) and the removal of water through pumping. Whereas from the nature of things these activities were located underground, the prime mover, usually a steam engine, was with infrequent exceptions and for compelling reasons located "at grass," the land surface adjacent to shaft mouth or pithead. Drainage of underground water was a widespread requirement in mining although the amount of water varied widely from region to region and from mine to mine. Pumps were the usual means of removal, save where the conditions of terrain favored natural drainage.

Whether of the bucket or plunger types, the barrel pumps employed were placed near the water-gathering sump at the bottom of the shaft. The reciprocating action of the pump was supplied by means of a string of wooden rods extending along the side of the shaft and taking its motion from the prime mover at the surface, either directly or through an angle bob (bell crank). As mine depths and water volume increased, the pump rod took on massive dimensions and weight and had to be offset at intervals along the mine shaft by balance bobs as counterweights. The whole assembly, or pitwork, which in the larger mines weighed scores and even hundreds of tons, was simply a huge connecting rod for conveying power from the source at the pithead to the pump hundreds or even thousands of feet below. This Cornish pumping system (named from Cornwall, the English county of origin) was the chief method of mine drainage during the prime decades of the first Industrial Revolution.

The winding rope or cable of the hoisting works served also in its different way as a kind of flexible connecting rod, varying in length as the cable or cage with its cargo rose and fell in the mine shaft. Although a slight affair compared to its shaft mate the pump rod, the hoisting rope in the deeper mines took on a weight comparable to and sometimes greater than the load it lifted. Moreover, this weight was variable, ranging from as much as several tons when extended to the shaft bottom to almost nothing at the end of the hoist, adding materially to the size and inefficiency of the winding engine. In time these difficulties were substantially mitigated by such expedients as conical winding drums, tapered ropes, and "balanced" hoisting. In a kind of horizontal extension of the use of the winding rope, power eventually was applied to underground haulage of minerals in the larger mines, especially coal mines. Difficulties in the transmission of power underground in mining operations proper—the opening of passages and extraction of minerals—effectively discouraged mechanization of mining on a significant scale before the closing decades of the nineteenth century.

New Conditions and New Needs

With the progressive advance of industrialization, industrial establishments assumed protean forms that often did violence to the conditions of compact and orderly arrangements most favorable to the use of conventional millwork. Such varied establishments as railroad shops, shipyards, oil refineries, chemical works, and plants making heavy machinery and equipment presented problems in power distribution that were not readily or effectively met by means of line shafts, pulleys, and belting. Of particular importance were those branches of industry in which the character of the materials, processes, and equipment required extensive sites and numerous structures. In many instances neither the production sequences, the character of the processing, nor the equipment used was adapted to power distribution from a central plant. As we shall see in a later volume, decentralization of power supply with a scattering of small steam engines provided a stopgap solution.

Between flouring mills on the Evans plan and the early textile mills there was little difference in the millwork. The textile mill had more of it, the dimensions were larger, and the ancient wooden cogwheels of the crown, spur, and lantern or trundle types had been replaced by cast-iron spur and bevel gear wheels.[32] Zachariah Allen of Rhode Island later recalled the method of transmitting power when he built his first cotton mill at Allendale in 1822:

[32] See Evans, *Miller's Guide,* on gear wheels, and *Knight's American Mechanical Dictionary* (1877), s.v. "Gearing."

It was deemed necessary to locate a mill directly over the waterwheels; so that a main upright shaft might be arranged upward through the several stories, to transmit the power more directly to a main horizontal shaft in each room, to distribute the power to each machine.

The shaftings were all made square, to receive the cast-iron wheels fastened by wedges. The pulleys were made of wood, by clamping together pieces of joists, notched to fit the shafts, by means of screw-bolts. Instead of the numerous light pulleys, now used, long wooden drums were built around the shafts, and made of boards nailed upon circular plank heads. With the slow speed of forty or fifty turns per minute, some of these drums were necessarily made three or four feet in diameter and several feet long, darkening the rooms by their ponderous magnitudes, and requiring very high ceilings to admit them.

These great drums being frailly nailed together, and unbalanced, could not be used with quick revolving movements, without shaking them to pieces, and also shaking the floors intolerably.[33]

The structural vibration reported by Allen came generally to be recognized as wasteful not only because of the power absorbed in the movement of the building but from the resulting distortion in the alignment of the shafting, accompanied by increased friction at the points of bearing.

British Innovations in Millwork

American millwork in the early years did little more than repeat, with a generation lag, that used in England. The typical English cotton mill of 1800 was equipped with main shafting four to six inches square and moving at 30 to 40 rpm to carry the power developed by a 50-horsepower steam engine.[34] It was not until some years later that what inevitably came to be called "the shafting revolution" began to gather momentum in Britain.

The incentive to change arose first in cotton textiles, where the great advances in spinning machinery brought demands for improvement in supporting and auxiliary equipment.[35] The drive for greater pro-

[33] Zachariah Allen, "The Transmission of Power from the Motor to the Machine," *Proceedings of the New England Cotton Manufacturers' Association* 10 (1871): For other references to early millwork, see idem, *Science of Mechanics,* pp. 249 ff.; discussion following Robert Allison, "The Old and the New," *TASME* 16 (1894–95):747 ff.; *The Locomotive,* Oct. 1883, pp. 169 ff.; Robert Brunton, *A Compendium of Mechanics,* ed. James Renwick (New York, 1830), pp. 95 ff.

[34] Ure, *Philosophy of Manufactures,* pp. 32 ff.

[35] See George Rennie's preface to Buchanan, *Practical Essays,* p. xviii.

duction and profits in British cotton mills required higher machine speeds, and these in turn led to demands for millwork capable of supporting such speeds. Wood in millwork was the first to go; it was replaced by cast iron first in wheelwork, especially gearing, and then increasingly in main drive shafts and the line shafts they placed in motion. Wooden shafting and wheelwork could not stand up under the loads imposed and the speeds required by an advancing industry. The replacement of wood by iron machine components began with the work of John Smeaton at the Carron Iron Works about 1769 and of John Rennie at the Albion flour mills in London during 1784-85.[36] With the limitations on speed imposed by the crude wooden wheelwork removed, attention was directed to shafting. The general replacement of timber shafting by cast-iron shafting, declared Robertson Buchanan in 1814, opened a new era in the history of millwork.[37]

In the United States some cast-iron gear wheels were used in flour mills before 1800, and by 1820 cast-iron bevel wheels had generally replaced the clumsy crown wheels used with the lantern wheels—wooden pin teeth meshing awkwardly with wooden staves—to change the direction and speed of motion in flour mills.[38] By the 1830s, at least in the older and more settled parts of the country, the shift from wooden to iron wheelwork in mills and factories was well advanced, as shown in the drawing of the Evans flour mill of 1826 (Fig. 110).[39] The brittle character of castings was offset in part by making them large and heavy. Even so the readiness with which gear wheels and other mill irons could be obtained from local foundries stocked with patterns of the styles and sizes most in demand led to their wide introduction.

The substitution of iron for wood in wheelwork, with shafting and other millwork to follow, was part of the larger transition from a technology of wood to one of iron basic in the process of industrialization. The strength and durability of iron made possible the use of power in larger amounts, in a more concentrated manner, and at far higher speeds than were feasible with wood. As we have noted, the compactness of iron, the intricate forms it could be given by casting, forging, and machining, and the dimensional precision possible in its shaping

[36] Ibid., pp. xii ff. See also Evans's 1806 memorandum in Bathe and Bathe, *Oliver Evans,* p. 126; Rees, ed., *The Cyclopaedia,* s.v. "Mill-Work."

[37] Robertson Buchanan, *Essay on the Shafts of Mills* (London, 1814), pp. 15 ff.

[38] Evans, *Miller's Guide* (1795), p. 55, and manuscript schedules of the 1820 federal census.

[39] Evans, *Miller's Guide* (1834), pp. 373 ff.; Allen, *Science of Mechanics,* pp. 255 ff.

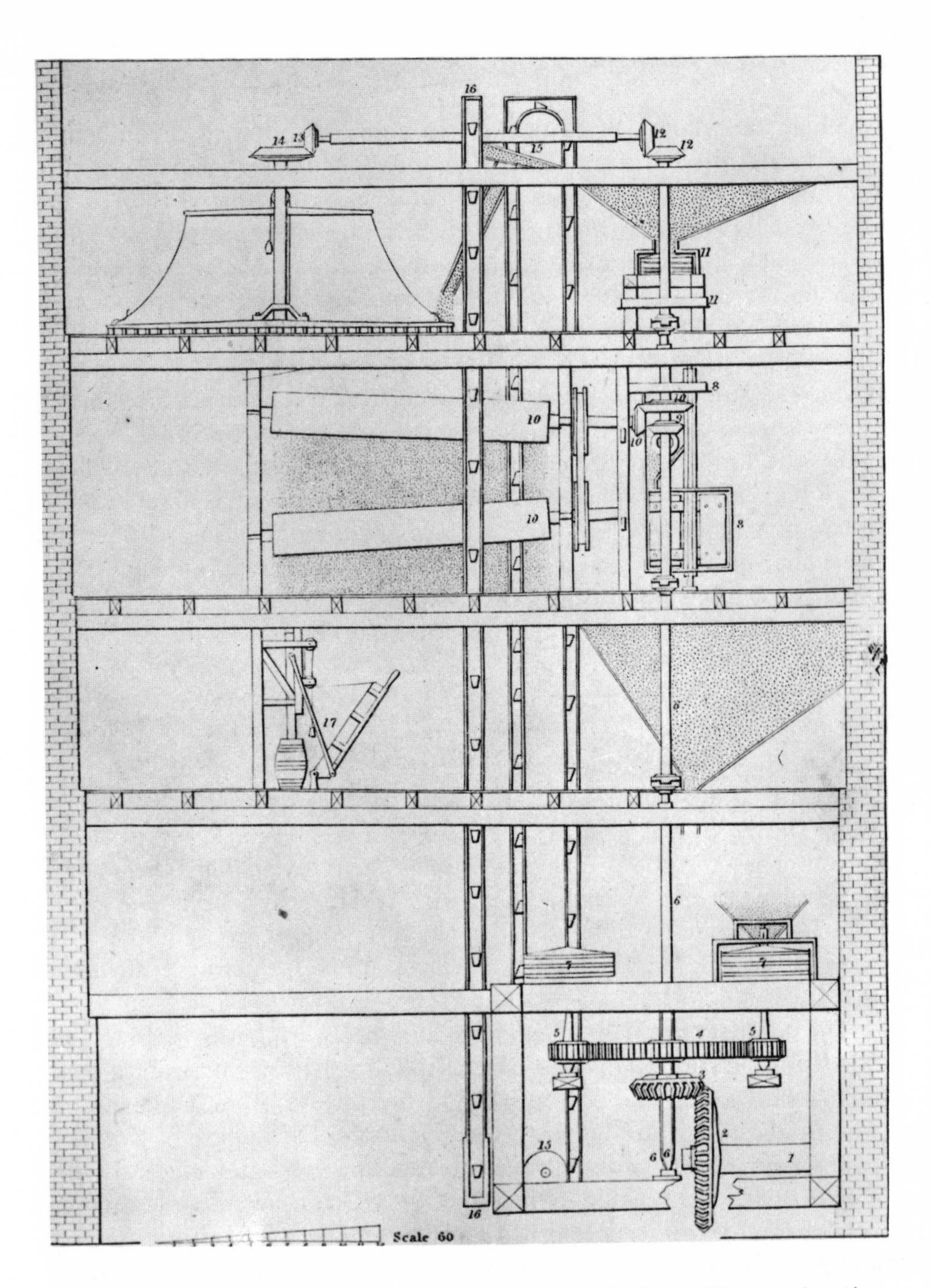

Fig. 110. Merchant flour mill, gears driving four pairs of five-foot millstones. A major advance in transmission by millwork is illustrated above in a mill designed by two sons of Oliver Evans. Built some thirty years after the initial publication of the *Miller's Guide* (1795). Here the wooden shafting and cogwheels were replaced by cast-iron shafts and wheelwork. Mills and factories equipped in the manner illustrated here mark the transfer to America of the transition from wood to iron in the millwork begun in Britain a half century earlier by John Smeaton and John Rennie. (Oliver Evans, *The Young Millwright and Miller's Guide,* 1850 ed.)

made it indispensable for the general introduction of machinery and power. With the new iron gearing leading the way, the design and fabrication of millwork were transferred from the millwright and mill carpenter to patternmakers, foundrymen, and, shortly, machinists. But these craftsmen were little better equipped than their predecessors to cope with the problems of gear-tooth and wheelwork design requiring for their solution the talents of mathematicians, geometers, and engineers.[40]

Millwright handbooks of the Evans type with tables of dimensions and proportions and rule-of-thumb directions were sadly out of date in the emerging age of industry and iron. The scientific foundations of gear design had been laid on the Continent in the seventeenth and eighteenth centuries, but it was not until the second quarter of the nineteenth century that British mathematicians and engineers bridged the gap between theory and application. While such European authorities as Philippe de la Hire, Robertson Buchanan, and Robert Willis were often cited, summarized, or quoted in such technical publications as the *Journal of the Franklin Institute* and considered at some length in Rees's *Cyclopaedia* and *Appleton's Dictionary of Machines,* the short-run influence upon American foundry and machine shop generally was probably slight.[41] Down to the Civil War years at least, reliance was chiefly upon cast-iron gearing with traditional design, used as taken from the casting flasks with some hand trimming and finishing. It was only too commonly assumed that gear teeth would wear into shape by using no lubricant, by adding emery to the lubricant in a kind of grinding compound, or by giving an uneven ratio to the number of teeth in the driving and the driven gear wheels. As late as 1876 we are told that the teeth of gearing "are too often merely smoothed with a file, or even left as they come from the foundry."[42] Derogatory contemporary comments on the quality of much mill gearing are by no means to be taken as evidence that American foundries and machine shops were necessarily laggard in standards and performance. Rather it would seem that the great majority of manufac-

[40] See Rennie's preface to Buchanan, *Practical Essays,* pp. vii–xxx.

[41] The article "Geering" in *Appleton's Dictionary of Machines,* "has been taken from *The Engineer and Machinist's Assistant* and is generally admitted to be the best yet published."

[42] Benson J. Lossing, *The American Centenary* (Philadelphia, 1876), p. 477; see also *Appleton's Dictionary of Machines,* 1:796; Evans, *Miller's Guide* (1860), Art. 82, pp. 201–2; Nicholson, *Operative Mechanic,* p. 44; *Scientific American* 14 (31 Mar. 1866): 209. On the wisdom of the cold-shaping of gear teeth with file and chisel, see Allen, *Science of Mechanics,* pp. 259–60; also Rees, ed., *The Cyclopaedia,* s.v. "Mill-Work."

tories, small in scale and limited in resources, felt no pressing need for precision machining either in millwork or in most of the products made. Precision work was costly. In millwork and other simple metal products common castings gave adequate service, and cheapness had a compelling appeal that most buyers of equipment were unable to resist in the capital-scarce American economy.

In England the first production gear-cutting machines, indispensable for obtaining precision of tooth form and dimensions, date from about 1820.[43] Robert Hoe, builder of printing presses and other machinery in New York, is said to have secured as early as 1830 "the first and for many years the only large machine in our country for cutting to correct theoretical shape the teeth of gearing." In 1840 one of the earliest American gear-cutting machines, invented by W. M. Hartshorne, was awarded a premium by the Franklin Institute. By mid-century gear-cutting machines had been developed by the Lowell Machine Shop and by George H. Corliss. To many engineers and mechanics visiting the great machinery hall at the Philadelphia Centennial Exhibition, 1876, the striking feature of the huge Corliss installation was not so much the spectacular but technically undistinguished double steam engine as the "nests" of massive machine-cut bevel gears, six feet in diameter, by which power was taken at four points from the main drive shaft for distribution throughout the exhibition hall. This was possibly as elaborate an installation of millwork as was to be seen in the Western world. The main drive shaft was 352 feet long; the four lateral shafts were 108 feet each. Through belts 30 inches wide and 70 feet long, they placed in motion more than 1.5 miles of overhead shafting supplying with power the thousands of machines on exhibition.[44]

There were frequent comments in technical journals in the late 1860s upon gears that rattled and jarred, often producing an almost deafening clatter as the teeth ground upon each other. The principles of gear cutting were ignored, declared one editor, and for every gear running almost noiselessly at high speed there was a multitude so

[43] Robert S. Woodbury, *History of the Gear-Cutting Machine* (Cambridge, Mass., 1958), pp. 65 ff.

[44] Lossing, *American Centenary,* p. 477; *JFI* 30 (1840):324; *Appleton's Dictionary of Machines,* 1:777–84. For the Lowell and Corliss machines, see *JFI* 103 (1877):4–5; and Samuel S. De Vere Burr, *Four Thousand Years of the World's Progress* (Hartford, Conn., 1878), pp. 270–75. See also reference to the high quality of gear wheels and gear-cutting machinery in J. F. Radinger, "Spur Fly-Wheels in the United States," abstract from *Dingler's Polytechnisher Journal* 229 (1878): 114. A description of a British gear-cutting machine, invented by Mr. Davies, appeared in *JFI* 29 (1840):202.

noisy as to prevent ordinary conversation nearby.[45] The great expense of gear-cutting machines was the main objection to their use; many machine shops evidently depended on contracting out such work when it could not be avoided. Relief was obtained in larger establishments through two expedients. One was the widely adopted practice of giving one of a pair of matching gear wheels wooden cogs in what was known from the manner of their insertion as the mortise wheel.[46] This arrangement gave smoother and quieter running. The other was the practice of much wider implications, to be discussed below, of replacing gear wheels in the main drive train from the prime mover by belts of appropriate size and strength.

Geared wheelwork might be termed the backward branch of American millwork, yet for many years it played an important part at the lower middle levels of industry. For all the noise, vibration, and friction accompanying its use, toothed gearing possessed two persuasive merits: it worked, and, in the common cast form, it was cheap. Yet this wheelwork was only one of the accessories of millwork, and as was demonstrated in this country from the early 1830s, the least essential. Its function was that of turning corners and changing speed ratios, a role that could be and, as we shall see, was taken by other means. As demonstrated at Lowell and elsewhere, smooth-faced pulleys and belting could for most purposes do these jobs more satisfactorily. As industrialization gathered momentum, the expanding scale of plant operations soon demanded a *system* of power distribution rather than merely workable arrangements for getting power from prime movers to machines. Under such conditions line shafts, pulleys, and belting and their accessories played the central role in millwork. Line shafts, as we have seen, were the prime arteries for conveying power. The course of their evolution was simple and clear. The trend in material was from wood to cast iron to wrought iron and eventually to mass-produced low-carbon steel. In operation the trend was toward ever higher rotating speeds, accompanied by proportionately smaller diameters. In fabrication the trend was toward ever greater concern with accuracy of form and dimension. There was a growing recognition

[45] *American Artisan,* n.s. 5 (9 Oct. 1867); *Scientific American* 14 (31 Mar. 1866):209; *Van Nostrand's Engineering Magazine,* 1:658; Thomas Dixon, *The Practical Mechanic and Engineer's Ready Reckoner* (Philadelphia, 1868), pp. 7 ff., 39.

[46] See description in *Appleton's Dictionary of Machines,* 1:785–86. See also Rees, ed., *The Cyclopaedia,* s.v. "Mill-Work"; Allen, *Science of Mechanics,* p. 262. The use of mortise gears, about 1850, was infrequent and experience with them not very satisfactory (remarks of Scott in Allison, "The Old and the New," p. 751).

that precision was hardly less essential in the means of distribution than in the machinery for generating and applying power. A factory's millwork, in short, was in the aggregate a great machine itself in the production and maintenance of which care and accuracy were demanded. Although anticipated earlier in New England textile circles, not until well after the Civil War did this view begin to obtain wide acceptance.

The Shafting Revolution of Fairbairn and Lillie

Owing to the far greater cost of wrought iron, mill shafting had long been made of cast iron. Low tensile strength and the slow speeds in general use required large diameters and heavy weight. The square or polygonal cross section of wooden shafting was taken over to accommodate the attachment and centering of gear wheels and pulleys by wedging.[47] Cast in relatively short lengths, this shafting required numerous bearings and couplings, adding to first cost, power waste, and maintenance expense. William Fairbairn, who with his partner, James Lillie, led the way in the shift from cast-iron to wrought-iron shafting, declared that the former at times absorbed almost as much power as the machinery driven. Ure cited an English mill in which shafting ranging from eight to four inches in diameter was used to distribute the power developed by a 50-horsepower engine. Zachariah Allen recalled his experience in 1857 with cast-iron shafting in a large New England cotton mill. Distribution of power from four large waterwheels was initially made through line shafting twelve inches in diameter in ten-foot lengths. Each length weighed 3,500 pounds, and the couplings for joining them weighed 3,800 pounds each. "So great was their weight and rigidity," declared Allen of these shafts, "that the settling of the foundations, changes of temperature, and the continual jar caused them to break so frequently as to render necessary the replacement of them by new wrought iron shafts."[48]

[47] Two discussions of millwork by Zachariah Allen separated by over forty years are of particular interest: *Science of Mechanics,* pp. 249 ff., and 'Transmission of Power," pp. 15 ff., 72 ff. See also Fairbairn, *Treatise on Mills and Millwork,* 2 pts. (London, 1861–63), pt. 2, pp. 104 ff.; and Ure, *Philosophy of Manufactures,* pp. 32–37.

[48] Fairbairn, *Treatise,* pt. 2 (1863), p. 72; Ure, *Philosophy of Manufactures,* pp. 36–37; Allen, "Transmission of Power," p. 17. Allen notes the persistence of the obsolete but does not explain it. Less surprising was a similar persistence on what might be termed the industrial frontier in Michigan. An engineer recalling his millwright apprenticeship in the early 1850s stated: "The shafting was six-sided cast-iron, and

The heavy shafting was matched in bulk if not in weight by the large, drumlike pulleys by which the power was conveyed through belts between line shafts and countershafts and thence to individual machines. Slow shafting speeds had to be offset by giving large diameters to the drive pulleys to secure the desired machine speeds. Roughly made of planking and boards, these "drums" often occupied the greater part of the shafting for convenience in the placement of machinery. They were difficult to place in position and to move, occupied much space, cut off badly needed light, and with the shafting on which they were mounted occasioned much friction and power waste.[49] As Zachariah Allen remarked, "Before the systematic balancing of pulleys commenced, no inconsiderable portion of the power was transmitted to shake the floors and even to cause some of the mills to rock to and fro."[50]

Taking the rude, cumbersome millwork of wood and cast iron as a baseline, three important advances were made in conventional transmission equipment before it was rendered obsolescent by the transmission revolution of the late nineteenth century. First and most important was the widespread introduction of lathe-turned wrought-iron shafting of small diameter running at high speed in place of the ponderous, slow-moving shafting of the early industrial years. This innovation was first made in England about 1820 and by 1840 was finding some acceptance in the United States. The second and distinctively American contribution to the new millwork was the replacement in the main drive train—linking prime mover with line shafting—of heavy shafting and gear wheels with belting and pulleys. The third and subordinate, though indispensable, advance was in raising the auxiliary equipment, such as hangers, couplings, and pulleys, to the mechanical level of the high-speed shafting. These innovations made available to the well-designed factory a system of power trans-

the eyes of the pulleys and gears were the same shape and these were fitted to the shafts with iron wedges. . . . It required very skillful workmanship to fit these keys or wedges with a cold chisel and file so they were true upon the shaft" (Emory F. Skinner, *Reminiscences* [Chicago, 1908], p. 27). In 1850 a New England cotton mill of 5,000 spindles and 70 horsepower used a cast-iron vertical drive shaft to carry power from the prime mover to machinery on the floors above. The line shafts were of wrought iron, ranging from 3.000 to 2.125 inches in diameter (Baird, *The American Cotton Spinner,* p. 36).

[49] See Webber and Worthen in discussion of Allison, "The Old and the New," p. 745; Evan Leigh, *The Science of Modern Cotton Spinning,* 2d ed., 2 vols. (Manchester, 1873), 2:273 ff.; "Notes on Mill Shafting," *The Locomotive,* Oct. 1883, pp. 169 ff.

[50] Allen, "Transmission of Power," pp. 15–18.

mission comparable mechanically to the prime mover and the machinery served. Important as these advances were, they were simply refinements of the basic millwork elements present in the Evans flour mill.

In his 1814 essay on the shafts of mills, Robertson Buchanan summed up the progress of the previous generation: the use of cast iron in place of wood in shafts and other millwork was almost universal. Its introduction had marked "a new kind of era in the history of mills," since in its absence the existing labor force in Britain could not have built one-tenth of the machinery erected in recent years.[51] The year 1815 was the point from which William Fairbairn later dated the beginnings of the new system of light and fast-moving shafting with which he and his partner, James Lillie, were so closely identified. Main shafts that thirty years earlier had moved at only 30 to 40 rpm by 1835 were turning at 120 to 150 rpm and occasionally in throstle spinning as fast as 200 rpm. The result was a 50 percent reduction of the mass of the millwork. Such practice, made possible with lathe-turned wrought-iron bars, permitted the shafting to be "strung like wires along the ceiling, and with small pulleys transmit the motion to the machinery without crowding the room or obstructing the light."[52] Indeed Ure reported of visits to millwright factories that he had difficulty in knowing "whether the polished shafts that drive the automatic lathes and planing machines were at rest or in motion, so truly and silently did they revolve."[53] In the case of some old mills remounted with the new millwork by Fairbairn and Lillie, an increase in power of fully 20 percent was obtained from the same prime mover. Redesigned and better made gear wheels, pulleys, hangers, and bearings were built to support and supplement the new shafting; and particular emphasis was placed upon securing and maintaining accurate alignment of the shafting.[54]

[51] Buchanan, *Practical Essays on Millwork,* pp. 176–77.

[52] Ibid., ed.'s note, p. xix, quoting letter from Fairbairn.

[53] Ure, *Philosophy of Manufactures,* pp. 32–37.

[54] Fairbairn, at first in partnership with Lillie and then by himself, engaged in the large-scale production of millwork. In an obituary *The Engineer* noted that his forte lay in millwright work. Not content with marked improvements in waterwheels, "he revolutionized the art of making mills. . . . He gave the milling world new shafting, new couplings, new gear accurately made, and properly proportioned to the work to be accomplished. . . . Fairbairn found millwrighting a second-rate trade. He abolished the millwright, and introduced the mechanical engineer" (quoted in Pole, *Sir William Fairbairn,* p. 436).

The New System in the United States

That the capacity of shafting to convey power increased with the speed of rotation was a familiar fact to all readers of Buchanan's essay of 1814 and Allen's *Science of Mechanics* of 1829.[55] The water shops built at the Springfield Armory in 1831–32 were equipped with shafting of but half the diameter and running at a far higher speed than that previously in use.[56] The major impulse to change in this as in other matters relating to motive power came from the new large-scale textile industry of Massachusetts. Hamilton Mills at Lowell, undoubtedly representing the best in traditional millwork practice about 1830, was powered by massive breast wheels, 13 feet in diameter by 14 feet across the face. These made but 6 rpm, a speed that was increased through segmental gearing to give the first-motion drive shafts 22 rpm. This speed was increased successively to 33 rpm in the heavy vertical cast iron shafts serving the upper floors and to 83 rpm in the line shafts taking their motion from the latter through bevel gearing and conveying it through belting and pulleys to the individual machines with whatever further increase in speed was appropriate to each class.[57]

The new order was already in the making. In his travels through England's industrial regions in 1825 Zachariah Allen became acquainted with the new practice of employing precision-made, fast-running shafting. Four years later he reported that in some of the more recently built cotton mills in this country, the speed of the upright shafting had been more than doubled, reaching 80 rpm.[58] In the Appleton Mills at Lowell, Paul Moody's pioneer installation of a main drive belt in place of shafting and gear wheels employed some-

[55] Allen, *Science of Mechanics,* pp. 249 ff. See also table in Brunton, *Compendium of Mechanics,* pp. 97–98, giving diameters of shafts required to carry speeds of 10 to 105 rpm. For a brief discussion of the limits on the extent to which this practice could be carried, see Richards, "The Principles of Shop Manipulation," 96:395 ff., 97:31 ff.

[56] The new shafting was designed by Nathaniel French, a mechanic who came to the armory from S. North's pistol factory at Middletown, Connecticut (Charles H. Fitch, "Report on the Manufacture of Engines and Boilers," p. 35).

[57] Beard, "Practical Observations on the Power Expended," pp. 6 ff.

[58] Zachariah Allen, Memorandum, Europe, 1825, Rhode Island Historical Society, Providence (courtesy of John Crnkovich). Idem, *Science of Mechanics,* pp. 261–64. Allen noted it was desirable "to increase the velocity of mill gearing gradually, beginning with the first motion rather slow by means of wheels and pinions of large diameter" (ibid., p. 262).

what lighter and faster-running line shafts.[59] When the new system was introduced at the Allendale cotton mill in 1839, the plan was to increase line-shaft speeds from 90 rpm to more than 200 rpm in order to reduce the size and weight of shafting and pulleys in nearly corresponding ratio. The goal: economy of costs and motive power. Allen in 1871 described some of the necessary innovations:

> A wheel pit was requisite outside of the mill building, in a separate wheel-house, for the double purpose of obtaining more space for larger cog-wheels, to get up the requisite speed, and of excluding the noxious steamy dampness arising from all water-wheels shut up within the mill walls. The pulleys in previous use, with ground or turned surfaces, would not operate quietly, without being turned inside as well as outside to balance the rims, and prevent the tremor consequent on the use of all unbalanced pulleys revolving rapidly. This improvement also reduced the weight of the pulleys to correspond with the reduced weight of the shafting and wheels, made of half the previous diameters. . . . This experiment, deemed somewhat wild at the time, proved successful. It has gradually been adopted as an economical system of transmitting power from motors to machines by high velocities of shafts and belts.[60]

Higher speeds made possible lighter shafting and pulleys but imposed much greater care in the manufacture of this equipment and in its installation and maintenance. The wrought-iron bar as it came from the rolling mill was neither very round nor wholly straight. It gave trouble at the higher speeds and had to be lathe-turned to precise dimensions, adding to its cost and limiting its production to machine shops with the special lathes required, a problem noted as early as 1829 by Allen.[61] Thomas Thornycroft, an English engineer, noted about 1850 that "previous to the introduction of the slide lathe, the shafting employed in spinning manufactories was a constant source of vexation and expense; the want of that perfect parallelism which is now obtained exposed the shaft to vibration or bending at every revolution; the consequence was constant fractures." The same problem was met with in shafting at ironworks, in waterwheels, in

[59] George S. Gibb, *The Saco-Lowell Shops: Textile Machinery Building in New England, 1813–1949* (Cambridge, Mass., 1950), pp. 79–80.

[60] Allen, "Transmission of Power," pp. 17–18.

[61] Allen, *Science of Mechanics,* pp. 249 ff.; see also idem, "Transmission of Power," pp. 16 ff.; Leigh, *Science of Modern Cotton Spinning,* 2:273 ff. A millwright's handbook noted in 1882 the problems of proper size and avoiding defects. Shafting turned within tolerances of a hundredth of an inch were available in any good shop and as low as one two-hundredth of an inch at others (Robert Grimshaw, *The Miller, Millwright and Millfurnisher* [New York, 1882], pp. 188–89).

marine engines, and in railway axles.[62] By 1850 the use of turned shafting was making headway, and by the end of the Civil War it was in wide use.[63] The narrowing of the margin of cost between cast and wrought iron favored the use of the latter. According to an account of 1888, over the previous thirty years the cost of wrought-iron engine shafts had fallen from several times that of cast-iron ones to only double the cast-iron cost.[64]

Further marked improvement came with the introduction about 1860 of "cold-rolled" shafting, made first of wrought iron and later of steel. In the finishing process this shafting was passed cold through special rolls "by which the metal is condensed and its strength increased, while it becomes remarkably straight, true, and uniform in size," with a highly polished surface. Owing to its greater tensile strength, cold-rolled shafting was reported to bring economies of about one-third in cost, weight, and power consumption as compared with ordinary turned iron shafting.[65] In their enthusiasm for the new system of millwork American mill engineers, in the textile industry at least, carried it further than was done in Britain, the country of origin. A New England textile manufacturer who had returned in 1867 from a trip to observe British mill practice reported much slower speeds of shafting and a much heavier weight of iron in the millwork of British cotton mills: "Our new mills contain scarcely one-fourth of the amount used there." While the length of line shafts in American textile mills did not often approach that found in some of the larger En-

[62] "On the Form of Shafts and Axles," reprinted in *JFI* 51(1851):56 ff.

[63] Grimshaw, *Miller, Millwright and Millfurnisher,* p. 188; Egbert P. Watson, *The Modern Practice of American Machinists and Engineers* (Philadelphia, 1867), p. 85; William Sellers and Company, *A Treatise on Machine Tools,* 4th ed. rev. (Philadelphia, 1877), pp. 184 ff. Specifications for millwork drawn by the Corliss Steam Engine company in the late 1840s called for shafting "turned and polished" (Order Book, 1847–50, Corliss Papers, John Hay Library, Brown University, Providence).

[64] Emory Edwards, *The American Steam Engineer* (Philadelphia, 1888), pp. 160–62.

[65] *Manufacturer and Builder* 10 (1878):273; W. H. Barlow, in discussion of H. Robinson, "The Transmission of Power to Distances," *PICE* 49 (1876–77):52–53. See also *Scientific American* 17 (17 Nov. 1867):305; R. H. Thurston, *Report on Cold Rolled Iron and Steel Shafting as Manufactured by Jones and Laughlin's American Iron Works, Pittsburgh* (Pittsburgh, 1878), pp. 6–7, 51 ff., 98; *Appleton's Cyclopaedia of Applied Mechanics,* s.v. "Shafting." Hollow iron (pipe) shafting was introduced in this country to a limited extent; in some instances it was made up to 12 to 15 inches in diameter, serving as a continuous drum taking belts at any point. It was used in the Georgia Mill at Providence, Colt's Armory at Hartford, and the Wheeler and Wilcox plant at Hartford. See *Van Nostrand's Engineering Magazine,* 1:333; Samuel Webber, "The Frictional Resistance of Shafting in Engineering Establishments," *TASME* 7 (1885–86):147–49; and Leigh, *Science of Modern Cotton Spinning,* 2:277.

glish mills, the size of shafting was carried to greater extremes. An extreme instance was reported by James B. Francis. Owing to the rapid increase in power requirements and the necessity of conserving the limited waterpower resources, the textile mills at Lowell were using smaller shafting at higher velocities, using lines but one and a quarter inches in diameter.[66]

The American Contribution: Belting in Main Drives

The most distinctive and important American contribution to power distribution within factories was the substitution of belting for the heavy shafting and toothed gearing previously used in Britain and this country in the main drive train of the mill. The use of endless bands, straps, or belts made of gut, rope, leather, or canvas for the purpose of conveying motion goes back far in the history of manufactures. The early use of belts appears to have been as a component of the machine itself; their employment to obtain motion from a power source was a natural extension of the practice. The use of belts in the Evans flouring mill has been noted. In early textile mills the belt became the common method of bringing motive power from the line shafting to the machines.[67] To extend its use from the service of individual machines to the main drive of an establishment was a far more ambitious undertaking. Heavy spur and bevel gearing combined with proportionately heavy shafting from the wheel shaft of the prime mover to the line shafts on the several floors had long been used for this purpose. From serving the individual machine with a fractional horsepower to bearing the entire power load of the establishment was a quantum jump. This use of what must have seemed a slight ribbon of leather must have required courage no less than imagination. Even with its relatively slow motion the older millwork, a major source of industrial accidents, was something to be handled with care and caution.

[66] A. D. Lockwood at Boston, 17 July 1867, in "English Cotton Manufactures," *Proceedings of the New England Cotton Manufacturers' Association* 11 (1872): 21; William Sellers and Co., *Treatise on Machine Tools,* pp. 207–8. The whipping effect—constant twisting and uncoiling of long lines of slender shafting, especially in loom sheds—was often countered by installing balance or flywheels at intervals. See Allen, "Transmission of Power," p. 27; *Proceedings of the New England Cotton Manufacturers' Association* 11 (1872):8–10; Fairbairn, *Treatise on Mills and Millwork,* pt. 2 (1865), p. 107; Buchanan, *Practical Essays,* p. 286.

[67] See Gibb, *Saco-Lowell Shops,* pp. 79–80.

There was occasional use of belts or "bands" in main drives in this country even before 1800, but only in establishments not much larger than country mills, for the most part in commercial sawmills, known as "band mills."[68] As early as 1825 the practicability of using belts in main factory drives was being urged in the East. "Power to almost any extent may be communicated through [leather bands] . . . . it is only necessary to increase their widths under proper tension proportionately to the increment of power required to be imparted." The use of "bands" greatly diminished friction, eliminated "the disagreeable noise which attends the movement of toothed machinery," and saved "considerable expense in construction and repairs."[69] The Philadelphia firm of Sellers and Pennock noted two further advantages: greater convenience in the arrangement of machinery and elimination of the unpleasant and often injurious jarring of machinery when toothed wheels were thrown into gear.[70] A Pittsburgh foundry in 1832 reported a preference for straps over gear wheels in western Pennsylvania sawmills, which were probably equipped either with the quick-moving muley saw or the recently introduced circular saws. Straps, according to this firm, are "more simple, easier worked and more easily repaired in case of accident. Our best mechanics prefer straps, and if made strong enough they will stretch but little and are not liable to break."[71]

These early instances of belt drive are illuminating if not in themselves very impressive. An event in 1828 in New England was of far greater significance. Belting took the place of wheelwork in the main drive of the newly built Appleton Mills at Lowell, probably the largest textile manufactory in the country at the time. Paul Moody, the millwright in charge of construction, in adapting the belt to the main drive "did away with the heavy English-type gearing which had creaked and groaned in every American mill from 1790 on . . . [and] touched off a new American style which soon came to constitute an important distinction between English and American mills."[72]

[68] See *Dictionary of American English,* s.v. "band mills"; *American Mechanics' Magazine* 1 (4 June 1825):300–1, communication headed "Leather Bands" by Sellers and Pennock, Philadelphia.

[69] *American Mechanics' Magazine,* 1:300–1. See also the brief item by Joseph Backwell, "On the Comparative Advantages of Employing Band-Wheels, Spur-Wheels, and Bevel-Wheels in Machinery," reprinted ibid. 2 (1826):7.

[70] Ibid., 1:300–1.

[71] Kingsland, Lightner, and Company, Letterbook no. 2, pp. 221, 239, Western Pennsylvania Historical Society, Pittsburgh.

[72] Gibb, *Saco-Lowell Shops,* p. 79. See also Samuel Batchelder, *The Introduction and Early Progress of the Cotton Manufacture in the United States* (Boston, 1863). "Though not

Fig. 111. British cotton mill, main drive by gear wheels and shafting. (William Fairbairn, *Treatise on Mills and Millwork,* pt. 2 [London, 1863].)

The following elevation and plan will represent a mode of gearing with belts in which the stress is as nearly equalized as is practicable, and causing the least possible waste of power and wear of journals by friction, the stress upon one main belt being counteracted by that upon another in an opposite direction, and thereby throwing the power of one belt into its opposite, instead of the bearings of the main line of drums from which the whole power of the mill is taken.

Elevation of a Cotton Mill.

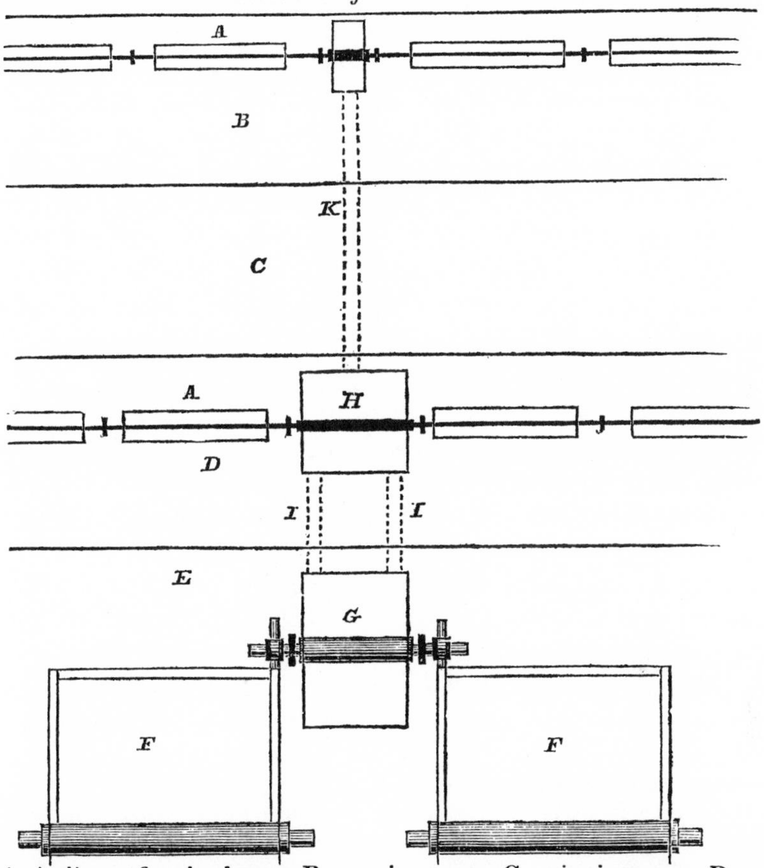

A A, lines of main drums; B, weaving room; C, spinning room; D, carding room; E, basement; F F, water wheels; G, main pulley; H, carding and spinning main pulley; I I, belts; K, belt to drive weaving and dressing.

In this elevation the two main belts, from the main pulley, sustain the whole power of the mill. Two such belts from 12 to 15 inches wide, made of the best of stock, are capable of operating 4,000 throstle spindles, with all the accompanying machinery for manufacturing coarse, heavy cotton goods.

Fig. 112. American cotton mill, main drive by belts and pulleys. (I. H. Beard, "Gearing Mills for the Manufacture of Cotton and Woolen Goods," *JFI* 23 [1837].)

The adoption of the belt main drive in factories of the largest size made power history. It prepared the way for the general introduction of the system in American industry and led to the eventual spread of belt main drives to Europe. It stimulated the replacement of gearing by belts in lesser mechanical applications, such as placing in motion the front rollers of spinning frames and, much later, in operating drilling machines.[73] Within ten years of the pioneer installation at Lowell the shift from toothed gears and shafting to belts and pulleys in the main drive was well advanced in the cotton textile mills. In 1840 a British engineer declared that in the United States "generally all the new [textile] mills use belts for gearing." By mid-century many of the older mills were joining the process by converting to belt drives.[74] The common arrangement of a belt drive was to have a separate belt for each floor, running from one wide or several narrow pulleys on the shaft of the prime mover (Fig. 112). A less approved arrangement, how widely used it is difficult to say, employed a single long belt "to run the whole round, from bottom to top of mill, turning every main [line] shaft and working its way to and fro."[75] The immense length and breadth of such a belt, its weight, and the difficulty of handling it, since the entire mill had to be stopped in the event of a need for adjustment or repairs, were strong objections to this arrangement.

Engineers differed over the value of the belting innovation. While conceding the successful operation of the first belt-driven mill, which he visited and examined, Zachariah Allen questioned its economy and mechanical desirability. On the basis of ten years' experience with both belt and gear drives, another American engineer held that belt drives ran lighter and consumed more power than the older type.[76]

---

to be called an invention, this proved to be a very important improvement, and was entirely original in its application to the transmission of fifty or a hundred horse-power by a single belt, and has been very generally adopted in the mills of New England" (p. 88).

[73] Oliver Dean to Mr. Gay, 11 Jan. 1831, to Mr. Sayles et al., 31 Jan. 1831, Amoskeag Manufacturing Company, Letterbook, 1826–31, Manchester Historic Association, Manchester. See also *Reports on the Philadelphia International Exhibition of 1876,* 1:231.

[74] Montgomery, *Cotton Manufacture of the U.S.,* pp. 24–25; *Scientific American* 3 (8 Apr. 1848):258. See also I. Beard, "Remarks on the Mode of Gearing Mills for the Manufacture of Cotton and Woolen Goods," *JFI* 19 (1837):451 ff.; M. Morin, "On belts," trans. Ellwood Morris, ibid. 38 (1844):22–23n.

[75] Leigh, *Science of Modern Cotton Spinning,* quoted in John H. Cooper, *A Treatise on the Use of Belting for the Transmission of Power* (Philadelphia, 1878), pp. 24–25; Beard, "Remarks on the Mode of Gearing Mills," pp. 452–53, with drawing.

[76] Allen, *Science of Mechanics,* p. 258. Beard, "Remarks on the Mode of Gearing Mills," pp. 451 ff.

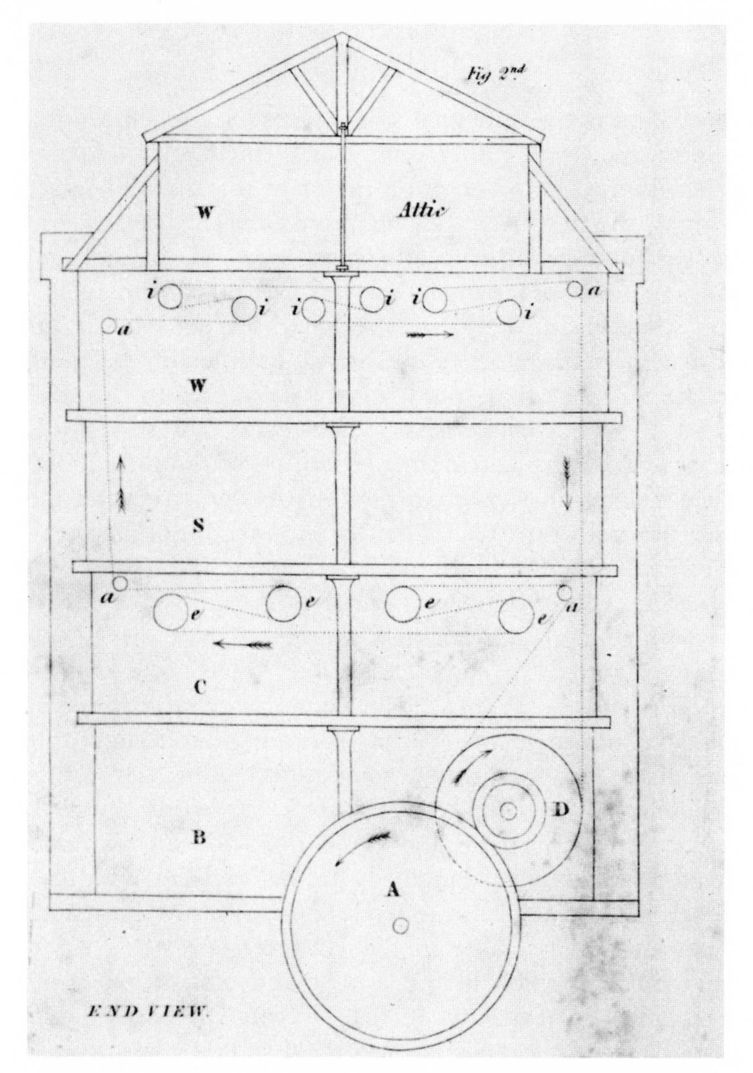

Fig. 113. American cotton mill equipped with continuous belt drive. (James Montgomery, *A Practical Detail of the Cotton Manufacture of the United States of America* [Glasgow, 1840].)

William Fairbairn disapproved of the American practice as wasteful of power because of its greater friction. Evan Leigh, his compatriot, on the other hand, admired this bold break with tradition: "Where capital is abundant and prejudice rife, men follow the beaten track. . . . Give a man a certain thing to do with limited means, and his ingenuity suggests a way of doing it; so in America heavy gearing is almost entirely discarded, and broad double belts or straps substituted."[77]

[77] Fairbairn, *Treatise on Mills and Millwork,* pt. 2 (1865), pp. 1 ff.; Leigh, *Science of Modern Cotton Spinning,* 1:37 ff.

Efficiency apart, the practical advantages of the innovation were obvious and a matter of rather general agreement. As compared with the typical wheelwork and shafting drives of the time, at least in the United States, the belt drive ran far more smoothly and quietly, cost less, and was more readily installed and repaired. Breakdowns were fewer and shutdown times much shorter. Less precise in their adjustment than wheel drives, belt drives required less substantial and rigid building structures for effective operation. The flexible belts absorbed shocks in the millwork that could be very damaging to the unyielding gearwork of the traditional drive.[78] Belts were less susceptible than gear wheels to misalignment from the settling of building foundations and change from shrinking or warping of wooden structural members supporting the millwork. Undoubtedly the insubstantial construction of factories in the United States no less than the indifferent quality of toothed gearing was an important factor in the shift to belt drives.

The smooth and almost noiseless motion of belt drives was their most striking feature. "In America," wrote Leigh, "the main driving belts are open straps . . . neatly boxed up, so that nothing is seen, nothing heard; whilst in this country [Britain], the disagreeable rumbling noise of the heavy gearing of some mills can be heard, in country places, a mile off."[79] Despite Fairbairn's contention that "geared wheels, when properly made and fitted up, will run as quietly and smoothly as belts," the fact remained that before 1850, with occasional exceptions in the more advanced machine shops of the East, such gearing was simply not available in the United States. Belt drives, too, represented but a fraction of the weight and cost of wheelwork and shafting of equivalent capacity. In 1848, to carry 60 horsepower thirty-six feet from the prime mover to the third story by cast-iron vertical shafting required more than four tons of metal, whereas 600 pounds of belting would accomplish the same purpose.[80] In large mills support of heavy upright shafting presented difficult structural problems; in one English cotton mill the upright drive shaft with its bevel wheels and other accessories weighed twelve tons.[81]

[78] Robert H. Thurston, *A Manual of the Steam Engine,* pt. 2, *Design, Construction, and Operation* (New York, 1891), pp. 528–29.

[79] Leigh, *Science of Modern Cotton Spinning,* 1:37 ff.; Fairbairn, *Treatise on Mills and Millwork,* pt. 2 (1865), pp. 1–2; Montgomery, *Cotton Manufacture of the United States,* pp. 24–25; Cooper, *Treatise on the Use of Belting,* pp. 277 ff. See also Richards, "Principles of Shop Manipulation," 97:167; Allen, *Science of Mechanics,* pp. 262–63.

[80] W. Montgomery, "Economy of Power in Cotton Factories," *Scientific American* 3 (6 Apr. 1848): 229.

[81] Leigh, *Science of Modern Cotton Spinning,* 1:26 ff.

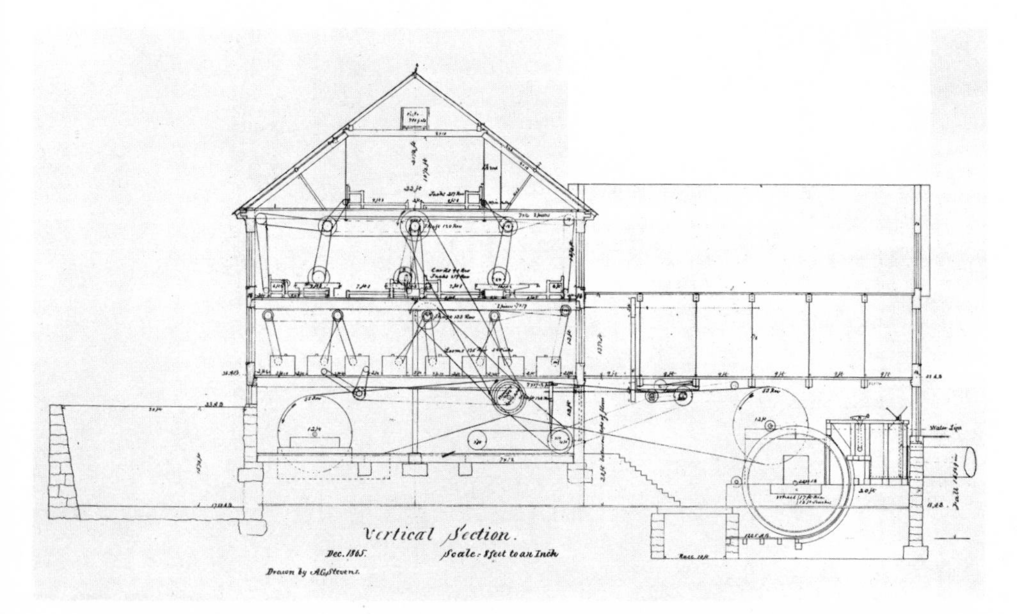

Fig. 114. Power transmission equipment: cotton mill, 1865. Except for the toothed segment of gearing between the waterwheel and main drive pulley, the power is distributed by a system of line shafts and counter shafts (ends only seen) driven by belting on pulleys. (Courtesy of the Merrimack Valley Textile Museum.)

Textile mills were simply the vanguard in a procession joined in time by large establishments, at least in most branches of manufacturing. The *Scientific American*, 4 February 1860, declared that wherever extensive machinery was used, the power to drive it, with very few exceptions, was transmitted by belting.[82] An American engineer announced in 1875 that in Europe the greater part of the power was transmitted by cogwheels, "but in this country ninety-nine percent is transmitted by belting."[83] Evidently the last percent was a large one, since Robert H. Thurston fifteen years later reported that gearing was "continually less and less used."[84] Advocates of belt drives conceded only one deficiency: they were not "positive." "If you start from the motor

[82] *Scientific American,* n.s. 2 (4 Feb. 1860):84; see also ibid. 33 (11 Dec. 1875):375. For illustrated examples of belt drives, see Cooper, *Treatise on the Use of Belting,* 4th ed. (Philadelphia, 1891), chap. 2, Arts. 101–6.

[83] John W. Sutton, quoted in *Scientific American* 32 (20 Feb. 1875):120; see also *Manufacturer and Builder* 10 (1878):156; Stephen Roper, *The Engineer's Handy-Book* (Philadelphia, 1881), pp. 637 ff.

[84] Thurston, *Manual of the Steam Engine,* pt. 2, p. 154.

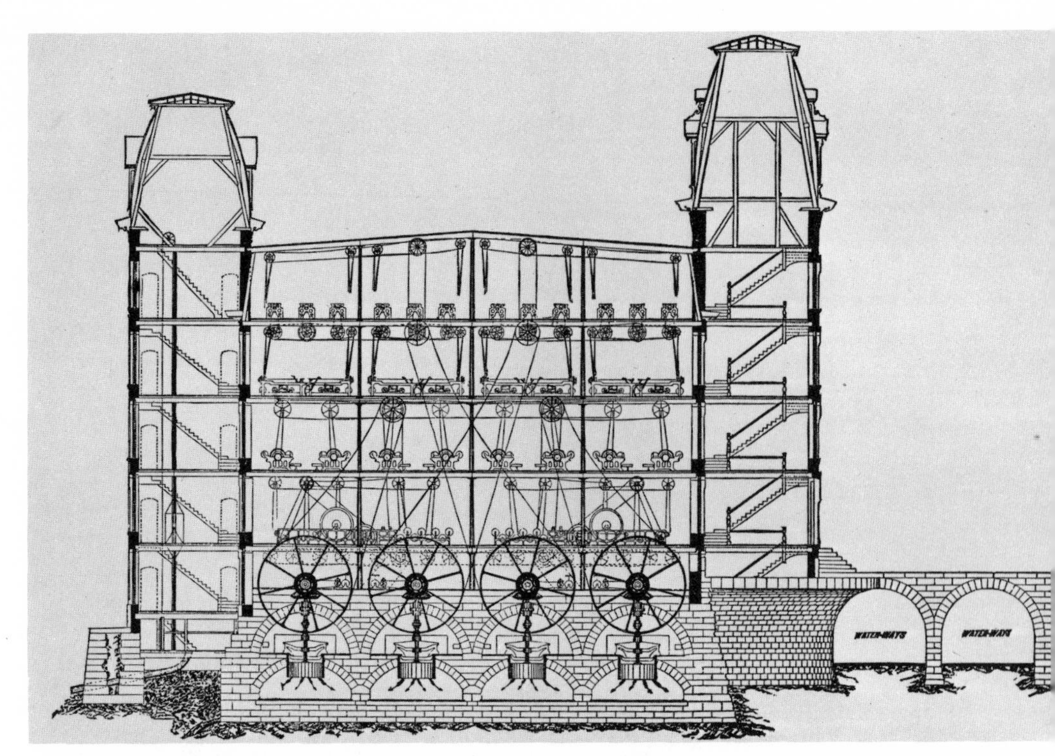

Fig. 115. No. 3 Mill, Manville Co., Manville, R.I., 1874. Transverse section. Capacity 130,000 spindles. Driven by four 84-inch, 400-h.p. Leffel turbines under 18–½ ft. head, Blackstone River. Power was carried from turbines by gearing to four main-drive flywheel pulleys 20 ft. in diameter and weighing ten tons each and was distributed throughout the mill by means of belting. An auxiliary steam plant of 800 h.p. capacity powered during low water. (Catalog, 1875. James Leffel Co., Springfield, Ohio. Courtesy of the Merrimack Valley Textile Museum.)

with a certain number of revolutions, you lose a portion of them with every belt used."[85] The number of applications of power where such slippage was undesirable, however, seems to have been relatively small.

The crowning achievement of the belt drive was its successful application to the most rugged of industrial operations, the rolling of iron and steel. The massive power requirements of rolling the heavier shapes combined with slow speeds of large steam engines to compel the use of huge gear wheels to obtain the necessary roll-train speeds. In a rolling mill erected during the Civil War, Jones and Laughlin of

[85] Sutton in *Scientific American,* 32:120; Allen early questioned the use of belting on this ground (*Science of Mechanics,* p. 268). See the comment of James B. Francis on Atlantic Cotton Mill No. 2, Apr.–May 1850, from "Exp. and Trans., 1845–57," LCP.

Pittsburgh took the bold step of replacing the traditional massive wheelwork gearing with belting. With the exception of a single roll train attached directly to the flywheel shaft of the 600-horsepower engine, the entire mill was driven by belts alone. The flywheel shaft measured 18 inches in diameter. The 68-inch-wide rim of the 40-ton flywheel, 25 feet in diameter, carried two belts by which the main line of shafting was driven. Each belt was 32 inches wide and 140 feet long. Plans for this belt drive were greeted with the skepticism that historical folklore seems never at a loss to supply for such events. "It never had and it never could be done." "Who ever heard of a rolling mill being driven by a belt?" Since in America failures are decently buried and forgotten, the outcome was necessarily a great success. In contrast with the repeated out-of-order condition of a companion geared mill, the belt-driven mill at the end of five years had yet to lose a belt. An observer remarked with surprise that one could converse without effort even with the mill in full operation.[86] Within little more than a decade nearly all of Pittsburgh's rolling mills were being driven by belts, and the system spread widely through the country.[87] Belting with its elasticity absorbed much of the heavy shock characteristic of rolling operations. Both engine and machinery were relieved of much wear and tear; there was a sharp reduction of breakages and shutdowns. Both first cost and operating expense were lowered.[88]

Millwork Fittings and Equipment

The American system of belt drives was slow to be taken up in Europe, especially in Britain, where the contrast between American and British practice had been noted as early as 1841.[89] In spite of the demonstrated success of the system, a long generation was to pass before the new plan began to obtain acceptance in the Old World.

[86] See *Scientific American* 20 (23 Jan. 1869):50 and 27 (3 Aug. 1872):72.

[87] Ibid. 31 (23 Jan. 1875):52 and 36 (20 Jan. 1877):42; Cooper, *Treatise on the Use of Belting* (1878), p. 94. See also description of belt used to drive a continuous rod mill about 1876, in William Hewitt, "The Continuous Rod Mill of the Trenton Iron Company," *TASME* 2 (1881):70 ff.

[88] Cooper, *Treatise on the Use of Belting* (1878), pp. 94 ff. See also Henry Pallett, *The Miller's, Millwright's and Engineer's Guide* (Philadelphia, 1866), p. 82.

[89] Robert Willis, *Principles of Mechanism* (London, 1841), quoted in Sellers, "Transmission of Motion," p. 243. On the persistence of geared drives in Europe, see Peter A. Kozin, *Flour Milling,* trans. M. Falkner and Theodor Fjelstrup (New York, 1917), pp. 296 ff.

A visiting American engineer in 1867 reported that the British "cannot conceive of our running a cotton mill by a 'strap.' "[90] The high quality of British gearing, the solid character of their mill structures, and a certain conservatism natural in an older and successful industrial nation discouraged innovation. As an English authority pointed out about 1873, fifty years had passed without material change in the revolutionary system of millwork introduced by Fairbairn and Lillie. The new conditions and requirements of industry called for changes of comparable magnitude.[91] In short, the outmoders were now outmoded. In Britain and on the Continent the use of belt drives increased rapidly from the 1870s, and by 1900 the practice of "the American system" was becoming general.[92] At bottom the acceptance of this innovation, in the United States as abroad, simply reflected the steady rise in machine speeds dictated by the drive for increased productivity and lower unit costs. Prime movers, as noted elsewhere, followed suit, and the millwork joining the two inevitably conformed. Toothed gearing, an English engineer remarked in the 1890s, "is an unsuitable medium for considerably increasing the speeds of shafts, as the teeth grind and vibrate and the wear, breakage and noise are excessive."[93]

The success of belt drives and the ever rising speed of shafting that it

[90] Lockwood, in "English Cotton Manufactures," p. 21. Lockwood remarked further that the English "insist on the superiority of their gearing system, as making the mill much more permanent, and less subject to accidents." But accidents closing mills for several days were met at three of the first mills visited and repairs were more difficult to make than with the American system.

[91] Leigh, *Science of Modern Cotton Spinning,* pp. 273 ff. There had been many difficulties in the American development of the belt main drive, as Allen had pointed out ("Transmission of Power," p. 20), and there were many American manufacturers and millowners, Sellers reported, "whose conversion [to the American system of mill shafting] was brought about through their pockets" (*Scientific American* 27:371). As late as 1869 a British treatise on millwork, Thomas Box's *A Practical Treatise on Mill-Gearing, Wheels, Shafts, Riggers . . . for the Use of Engineers* (London, 1869), contained no discussion of belting.

[92] "The mode of driving mills, factories, etc., by means of gearing seems now to be in a fair way of being abandoned in Europe, and that adopted in this country many years ago, of driving by main belts, substituted in its place" (*Reports of the U.S. Commissioners to the Paris Universal Exposition, 1878,* House Executive Document 42, 46th Cong., 3d sess., 5 vols. (Washington, 1880), 4:401 ff.

[93] W. P. Bale, *Modern Shafting and Gearing and the Economical Transmission of Power* (London, 1893), p. 57. Another British engineer noted that many belts transmitting 300 to 400 horsepower "are driven up to 5000 feet per minute and beyond with ease, certainty and long endurance in contrast to gearing with its noise, inelasticity, and risk of destruction" (Arthur Rigg, *A Practical Treatise on the Steam Engine,* 2d ed. [London, 1894], p. 184).

made possible were owing in large part, as with the earlier achievement of Fairbairn and Lillie, to the redesign and systematic production not only of the shafting itself but of the auxiliary equipment ranging from hangers, couplings, and bearings to the pulleys and belts forming the first and final links between the motor and the machine. Many early attempts to introduce belt drives had failed because of the low belting speeds adopted and the inadequate supporting components. "Pulleys had not been made sufficiently light and well balanced for anyone to adventure to use them with the high speed required for leather belts to operate advantageously," Zachariah Allen pointed out. "With the slow speed, it was necessary to strain the belts so tightly on the pulleys, to produce sufficient adhesion, without slipping around on the smooth surfaces that the lacings and texture of the leather yielded; and so frequently repairs were required, that the superintendents of mills nearly all abandoned the use of them for transmitting the power from the motors to the mill shafting."[94]

In addition to pulleys for taking and receiving motion from the line shafts there were hangers or brackets for supporting shafts and countershafts, couplings, bearings, and devices such as fast-and-loose pulleys and clutches for engaging and disengaging the driven machines, step or cone pulleys for changing speed ratios at the machines, lubricating devices, and the like. The key to the development of these items was found in their role as components in what Coleman Sellers termed "a machine to transmit motion . . . most frequently the largest in the establishment."[95] This key was found in the manufacture of millwork with much of the same accuracy and precision required in the motor and the machinery driven. The chief difficulties in building and maintaining this "transmission machine" lay in the attenuated character of its numerous arms, the line shafts extending hundreds or even thousands of feet throughout the establishment, and in the character of the machine's framework, the building structure itself. Apart from foundations and main walls, frequently of brick or other masonry, mill structures in this country were typically framed of wood. The shrinkage or warping of structural timbers, the yielding under changes in floor loads, and the not uncommon settling of foundations often played havoc with the proper alignment of shafting and bearings. Seasonal change in weather alone might disturb alignment badly. The unbalanced pull of belting from opposite sides could produce a "springing" of the shafting with similar consequences. Lubrication of

[94] Allen, "Transmission of Power," p. 20.

[95] Sellers, "Transmission of Motion," p. 233.

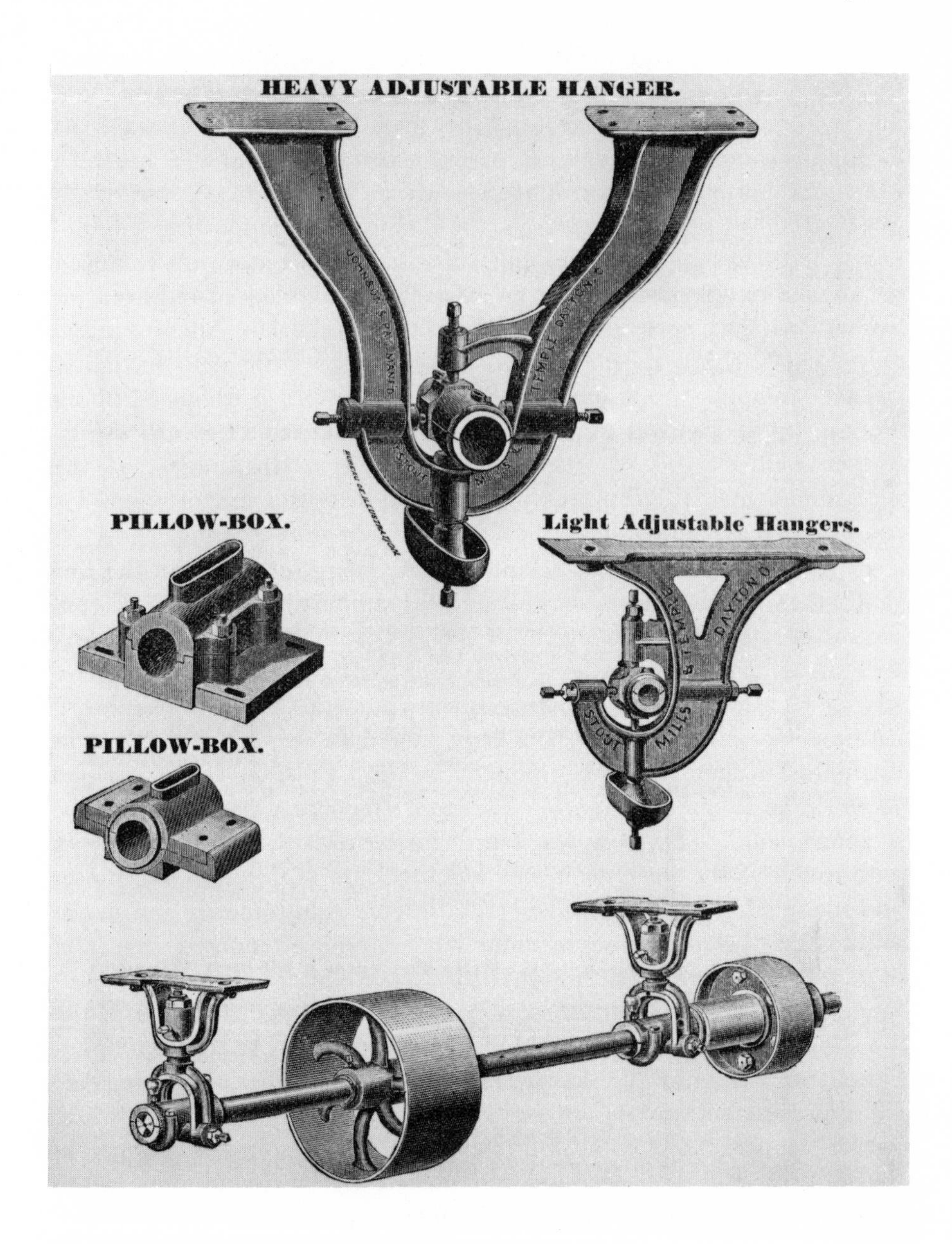

Fig. 116. Patented line-shafting hardware offered by a turbine manufacturer. (Globe Iron Works, *The American Turbine* [Dayton, Ohio, 1875].)

Fig. 117. A millwork manufacturer's advertisement of the 1870s. (*The Manufacturer and Builder* 7 [Mar. 1875].)

hundreds of awkwardly reached bearings could become the maintenance millwright's nightmare. It is significant that the figurative if not literal inheritor of the mantle and fame of the Manchester firm of Fairbairn and Lillie, who entered millwork production by way of millwrighting and mill building, was the Philadelphia firm of William Sellers and Company, builders of machine tools, the most demanding of machines in respect to precision in building and operation.

The Sellers firm had its beginning in 1848 when two machinists of Providence, Rhode Island, joined forces as Bancroft and Sellers to establish a manufactory of machine tools and millwork in Philadelphia. About 1845 Edward Bancroft had invented the self-adjusting swivel hanger in which the bearing accommodated its position to minor changes in the hanger's position occasioned by the circumstances mentioned above. He failed to interest leading New England machine-tool builders in manufacturing a hanger that they regarded as a needless refinement too costly for general use. The success of the Bancroft hanger, introduced by the Sellers Machine Tool Works, was followed by 'the development of a complete system of millwork. This included not only precision-built shafting designed to run at speeds as high as 400 rpm but a double-cone coupling that virtually eliminated the careful individual fitting on the shafting hitherto necessary and a light, self-centering, and carefully balanced pulley. Through an extensive series of experiments the Sellers firm determined what strength and consequently what weight and proportions were required for each millwork item. Using special machine tools and gauges and a carefully systematized organization of production, the firm turned out immense numbers of interchangeable parts at costs and prices below those of the common sort of millwork sold typically by the pound.[96]

The official British parliamentary reports on the Philadelphia Centennial Exhibition cited the Sellers firm as an example of the American practice of machine-tool builders themselves manufacturing light mill gearing, "thereby giving a tone and efficiency to this important branch of manufactures." Sellers and Company were "the great authorities in this class of work in the United States. One of the partners . . . has taken up the subject of the transmission of power at the point where it was brought up to by the late Sir William Fairbairn."[97] The parliamentary report also summarized the chief points in the Sellers "mill gear system," which was distinguished by its efficiency, simplicity, and economy.

Americans, declared a French engineer, "aim at achieving the lightest possible weight and the highest possible speed. Hence the universal substitution of belting for gearing, and the general adoption of light shafting and small pulleys. . . . The ease and durability of the system would astonish advocates of gearing."[98] One general rule covered the whole ground, declared an American publication: "Make your shaft as small as you can, and run it as fast as you can." Returning from the Vienna Exposition of 1873, R. H. Thurston joined the chorus: this mode of millwork was essentially an American system, seldom seen in Europe.[99]

[96] The two lectures given by Sellers before the Stevens Institute of Technology in 1872 and published in *JFI* 94:233–43, 307–19, illustrated with simple drawings, are the best single account of the subject, given by a man who was without doubt the leading figure in the United States in millwork design and construction. See also *DAB*, s.v. "Sellers, Coleman," and Fitch, "Report on the Manufacture of Steam Engines and Boilers," pp. 34–35.

[97] *Reports on the Philadelphia Centennial Exhibition of 1876,* 1:232–33.

[98] Quoted in *Manufacturer and Builder* 7 (1875):70.

[99] *The Locomotive,* Oct. 1883, pp. 169 ff.; Robert H. Thurston, "Report on Manufactures and Machinery," *Reports of the Commissioners to the International Exhibition Held at Vienna, 1873,* ed. Robert H. Thurston, 4 vols. (Washington, D.C., 1876), 3:212–13.

Fig. 118. Main machine room of a pipe-fitting manufacturer, 1881. (*Scientific American*, Oct. 22, 1881.)

Although belts at the practical working level were a subject of endless discussion, they were slow to receive systematic engineering attention.[100] It was generally agreed that they were low in cost and high in efficiency, gave good service most of the time, and were readily repaired and maintained. Opinion differed, however, on such matters as the size and speed of belting to serve a given purpose, the materials and manner of construction, and the details of installation and maintenance. Such experimental investigations as were made in the early years gave such variable results as to provide little sure guidance to millwrights and engineers in selection or installation, but these uncertainties were but minor deficiencies.[101] Belting made by the millowner or the local saddler was apt to be uneven in thickness and irregular in operation; the parts of the hide from which it was made were sewn or laced together with much overlapping. Belt making as a special branch of enterprise, which brought great improvement, did not assume much

[100] A useful source is Cooper's *Treatise on the Use of Belting*, a collection of articles, excerpts, tables, and lengthy quotations from a variety of sources, including many drawings. A useful article is Charles A. Schieren, "Leather Belting–Its Origin and Progress," *Electrical Engineer*, Mar. 1888, pp. 112–14. See also *Appleton's Cyclopaedia of Machines*, s.v. "Belts," supplemented in *Appleton's Modern Mechanism*, pp. 43–51. An illuminating reminiscent account is Fred H. Colvin, *Sixty Years with Men and Machines: An Autobiography* (New York, 1947), pp. 17–24.

[101] See *Appleton's Modern Mechanism*, pp. 43–45.

importance before the 1840s. To combine the select portions of cattle hides into belts of uniform thickness from a few inches to several feet wide and as much as 200 feet long was a remarkable achievement, especially the making of the large belts required for main drives and the high speeds increasingly used. With proper care leather belts, and under some conditions the rubber ones, which attained a certain popularity from the 1850s, lasted for years.

The Limitations of Millwork

By the 1870s the important innovations in power transmission associated with the first Industrial Revolution and the use of power-driven machinery had been made. The substitution of speed of motion for volume and weight in the means of transmission—shaft, wheel, or belt—was the key to advance: power equals force in pounds times velocity in feet per minute. Replacing size of shaft or width of belt with speed of rotation lowered first cost, reduced friction, and gave economy in power. It also required greater precision in design and fabrication and more care in maintenance and lubrication. At the time the savings in first cost and in the consumption of power were impressive. Yet as industrial establishments grew in scale and in complexity of operations, such gains were offset by difficulties inherent in a system of power distribution which became more cumbersome and less efficient with greater extent and which imposed a straitjacket upon the organization of production. The requirements of power transmission before the coming of electricity imposed a heavy burden on engineers in mill design and construction. After the machinery layout had been determined, shafting plans had to be made covering every part of the plant and each individual machine. These plans had to be made in minute detail with precise calculations for obtaining the speeds for which each machine was designed:

> Most rooms, with concentrations of machinery had the ceilings pretty well covered with lines of shafting, pulleys and hangers, except that for economy two floors of machinery would often be driven from one set of shafting . . . the lower set of machinery by means of belts from over head and the upper set by belts from below coming up through the floor. . . . [Mill engineers] had to schedule each and every pulley—not only on the machines but also on the shafting—as well as all hangers, couplings, etc., and all belting.

Yet the headaches were not over even after the shafting and connections were laid out; there was the matter of supports and bracing to consider. The main

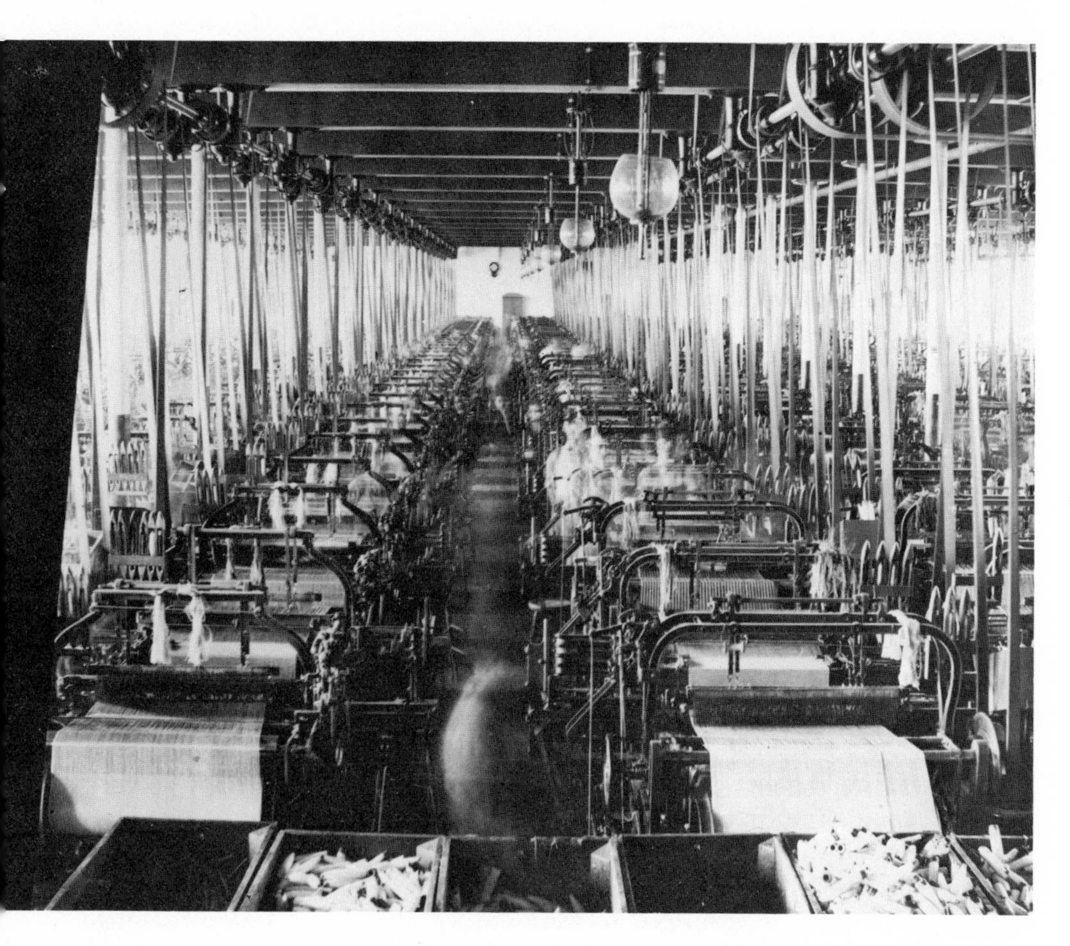

Fig. 119. Line shafts and belting: a weaving room in the Amoskeag Mills, Manchester, N.H. These shafts not only conveyed power to the machines visible in the photograph, but belting from the same shafts conveyed power through the ceiling to machines on the floor above. (Courtesy of the Smithsonian Institution.)

lines of shafting which received power from the water wheels or engines would necessarily run into large sizes in the case of a large mill. Sizes of main-line shafting up to ten inches in diameter were not unusual, and required special construction of that part of mill buildings and foundations which had to support such heavy weights—and in addition the structures had to withstand the pull of the huge belts or rope drives involved.[102]

[102] Samuel B. Lincoln, *Lockwood Greene* (Brattleboro, Vt., 1960), pp. 140 ff.; see also ibid., pp. 37 ff.

From an operating viewpoint the deficiencies of the conventional millwork centered in the maintenance burdens it placed upon shop management. Even so routine a matter as lubrication and cleaning of bearings was a laborious and often neglected task owing to the awkward position of the ceiling hangers—one to every eight or ten feet—by which the long lines of shafting were supported. Overdone, lubrication was a source of oil drippings, work spoilage and waste; carelessly done, it led to lost work in friction, overheating, damaged bearings, and occasionally fires. A more difficult, demanding, and even more neglected maintenance task was that of keeping shafting and hangers properly aligned, pulleys balanced, and belts adjusted to correct tension. In all but the most substantial mill structures the lack of rigidity in walls and floors added to the shake occasioned by poorly installed and carelessly maintained millwork. Changes in the position of machinery, additions to or changes in the distribution of floor load, and marked changes in temperature and humidity produced deflections in alignment of the millwork. Finally, in addition to occupying excessive space and interfering with lighting and handling of materials, the long lines of overhead shafting with their whirling pulleys and rows of flapping belts in swift motion were a hazard to life and limb and a source of frequent accidents.[103]

As a system of power distribution the millwork of shafting, belting, and pulleys met the needs of such industries as textiles, whose processes required multiple machines of the same kinds, which could be ranged in long rows paralleling the lines of shafting overhead. But millwork was largely lacking in the flexibility necessary to meet the varied needs of industries whose operations were not adapted to the orderly arrangements of a textile mill, raised story upon story. By its compulsion toward linear, rectangular, and symmetrical arrangements, millwork placed serious restrictions upon factory layout, discouraging when it did not actually prevent the rationalization of industrial operations. The vital factor of power supply, distributed through conventional millwork, began as a servant but in time tended to become a master from which industry had to be freed.

[103] On the difficulties in adapting power transmission to the rapid increase in the use of machinery in large establishments during the 1890s, see H. J. Westover, "Mechanical Power Transmission," *JWSE* 3 (1898):1021 ff. See also Colvin, *Sixty Years with Men and Machines,* pp. 17 ff., and James F. Hobart, *Millwrighting* (New York, 1909), chaps. 11–13. According to Pole, the new millwrighting firm of Fairbairn and Lillie was harassed by the common Sunday work resulting from frequent breakdowns of the old-style wooden and cast-iron millwork, from which they were largely to be freed by their development of fast-running, precision-made shafting and associated components (*Sir William Fairbairn,* pp. 113–15).

10
The Decline of Direct-Drive Waterpower

THE GENERATIONS FOLLOWING the Civil War witnessed the progressive decline of waterpower from its long-dominant role in the American economy. Of the increase in capacity in manufacturing nationwide, from 2.4 million to 6 million horsepower from 1869 to 1889, steam power comprised 96 percent. Waterpower reached its lowest ebb from 1879 to 1889, when its increase was less than 3 percent in a period of great expansion of industrial capacity and output nationally. A substantial recovery in growth during the 1890s and early 1900s failed to halt the further relative decline compared with steam power. By 1900, thanks to hydroelectricity, waterpower obtained a new, greatly enlarged, and indefinitely extended lease on life; but the years of continued direct-drive use were numbered. The renewed vigor of this ancient, beneficent, and in many respects admirable source of motive power, in important ways never matched by steam power, depended almost wholly upon the distant transmission of power by means of electricity. The federal census of 1900 reported that waterpower, long the primary determinant of industrial location, had ceased to exercise an appreciable influence upon the geographical distribution of industry.[1]

[1] Carroll R. Daugherty, "The Development of Horsepower Equipment in the United States," in *Power Capacity and Production in the United States,* U.S. Geological Survey, Water-Supply Paper no. 579 (Washington, D.C., 1928), p. 49. Also, U.S., Census Office, Eleventh Census, 1890, vol. 6, *Report on Manufacturing Industries* (Washington, D.C., 1895), pt. 1, pp. 748–49; U.S., Bureau of the Census, Twelfth Census, 1900, vol. 7, *Manufactures,* pt. 1, *U.S. by Industries* (Washington, D.C., 1902), p. cccxx, table 4, cxc. In absolute terms increases in direct-drive waterpower, 1869–1909, were: United States, 61 percent; New England, 109 percent, with increases each decade; Middle Atlantic states, 25 percent, with declines in three decades, increases in two; North Central states, 60 percent overall. See H. H. Bennett, "Utilization of Small Water Powers," *Transactions of the Third World Power Conference, 1936,* 10 vols. (Washington, D.C., 1938), 7:481, table 4. The Thirteenth Census, 1910, showed that while direct-drive waterpower had increased 692,457 horsepower, 1869–1909, its share in the national total decreased to 9.8 percent (U.S., Bureau of the Census, Thirteenth Census, 1910, vol. 8, *Manufactures* [Washington, D.C., 1913], pp. 330–31). The decennial census did not report hydroelectric capacity in horsepower at this date. Separate reports of central electric light and power stations were issued quinquennially beginning in 1902.

Factors in the Change

Of the many factors contributing to the passing of direct-drive waterpower, three may be noted as fundamental: the progressive industrialization of the economy, the extension and completion of the railroad network nationally, and the urbanization of manufacturing. Industrialization was the complex of economic, technological, and social forces set in motion by the first Industrial Revolution. With roots deeply anchored in the past, industrialization attained a certain maturity in the generation following the Civil War. Under the stimulus of the institutions of private enterprise its advance was marked by an accelerating drive toward ever larger size and capacity of firms, the rationalization of production, and the outward thrust of markets. Industrialization brought about not only the progressive takeover of manufacturing as traditionally conducted in the home, on the farm, and in the craft and workshop industries but the increasing subordination of agriculture and trade within the economy. The takeover was accompanied by the extension of mechanization, the replacement of hand by machine operations within established industries, and the proliferation of new industries engaged in the provision of an unending variety of consumer and producer goods and services.

Transportation facilities were basic to overall economic development and industrial growth. For three-quarters of a century from its active beginnings in the 1840s, railways were the key agency upon which the advance of industrialization depended. Railways freed the national economy from the limitations of a system of inland waterways that had served the American people so well between Independence and Civil War. After a century of federal improvement at an outlay of a half-billion dollars, this system of rivers, canals, and inland seas at its peak afforded no more than 30,000 miles of navigable waterways, nearly half of which with a depth of less than four feet could be used only by the smallest class of steam vessels. For all the benefits in advancing the frontiers of settlement and the regional interchange of farm products and manufactured goods, the deficiencies of the system, felt increasingly from mid-century, were threefold. Water transport held the national economy in the grip of seasonal embargoes upon transportation and trade by low water and ice during one-third or more of the year. The water routes, both natural and man-made, were circuitous and lengthy, affording service that while low in cost was lacking in system and regularity and could not be appreciably extended. The most serious deficiency was the limitation of the direct benefits to but a small population of the country, the communities within two or three days wagoning distance—perhaps twenty-five to thirty-five miles—by common dirt roads to a navigable waterway.[2]

The railway changed all this. Occasional and brief interruptions by storm and flood apart, railways freed the economy from the periodic and extended seasonal embargoes on transportation service. They not only provided more direct and faster service but in time came to supply a regularity and reliability of service only too commonly wanting on many of the trunkline water routes. The construction of railways proceeded at an extraordinary and accelerating pace. By 1860 railway mileage had passed the 30,000-mile mark, exceeding the length of inland waterways at their peak. The 53,000 miles of the national network in 1870 had doubled by 1882; and it doubled again by 1904. This remarkable territorial extension was matched by the number and power of locomotives placed in operation.[3]

The growth in railway facilities and motive power was accompanied by a reduction in railway rates, which in time only the more favorably situated inland waterways could better. In the critical matter of road transport upon which most of the agricultural population in the first instance depended, rates per ton-mile in the years before 1860 were cited as follows: 20 cents on highways in good condition; 40 cents on poor roads or during bad weather. Railway rates about 1840 ranged between 4 and 10 cents per ton-mile. By 1870, the figure had been reduced to 1.5 cents and by 1900 was down to 0.5 cents.[4]

The extension of the railway network and the provision of transportation service virtually free from the vicissitudes of weather and the seasons gave rise to national markets, or at least extensive regional mar-

[2] These figures are based on the records of the U.S. Army Corps of Engineers, as reported in 1908 by the Inland Waterways Commission and later updated to 1925 with little change. They do not include the Great Lakes. See Bureau of Railway Economics, *An Economic Survey of Inland Waterway Transportation in the United States* (Washington, D.C., 1930), pp. 7-23. The seasonal limitations on water transportation in the northeastern United States are reviewed with particular reference to the Ohio Valley in Louis C. Hunter, *Studies in the Economic History of the Ohio Valley* (Northampton, Mass., 1934), pp. 5–49; see also Douglass C. North, *Economic Growth of the United States, 1790–1860* (Englewood Cliffs, N.J., 1961), pp. 142–45.

[3] Kent T. Healy, "American Transportation before the Civil War" and "Development of a National Transportation System," in Harold F. Williamson, ed., *The Growth of the American Economy,* 2d ed. (Englewood Cliffs, N.J., 1951); Daugherty, "Development of Horsepower Equipment," table 2, p. 46.

[4] Healy, "Development of a National Transportation System," esp. fig. 3, pp. 380–82. See also Chester W. Wright, *Economic History of the United States* (New York, 1941), pp. 353–55. Prices are in current dollars.

kets, for an ever widening range of products. It created a competitive situation in which the uncertainties of a power supply resting upon falling water became increasingly unacceptable. The progressive penetration of innumerable local markets hitherto protected from outside competition by transport and marketing costs gradually undermined the position of small industrial establishments serving as the last bastion of waterpower in its traditional form. The ground was thus cleared for two related developments of prime importance: the concentration of production in establishments of ever larger size to obtain economies of scale and to meet the challenge of widening competition; and the centralization of the manufacturing industry at those points where conditions were not only most favorable for access to markets and raw materials but where other circumstances necessary for the conduct of business enterprise were present. The absence of waterpower in most commercial centers, long regarded as a major deterrent to the conduct of manufacturing operations in urban communities of any size, had ceased to be a relevant factor. Steam was increasingly recognized as the power source upon which all but a limited range of traditional local industries had to depend. Power supply thus gave way to transportation facilities as the most critical factor in industrial location.

So far as the larger fields of competitive enterprise were concerned, there gradually emerged a wide range of locational factors bearing upon the viability of an enterprise, little known or appreciated in earlier times. In reviewing the circumstances in which Boston capitalists moved into the largely undeveloped cotton textile field in the 1820s and 1830s, John Coolidge has remarked: "Proximity to capital, to markets, to raw materials, to labor, was not considered. . . . Water power, and plenty of it, was all they sought."[5] For a generation to come this approach to industrial location was widely shared and acted upon. After the Civil War there was a growing awareness and recognition that in the emerging industrial economy motive power was but one among many locational factors to be weighed and more commonly than not a minor one, accounting for no more than 5 to 10 percent of total production costs. By this time such cost advantage as waterpower still possessed over steam power might readily be overruled by other cost and advantage considerations.

Industrialization was accompanied by the concentration of manufacturing and industrial and commercial operations generally, not only in firms of ever larger size and resources but in communities

[5] John P. Coolidge, *Mill and Mansion: A Study of Architecture and Society in Lowell, Massachusetts, 1820–1865* (New York, 1942), p. 167.

where conditions for the conduct of production and business enterprise were most favorable. Increasingly evident from mid-century, urbanization and industrialization proceeded hand in hand. Access to markets, raw materials, capital, labor supply, and various service facilities made the larger urban communities increasingly attractive to manufacturers, especially growing cities situated at nodal points on channels of transportation, trade, and communications. Originally deriving their existence and growth from favorable means of water transportation, such commercial centers contributed to further growth by promoting and supporting improvements in overland transport—roads, canals, and from the 1830s and 1840s the extension of railroad lines, their consolidation into systems, and interline coordination of services. The accelerating concentration of manufacturing in cities was the principal factor in the drastic reversal of the status of waterpower in the postbellum generation. The availability of steam power at a fair level of proficiency and economy was the principal technological condition of the changeover.

Urbanization and Power Supply

The industrial and commercial utility of water courses, as observed in an earlier chapter, was directly opposed. The marked slopes, the interruption of streambeds by falls and rapids, and the limited depth and breadth of flow characteristic of millstreams effectively prevented their use for commercial navigation. In the prerailway years the primary condition of commercial development in centers of any importance was access to navigable waterways. As a rule the only manufactures to thrive in commercial centers were the craft and workshop industries serving the requirements of trade and shipping and of the indigenous population. The use of machinery was almost entirely confined to such simple kinds as could be driven by hand cranks, human treadmills, or by draft animals similarly harnessed. As discussed in the succeeding volume, when steam power from the 1820s made its way slowly into industrial use, it was chiefly concentrated in the larger commercial centers and was applied principally by engines of quite small capacity. As late as 1860 barely a dozen cities of more than 10,000 population could be described as having a substantial industrial base, with, say, at least a quarter of the population engaged in manufacturing. See the accompanying tabulation based on census data.

	Population	*Persons engaged in manufacturing*	*Percent of total population*
New Bedford, Mass.	22,300	11,297	51
Lynn, Mass.	19,083	9,588	50
Lawrence, Mass.	17,639	7,150	41
Lowell, Mass.	36,827	13,206	36
Manchester, N.H.	20,107	7,000	35
Fall River, Mass.	14,026	4,621	33
Smithfield, R.I.	13,283	3,801	29
Gloucester, Mass.	10,904	3,095	28
Newark, N.J.	71,941	18,851	26
Nashua, N.H.	10,065	2,542	25
Waterbury, Conn.	10,004	2,502	25
Bridgeport, Conn.	13,299	3,269	25

Of the twenty leading cities in the United States in 1860—those with populations of 45,000 or more—only Newark fell within the category of industrial cities so defined. In contrast, only one-eighth and one-fifth of the populations of the nation's largest cities, New York and Philadelphia, were engaged in manufacturing. Less than a tenth of the inhabitants of such cities as Baltimore, Brooklyn, Buffalo, Chicago, Louisville, Milwaukee, and Saint Louis were so employed.[6]

In 1860, with waterpower still the chief source of motive power in manufacturing in the estimated ratio of fifty-nine to forty-one, only five of the foregoing twelve industrial cities depended primarily on this traditional source of power. They were the Merrimack River centers, Lowell, Lawrence, and Manchester; nearby Nashua, New Hampshire; and Fall River, Massachusetts, which by 1860 was obtaining only about 25 percent of its industrial power from steam. Of the 102 cities in 1860 having 10,000 or more inhabitants, less than a score made appreciable use of waterpower; most of these fell within the lower half of the list in order of size. They accounted for but 8 percent of the aggregate population of the 102 cities and 13 percent of the hands employed in manufacturing. The industrial employees of these 102 cities comprised only 43 percent of the total such workers reported for the nation. The greater part of industrial employment in 1860 thus was found in some 300 cities of 2,500 to 10,000 population and in the innumerable smaller towns, villages,

[6] U.S., Census Office, Eighth Census, 1860, *Statistics of the United States* (Washington, D.C., 1866), pp. xviii–xix. See also volume on steam power, in preparation.

and hamlets distributed throughout the nation. It was especially in this last and largest group that waterpower had its principal use and strength.

If we use the census definition of a city as 2,500 or more inhabitants, the number of cities nationwide rises from 131 in 1840 to 392 in 1860, to 939 in 1880, and to 1,737 in 1900. The total urban population in these same years increased from 17 million to 31 million to 50 million and 76 million respectively, and accounted for 11, 20, 28, and 40 percent of the aggregate population during the same period. Here was an impressive base for industrialization, depending of necessity almost entirely upon steam power, which in turn was aided by important advances in new engine types from 1850. In 1875 Robert Thurston remarked upon the reversal of circumstances that had long favored waterpower over steam power. Fuel consumption in the best engines had been reduced from 10.0 to 2.5 pounds of coal per horsepower hour. Initial costs of steam plants were down, in part owing to the lower expense of transportation from engine builders. Maintenance and repair costs were also much lower. Large steam mills were erected where some years earlier the sharp competition from waterpower sites offered slight hope of success. An 1897 review of steam-power trends since 1870 developed the same theme: engines and boiler plant equipment were better designed, better built, and lower in first cost by nearly 20 percent; operating costs were down by 50 percent.[7]

The decennial censuses documented the marked growth of manufacturing in cities. "No one will hesitate," declared Superintendent Francis Amasa Walker of the 1880 census, "to assent to the proposition that the growth of the cities of the United States since 1850 has been due in far greater measure to their development as manufacturing

[7] U.S., Bureau of the Census, *Historical Statistics of the United States from Colonial Times to 1957* (Washington, D.C., 1960), p. 14. Robert H. Thurston, "Report on Manufactures and Machinery" *Reports of the Commissioners to the International Exhibition Held at Vienna, 1873,* ed. Robert H. Thurston, 4 vols. (Washington, D.C., 1876), 3:174–75. F. W. Dean, "Reduction in the Cost of Steam Power from 1870 to 1897," *TASME* 19 (1898):301–29: Figs. on p. 323 relate to plants of some 1,000 horsepower and were not applicable to plants of small capacity. According to Thurston: "The usual dominant consideration in locating any manufacturing establishment," in consequence of the reduced expense of steam power, "is proximity to the market and convenience of transportation" ("Report on Manufactures and Machinery," p. 175). An early discussion of industrial location and the advantages of an urban situation, primarily for engineering works, is Charles G. Willcox, "Iron Works—Their location, Arrangement and Construction," *Scientific American,* n.s. 3 (15 Sept. 1860): 180. See also John C. Merriam, "Steam," in *Eighty Years' Progress of the United States* (Hartford, Conn., 1868), p. 271.

centers than to their increased business as centers for the distribution of commercial products."[8] Of the twenty leading cities of the nation in 1860, the greater number by 1880 had increased substantially the proportion of their now greatly enlarged populations engaged in manufacturing. Since 1860 Chicago, Detroit, Milwaukee, and San Francisco tripled or better the percentage of their populations in industrial occupations; Baltimore, Cleveland, and Saint Louis practically doubled theirs; and New York, Boston, Buffalo, and Louisville increased theirs by 50 percent. In 1880 only four of the twenty leading cities had fewer than 10 percent of their inhabitants engaged in manufacturing as compared with thirteen in 1860.[9] As Adna F. Weber later noted, the trend in the concentration of manufacturing was particularly marked in the largest cities. In 1860 the ten cities with 50,000 or more population had a per capita product less than half that in ten cities of lesser population. By 1890 per capita product was proportioned, roughly, according to size, ranging from $455 in the twenty-eight "great cities," to $355 in 137 cities of 20,000 to 100,000 population and $58 for the "remainder of the country."[10] The high point in the relative concentration of manufacturing in cities was reached about 1890, when the hundred largest cities of the nation, with 20 percent of the population, accounted for 60 percent of total manufactured product.[11]

Regional and National Trends

The trend to steam power in industry was most pronounced in the trans-Appalachian West. Drawing strength from the westward advance of population and settlement and protected from eastern

[8] U.S., Census Office, Tenth Census, vol. 2, *Report on Manufactures of the United States* (Washington, D.C., 1883), p. xxii. Hamilton Smith made the same point with greater assurance if less evidence thirty years earlier (James DeBow, *Industrial Resources, Etc., of the Southern and Western States,* 3 vols. [New Orleans, 1852–53], 2:122–23).

[9] U.S., Census Office, Tenth Census, 1880, *Compendium of the Tenth Census,* 2 pts. (Washington, D.C., 1883), pt. 2, p. 1218.

[10] Adna F. Weber, *The Growth of Cities in the Nineteenth Century* (New York, 1899), pp. 201–2, citing E. Laspeyres, "Die Gruppirung der Industrie innerhalb der Nordamerikanischen Union," *Vierteljahrschrift für Volkswirtschaft und Kulturgeschichte,* 34:17. See pp. 228 ff. for Weber on the relation of industry to urban trends.

[11] Twelfth Census, *Manufactures,* pt. 1, pp. ccxviii–ccxix. A reversal of the centralization trend began in the 1890s so that by 1900 the 164 largest cities produced practically the same proportion of the national industrial product as the one hundred largest ten years earlier.

competition by transportation costs, manufacturing expanded vigorously following an early foothold in the upper Ohio Valley, especially in the northwest. The center of manufacturing in the United States, which in 1850 was located in central Pennsylvania, had by 1900 moved some 300 miles westward into Ohio. Between 1850 and 1880 manufacturing by value of output increased twice as fast for the states north of the Ohio River as for the nation at large. While the North Atlantic states' share of the national product in manufactures declined from 66 to 62 percent, 1860–80, the North Central states' share rose from 17 to 26 percent. By 1890 the gap between the two regions dominating the national scene had narrowed further, standing at 52 and 33 percent, respectively.[12]

The greatest gains of steam power in the post–Civil War decades were understandably in the regions where waterpower resources were least abundant. These included not only the middlewestern states where manufacturing was most advanced—Ohio, Indiana, Illinois, and Michigan—but the South Central states—Kentucky, Tennessee, Alabama, and Mississippi—where manufacturing beyond the country-mill stage had made slow progress. By 1870 industry had made little more than a beginning west of the Mississippi; in most states steam engines supplied from 65 to 75 percent of the power in use.[13] In Missouri, the most advanced state industrially in this group, about 90 percent of the power employed in manufacturing in 1870 was steam power. Except in certain mining regions, industrial development in the Far West and on the Pacific coast likewise depended largely upon steam power.[14]

The urbanization of manufacturing, largely responsible for the wholesale shift to steam power, was simply the most striking and visible of the larger processes of the expansion and industrialization of the American economy. Table 18 summarizes the growth in the leading sectors of the economy by value added by production. In the fifty-year period 1849-99 (to avoid exaggeration), there was more than a threefold increase of population, including an eightfold increase in urban population (not shown); a fourfold increase in agriculture; and a thirteenfold increase in manufacturing, supported by a tenfold increase in horse-

[12] Tenth Census, *Manufactures,* pp. xii ff.; Eleventh Census, *Manufacturing Industries,* pt. 1, pp. 8 ff. On the westward shift in the center of manufactures, see Twelfth Census, *Manufactures,* pp. clxx–clxxi.

[13] U.S., Census Office, Ninth Census, 1870, vol. 3, *Statistics of Wealth and Industry* (Washington, D.C., 1872), pp. 392 ff.

[14] Ninth Census, *Statistics of Wealth and Industry*, pp. 392 ff.

power capacity. Construction figures show a modest sixfold increase, but mining and coal, if graphed, would leap off the page.

The impact of mechanization, the shift from hand to machine production for the years 1870–1900, is shown in Table 19. There are wide variations both in the power requirements among the eleven industries covered and in the amount of increase. In only four instances did the increase in horsepower per wage earner fall below 100 percent; in several cases the increase ranged from nearly threefold to fivefold.

Power Trends by Major Industries

Table 20 reviews the varying response of a representative group of manufacturing industries to the alternatives of steam power and waterpower, 1870-1900. Among the eleven industries covered, only one, paper and wood pulp, failed to join the pervasive move from waterpower to steam power. The paper and wood pulp industry in 1870 obtained 72 percent of its power from waterwheels and as late as 1909 still employed 60 percent waterpower.[15] The explanation lies in the need for large amounts of both power and water in production, especially in pulp grinding and waste disposal, favoring waterpower with its low unit costs and placing a premium on river sites. At the other extreme was the iron and steel industry, which from the 1860s was the most striking example of a rapid and thorough conversion from waterpower to steam power within some thirty years. In the 1820s and 1830s the iron industry, as we noted in another place, closely resembled other early small rural mill industries. In 1838 barely 10 percent of the forges, blast furnaces, and rolling mills in the

[15] U.S., Census Office, Eighth Census, 1860, *Manufactures of the United States in 1860* (Washington, D.C., 1865), pp. cxxi–cxxix; U.S., Census Office, Tenth Census, 1880, vols. 16–17, *Reports on the Water-Power of the United States* (Washington, D.C., 1885, 1887), pt. 1, p. 328n.; Thirteenth Census, *Manufactures,* p. 333. In 1909 the paper and wood pulp industry was the only industry among the 102 leading industries in which waterpower was the principal source of motive power (ibid., table 3). This industry between 1869 and 1899 increased its power requirements per hand and per establishment more than fivefold, leading the industrial field in this respect ("Power Employed in Manufactures," Twelfth Census, *Manufactures*, pt. 1, pp. cccxxx, cccxvii). Steam power was not introduced in the paper mills at Holyoke, Massachusetts, until the early 1880s (Tenth Census, *Water-Power,* pt. 1, p. 224). On the heavy power demands of the paper industry, see Andrew Ure, *Dictionary of Arts, Manufactures, and Mines* (Philadelphia, 1848), pp. 929 ff.; also JBF to Clemens Herschel, 11 Aug. 1880, LCP, vol. DB-10, 325.

Table 18. Value added by production in major sectors of the economy, with figures for population and horsepower, 1839–99
(1879 prices; dollar figures in millions)

	Population (millions)	Agriculture	Mining	Manufacturing	Mineral coal	Construction	Horsepower in manufactures (1000s)
1839	17.1	787	7	190	2.4	110	–
1849	23.2	989	17	488	8.4	163	1,100
1859	31.4	1,492	33	677	17.0	302	1,600
1869	38.6	1,720	70	1,078	41.0	403	2,346
1879	50.2	2,599	153	1,962	84.0	590	3,411
1889	62.9	3,238	346	4,156	172.0	919	5,970
1899	76.0	3,918	551	6,282	298.0	1,020	10,720

SOURCES: National Bureau of Economic Research, *Trends in the American Economy during the Nineteenth Century* (New York, 1960), p. 43; Carroll R. Daugherty, "The Development of Horsepower Equipment in the United States," in *Power Capacity and Production in the United States*, U.S. Geological Survey, Water-Supply Paper no. 579 (Washington, D.C., 1928), pp. 48–49.

Table 19. Increase in horsepower equipment in selected industries by wage earner and by $1,000 value of product, 1870–1900

	HP per wage earner				HP per $1,000 value of product			
	1870	1880	1890	1900	1870	1880	1890	1900
Agricultural implements	1.0	1.1	1.3	1.7	0.5	0.7	0.6	0.8
Boots and shoes, factory product	—[a]	0.1	0.2	0.4	—[a]	0.1	0.1	0.2
Cotton goods	1.1	1.5	2.1	2.7	0.8	1.3	1.7	2.4
Flouring and gristmill products	9.9	13.2	15.9	27.4	1.3	1.5	1.5	1.8
Hosiery and knit goods	0.4	0.4	0.6	0.7	0.4	0.4	0.5	0.6
Iron and steel	2.2	2.8	5.0	7.4	0.6	1.3	1.6	2.0
Lumber and timber products	4.3	5.6	3.1	5.7	3.1	3.5	2.2	2.9
Paper and wood pulp	3.0	4.8	9.6	15.4	1.1	2.3	3.1	6.0
Silk and silk goods	0.3	0.3	0.6	0.9	0.2	0.2	0.3	0.6
Woolen goods	1.1	1.2	1.6	2.0	0.6	0.7	0.9	1.2
Worsted goods	0.6	0.9	1.3	1.7	0.4	0.5	0.7	0.8

SOURCE: U.S., Bureau of the Census, Twelfth Census, 1900, vol. 7, *Manufactures* pt. 1, *U.S. by Industries* (Washington, D.C., 1902), p. cccxxx, table 8.

[a] Less than 0.1 percent.

Table 20. Waterpower and steam power used in selected industries, 1870 and 1900

(In thousands of horsepower)

	1870		1900	
	Water	Steam	Water	Steam
All industries	1,130.0	1,216	1,727.0	8,742
Agricultural implements	10.0	16	7.0	61
Boots and shoes, factory product	0.2	3	2.4	35
Cotton goods	99.0	47	252.0	532
Flouring and gristmill products	408.0	169	451.0	534
Hosiery and knit goods	4.0	2	15.0	40
Iron and steel	17.0	154	9.0	1,582
Lumber and timber products	327.0	315	201.0	1,402
Paper and wood pulp	42.0	12	505.0	256
Silk and silk goods	0.8	1	7.0	46
Woolen goods	53.0	32	52.0	83
Worsted goods	5.0	3	20.0	73
All other industries	165.0	462	206.0	4,100

SOURCE: U.S., Bureau of the Census, Twelfth Census, 1900, vol. 7, *Manufactures*, pt. 1, *U.S. by Industries* (Washington, D.C., 1902), p. cccxxviii, table 6.
NOTE: "All other power," first reported in 1890, rose from 29,283 to 830,407 horsepower, 1890–1900.

nation employed steam power.[16] Yet twenty years later, when John Peter Lesley compiled his *Iron Manufacturers' Guide* (Philadelphia, 1859), nearly 50 percent of the ironworks of all types were driven by steam power, the proportion rising with the larger and more modern types of plants. Thus some 85 percent of the small, old-style bloomeries and forges making wrought iron directly from the ore depended

[16] Victor S. Clark, *History of Manufactures in the United States, 1607–1860,* 3 vols. (New York, 1929), 1: chap. 29; J. P. Lesley, *The Iron Manufacturers' Guide* (New York, 1859), pp. 747 ff.; Frederick Overman, *The Manufacture of Iron in All Its Various Branches* (Philadelphia, 1850), pp. 460 ff.; Louis C. Hunter, "Influence of the Market upon Technique in the Iron Industry of Western Pennsylvania up to 1860," *JEBH* 1 (1929):241 ff. See also volume 2, in preparation. Sixty-one basic metalworks in 1838 used steam engines with an aggregate of 2,917 horsepower. In 1840 there was a total of 795 ironworks in the United States–forges, blast furnaces, and rolling mills—and also 804 foundries, compared to 93 foundries reported as using steam power in 1838 (*U.S., Secretary of the Treasury, Report on the Steam-Engines in the United States,* House Ex. Doc. 21, 25th Cong., 3d sess. [Washington, D.C., 1839]; U.S., Census Office, Sixth Census, 1840, *Compendium of the Enumeration of the Inhabitants and Statistics of the United States* [Washington, D.C., 1841], p. 358).

upon waterpower, as did nearly 50 percent of the charcoal blast furnaces. But 90 percent of the coal and coke blast furnaces and about 66 percent of the rolling mills, by far the largest types of ironworks and the heaviest consumers of power, used steam power.

The shift to steam power was slowest in the East, furthest advanced in the West. According to Lesley's *Iron Manufacturers' Guide*, in the Atlantic seaboard states the ratio of water-powered to steam-powered ironworks of all types was about two to one; west of the Appalachians this ratio was more than reversed, standing at one to two and one-half. In the Ohio Valley, that early bastion of steam power, 75 percent of all ironworks and all but a handful of mineral coal blast furnaces and rolling mills were driven by steam engines.[17] By 1870 the transition to steam power in the industry was nearly complete, with 90 percent (96 percent in 1880) of the horsepower in all branches of the industry consisting of steam power. The new and fast-growing Bessemer and open-hearth steelworks employed no waterpower at all. During the 1870s the water horsepower in the industry remained virtually at a standstill while steam horsepower rose from 154,000 to 381,000 horsepower.[18]

This early, rapid, and, by 1880, virtually complete change to a steam-power base in the basic iron industry was owing in the first instance to the attractions of urban situations, with access to markets and raw materials, of which Pittsburgh was perhaps the earliest and most striking example. This concentration was marked especially in the West by the transition from small-scale rural charcoal blast furnaces depending upon local ores and fuels to massive coal and coke blast furnaces using the rich ores from distant Missouri and Lake Superior fields made available by low-cost rail or water transport. Waste heat and gases from the new closed-top, hot-blast furnaces, combined often with puddling furnaces as an adjunct to rolling mills and nailworks, were used for steam raising and eliminated the traditional cost advantage of waterpower over steam power. Finally, the heavy and rapidly mounting power requirements of ever larger capacity blast furnaces and rolling mills could not possibly have been met or concentrated mechanically at the points of use so far as waterpower was concerned.[19]

[17] See tables based upon Lesley, *Iron Manufacturers' Guide* in Louis C. Hunter, *Studies in the Economic History of the Ohio Valley* (Northampton, Mass., 1934), pp. 39–41.

[18] Herman Hollerith, "Statistics of Steam and Water Power Used in the Manufacture of Iron and Steel," in Tenth Census, 1880, vol. 22, *Report on Power and Machinery in Manufactures* (Washington, D.C., 1888), pp. 1–5.

[19] Clark, *History of Manufactures,* 1:chap. 19; also vol. 2; Overman, *Manufacture of*

In textiles, long depending largely upon waterpower, steam power provided the only feasible basis for the accelerating expansion of production after 1850. By 1860 Lowell mills were in some instances compelled to employ auxiliary steam power and by 1880 more steam power than waterpower was in use here, despite the reluctance of some of the textile corporations to resort to steam. As the following tabulation from Thomas R. Smith's *The Cotton Textile Industry of Fall River, Massachusetts* shows, in New England, accounting as late as 1880 for over 50 percent of the national output of cotton textiles, steam power by 1900 outranked waterpower two to one, whereas thirty years earlier waterpower had outranked steam three to one.

	Waterpower		*Steam power*	
	Horsepower	*%*	*Horsepower*	*%*
1870	80,271	75	26,763	25
1880	116,854	56	90,521	44
1890	145,563	49	154,286	51
1900	162,619	33	324,162	67

Nationwide the cotton textile industry, and woolen textiles in similar fashion, followed much the same course, accompanied in New England by a major shift from inland to coastal locations. Fall River's striking rise as a cotton textile center underscores the role of steam power. Based originally upon the small but excellent power of the Quequechan River, the Fall River cotton industry in 1843 obtained its first steam mill and from the early fifties depended entirely upon steam power for its rapid growth. In 1870 Fall River had more spindles than Lowell and by 1875 led the nation in capacity and output with twice the number of spindles as the longtime leader on the lower Merrimack.[20] The pattern of change was much the same in other textile industries and in industry generally throughout the nation, as shown in Table 20. In the flour and grist milling and the lumber and timber industries the substantial conversion to steam power usually went

Iron, pp. 460 ff.; Walter Johnson, *Notes on the Use of Anthracite in the Manufacture of Iron* (Boston, 1841), pp. 28–31; Samuel H. Daddow and Benjamin Bannon, *Coal, Iron, and Oil; or the Practical American Miner* (Pottsville, Pa., 1866); John Fritz, *The Autobiography of John Fritz* (New York, 1912); Edmund C. Pechin, "The Position of the American Pig-Iron Manufacture," *TAIME* 1 (1871–73): 280 ff.

[20] Thomas R. Smith, *The Cotton Textile Industry of Fall River, Massachusetts* (New York, 1944), chap. 2. See also the account of Fall River in John Hayward, *A Gazetteer of the State of Massachusetts*, rev. ed. (Boston, 1849).

hand in hand with modernization of equipment, increase in scale, and relocation with reference to access to materials and rail transportation. In the case of lumber and timber, the use of waste wood for fuel at no cost save handling eliminated a nuisance and brought the marked shift from waterpower to steam power.

Motive Power and Two Economies

The broad changes in the locational patterns of industry reflecting the extension of the railway network brought about a gradual but decisive reversal of the traditional relationships between the two sources of mechanical power. During the years of transition the two kinds of motive power seemed almost to become identified with different industrial economies, the one associated with the older industrial regions and the mill industries and industrial villages characteristic of the past and the other with the emerging dynamic and diversified economy of the future, which gave to the multiplying commercial centers of the nation an increasingly industrialized and urban character. The new industrial America is disclosed by the decennial censuses of manufacturing. These surveys document forcefully the accelerating shift from waterpower to steam power from the 1860s. Against this picture of progress may be placed the carefully delineated if partial view of an older industrial America revealed by the special 1880 census survey of waterpower. The industrial economy that emerges from the 1880 survey is one dominated by the small-scale, rural-oriented industries serving market areas of limited extent. This economy was identified above all with the basic flour, grist-, and sawmills that for over two centuries had played so vital a role in an expanding and pioneering agricultural America.

If we except some half-dozen major drainage regions in the Northeast—chiefly the coastal river basins from eastern Maine to New Jersey, in capacity led by the Merrimack, the Connecticut, and the Hudson river systems—we find that gristmills, flour mills, and sawmills accounted for the greater part of the waterpower in use in the United States in 1880. In only one Atlantic seaboard drainage basin south of New Jersey did this group of mills employ as little as 60 percent of the waterpower in use. In the remaining drainage basins the percentage ranged from 71 to 94 percent. West of the Appalachians only on "the streams tributary to Lake Ontario" did such mills account for less than 60 percent of the waterpower in use. In only two other western river basins, the Beaver River, a minor stream in southwestern Pennsylvania (74 percent), and the Miami, a substantial river in southwestern Ohio (66 percent), did the traditional milling industries employ less than 88 percent of the waterpower reported in use. The average figure was between 90 and 95 percent. In the nation as a whole in 1880 flour and gristmills together with lumber mills employed 61 percent of the waterpower used in manufacturing; cotton and woolen goods and paper raised the figure to 85 percent. Thus, five industries, all among the oldest in the country, employed much the greater part of the waterpower used in manufacturing in 1880. The remainder of the developed waterpowers in the drainage basins dominated by the traditional milling industries was employed largely—and characteristically by establishments of quite modest size—in woodworking, metal and metalworking, and textile enterprises.[21]

Traditional Industries in Decline

For much of the older industry based on waterpower the future was to hold little promise; the great majority of individual establishments as well as the greater number of small industrial communities were destined slowly to wither on the vine. "All through the United States," declared James Emerson, the veteran hydraulic turbine tester of Massachusetts about 1890,

> may be seen relics of old mills, wheelwright shops, etc., located upon streams of little capacity except during the spring of the year or during heavy rains. These were very useful in early times, but now almost every want is supplied by large manufacturers at a lower price than the raw material would cost at these isolated places, hence few of them continue in operation, and such as do bear the marks of a lingering old age going to seed.[22]

[21] Figures compiled from Tenth Census, *Water-Power,* pts. 1 and 2, esp. pt. 1, p. xvi. By contrast with the situation in most parts of the country, flour, grist- and sawmills employed only 17 percent of the horsepower in use in the Merrimack basin, 16 percent for "the coast streams south of the Merrimack," 33 percent for the Connecticut River and tributaries, 28 percent for "the streams tributary to Long Island Sound," and 42 percent for the Hudson River and tributaries. Figures for other parts of the country are also based on data from this source.

[22] James Emerson, *Treatise Relative to the Testing of Water-Wheels and Machinery,* 4th ed. (n.p., 1892), p. 27.

The traveler through the older portions of Ohio, observed a geological report in 1904,

has his attention arrested by the desolateness of the old water-power mills. On every hand there are to be found the ruins of some once busy mill, its dam broken and its race clogged and overgrown with weeds. At no very distant time these small mills entered very largely into the daily life of a good portion of the people, not from the importance of the mills themselves, but from the fact that they supplied the daily necessities of life, especially in the more isolated districts. The usual grist and saw mill was here supplemented by small woolen mills, cooper shops, an occasional machine or wagon shop, and many numerous cider presses and cane mills.

. . . At one time in the Muskingum basin water powers were to be found on many small streams, the back water of one dam reaching to the tail-race of the power above. These numerous dams made the flow of the stream more uniform than ever, and frequently gave steady power throughout the year.[23]

A generation later another geologist reported that of the 253 waterpowers of the Muskingum Valley utilized for gristmills and sawmills in 1880, only ninety remained in 1904 and but fourteen in 1936.[24] At the turn of the new century the federal census commented with professional detachment:

Every day observation shows us that the smaller establishments in many lines of industry tend to disappear before the superior competitive facilities of larger enterprises. Abandoned mill sites are common on streams which were once relatively valuable water powers, but which have gradually shrunk until they can be depended upon only for a portion of the year. This is especially true of saw and grist mills . . . . a constant shifting of industries from undesirable localities to places which offer better facilities of one kind or another is in progress throughout the country as one phase of the steady adjustment which is taking place.[25]

By 1900 the rapid decline of the milling industries that had long stood at the head of the nation's industrial establishments was a matter of official comment. Decline was not confined to water mills but

[23] B. H. and M. Flynn, *The Natural Features and Economic Development of the Drainage Areas in Ohio of Sandusky, Maumee, Muskingum and Miami Valleys,* U.S. Geological Survey, Water-Supply Paper no. 91 (Washington, D.C., 1904), pp. 34–35.

[24] H. H. Bennett, "Utilization of Small Water Powers," p. 483.

[25] Twelfth Census, *Manufactures,* pt. 1, pp. lxiv–lxv. In the mid-1930s there was an interesting exception, as reported by Bennett, to the widespread decline of small waterpowers: the Tennessee Valley. Under the fostering care of the new Tennessee Valley Authority some 1,700 waterpowers of 100 horsepower capacity or less were in use ("Utilization of Small Water Powers," pp. 490–91).

extended progressively to the village workshop and handicraft industries, regardless of the kind of power, if any, employed; it was incident to the commercialization of farming and the eventual decline of rural populations. Yet thousands of country mills, especially gristmills supplying local needs for meal, flour, and feed, persisted well into the twentieth century. Their role and the aggregate power employed were of minor if not negligible importance in the new industrial age.[26]

Unutilized and Underutilized Waterpowers

In the competitive struggle between the old industry and the new, power supply was but one element and rarely the decisive one. With an active, growing establishment the limitations of waterpower could almost always be overcome by adding auxiliary steam power, which, we will note later, became a practice of common, if typically reluctant, resort. On a regional basis it was not for want of waterpowers of a capacity and character otherwise suited to development that the shift to steam power proceeded so rapidly and decisively during the post–Civil War decades. The massive 1880 census survey of waterpower contains abundant evidence of both unused and underutilized waterpowers throughout the greater part of the northeastern United States. Even in highly industrialized southern New England the millstream presenting an almost uninterrupted succession of developed waterpowers was in many districts matched by little used millstreams. In addition to innumerable and widely distributed undeveloped waterpowers up to 100 horsepower in capacity, the 1880 survey revealed 127 undeveloped powers of 2,000 or more gross horsepower, of which fifty-three had capacities over 5,000 and twenty-eight over 10,000

[26] The decline is discussed for our northern neighbors in John MacDougall, *Rural Life in Canada: Its Trends and Tasks* (1913; reprint ed., Toronto, 1973), pp. 57 ff. For details respecting the larger undeveloped powers, see Tenth Census, *Water-Power,* pt. 1, pp. 33–35, and under the appropriate river basins. The total potential waterpower of Maine about 1902, as estimated by the U.S. Geological Survey, was some 1.5 million horsepower as compared with 350,000 horsepower in actual use (Maine, State Water Storage Commission, *Third Annual Report* [Waterville, 1913], pp. 32–33). The hydrographic survey of New York, reported stream by stream in 1919, concluded that the undeveloped waterpower which could be made available with the aid of reservoirs was twice as great as that in use (*Water Power Resources of New York* (Albany, 1919), pp. 37 ff. See also C. M. Leighton, "Underdeveloped Water Powers," in *Papers on the Conservation of Water Resources,* U.S. Geological Survey, Water-Supply Paper no. 234 (Washington, D.C., 1909), pp. 51 ff.

gross horsepower (see Table 21). In New England alone there were sixteen such large undeveloped powers. The Middle Atlantic states accounted for sixteen more. With all but two of its eighty-three large waterpowers in the undeveloped class, development in the South Atlantic states lay almost wholly in the future. Nonuse of large, as of lesser, powers was owing to remoteness from markets, inadequate transportation facilities, and the limited development of manufacturing owing to these and other factors.[27]

Expedients for Enlarging Power Supply

Idle waterpower in periods of expanding industry and the increasing adoption of steam power underscored the limitations of direct-use waterpower. "I have no doubt you have a large and valuable property," Superintendent Francis of Lowell's Locks and Canals replied in 1884 to an effort to interest him in an upstate New York waterpower, "and if you could transport it here we should be glad of it at a big price, as our water power is exhausted and we are not near the coal-fields."[28] Except for distances usually limited to some hundreds of yards by means of costly canals, waterpowers were not portable. Removal of plant and equipment to a power site of greater capacity was a solution evidently adopted with considerable reluctance and was evidently not a frequent practice. Eli Whitney's move from New Haven to Whitneyville to secure needed power and that of the Yale brothers from their original base to the water privilege around which developed the industrial community of Yalesville were not typical. When the Stamford Manufacturing Company, incorporated in 1844, required additional capacity for the production of dyewood extracts, additional mills were established in turn at several other points in Connecticut and New York, a process reversed in later years when the adoption of steam power permitted a gradual reconcentration of operations in the home plant.[29]

In an earlier chapter attention was called to the frequent practice of a single manufacturer securing control of successive adjacent privileges on the same stream, usually at some inconvenience and additional

[27] See George F. Swain in Tenth Census, *Water-Power,* pt. 1, pp. 667 ff., 823–24. There is little explicit discussion of the industrial potential of this region.

[28] JBF to S. B. Rice, 1 Jan. 1884, LCP, DB-12, 286.

[29] William T. Davis, ed., *The New England States,* 4 vols. (Boston, 1897), 2:944 ff., 983–84.

Table 21. Developed and undeveloped waterpowers in the United States, 1880

	Number of developed powers, utilizing in 1880			Number of undeveloped powers, with an available power of			Total number of powers, with an available power of		
	over 10,000 HP[a]	over 5,000 HP	over 2,000 HP	over 10,000 HP	over 5,000 HP	over 2,000 HP	over 10,000 HP	over 5,000 HP	over 2,000 HP
New England									
Massachusetts	3	4	4	0	0	0	4	4	4
New Hampshire	1	1	3	1	1	3	2	5	9
Vermont	0	1	1	0	0	0	0	1	1
Between New Hampshire and Vermont	0	0	0	0	2	2	0	2	2
Maine	1	1	4	2	6	9	5	15	19
Connecticut	0	0	1	0	1	2	1	2	3
Total	5	7	13	3	10	16	12	29	38
Middle Atlantic states									
New York	0	3	8	2	4	7	9	12	26
New Jersey or between New Jersey and Pennsylvania	0	0	1	0	2	2	1	4	5
Pennsylvania	0	0	0	2	2	2	2	2	2
Between Maryland and Virginia	0	0	0	2	4	4	2	4	4
Virginia	0	0	1	—	—	—	—	—	—
West Virginia	0	0	0	1	1	1	1	1	1
Total	0	3	10	7	13	16	15	23	38
South Atlantic states									
North Carolina	0	0	0	3	4	6	3	4	6
South Carolina	0	0	0	4	9	13	4	9	13
Georgia	0	0	2	0	2	31	2	4	33
Between South Carolina and Georgia	0	0	0	2	3	4	2	3	4
Alabama	0	0	0	1	1	26	2	2	27
Total	0	0	2	10	19	80	13	22	83

Table 21. (cont.)

	Number of developed powers, utilizing in 1880			Number of undeveloped powers, with an available power of			Total number of powers, with an available power of		
	over 10,000 HP[a]	over 5,000 HP	over 2,000 HP	over 10,000 HP	over 5,000 HP	over 2,000 HP	over 10,000 HP	over 5,000 HP	over 2,000 HP
Northwest									
Minnesota	1	1	1	3	3	3	5	5	5
Wisconsin	0	0	1	5	8	12	7	11	15
Kansas	0	0	0	0	0	0	0	0	1
Total	1	1	2	8	11	15	12	16	21

Source: George F. Swain, "Statistics of Water Power Employed in Manufacturing in the United States," *PASA*, n.s. 1 (1888–89): 44.

[a] HP refers to theoretical horsepower during twelve hours, in dry seasons, supposing that the water could be stored, and not wasted, during the other twelve hours.

expense in operations, as in the cases of ironworks in Boonton, New Jersey; Troy, New York; and Weymouth, Massachusetts. Many similar examples would be found by combing the stream-by-stream reports on water privileges in the two-volume 1880 census survey. One of the more impressive of such arrangements was the West Warren Cotton Mills, with four mills occupying successive privileges with falls ranging between twelve and sixteen feet over a half-mile stretch on the Quaboag River in western Massachusetts. Developed at considerable outlay in grading at millsites, this establishment comprised in the aggregate some 32,000 spindles (compared with 45,000 and 59,000 spindles of two representative companies at Lowell), 650 hands, and, at three of the mills, 650-horsepower capacity.[30]

More characteristic was the situation of the Springfield Armory, not far distant from West Warren, where several workshops were strung along the Mill River for some two-thirds of a mile. Faced in the mid-1820s with the alternatives of further dispersal of operations in new shops, removal at considerable expense to a new site, or the

[30] Tenth Census, *Water-Power,* pt. 1, p. 270. Instances may be cited of a similar distribution of works at powder manufactories, but safety considerations may have been primary here (ibid., p. 618 [DuPont Company, Del.] and pp. 252–53 [Hazard Powder Company, Conn.]).

unwelcome addition of steam power, the armory authorities reluctantly accepted the continued dispersal of operations with the drawbacks of expense of materials handling between the several shops and the problems of coordinating power use. As in all similar situations, the shops located downstream were dependent upon the upper ones for the release of water—used or "wasted"—to drive their wheels.[31] The vigorous growth needs of such companies as Ansonia Brass and Copper in Ansonia, Connecticut, and Washburn and Moen in Worcester, Massachusetts, were for some years met by building additional facilities on scattered waterpowers within the same community.[32]

The most generally available expedient for relieving power shortages was elimination of waste in the existing plant: stopping waste flow over or through the dam; checking and closing leaks at gates, penstocks, races, wasteways, and wheels; replacing worn and inefficient wheels with new and better ones; and, less directly, reducing power losses in transmission and machinery. On the basis of fragmentary evidence, such methods of relief were not carried to excess in the millwright stage of engineering. The vigor with which such large and ably managed waterpower companies as Lowell's Locks and Canals, under Superintendent Francis, applied itself to this task and the indifference and even hostility at times from the mill managements suggest that ignorance and neglect were the common rule. As a consultant, Francis informed the management of the Perkins textile mills of Chicopee, Massachusetts, that a change in the position of existing waterwheels alone could increase power more than 40 percent, with additional gains possible through replacing existing wheels with turbines and other measures.[33] A New Hampshire waterpower commission reported in 1870 that in many cases less than a third of the available power was in actual use: "Many mills are operated by leaky dams, and by water wheels which properly belong to the age of pod-augers and wooden plows.[34] According to Charles T. Main, "A certain

[31] See Col. Roswell Lee to Col. G. Bomford, 23 Mar. 1824 and 18 Sept. 1826, and to S. Lathrop, 11 Jan. 1823, Springfield Armory Records, National Archives.

[32] Davis, ed., *The New England States,* 2:907–8; C. E. Goodrich, *The Story of the Washburn and Moen Manufacturing Company, 1831–1899* (Worcester, Mass., 1935), pp. 4 ff., 68 ff.

[33] JBF to F. H. Story, 31 Jan. 1853, LCP, A-18, no. 90.

[34] New Hampshire, Water Power Commission, *Report on the Preliminary Examination of the Water Power of New Hampshire* (Manchester, 1870), p. 8. On the economy of obtaining turbines of high guaranteed efficiency, see Joseph P. Frizell, *Water-Power: An Outline of the Development and Application of the Energy of Flowing Water,* 2d ed. (New York, 1901).

amount of water is unavoidably wasted over the dam and by leakage through the various parts of the plant. This allowance of leakage and waste I usually place at 10 per cent of the flow which could be theoretically stored and used."[35]

Ignorance, inertia, and expense delayed in many establishments the replacement by hydraulic turbines of the old-style waterwheels of low efficiency.[36] An 1866 survey of New England cotton mills showed that in forty-seven mills the ratio of breast and overshot wheels to turbines was 88 to 119. Ten years later the Massachusetts state census showed that nearly 80 percent of the waterwheels used in manufactures were turbines.[37]

Power shortages often were a consequence of errors in the initial estimate of the available capacity at a given site. Zachariah Allen early remarked a common tendency to choose a waterpower and millsite during the ample spring flow, only to discover following the completion of the mill that it must stand idle during the low-water months of summer "when the days are longest and the expenses of light and fuel are not required."[38] Yet in the absence of streamflow records of some duration, local report was only too commonly unreliable as a guide. As late as 1880 there were probably not over half a dozen streams in the country that had been regularly and carefully gauged over a period of years. The systematic collection of streamflow data by the United States Geological Survey did not begin until 1888. The engineers engaged in the federal waterpower survey of 1880 were compelled to resort to the crudest of "guesstimates" on this critical element in

[35] "Computation of the Values of Water Powers," *TASME* 26 (1904–05): 73.

[36] The letters of Henry Burden of Troy, New York, to Col. Roswell Lee of the Springfield Armory call attention to a gain in power of 50 to 100 percent from the replacement of undershot by overshot wheels. See also Lee to Crocker, Richmond, and Company, 25 Nov. et seq. 1825, Springfield Armory Records.

[37] *Statistics of Cotton Manufacturers in New England* 1 (1866); Massachusetts, Bureau of Labor Statistics, *Census of Massachusetts, 1875*, 3 vols. (Boston, 1877), 2:325.

[38] Zachariah Allen, *The Science of Mechanics* (Providence, 1829), pp. 197–98. Allen added, "These mills may be seen in different parts of the country abandoned and desolate, standing as monuments of the folly of those who erected them." See also U.S., Secretary of the Treasury, *Documents Relative to the Manufactures in the United States*, House Ex. Doc. 308, 22d Cong., 1st sess., 2 vols. (Washington, D.C., 1833). More than a half century after Allen's comment, a prominent hydraulic engineer of Lawrence, Massachusetts, declared that one of the principal sources of disappointment in the use of waterpowers was "the placing of more machinery on a given water privilege" than could be carried with its capacity at minimum flow (Charles T. Main, "On the Use of Compound Engines for Manufacturing Purposes, *TASME* 10 [1888–89]:78).

their reports.[39] As we have noted, not until the rapid rise of interest in hydroelectricity did federal and state governments provide essential stream gauging and recording facilities.

Irregularities in Streamflow

The more serious problem for waterpower in the long run arose from seasonal variations in streamflow and the resulting interruptions in the direct-drive power supply. In the first part of the nineteenth century millowners and manufacturers did not, in general, expect a continuous supply of streamflow and power. The amount of machinery employed was proportioned to what was variously termed "the milling" or "ordinary" stage of the millstream, defined sometimes simply as "most of the year," but usually meaning some eight or nine months.[40] As we have seen, a period of idleness with minor variations was accepted and adjusted to, but as manufacturing took on a larger and increasingly "commercial" scale, efforts were made to reduce the length of these fairly regular interruptions by such means as upstream reservoirs. More serious were marked irregularities and prolonged interruptions of streamflow attended by losses in production and income. Irregularities and unreliability of flow were usually most pronounced on the smaller streams, on which the great majority of water mills were dependent.[41] In the lumber industry from Maine to Wisconsin the lack of power attending severe weather limited the sawing season to about half the year.[42] The small average output of rural blast

[39] Tenth Census, *Water-Power,* pt. 1, p. 171. The engineers found that local accounts of streamflow were usually vague, imprecise, and often clearly in error. For the purposes of the survey, they attempted to supply data under such unsatisfactory phrases as "absolute minimum flow," "minimum low seasons flow," "low seasons flow in dry years but not the driest," "flow available ten months in the year," and "maximum-flow available with storage." See, e.g., pt. 1, pp. 56–57, 171 ff., 529 ff., 678 ff. On the streamflow gauging activities of states, associated usually with other programs, such as reservoirs in support of waterpower development and conservation, see Maine, State Water Storage Commission, *Second Annual Report* (Waterville, 1912); W. R. King, *The Surface Waters of Kentucky* (Frankfort, 1924); W. N. Gladson, *The Geological Survey of Arkansas: Water Powers of Arkansas* (n.p., 1911).

[40] D. M. Greene, "On Gauging of Streams," *TASCE* 5 (1876): 251; George A. Kimball, "Water Power—Its Measurement and Value," *JAES* 13 (1894):110–11. This matter is discussed in Hunter, *Studies in the Economic History of the Ohio Valley,* pp. 5–49.

[41] *Scientific American* 2 (24 Oct. 1846):37; also W. C. Howells, *Recollections of Life in Ohio, 1813–1840* (Cincinnati, 1895), p. 136.

[42] *DeBow's Review* 14 (1853):333–34; R. G. Wood, *A History of Lumbering in Maine, 1820–1861* (Orono, 1935), p. 159. The reliance upon streams for floating logs to mar-

furnaces, long the main supply of puddling furnaces and rolling mills, was in part owing to the same cause.[43] As late as 1858 three-fourths of the 750 paper mills in the nation were reported limited by power supply to eight months' operation each year.[44]

The vagaries of power supply became increasingly burdensome with the growth in scale of operations, and a rising proportion of industry sought relief and competitive advantage by adopting steam power, in part or as primary supply. By the 1850s references to the restrictions imposed by seasonal interruptions of supply were becoming common. "Many of the manufactures of this place are crippled by loss of water power through the drought." "Many of the mills have to stop during the dry seasons." "The river is too uncertain to admit of works requiring a constant power supply." "Our water fails us two or three months in the year."[45] An official survey of Maine waterpowers in 1869 boasted that the state's rivers "to a remarkable degree enjoy immunity from those ruinous drawbacks to water-power manufacturing: *water-dearth* and *water-freshet.*" Yet the descriptions of waterpowers in the basin-by-basin survey that followed contain frequent references to seasonal interruptions.[46]

Zachariah Allen noted in 1829 that the loss to millowners from seasonal interruptions was much greater than that suggested by a complete suspension of activity, since at times it was necessary to run at reduced speed. "The actual stoppages take place at different hours of each day and the labor is desultory, while the operatives cannot undertake any other temporary business for their support."[47] Thirty years later the same complaint came from the superintendent of one of the

---

ket and mills compounded difficulties for the lumber industry (William G. Rector, *Log Transportation in the Lake States Lumber Industry, 1840–1915* [Glendale, Calif., 1953], p. 28).

[43] Thomas Chambers, "The Iron Trade," *National Magazine and Industrial Record* 1 (July 1845): 147; see also *Documents Relating to the Manufactures in the United States,* 2:864–65.

[44] See Hunter, *Studies in the Economic History of the Ohio Valley,* pp. 34–37.

[45] *Scientific American* 4 (9 Oct. 1858):14, 33.

[46] Maine, Hydrographic Survey, *The Water-Power of Maine, by Walter Wells* (Augusta, 1869), pp. 49–50. See also *Report on the Preliminary Examination of the Water Power of New Hampshire,* p. 9.

[47] Allen, *Science of Mechanics,* p. 201. Allen cited an instance in which the millpond was so limited in size that during low water work was suspended regularly once an hour, to be resumed after thirty minutes when the pond refilled from streamflow. Writing some fifty years later, he stressed the deprivation and distress of the poor resulting from such interruptions to work and compelling reliance on the credit or charity of the mill owners ("Sketch of the Life of Zachariah Allen," Rhode Island Historical Society, Providence).

largest waterpowers in northern New England. To outright stoppages of twenty-five days during the fall season were added the burden of low-speed operations much of the time. As a result, the superintendent said, "our men became uneasy, some grumbled and many went home, and others procured employment elsewhere, and all suffered more or less. This state of things broke up our organization and it was many weeks after we had plenty of water before our complement of men could be got together again and all working to as good advantage as before we were stopped."[48] It was most unusual, everyone agreed, for the waters of Goose Creek to fail as they did at Harrisville in southern New Hampshire in the fall of 1880, resulting in a three-month stoppage of the mills. "Woolens in process could not be completed, and business was lost. A large number of operatives left town to seek employment elsewhere, and for those who stayed it was a bleak winter." With this, the second such failure in a row, labor recruitment became a major concern for two years. Reliance on a limited local labor supply, gathered to the village or town serving the mills and subject to depletion at such times, was a handicap not present in urban centers with their labor pools on which all industry might draw.[49]

Severe power shortages had affected the textile industry in New England in 1866 and 1870, and beginning in 1879 an extreme drought

[48] Quoted in George S. Gibb, *The Saco-Lowell Shops: Textile Machinery Building in New England, 1813–1949* (Cambridge, Mass., 1950), pp. 375–76. Various mills at Richmond, Petersburg, and Lynchburg, Virginia, had to suspend operations for some time during dry-season shortages (Tenth Census, *Water-Power,* pt. 1, pp. 534 ff. While two New England textile companies succeeded in increasing their water supply sufficiently to run their mills during the summer and fall droughts, for many years it was necessary to curtail operations during the summer. During the dull market years of the 1860s, moreover, the layoff gave many operatives the opportunity to visit their former homes in Canada (Evelyn H. Knowlton, *Pepperell's Progress: A History of a Cotton Textile Company, 1844–1945* [Cambridge, Mass., 1948], pp. 8–9). Francis reported a similar effect on labor earnings and morale at Lowell when diminishing streamflow reduced machine speeds (JBF to T. Lyman, 28 Oct. 1865, LCP, DB-3, 719). Seasonal interruptions of industrial activity were by no means limited to manufactories dependent upon waterpower. The reliance of inland commerce on water transportation before the Civil War imposed a seasonal pattern on commercial and industrial activity alike in much of the nation. In the trans-Appalachian West periods of protracted low water on the Ohio River not only virtually halted navigation but in some years imposed an embargo on the coal traffic, which literally brought steam engines at Cincinnati and Louisville to a stand. See Hunter, *Studies in the Economic History of the Ohio Valley.*

[49] John B. Armstrong, *Factory under the Elms: A History of Harrisville, New Hampshire, 1774–1969* (Cambridge, Mass., 1969), pp. 132–33. The same point was stressed by Main, "On the Use of Compound Engines," p. 87.

extended throughout a large part of the Northeast and the Middle Atlantic states. Long stoppages of mills and factories seriously affected industrial communities from Maine into Virginia. Canal navigation and associated waterpowers were hampered at many points. The crisis even extended to urban water supply, ranging from York, Kittery, and Wells, Maine; Nashua and Manchester, New Hampshire; to Kingston, Manhattan, and Brooklyn, New York; Reading and Altoona, Pennsylvania; Baltimore; and Petersburg and Richmond, Virginia.[50]

The response of Charles E. Emery in 1883 to such a series of disrupting and burdensome droughts reflects the changes that had taken place in the industrial scene in the later decades of the nineteenth century. Among considerations affecting the value of a waterpower Emery placed foremost "the reliability of the power throughout the entire year and continuously for a number of years."

For a country saw-mill operated by men having other duties at times, a simple torrent, dry or nearly so at times during the summer, will answer very well. The farmers' boys can get in their harvest in the dry season, get out and haul logs before the snow melts, and saw them by water power when the stream is full. Everything is of the simplest description; no labor is left idle, and the cost of the whole plant is so small that the mill may be unused for long periods without a loss worthy of consideration.

Such a state of things is not possible with the manufacturing interests of modern times. Large contracts are to be filled; large numbers of operatives are employed, skilled only in particular branches of the particular work done; hundreds of thousands of dollars of capital are invested—the mills cannot be stopped if the owners hope to compete with others doing business in a business way. If the water power fails for a season, steam is employed; and indeed, from its greater reliability, steam is used exclusively in many cases in successful competition with water power.[51]

[50] *U.S. Economist,* 3 Feb. 1866, quoted in Cincinnati *Gazette,* 7 Feb. 1866; see similar references ibid., 1 Oct. 1870; J. Leander Bishop, *A History of American Manufactures from 1608 to 1860,* 3d ed., 3 vols. (Philadelphia, 1868), 1:101–4, 127–32; A. Fteley, "The Flow of the Sudbury River, Massachusetts, for the Years, 1875–1879," *TASCE* 10 (1881):225 ff., 235–36. Commenting upon an unusual drought at Rochester, New York, onetime famed "water-power city," an engineer confided to his colleagues: "I tell you the honest truth when I say that parties are relinquishing their water rights and running by steam. Just now we have got no water at all in the Genessee" (discussion of Charles T. Main, "The Value of a Water Power," *TASME* 13 [1891–92]:156). See also discussion of idem, "On the Use of Compound Engines," p. 77.

[51] Charles E. Emery, "The Cost of Steam Power," *TASCE* 12 (1883):429–30. See also George F. Swain, "Statistics of the Water Power Employed in Manufacturing in the U.S.," *PASA*, n.s. 1 (1888–89): 26–27, 36; W. M. Barr, "Hydraulic Generation and Application of Motive Power," *Engineering Magazine* 29 (1905):867; C. T. Main, discussion of Kimball, "Water Power," p. 89.

William Worthen, whose experience as a hydraulic engineer had been mainly in New England textiles, in 1887 remarked upon the extraordinary upsurge of steam power in manufacturing during the 1870s, 80 percent compared with 8 percent in waterpower from a slightly larger base. He explained the disparity in terms of "a convenience of access to business and labor centers" that largely favored the increased use of steam power. "And although there are immense water powers yet undeveloped, and the cost of steam power is largely in excess of that of water power," he continued, "yet position and its relations have decided in favor of steam." Where large manufactories have depended upon waterpower and this source has been exhausted or is very irregular, "business has been extended by the use of steam, rather than by a diversion of it to a new place for water."[52] Without steam power and with irregular streamflow at times compelling plant stoppages, declared Charles T. Main in 1891, waterpower had little value "except for a very limited range of business. No business, employing any amount of labor, carried on in such a way, could compete successfully with concerns which have a continuous run."[53]

Steam power was long held an expedient of last resort; often the initial outlay eliminated it as a practical option. Much could be done to postpone what to many was an unacceptable alternative. Steam power raised the problems not only of cost, usually in hard and scarce money, and availability for purchase but of maintenance and repair. There were also the obstacles of ignorance, prejudice, and dread of the unfamiliar. On the other hand, as earlier remarked, a rundown, dilapidated water privilege could be put in order and improved. Old and inefficient wheels could be repaired or replaced, dams repaired and their height raised; additional flowage land obtained, and pondage increased. Adjacent privileges on the stream might be available for purchase and brought under development, individually or perhaps consolidated to advantage. Losses in headraces might be corrected by checking leakage, regrading, or relocation; tailraces could be deepened, and the fall increased. An intelligent millwright with the aid of a surveyor might carry improvement well beyond maintenance and repair, substantially increasing capacity.

[52] William E. Worthen, "Address of the President," *TASCE* 17 (1887): 4.

[53] Main, "The Value of a Water Power," pp. 141, 149. Main pointed out that a supplementary steam plant was necessary for most waterpowers; with many days of insufficient water, the steam plant had to equal the water plant in capacity, an expensive business.

Storage Reservoirs

Upstream reservoirs and, in time, auxiliary steam power were the most common means of relief from the periodic paralyses or slowdowns imposed by low water. The first was the more readily accepted remedy, its practice falling within the framework of things familiar and its cost within acceptable limits, especially if shared by a number of millowners as beneficiaries. The upstream reservoir was simply a millpond multiplied a hundredfold or more. Whereas the millpond usually did no more than store nighttime flow for daytime use, one or more reservoirs located at upstream points could capture and store enough surplus wet-season flow to offset the greatly diminished flow and power supply of the seasons of drought, most commonly midsummer through early fall. Reservoirs of the early waterpower years were nearly always based upon one or more natural lakes or ponds of the desired capacity having outlet in the millstream or one of its tributaries, possibly at a point many miles distant from the mill or mills drawing upon them. Such bodies of water, numbered by the thousands in New England, upstate New York, and the upper Mississippi Valley, served in their natural state as accumulators of water during wet seasons for continuous but gradual release through stream outlets in subsequent months. "Our lakes are so many basins of reserve power," declared a New Hampshire waterpower survey report of 1870. "They are like a spring wound up for use during rainy weather to be let down if other sources fail or prove insufficient."[54] By placing dams with regulating gates at the lake outlets, their level could thereby be raised, storage capacity greatly enlarged, and the low-water streamflow to the mills below substantially increased. In the rough hill country often characteristic of upper drainage basins, the necessary land could be purchased or flowage awards to landowners met, and construction, maintenance, and management provided at acceptable expense for the benefits received. On the busier millstreams, with millowners cooperating in corporate or informal association, improvement and regulation of streamflow could be carried out to mutual advantage and at moderate cost. Large establishments often found it possible and preferable to provide their own upstream reservoir capacity, the better to adapt streamflow to the needs of their own operations as and when business required, no doubt usually to the advantage but sometimes to the annoyance of other millowners on the reservoired stream.

[54] *Report on the Preliminary Examination of Water Power of New Hampshire,* p. 12.

There were Old World precedents for power-supply reservoirs and at least two instances of their use here during the colonial period.[55] The earliest example to come to my attention was the reservoir established about 1645 in connection with the pioneer New England ironworks near the site of Lynn, Massachusetts. A second example was that provided by Peter Hasenclever, the ironmaster, in connection with the ironworks built about 1767 at Charlottenberg on the Poquanock River in New Jersey. Here several ponds were converted into reservoirs by means of dams at their outlets. The water supply thus obtained was used to maintain power during periods of drought.[56]

The effective introduction of reservoirs after 1800 appears to stem from the pioneer installations in Rhode Island, established in 1822 on the initiative of the imaginative and inventive industrialist, Zachariah Allen, in cooperation with other millowners on the headwaters of the Woonasquatucket River. This project, according to Allen, was carried out under an act of incorporation, which was "the first ever granted in New England for the special object of constructing reservoirs." This innovation, later supplemented by additional reservoirs, transformed a minor stream but twelve miles in length into an important power source, eventually supplying a dozen textile mills.[57] Success led to

[55] See John F. Bateman, "Description of the Bann Reservoirs, County Down, Ireland," *PICE* (1841): 168 ff.; *Library of Universal Knowledge* (1880 ed. of *Chambers's Encyclopaedia)*, s.v. "Water Power." In Britain John Smeaton employed reservoirs as early as the 1760s to supplement streamflow in dry seasons (John Farey, *A Treatise on the Steam Engine* [London, 1827], pp. 273–74).

[56] Edward N. Hartley, *Ironworks on the Saugus* (Norman, Okla., 1957), pp. 182–83; John B. Pearse, *A Concise History of the Iron Manufacture of the American Colonies up to the Revolution* (Philadelphia, 1876), pp. 60–65. There was a misguided attempt in Orleans, Vermont, in 1810 to convert a sizable pond into a reservoir to supply mills below on the Barton River. When a ditch from pond to stream was opened, the water rushed out en masse, creating havoc and entirely emptying the pond in some fifteen minutes (Zadock Thompson, *A Gazetteer of the State of Vermont* [Montpelier, 1824], p. 131). For a similar episode in another part of Vermont in 1830, see Hamilton Child, ed., *Gazetteer of Orange County, Vermont, 1762–1888* (Syracuse, N.Y., 1888), p. 511.

[57] Allen, *Science of Mechanics,* pp. 199–200. See also Amos Perry, *Memorial of Zachariah Allen, 1795–1882* (Cambridge, Mass., 1883), pp. 49, 56, 61; Allen, "The Transmission of Power from the Motor to the Machine," *Proceedings of the New England Cotton Manufacturers' Association,* 10 (1871): 35. By 1860 eight corporations in all had been chartered in Rhode Island for establishing reservoirs in the Woonasquatucket and nearby river basins (Peter J. Coleman, *The Transformation of Rhode Island, 1790–1860* [Providence, 1963], p. 90). The Watuppa Reservoir Company for controlling the flow of water to mills on the Quequechan River at Fall River was incorporated by the Massachusetts legislature in 1826 (Fall River, Mass., Watuppa Ponds and Quequechan River Commission, *Report . . . to the City Council* [Boston, 1915], pp. 16–17).

imitation; many similar companies were established in this busy industrial state and in adjacent areas in Massachusetts, sometimes termed the cradle of the American textile industry, a region where at the peak of its development the use of waterpower per square mile was the highest in the nation.[58] By the 1830s the use of storage reservoirs for regulating power supply was a familiar and growing practice in New England. By 1832, as described by Samuel Slater, a textile pioneer, the system had been extended to the other main rivers of Rhode Island. Although none of these streams was by western standards more than a creek, as Slater pointed out, they and their tributaries furnished in their short runs "innumerable cascades, and a power of propelling machinery almost incalculable in amount." Improved by such means, many of these streams whose natural flow "would drive the works upon them only ten or eleven months of the year, had been made to operate throughout the year."[59] Except during the most unusual droughts it became common for mills using stored water to operate at close to normal levels at all times.

As revealed by the federal waterpower survey of 1880, the use of storage reservoirs had become widespread in New England, in many instances established "at merely nominal expense," often no more than a few thousand dollars.[60] Many secondary rivers of the region, including such substantial streams as the Saco, Penobscot, Salmon Falls, Hoosic, Willimantic, Naugatuck, and Housatonic rivers, were materially improved as sources of power by this means. Although reservoirs were talked of for such major rivers as the Connecticut, the upper Hudson, and, much later, the upper Mississippi rivers, such projects were too ambitious to be undertaken except under government sponsorship, which was not forthcoming before the age of hydroelectricity. A competent geologist later estimated that a probable 10 percent of the floodwaters of New England and New York were saved by storage in artificial reservoirs and ponds.[61]

[58] Swain, "Statistics," p. 38, table 9.

[59] Samuel Slater, in *Documents Relating to the Manufactures in the United States,* 1:927. The chief source of information on waterpower reservoirs is the Tenth Census, *Water-Power,* esp. pt. 1, pp. 181–84, 346–48, 364, 434, 619, 625, pt. 2, pp. 104, 131–32, 219, 226, 467. Maine, Hydrographic Survey, *Water-Power of Maine,* pp. 158–59, discusses the value of reservoirs.

[60] Tenth Census, *Water-Power,* pt. 1, pp. 182–84; see also Maine, Hydrographic Survey, *Water-Power of Maine,* p. 159.

[61] M. O. Leighton, "Floods," in *Papers in the Conservation of Water Resources,* U.S. Geological Survey, Water-Supply Paper no. 234 (Washington, D.C., 1909), pp. 26–27. For reservoir proposals for larger rivers, see Tenth Census, *Water-Power,* pt. 1, pp.

The most impressive use of storage reservoirs was that initiated in the 1840s by the Locks and Canals company of Lowell, described in chapter 6 above. This extensive project, carried out with the support of the waterpower companies serving the Merrimack River textile cities of Lawrence and Manchester, involved the conversion into reservoirs of Lake Winnipesaukee and other smaller New Hampshire lakes, thus enabling the Locks and Canals to increase by slightly more than half the basic, or "permanent," power guaranteed the lessee mills. The lake reservoirs amply justified the substantial expenditures involved and the large responsibilities their maintenance and management entailed. Superintendent Francis of the Locks and Canals remarked upon their wider significance: "The tendency is in the direction of a more uniform flow through the year, from the extension of the reservoir system in the smaller branches of the river, of which, of course, we get the benefit by rendering available for manufacturing purposes [here] a larger proportion of the surplus [river flow.]"[62]

The use of upstream reservoirs as a means of overcoming seasonal shortages and, less commonly, of enlarging year-round power supply was far from general and appears to have been largely dependent upon two favoring circumstances. These were, first, the presence in the upper drainage basins of natural lakes or sizable ponds that could be converted to reservoirs at acceptable cost; second, a sufficient number of mills and factories dependent upon a millstream to encourage formation of associations to share the expense involved in reservoir systems. Outside of New England, upstate New York, and parts of the upper Mississippi Valley, the first and basic condition was not often present.[63] Where nature did provide lakes and ponds, joint

346–47; pt. 2, pp. 170–71, 226. For a proposed reservoir system on the upper Schuylkill River to benefit the Fairmount Water Works in Philadelphia, see ibid., pt. 1, p. 620. On the discussion and introduction of reservoiring on the upper Mississippi River to improve both navigation and power supply, see Lucille M. Kane, *The Waterfall That Built a City: The Falls of St. Anthony in Minneapolis* (Saint Paul, 1966), 1, pp. 128–33.

[62] JBF to T. Jefferson Coolidge, 2 Oct. 1868, LCP, DB-5, 312–315.

[63] Another early example of reservoiring in the upper Mississippi Valley was that reported in Tenth Census, *Water-Power,* pt. 2, pp. 226 ff., for the Rock River basin of Illinois and southern Wisconsin; see also ibid., pp. 26 ff., 467. The abundant waterpower of the lower Fox River in Wisconsin was served by Lake Winnebago as a natural reservoir (ibid., p. 35). In such lumbering regions as Maine and Wisconsin, the use of reservoirs on major logging steams to supply water for flushing the winter cut of logs in spring drives often led to conflicts with the power needs of milling interests. The result at times was extended controversy, legislative contests, and litigation, in the happier instances leading to cooperative arrangements for reservoir maintenance and management. See Maine, State Water Storage Commission, *Third Annual Report* (Waterville, 1913), pp. 31, 54–55; Rector, *Log Transportation in Lake States.*

action by the benefiting millowners was for many years the more common method by which reservoiring was carried out, especially in New England. First cost and maintenance expenses were shared usually in proportion to the power used by each millowner. Under favorable conditions, some large concerns preferred independent action and the advantage of being free to release reservoir water in accord with their own requirements. In such cases other power users located on the stream between owner and reservoir obtained at least some benefits at no cost save that of being obliged to adapt their own operations to those of the controlling company during periods of low water. Even in New England, however, the establishment of reservoirs independently by a single large manufactory does not appear to have been common.[64]

The building of artificial reservoirs by throwing dams across river valleys at narrows was a practice generally reserved for the greater opportunities and resources of the hydroelectric period ahead. In time, too, the increase of population, the extension of agriculture, and the rise of land values tended to discourage reservoiring in otherwise favorable regions by adding to its cost.[65] In sum, although reservoirs often provided welcome relief from seasonal shortages and even added to the year-round power supply, they were a cumbrous method of enlarging capacity. The widespread use of auxiliary steam power at manufactories dependent upon waterpower, reported by the Tenth Census waterpower survey of 1880, indicates a preference for this alternative method of enlarging power supply. The regular and systematic use of storage reservoirs to increase capacity begins with the use of hydroelectricity. From the 1890s engineering studies of site feasibility for large hydroelectric ventures nearly always included surveys of upstream storage possibilities.[66]

[64] See, e.g., Tenth Census, *Water-Power,* pt. 1, pp. 112–14, 120–21, 187, 197–99, 204–13. For examples of joint ownership and management, see ibid., pt. 1, pp. 61–71, 90, 195–99, 206–8, 241, 274–79, 310. For pros and cons of reservoirs from the viewpoint of individual mills and factories, see ibid., pt. 1, pp. 270 ff., 317–18, 325, 347–48.

[65] See *The Water Power Resources of New York,* pp. 5 ff.; Tenth Census, *Water-Power,* pt. 1, pp. 186, 196, 202; G. B. Leighton, *Report of the* [*New Hampshire*] *Commission on Water Power Conservation and Water Power, 1917–1918* (n.p., 1919), pp. 24–26, 47.

[66] Following the highly publicized hydroelectric developments at Niagara Falls and the rapid buying up of attractive power sites by large holding companies, state governments moved to establish controls over hydraulic power development within their boundaries. One of the first steps taken was to make surveys of the waterpowers within the state, including studies of the storage capabilities. See, e.g., the first three annual reports of Maine's State Water Storage Commission, 1910–13, esp. *Second Annual Report,* pp. 3–5, and *Third Annual Report,* pp. 19 ff., 25. For discussion of storage reservoirs used with hydroelectric development, see Daniel W. Mead, *Water Power*

Auxiliary Steam Power

Steam power was adopted almost as early as storage reservoirs for overcoming seasonal deficiencies of streamflow and eventually obtained general acceptance as the most feasible alternative. Little more than twenty years after its founding, the pioneer Boston Manufacturing Company in 1836 installed steam power in its Waltham mills to meet power needs during droughts. By this date, too, several Fall River manufactories had adopted steam power for the same purpose.[67] It was as insurance against interruptions in waterpower supply that the Springfield Armory reluctantly moved in the 1830s to add steam power to their plant. Before 1850 such thriving industrial communities as Worcester, Massachusetts, and Naugatuck, Connecticut, had outgrown the capacity of the numerous small privileges on the streams in their midst and turned to the steam engine as the essential base for further growth.[68] One of the largest industrial establishments in Maine, the Saco Water Power Company, in 1860 suffered outright stoppage of its mills for twenty-five days and low-speed operations during most of the autumn months. With steam power installed, the company saved fifty days of production during 1861. Following one of the driest summers on record, 1882, leading to the first restrictions on water use by the power company, the paper mills at Holyoke, Massachusetts, began adding steam power.[69]

Engineering: The Theory, Investigation, and Development of Water Power (New York, 1908), chap. 26; Joseph P. Frizell, *Water-Power,* 3rd ed. (New York, 1903), chap. 24; and George F. Swain, *Conservation of Water by Storage* (New Haven, 1915), esp. chap. 6.

[67] William R. Bagnall, *Sketches of Manufacturing Establishments in New York City and of Textile Establishments in the Eastern States,* ed. Victor S. Clark, 3 vols. (New York, 1908), 3:2045; *Report on the Steam-Engines in the United States,* p. 57; see also John Hayward, *The New England Gazetteer,* 6th ed. (Boston, 1839), s.v. "Exeter."

[68] Felicia J. Deyrup, *Arms Makers of the Connecticut Valley* (Northampton, Mass., 1948), pp. 147–48; Charles G. Washburn, *Industrial Worcester* (Worcester, Mass., 1917), pp. 31 ff.; Constance McLaughlin Green, *History of Naugatuck* (Naugatuck, Conn., 1948), pp. 156 ff.

[69] Gibb, *Saco-Lowell Shops,* pp. 375–76; Tenth Census, *Water-Power,* pt. 1, p. 224. See also Constance McLaughlin Green, *Holyoke, Massachusetts: A Case History of the Industrial Revolution in America* (New Haven, 1939), p. 80. By 1880 the largest textile company at Biddeford-Saco was the Pepperell Manufacturing Company. It used as much as 2,000 water horsepower during eight months of the year but during the remaining months often could get no more than 300 to 400 horsepower. It was equipped with a steam power capacity of 1,500 horsepower (Tenth Census, *Water-Power,* pt. 1, p. 120).

Rising overall power shortages after the Civil War combined with seasonal needs to accelerate the changeover to steam power in large and lesser establishments alike. Even in the case of so large and carefully developed and managed a waterpower system as Lowell's, the splendid lake reservoirs at the headwaters of the Merrimack served only to mitigate power shortages. "Nearly all of the water-mills in operation today," declared a speaker at an 1880 gathering of the New England textile manufacturers, "have outgrown the capacity . . . of the streams where they are located." With variations in detail the story was much the same everywhere in the older waterpower regions. The dead end, as it seemed, of full utilization of waterpower was eventually reached. Despite its many reservoirs, Connecticut's Quinebaug River could no longer supply the growing demands resulting from the owners of the numerous mill privileges "adding machinery to their mills and driving it at a greater speed than formerly." The chief source of disappointment with waterpower, declared T. J. Borden of Fall River, was "the placing of more machinery on a given water privilege" than the stream during periods of minimum flow was capable of driving.[70]

The 1880 waterpower survey showed that auxiliary steam power was widely adopted in the principal river basins east of the Mississippi. The greater number of the 3,400 mills and factories located upon and obtaining power from the streams tributary to Long Island Sound, including the Connecticut River and its tributaries, depended on steam engines for nearly 20 percent of the total power used, although reliance on auxiliary steam power varied widely from industry to industry. In the various drainage basins under consideration, for instance, the ratio of steam power to aggregate power as reported by industries was approximately as follows: silk, 50 to 60 percent; other textiles (cotton, woolen), 20 to 30 percent; wood and metalworking, 10 to 30 percent; and grist- and sawmills, 2 to 5 percent.[71]

In the much more extensive Hudson basin the proportion of auxiliary steam power to total power was about the same as for streams tributary to Long Island Sound. On the streams of upstate New York tributary to Lake Ontario the proportion was much lower: about 5 percent. Throughout the eastern half of the United States the extent of reliance upon auxiliary steam power, wherever reported in the 1880

[70] Tenth Census, *Water-Power,* pt. 1, pp. 75, 85, 200–1; T. J. Borden in discussion of Main, "On the Use of Compound Engines," p. 78. See also description of the Naugatuck River and tributaries, Tenth Census, *Water-Power,* pt. 1, p. 318.

[71] Tenth Census, Water-Power, pt. 1, pp. 304 ff., 328 ff.

census survey, varied from one river basin to the next; rarely, save where gristmills and sawmills predominated, was the supporting role of steam power negligible. Even on such moderately developed power streams as the Allegheny, Monongahela, Big Beaver, Scioto, and Miami rivers the auxiliary power of steam engines ran as high as 25 percent of the aggregate power employed on the hundreds of mills and factories concerned. In the Miami basin, for example, the lower portions of which contained much manufacturing, 9,300 water horsepower and 3,000 steam horsepower were in use. On some of the smaller millstreams the proportion often ran higher, with steam power comprising the chief supply of mills once driven entirely by waterwheels. The improvement of steam engines, the decline in first cost, and the rising efficiency in the use of fuel stimulated the trend.[72]

Comparative Costs of Steam Power and Waterpower

The low cost of waterpower compared with steam power was long accepted as a matter so obvious as not to require discussion. How slow this favorable cost comparison was to give way is suggested by the article "Water-power" in *Appleton's Dictionary of Machines* (1852):

> The waterfall is rendered available comparatively without labor, and furnishes its supplies without the intervention of human aid. The energies of the steam engine, on the contrary, can be commanded in any situation, only by the influence of the miner; and in localities much removed from sources of fuel, can only be sustained at an expense which falls heavily upon the operations to which they are subservient. That expense, it is true, is continually being diminished, and by means of the steam engine itself, in its character as a carrier; but no happy discovery, no possibility, can reduce it to the minimum at which our water-runs are maintained.

More than thirty years later a prominent hydraulic engineer remarked upon the persistence of the long accepted but no longer meaningful view. There was an earlier period when there was no question of steam

[72] Ibid., pp. 400–401; pt. 2, esp. pp. 450–90. Auxiliary steam power was not reported for many of the river basins west and south of New York State. In the Ohio Valley drainage basins, the number of *places* at which auxiliary steam power was reported in use ran considerably higher than the ratio of steam to water horsepower, ranging from one-third to one-half of the places listed. A New Haven builder of large capacity high-speed engines in 1871 described his product as "peculiarly desirable as an auxiliary power" for manufacturers using waterpower, who from the past season's experience would understand the necessity of providing against "the stoppage of their works by drought in the future" (*Manufacturer and Builder* 3 [1871]:132).

power competing with waterpower under ordinary conditions, but time had brought about a very different set of conditions and a change in waterpower's unquestioned superiority to steam in cost. "I have nothing," declared Charles Main in 1889, "but respect and admiration for our predecessors who have done so nobly and who hold so tenaciously to the old doctrine, but they cannot but see, if they will, that the reverence with which water power has always been mentioned is gradually disappearing."[73]

The wide differences in the cost of waterpower and steam power, both in initial capital outlay and in operating expense, especially with the fuel and attendance costs remarked in the quotation above, together with unfamiliarity and problems of maintenance and repair, long ruled steam power out of consideration for most stationary applications. This situation underwent gradual but slowly accelerating change from about 1850 with advances in the design and construction of steam engines and the growth of manufacturing in the cities. The cost figures presented in the reports on waterpower prepared by the Tenth Census and arranged by Professor George F. Swain in a separate report provide a baseline for comparison with steam-power costs in the postbellum years (see Table 22). The unit generally employed in engineering discussions, as in commercial practice, was the horsepower year: the equivalent of one net horsepower delivered under head at mill or factory penstock during 308 ten-hour working days. To these figures must be added the expense borne by the consumer of providing and maintaining the wheel installations with appurtenant structures and equipment, estimated by Professor Swain at $9 per horsepower year. Swain's tabulation shows the very wide range within the eastern half of the continent in the cost of commercial waterpower, from as little as $11 to as much as $80 per horsepower per year.

The most familiar and widely cited rates were those in effect at the larger textile centers of New England on the lower Merrimack River, long the largest and most fully developed and utilized waterpowers in the country. The net rates alike at Lowell, Lawrence, and Manchester were almost identical at about $20 per horsepower year for the "permanent" power available the year round with the "surplus" power available seasonally charged at somewhat higher rates to discourage use. In 1881 the surplus-water rate at Lowell of $5 per mill power (65 horsepower) was described as "about the cost of the coal necessary to

[73] Charles T. Main, "Cost of Steam and Water Power," *TASME* 11 (1889–90): 125.

Table 22. Waterpower rates and cost in the United States, 1880 (Net effective horsepower per annum)*

Place, and conditions.	Cost of water per N. E. H. P. per annum.	Total cost of power per N. E. H. P. per annum.
Lawrence, Mass., for power originally granted...	$15.00	$24.00
" " " surplus up to 20%.........	20.61	29.61
" " " " 20 to 50%.........	41.22	50.22
" " " " above 50%.........	20.61	29.61
" " for permanent leases at present..	20.00	29.00
Lowell, Mass., for original leases...............	15.00	24.00
" " " surplus up to 40%............	25.75	34.75
" " " " " 40–50%.........	51.50	60.50
" " " " " 50–60%.........	103.00	112.00
" " " " during "backwater"..	5.15	14.15
Manchester, N. H., for surplus power............	25.75	34.75
Saco and Biddeford, Me., for surplus power.....	15.45	24.45
Lewiston, Me., ordinary leases.....................	2.50 to 12.50	11.50 to 21.50
Windsor Locks, Conn. (according to fall).........	18.00 to 27.00	27.00 to 36.00
Holyoke, Mass..............................	12.87	21.87
Turner's Falls, Mass............................	7.50	16.50
Bellows Falls, Vt...............................	7.50	16.50
Unionville, Conn.............................	16.20	25.20
Occum, Conn.....................................	20.00	29.00
Barrett's Junction, Mass........................	9.00	18.00
Birmingham, Conn., "permanent water.".......	27.00	36.00
" " 1st surplus....................	16.00	25.00
" " 2nd "	11.00	20.00
Ansonia, Conn., "permanent water."...........	28.50	37.50
" " surplus........................	12.00 to 24.00	21.00 to 33.00
Oswego, N. Y., 1st class..........................	19.00	28.00
" " 2nd "	13.50 to 16.10	22.50 to 25.10
" " surplus	9.35 to 12.50	18.35 to 21.50
" " (2nd privilege) 1st class..........	9.30 to 11.30	18.30 to 20.30
" " " 2nd and 3rd class...	4.70 to 5.65	13.70 to 14.65
Cohoes, N. Y....................................	21.00	30.00
Rochester, N. Y. (at one privilege)..............	25.00	34.00
Lockport, N. Y..................................	12.00 to 15.90	21.00 to 24.90
Niagara Falls, N. Y. (1.).........................	10.00	19.00
" " " (2.).........................	7.00	16.00
Passaic, N. J....................................	47.50	56.50
Paterson, N. J...................................	51.00	60.00
Trenton, N. J....................................	53.50 to 71.00	62.50 to 80.00
Fredericksburg, Va..............................	5.00 to 15.00	14.00 to 24.00
Manchester, Va..................................	42.00 to 60.00	51.00 to 69.00
Augusta, Ga.....................................	5.50	14.50
Hamilton, Ohio..................................	27.30 to 62.90	36.30 to 71.90
Middletown, Ohio................................	41.00	50.00
Franklin, Ohio...................................	42.00	51.00
Dayton View, Ohio..............................	52.00	61.00
Dayton, Ohio....................................	43.00	52.00
" " (from M. and E. Canal)............	35.30 to 40.00	44.30 to 49.00
Appleton, Wis. (with land), 500–1000 H. P........	1.00 to 2.00	10.00 to 11.00
" " " 100–300 H. P..........	3.00 to 4.00	12.00 to 13.00
" " " 50 H. P..............	2.00 to 3.00	11.00 to 12.00
Kaukauna, Wis. " 100–300 H. P..........	2.00 to 5.00	11.00 to 14.00
Lawrence, Kansas................................	20.00	29.00

*Efficiency of wheel assumed at 80 per cent. This is high, as an average, for actual wheels, but for a different per cent the cost can easily be calculated, being inversely as the efficiency.

SOURCE: George F. Swain, "Statistics of Water Power Employed in Manufacturing," *PASA*, n.s. 1 (1888–89): 33. Based upon the more detailed tables in Tenth Census, *Water-Power*, pt. 1, xxxv-xxxviii.

Note: The cost of water (left-hand column) was for water delivered at plant penstocks under head. The annual expense in fixed and operating costs per net horsepower of the wheel installations provided by the consumer of water (penstocks, waterwheels, wheel pits, etc.) were estimated by Swain at $9 per net effective horsepower as shown in right-hand column.

make the same amount of steam power."[74] Rates at the large Merrimack centers were among the lowest in the country, although admittedly available only to the limited number of very large textile corporations under lease. Commercial rates lower than these were available at only a few commercial powers, chiefly in the upper Connecticut Valley, upstate New York, and in other industrially less attractive situations as in Georgia and Wisconsin.

More striking were the instances where commercial waterpower commanded rates two and three times those of the New England textile centers, most noticeably at the New Jersey cities of Paterson, Passaic, and Trenton; Manchester, Virginia (across the James River from Richmond); and the southern Ohio cities of Dayton, Middletown, and Hamilton on the Miami River. Rates of $50 to $80 reflected a preference for waterpower over steam power based on something other than price. For many users waterpower was preferred for its cleanliness and the smaller space requirements of the equipment, and for freedom from the fire and explosion hazards of steam power, from the dirt and nuisance attending the handling and storage of fuel and ashes, and from the burden and expense of engine- and boiler-room attendance.[75] Decisions on power plant were affected, too, by process requirements for water, as notably in the paper and wood-pulp industry, combined in this instance with heavy power consumption in the basic operation of grinding; or, as in textiles, the large requirements of process heat supplied most conveniently and cheaply as a by-product of steam power in the exhaust steam.

In the competition with steam power, waterpower was favored by other considerations: it was the familiar power, strengthened by tradition and long usage. Ownership of the waterpower plant by the user was by far the predominant practice and may well have enjoyed a preference over commercial waterpower and probably a cost advantage. Waterpower had another advantage over steam power in the smaller plant capacities: unit costs tended to increase with the size of the plant. The reverse was nearly always the case for steam power; the unit cost of plants and operating expenses were higher for small than for large plants.[76]

From the beginning steam power labored under a decided cost disadvantage and found acceptance typically only in the absence or the

[74] JBF to W. H. Thompson, 21 Mar. 1881, LCP, DB-11, 81-82.

[75] Main, "Cost of Steam and Water Power," p. 119.

[76] Main, "On the Use of Compound Engines for Manufacturing Purposes," p. 87; see similar statement in Swain, "Statistics," pp. 30-31.

inadequate supply of waterpower. This is best illustrated in textiles, the first of the quantity-production mechanized industries to attain importance in this country. The resort to steam power was in most instances a matter not of choice but of necessity in the face of insufficient streamflow to accommodate expanding operations. The pioneer integrated cotton mill of the Boston Manufacturing Company, on the Charles River at Waltham, turned to steam power in 1836 for use in periods of drought. The Exeter Manufacturing Company in New Hampshire installed a steam engine in 1839 for use when the water failed. By this time steam power began to be introduced in the Lowell mills for the same reason. As superintendent James B. Francis of the Locks and Canals later recalled, steam power was in use at four of the ten textile corporations by 1847. When in 1871 some 9,000 water horsepower were in use at the Lowell mills, 2,500 steam power were in continuous use. Ten years later "about twice as much steam power as water power" was installed there.[77]

Other New England textiles followed much the same pattern. "Nearly all of the water-mills in operation today," ran a report in 1880, "have outgrown the capacity of the streams where they are located." "With the increase of machinery and low water in our rivers it is safe to say [the consumption of fuel and steam] will double in the next ten years." Following the driest summer on record, in 1882, and the introduction of restrictions on water use by the power corporation at Holyoke, some of the paper mills there began adding steam power. Although the Mill River, the tributary entering the Connecticut River at Northampton, Massachusetts, was fairly lined with water mills by 1850, no steam power was introduced until the late 1850s and then by new establishments. In less than twenty years thereafter steam power was supplying nearly one-half of the total power in use.[78]

This progressive shift to steam power was made despite far higher costs, gradually reduced from the 1850s, with the marked advances in steam engine efficiency and economy. Nonetheless, for many years steam-power costs ranged from two to three times higher than those at

[77] Bagnall, *Sketches of Manufacturing Establishments,* 3:2045; John Hayward, *A Gazetteer of New Hampshire* (Concord, 1839), s.v. "Exeter"; and JBF to J. H. Sawyer, 28 Nov. 1883, LCP, DB-12, 272, to L. Dean, 26 Dec. 1871, DB-6, 349, to R. L. Steele, ca. 1882–83, DB-12, 221.

[78] W. E. Parker, "Combustion of Fuel, with Relation to the Use of Different Sizes of Coals," *Proceedings of the 29th New England Cotton Manufacturers' Association 28* (1880): 30 ff.; Green, *Holyoke,* p. 80. Some steam power had been used in the city before this (Agnes Hannay, *A Chronicle of Industry on the Mill River* [Northampton, Mass., 1935–36], pp. 90–91).

the large commercial waterpowers cited above. At Lowell, Superintendent James B. Francis followed carefully the course of steam-power costs, concerned with rising power demands of the mills and the limits on streamflow upon which they depended, aided by information and advice from his friend George A. Corliss of Providence, Rhode Island, leading designer and builder of large mill engines. In 1854 Francis reported steam-power costs as three times those of waterpower as the latter was reckoned at the lower Merrimack textile centers. Ten years later the ratio had fallen to two to one, changing only slowly during the succeeding years. In correspondence with Walter Wells, a prominent hydraulic engineer in Maine, Francis declared that "the highest rates [of waterpower] I know of are about half the cost of steam power, say, $70 per horsepower per annum." The last figure was reported ten years later for several steam mills at coastal points in New England by Samuel Webber. Francis in 1876 and Corliss in 1879 reported steam power available at $45 and $49 per horsepower year, respectively.[79] With continued engine improvement, especially the shift from simple to compound engines and from noncondensing to condensing practice and the gradual raising of steam pressures, the trend toward higher efficiency and lower costs continued. The margin of waterpower's advantage narrowed and in the case of the largest and most efficient steam plants was to be reversed.

The earlier instances of steam-power costs cited above were based typically on somewhat random instances of performance in terms of

[79] JBF to Corliss and Nightingale, 6 Jan. 1854, memoranda of 13 Jan., 22 July 1854, s.v. "Power Mill, 1851," LCP, vol. A-2; JBF to Edmund Dwight, 20 Mar. 1864, DF-1, 144. The memorandum of 13 Jan. 1854 describes the visit of George Corliss to Lowell and gives details of first cost and operation of a 280-horsepower Corliss condensing engine. See also JBF to T. C. Avery, 6 Dec. 1858, DA-5, 58 ff. With estimates of the cost of waterpower at Lowell ranging from $15 to $25 per horsepower per year, with $20 to $22 the most common figure, the following are contemporary statements of steam power costs: for 1863, $27.90 for fuel alone, with a Corliss engine using coal at $6.00 a ton and 2.5 pounds per horsepower-hour (Samuel Batchelder, *The Introduction and Early Progress of the Cotton Manufacture in the United States* [Boston, 1863], pp. 91 ff.); for 1879 $70 per horsepower-year, including interest and depreciation, in four different steam cotton mills at New England coastal points; also an estimate of $48.66 per horsepower-year, including depreciation and interest, for a Corliss engine (Samuel Webber, *Manual of Power for Machines, Shafts, and Belts* [New York, 1879], sec. 2, pp. 79–81); for 1891, at Lowell, $27.50 per horsepower-year (310 twelve-hour days; coal at $5.50 per ton and consumption at 3.0 pounds per horsepower-hour) (Col. J. Francis, 22 Apr. 1891, LCP, DF-1, 144); for 1896, also at Lowell, about $31 per horsepower-year (*Engineering Record* 33 [16 May 1896]:418–19). In 1876 Francis estimated the cost of steam power at about twice the cost of "surplus" waterpower, or some $45 per horsepower-year (JBF to T. O. Selfridge, 1 Sept. 1876, LCP, DB-9, 16–17).

coal consumption per horsepower per hour and the prices charged by engine builders. During the 1880s cost figures at last became available resting upon systematic investigation by competent engineers, who in the light of their own experience drew upon all feasible sources of information, especially the records of engine builders, in some cases using detailed questionnaires, carefully conducted tests of engine performance, and the evidence of expert witnesses presented in litigation relating to public water supply. Detailed reports of cost and operating data were obtained not only for engines, boilers, and accessory equipment but for engine foundations and housing, chimneys, and installation costs. Included were not only figures on first cost, fuel, repairs, and attendance but also on interest or depreciation, taxes, and insurance. Particular attention was given distinctions between plants of varying capacity and to engines of different types. Table 23, based on the more detailed data presented by Robert H. Thurston in his *Manual of the Steam Engine* (1891), reveals the wide range of costs attending different capacities of plants and the simpler types of engines: portable, noncondensing, and condensing. A comparison of these cost figures with the somewhat earlier figures for waterpower cost shows the narrowing margins of costs with the larger and more efficient engines reported here. Thurston's discussion was directed to the generality of steam-power users, and his tabular presentation ignored the larger and more efficient plants employing multicylinder engines.

For much the greater part of industrial expansion during the post–Civil War years, the question of steam-power versus waterpower costs was irrelevant. The urban communities, large and small, in which, as noted earlier, manufacturing became increasingly concentrated, had no alternative to steam power. However rapid its relative decline, waterpower continued to increase in quantity. Despite the virtual standstill in capacity during the 1880s, in the period 1869-1909 water horsepower capacity increased in the nation as a whole 61 percent and in New England at nearly twice this rate—a 109 percent increase.[80] The one major industry in which the issue of relative power costs continued to be meaningful was textiles. In 1890 the division in cotton textiles between steam power and waterpower nationwide was, in horsepower capacity, virtually even, 51 and 49 percent respectively. Steam power had, years ago, lost its earlier role as auxiliary power; yet the traditional view of waterpower as the cheaper persisted, and at Lowell, for example, the capacity of turbine installations had been increased to more than double the "permanent" leased power, in

[80] Smith, *Cotton Textile Industry of Fall River*, p. 41.

Table 23. Operating and capital costs of steam-power plant with different types and sizes of engines

	Portable upright		Horizontal noncondensing		Condensing single-cylinder		
	5 hp[a]	15 hp	20 hp	50 hp	100hp	250 hp	500 hp
	(in dollars)						
Cost of engine, boiler, and amortization	781.00	1,800.00	2,399.00	6,455.00	11,148.00	24,732.00	43,856.00
Cost of buildings, chimneys	313.00	504.00	599.00	1,029.00	1,487.00	3,304.00	7,260.00
Annual interest (6%) on items above	65.64	138.24	179.28	449.04	758.10	1,682.16	3,066.96
Annual interest and taxes on items above	78.55	167.99	218.01	555.67	943.15	2,090.68	3,791.36
Interest, insurance, and taxation[b]	15.71	11.20	10.90	11.11	9.43	8.36	7.58
Operating and misc. expenses	131.13	52.25	40.92	28.09	17.04	8.50	7.64
Coal, per long ton	45.33	37.89	32.36	24.06	18.98	18.02	18.02
Total cost, per hp-year	192.17	101.34	84.18	63.26	45.45	34.88	33.24

SOURCE: Robert H. Thurston, *A Manual of the Steam Engine,* pt. 1 (New York, 1891), pp. 816–19.

[a] Engine sizes are for brake horsepower; indicated horsepower (IHP) are slightly larger as follows: 6.25, 18.29, 23.53, 56.82, 112.36, 276.24, 552.49. Consumption of steam and of coal per IHP per hour given in pounds as follows: *feedwater,* 42, 36, 34, 27, 23, 22.2, 22.2; *coal*: 5.60, 4.80, 4.25, 3.27, 2.61, 2.52, 2.52. *Friction* as a percentage of engine power: 20, 18, 15, 12, 11, 9.5, 9.5.

[b] *Insurance:* 0.5 percent on total valuation of engine, boiler, and all appurtenances installed, not including engine and boiler house or chimney. *Taxation*: rate of $15 per M of total value.

order to take advantage of the excess streamflow during many months in the year.

In successive meetings of the American Society of Mechanical Engineers the traditional view was sharply challenged in two papers presented by leading hydraulic engineers long identified with the New England textile industry and especially with the Merrimack River centers. Charles H. Manning was for years chief engineer and general superintendent of the Amoskeag Manufacturing Company, owner of the waterpower and of much of the textile mill capacity at Manchester, New Hampshire. Charles T. Main, onetime president of the American Society of Mechanical Engineers, was superintendent of the Lower Pacific Mills at Lawrence, Massachusetts. Main's paper dealt only with steam-power costs, but the significance of its conclusions for waterpower was the chief subject of the long discussion that followed.[81] It presented the results of an extended investigation into cost of plant and of the power produced in plants of 1,000-horsepower capacity, equipped with paired engines of three different types: single-cylinder noncondensing, single-cylinder condensing, and compound (two-cylinder) condensing engines, operating under common conditions and assumptions. Data were drawn wherever possible from actual installations. Particular care was given to the inclusion of all plant components, from engine and boiler installation to chimney, and to all items of expense, fixed and variable, including not only fuel, attendance, and repairs but depreciation, insurance, and taxes. As a fellow engineer was later to remark: "nine steam users out of ten never allow a tithe of the constant charges to enter into their calculations" and were "quite unaware of real power costs."[82]

A primary consideration entering into power costs in textile mills as examined in the Main and Manning papers was the factitious role played by heat requirements in processing—essential in such operations as boiling, bleaching, dyeing, drying, and, in the cold months, space heating. Such process heat was most economically supplied by taking the heat normally rejected by the engine as exhaust, the advantage of this being partly offset in the case of condensing engines by the sacrifice of the gain from a vacuum in the condenser.[83] In effect the engine was simply used as a pressure-reducing valve, at the same time supplying the required power, with the 80 to 100 psi brought down to the few pounds pressure conveniently manageable in processing. In his article Main included tables to show the daily and yearly cost per indicated horsepower with the three types of engines, and to summarize the cost of waterpower at Manchester, Lowell, and Lawrence (see Tables 24-27). The total yearly expense per indicated horsepower (IHP) for com-

[81] Main, "On the Use of Compound Engines," pp. 48–87.

[82] T. A. Taylor, "The Competitive Cost of Steam and Electric Plants," *Electrical Engineer* 24 (1897):551–52.

[83] Of such processing and space heat requirements Main declared: "Rarely ever in a textile mill would the amount [of such requirements] fall below 20 per cent of the total heat rejected by the engine" (Main, "Computation of the Values of Water Powers," p. 91).

Table 24. Approximate total yearly expense of steam power per I.H.P., 1890

TABLE I.—SHOWING APPROXIMATE TOTAL YEARLY EXPENSE OF STEAM POWER PER I. H. P.

Engine.	I. H. P. of Plant.	STEAM USED FOR POWER ONLY. Cost of Coal per Ton.				25% OF EXHAUST STEAM USED. Cost of Coal per Ton.				50% OF EXHAUST STEAM USED. Cost of Coal per Ton.				75% OF EXHAUST STEAM USED. Cost of Coal per Ton.				100% OF EXHAUST STEAM USED. Cost of Coal per Ton.			
		$ 3.00	$ 4.00	$ 5.00	$ 6.00	$ 3.00	$ 4.00	$ 5.00	$ 6.00	$ 3.00	$ 4.00	$ 5.00	$ 6.00	$ 3.00	$ 4.00	$ 5.00	$ 6.00	$ 3.00	$ 4.00	$ 5.00	$ 6.00
Compound.	500	22.35	24.82	27.28	29.74	20.73	22 84	24.95	27.06	19.10	20.86	22.62	24.38	17.42	18.83	20.24	21.65				
	1000	19.16	21.63	24.09	26.55	17.57	19.68	21.79	23.90	15.39	17.75	19.51	21.27	14.34	15.75	17.16	18.57				
	1500	17.38	19.84	22.30	24.76	15.81	17.92	20.03	22.14	14.23	15.99	17.75	19.51	12.62	14.03	15.44	16.85				
	2000	16.23	18.69	21.15	23.61	14.73	16.84	18.95	21.06	13.18	14.94	16.70	18.46	11.66	13.07	14.48	15.89				
Condensing.	500	24.83	28.35	31.87	35.39	22.03	24.93	27.83	30.73	19.06	21.36	23.66	25.96	16.28	17.96	19.64	21.32				
	1000	22.34	25.84	29.36	32.88	19.54	22.44	25.34	28.24	16.62	18.92	21.22	23.52	13.87	15.54	17.22	18.90				
	1500	21.01	24.53	28.05	31.57	18.81	21.71	24.61	27.51	15.28	17.58	19.88	22.18	12.48	14.16	15.84	17.52				
	2000	20.13	23.65	27.17	30.69	17.93	20.83	23.73	26.63	14.40	16.70	19.00	21.30	11.60	13.28	14.96	16.64				
High-Pressure.	500	27.16	31.39	35.62	39.85	23.64	27.06	30.50	33.94	20.07	22.72	25.37	28.02	16.47	18.32	20.17	23.02	12.89	13.95	15.01	16.07
	1000	24.82	29.03	33.25	37.48	21.26	24.70	28.14	31.58	17.72	20.37	23.02	25.67	4.16	16.01	17.86	19.71	10.54	11.60	12.66	13 72
	1500	23.69	27.92	32.15	36.38	20.16	23.60	27.04	30.48	16.59	19.24	21.89	24.54	3.06	14.91	16.76	18.61	9.38	10.44	11.50	12.56
	2000	22.79	27.02	31.25	35.48	19.26	22.70	26.14	29.58	15.69	18.34	20.99	23.64	2.16	14.01	15.86	17.71	8.48	9.54	10.60	11.66

COST OF STEAM AND WATER POWER.

SOURCE: Charles T. Main, "Cost of Steam and Water Power," *TASME* 11 (1890): 110.

Table 25. Ordinary running daily and yearly expenses of 1,000-horse-power steam plant using different engine types

TABLE III.—SHOWING ORDINARY RUNNING DAILY AND YEARLY EXPENSES. 1,000 H. P. PLANT.

Col. 1	2	3	4	5	6	7	8	9	10	11	12	13	14	15	16	17	18	19	20	21	22
Per cent. of exhaust steam used.	Lbs. coal per I.H.P. per hour.			Cost of coal per I.H.P. per day of 10½ hours @ $5.00 per long ton—2,240 lbs.			Attendance of boilers per I.H.P. per day.			Attendance of engine per I.H.P. per day.			Oil, waste & supplies per I.H.P. per day.			Total daily expense.			Total yearly expense —308 days.		
	Compound.	Condens'g.	H. P.	Compound.	Condens'g.	H. P.	Compound.	Condens'g.	H. P.	Compound.	Condens'g.	H. P.	Compound.	Condens'g.	H. P.	Compound.	Condens'g	H. P.	Compound.	Condens'g.	H. P.
				c.	c.	c.	c.	c.	c.	c.	c.	c.	c.	c.	c.	c.	c.	c.	$	$	$
0	1.75	2.50	3.00	4.00	5.72	6.86	0.53	0.75	0.90	0.60	0.40	0.35	0.25	0.22	0.20	5.38	7.09	8.31	16.570	21.837	25.595
25	1.50	2.06	2.44	3.43	4.71	5.58	.45	.62	.73	.60	.40	.35	.25	.22	.20	4.73	5.95	6.86	14.568	18.326	21.129
50	1.25	1.63	1.88	2.86	3.73	4.30	.38	.49	.56	.60	.40	.35	.25	.22	.20	4.09	4.84	5.41	12.597	14.907	16.663
75	1.00	1.19	1.31	2.29	2.72	3.00	.30	.36	.39	.60	.40	.35	.25	.22	.20	3.48	3.72	3.94	10.564	11.458	12.135
100	......	......	0.75	......	......	1.72	......	......	.23	......	......	.35	......	......	.20	......	......	2.50	......	......	7.700

COST OF STEAM AND WATER POWER.

SOURCE: Main, "Cost of Steam and Water Power," p. 113.

Table 26. Cost of plant, fixed charges, and total yearly expense per I.H.P. for different engine types

Col. 1	2	3	4	5	6	7	8	9	10	11	12	13	14	15	16	17	18	19	20	21	22	23	24
Engine.	Per cent. of exhaust steam used.	Engine and piping complete.	Engine house.	Engine foundations.	Total cost of engine plant.	Depreciation at 4 % on total cost.	Repairs @ 2 % on total cost.	Interest @ 5 % on total cost.	Taxation @ 1.5 % on ¾ cost.	Insurance @ 0.5 % on engine and engine house.	Totals of Cols. 7, 8, 9, 10, and 11.	Boilers complete, including feed pumps, etc.	Boiler house.	Chimney and flues.	Total cost of boiler plant.	Depreciation @ 5 % on total cost.	Repairs @ 2 % on total cost.	Interest @ 5 % on total cost.	Taxation @ 1½ % on ¾ cost.	Insurance at 0.5 % on total cost.	Totals of Cols. 17, 18, 19, 20, and 21.	Total ordinary running expense Cols. 20, 21, and 22, Table II.	Total yearly expense per I.H.P. Cols. 12+22+23.
		$	$	$	$	$	$	$	$	$	$	$	$	$	$	$	$	$	$	$	$	$	$
Compound.	0	25.00	8.00	7.00	40.00	1.60	0.80	2.00	0.45	0.165	5.015	9.33	2.92	6.11	18.36	0.918	0.367	0.918	0.207	0.092	2.502	16.570	24.087
	25	Constant.	"	"	"	"	"	"	"	"	"	8.00	2.50	5.66	16.16	.808	.323	.808	.182	.081	2.202	14.568	21.785
	50											6.67	2.08	5.15	13.90	.695	.278	.695	.156	.070	1.894	12.597	19.506
	75											5.33	1.67	4.60	11.60	.580	.232	.580	.131	.058	1.581	10.564	17.160
Condensing.	0	20.00	7.50	5.50	33.00	1.32	0.66	1.65	0.371	0.138	4.139	13.33	4.17	7.30	24.80	1.240	.496	1.240	.279	.124	3.379	21.837	29.355
	50	20.00	7.50	5.50	33.00	1.32	0.66	1.65	.371	.138	4.139	10.99	3.43	6.70	21.12	1.056	.422	1.056	.238	.106	2.878	18.326	25.343
	25	19.00	7.50	5.00	31.00	1.26	0.63	1.575	.354	133	3.952	8.69	2.72	5.92	17.33	.867	.347	.867	.195	.086	2.3[illegible]2	14.907	21.221
	75	19.00	7.50	5.00	31.00	1.26	0.63	1.575	.354	.133	3.952	6.35	1.98	4.92	13.25	.663	.265	.663	.149	.066	1.806	11.458	17.216
Non-Condensing.	0	17.50	7.50	4.50	29.50	1.18	0.59	1.475	0.332	0.125	3.702	16.00	5.00	8.00	29.00	1.450	.580	1.450	.326	.145	3.951	25.595	33.248
	50	Constant.	"	"	"	"	"	"	"	"	"	13.01	4.07	7.20	24.28	1.214	.486	1.214	.273	.121	3.308	21.129	28.139
	25											10.03	3.13	6.30	19.46	.973	.392	.973	.219	.097	2.654	16.663	23.019
	75											6.99	2.18	5.30	14.47	.724	.289	.724	.163	.072	1.972	12.185	17.859
	100											4.00	1.25	4.00	9.25	.463	.185	.463	.104	.046	1.261	7.700	12.663

SOURCE: Main, "Cost of Steam and Water Power," p. 114.

Table 27. Yearly expense of waterpower per horsepower on wheel shaft at Merrimack River centers

Charges for Water.			Attendance, oil, supplies, etc.	Fixed Charges on Cost of Plant.						Total Yearly Expense per H. P.					
				Cost of plant.						For plants costing					
Per mill power.		Per H.P. per year.		$50	$60	$70	$80	$90	$100	$50	$60	$70	$80	$90	$100
$300 per year...	a	$4.62	$0.72	$5.08	$6.10	$7.11	$8.12	$9.13	$10.15	$10.42	$11.44	$12.45	$13.46	$14.47	$15.49
	b	12.31								18.11	19.13	20.14	21.15	22.16	23.18
2 per day.......		9.48								15.28	16.30	17.31	18.32	19.33	20.35
4 " "		18.96								24.76	25.78	26.79	27.80	28.81	29.83
8 " "		37.92								43.72	44.74	45.75	46.76	47.77	48.79
10 " "		47.40								53.20	54.22	55.23	56.24	57.25	58.27
20 " "		94.80								100.60	101.62	102.63	103.64	104.65	105.67

SOURCE: Main, "Cost of Steam and Water Power," p. 118.

pound, condensing, and high-pressure noncondensing engines using no exhaust steam for processing ranged from $24.08 for compound, $29.36 for condensing, and $33.25 for high-pressure noncondensing engines. By progressive increases in the exhaust steam employed to 25, 50, 75, and 100 percent, power costs were reduced. With all the exhaust steam so used, power costs were $14.90, $16.81, and $12.66, respectively; the condensing engine, through loss of the benefits of a vacuum secured through condensation, appeared at a certain disadvantage. Such rates were to be matched by waterpower at few important sites.

The Manning paper presented comparative figures for steampower and waterpower costs.[84] The cost of "permanent" leased water at Lowell, Lawrence, and Manchester per horsepower year was virtually the same at each center and was taken as $10.50 (net horsepower on the wheel shaft, $14.00). Wheel efficiency and fixed and operating costs on wheel installations of 1,000 net horsepower were taken as the same at all three centers and gave a total cost of $22.62 per net horsepower per year. With surplus power reckoned under varying conditions as $5.00 and $2.00 per mill power (65 net horsepower) per diem, the cost of surplus waterpower per net horsepower per year was $27.63 and $16.20. Calculations for steam power similar to those by Main were made for two types of engines in plants of 1,000-horsepower capacity. For the compound engine, with no exhaust steam and 25 percent exhaust steam taken for process heat, the costs per net horsepower per year were $22.00 and $19.34; for the single-cylinder noncondensing engine with all exhaust steam taken the cost was $14.58. In short, taking the Manning estimate for waterpower cost at the three lower Merrimack textile centers as compatible with the figures cited in Table 22 above, and with more or less exhaust steam used in the engines cited, steam power had secured a certain margin of advantage over waterpower in cost at these longtime citadels of waterpower.

In the following year, 1890, Main published another paper making a direct comparison of steam-power and waterpower costs at the three Merrimack River textile centers, reaching conclusions in substantial agreement with those of Manning. There were, he added, certain advantages of each form of power that might outweigh differences in cost. Such advantages for steam power were reliability of power throughout the year; choice in the selection of sites with respect to freight rates, fuel costs, and favorable conditions of labor supply or

[84] Charles H. Manning, "Comparative Cost of Steam and Water Power," *TASME* 10 (1888–89):499 ff.

markets; or operation at a more uniform speed than possible with waterwheels. Waterpower afforded greater cleanliness, smaller space requirements, and abundant water when required for processing. Main concluded: "That the balance of advantages and cost combined is in favor of steam power for textile manufactures is proven at the present time by the erection of steam mills almost entirely, while there are still undeveloped water powers which are available."[85]

Even the generality of power users, to whom the economies of plants of 500 to 1,000 horsepower equipped with engines of high efficiency were unavailable, the unreliability of waterpower was commonly a matter of primary concern. In the discussion of Main's first paper Professor Thurston spoke forcefully to this point: "We cannot rely on having water power when we want it; and although the stream may give full power for nine months in the year, for three months it is apt to be of very uncertain flow as well as very small volume." Yet Thurston, too, ended on the note of steam power's greatly reduced cost: "It has become financially practicable today to put steam in a mill at Holyoke or Lowell, and to neglect the water power which lies right at hand."[86]

[85] Main, "Cost of Steam and Water Power," pp. 108–25. See pp. 115 ff. for the discussion of waterpower costs with specific examples of power plants at Lowell, Lawrence, and Manchester cited. A later and more detailed consideration of waterpower costs is found in Main, "Computation of the Values of Water Powers," pp. 68 ff., treating the complex problems arising from diversion of streamflow to extend public water supply systems. One other engineering authority to be consulted is Charles E. Emery, a mechanical engineer of wide experience and high distinction. See his papers "The Cost of Steam Power," in *TASCE,* pp. 425 ff. and *TAIEE* 10 (1893):119–62. It remained for Samuel Webber, another prominent New England textile and hydraulic engineer, to point out in 1893 the irrelevance of such calculations. Only by using engines of 1,000 or more horsepower and 120 pounds of pressure in carefully controlled ten-hour tests could the cost of steam be reduced to $20 per horsepower. Average annual cost in practical use was $40 to $50 per horsepower. (William Kent, *Mechanical Engineers' Pocketbook*, 9th ed. rev. [New York, 1916], p. 768).

[86] In discussion of Main, "On the Use of Compound Engines," pp. 79–80, 83–84. Another participant pointed out that if mill construction ceased at the point where the water was ample the year round, there would be large quantities of water running to waste two-thirds or even three-quarters of the year. When mills and factories needed to extend their facilities, there were important advantages in enlarging the existing plant rather than seeking a new location; the simplest answer, if often an expensive one, was the installation of an auxiliary steam plant. "In a large number of water power developments which I have examined," declared Main in 1905, "a very large percentage have been developed with wheel capacity sufficient to use all of the water from six to seven months in an average year, and during the remaining months water would go to waste. The economical development has been stated by some engineers to be nine months" ("Computation of the Values of Water Powers," pp. 81 ff. To obtain reliable data on streamflow throughout the year and over a sufficient number of years

An Indictment Agreed Upon

For generations the crucial power issue had been availability. With no alternative source, aside from beasts of burden, waterpower ruled the field—it was literally ne plus ultra. Then in the nineteenth century in response to conditions favoring the growth of manufactures at the sites affording transport facilities, first by inland and coastal waterways and in time by railways, the positions of waterpower and steam power, the last at an advanced state of development, were reversed. Waterpower's shortcomings of immobility, inflexibility, and unreliability, unfortunately projected in later discussions into the earlier years, became obvious and deserved stigma. In the typical urban situation steam was the only practicable source of power. Increasingly it became evident that its higher cost, especially in small- to medium-scale establishments, was greatly outweighed by the many advantages of an urban location. "In these days of sharp competition," declared Professor Swain in 1888, "location and commercial facilities of every kind are of far more importance than a few dollars more or less expended for power; while the advantages of a fixed and steady power afforded by a steam engine over the uncertain and fluctuating power furnished by a natural stream are becoming more and more recognized." In manufacturing, declared another engineer, "you are generally in pursuance of a contract or you have to get certain goods into market at a certain time. You have plenty of water at times when you do not need it, but you must have means of getting your goods out whenever they are called for, and there are certain seasons of the year when you have not got the power, and then you are obliged to supplement it by other means." A steam-power site, remarked another discussant, "can be selected with reference to the markets, to the low cost of fuel, and to the facility of procuring operatives." "It is cheaper to transport freight on a canal per ton mile," ran a variant of the general theme, "than it is on a railroad; but the railroads are drying up the canals. Why? Because time is a great element."[87] The engineer discussants vied with each other in heaping discredit on what had long been a major national resource. Transportation facilities increasingly occupied the critical role in industrial location so long held by motive power.

and to calculate the costs and gains attending the additional power capacity and of the auxiliary steam plant required to bridge the gap advantageously in dry months called for engineering resources available to relatively few.

[87] Swain, "Statistics," p. 36; Marshall M. Tidd and Main, in discussion of Kimball, "Water Power," p. 82; Professor De Volson Wood, in Main, "On the Use of Compound Engines," p. 81.

A New Contender for Stream Use

There were other elements bearing upon the choice of power, less measurable even when tangible. How weigh the hazard of occasional great floods—hardly an important millstream was without its tale of major, even catastrophic, losses from this cause—against the fire and explosion risks of steam power?[88] Much might be learned from the records of insurance costs and claims. What balance is to be struck in the nuisances attending each—the difficulties of coping with floating and anchor ice in winter and with the trash of flood seasons, on the one hand, and on the other the smoke, grime, and ashes of the boiler plant?[89] Much more important evidently, although the record here is

[88] Examples of major flood damage frequently appear in local histories. For instance, a violent summer freshet in Poultney, Vermont, in 1811 swept away nine mills, one woolen factory, and several other buildings (Thompson, *Gazetteer of Vermont,* p. 221). A few years later an August storm "of uncommon violence caused an immense destruction of mill-dams, mills, factories, forges and bridges, etc., upon the Atlantic seaboard, particularly in Philadelphia, Baltimore and their vicinities" (Bishop, *History of American Manufactures,* 2:240). In 1839 a heavy May freshet destroyed a large part of the recently completed dam on the Kennebec River at Augusta, Maine, and totally destroyed ten mills just erected to use the power (U.S., Census Office, Tenth Census, 1880, vols. 18–19, *Report on the Social Statistics of Cities* [Washington, D.C., 1886], pt. 1, p. 5). September rains in 1856 brought an unprecedented rise in the Ausable River in upstate New York; at Keeseville, Clinton, and Ausable many grist- and sawmills, machine shops, and rolling mills were destroyed or damaged (*Scientific American* 12 [25 Oct. 1856]: 54). In the 1870s recurrent freshets in different parts of New England caused extensive damage in many communities (Tenth Census, *Water-Power,* pt. 1, pp. 190 ff., 230, 254, 277). Sometimes man's ignorance or negligence played a part, as when reservoir dams gave way. Such were the Mill River and Ware River disasters in the Connecticut Valley near Northampton and Staffordville, with heavy destruction in buildings, mills, and other property—in the former instance, the loss of some 150 lives (ibid., pt. 1, pp. 271–75, and *A Full and Graphic Account of the Terrible Mill River Disaster* [Springfield, Mass., 1874]).

[89] For the ice problem, see Tenth Census, *Water-Power,* e.g., pt. 1, pp. 220–21, 343 ff., 369 ff. The most detailed and vivid information to come to my attention is numerous references in the correspondence and memoranda of JBF in LCP. At Lowell ice formed on the supply canals at night when unchecked by streamflow; in the morning the thin ice broke up and clogged the trash racks protecting the waterwheels, requiring the diversion of many hands each morning to clear ice from the canals through wasteways. At some mills "losses and troubles from this source were formerly frequent and intolerable." The perfect remedy, not readily applied, according to Francis, was "great depth and a current slow enough to let the canals freeze over as thick as it can, and have water way under the ice" (JBF to A. F. Whitford, LCP, DB-7, 384–85). The subsurface form known as "anchor ice" presented difficulties of a less tractable sort (JBF, "Anchor Ice," *JFI,* 1866). See also discussion of anchor ice in Tenth Census, *Water-Power,* pt. 1,

largely silent and the references fragmentary, was the degree of independence of power control. Here the advantage lay clearly with steam power. With engine and boiler the millowner could do as he pleased, virtually without let or hindrance, subject only to the eventual restrictions of inspection laws and ordinances. By contrast the owner and developer of a water privilege was hemmed in and at times harassed by a body of prescriptive rights, statutory law, and judicial decisions bearing upon the exercise of rights in streamflow. The rights, discussed in chapter 3 above, were those of usage, not of ownership, and were shared by other millowners, water users, and landowners along the same stream.

For the better part of three centuries the milling and manufacturing interests had had to cope with the rights and often vociferous demands of a succession of competing stream users—lumbermen and loggers, flatboat and rafting interests, riparian landowners generally and streamside farmers particularly, and seasonal fishing folk on coastal streams. The power users, too, as we have noted, were usually their own worst enemies, interfering with stream use by others and litigious to a fault. In the post–Civil War years some of the old-time challengers were gone or dwindling—the alewife gluttons, the flatboatmen and raft hands, the onetime rampageous, dam-busting loggers. If fellow millowners were no less contentious, they were diminishing in numbers. On the other hand, with the continued spread of settlement, land clearing, and cultivation, the farming interests had become more numerous and more aggressive in defense against mill flowage of rich bottomlands and meadows. Damages assessed for the latter in any event had become more costly to millowners in consequence of rising land and crop values.[90]

Symptomatic of the changing status and importance of waterpower was the mounting threat from a new and more formidable source: urban water supply. The problems of water for the cities, which had risen in Britain more than a half century earlier, moved into the fore-

---

p. 185. See also J. P. Frizell, *Water-Power* (3d ed.), pp. 254–55, 294–95. Emery remarked upon the discomforts and risks of waterpower management during New England winters "directing the movements of several hundred men in clearing the canals and races of ice. . . . cheap as water power is under normal conditions, the irregularities due to freshets, droughts and ice must be duly considered in striking a balance with steam power" ("The Cost of Steam Power." *TAIEE* 10:160–61).

[90] For a more recent protest, see Vermont, General Assembly, *Report of the Committee Designated by the Vermont General Assembly . . . to Study the Power Dams and Their Relations to Agriculture* (Montpelier, 1945).

ground of attention during the 1870s. "The water boards of our cities," declared an engineering paper of 1900, "are acquiring, by legislative processes, stream after stream now yielding water power for industrial purposes."[91] The situation first became acute in the Northeast. To meet the increasingly urgent domestic and sanitary needs for water, cities from Maine to Maryland were commandeering the flow of streams in their environs, reaching out in some instances scores of miles to meet their wants, as did Boston and New York. The manifest priority of urban needs on health grounds alone was everywhere acknowledged. The central issue typically was one of compensation to the millowners, yet there were conflicts of interests, as illustrated by the 1877 report of New Jersey's State Sanitary Association. The transition from an agricultural to an industrial economy revealed conflicting views as to the best utilization of streamflow in the leading river basins:

> The one, which is held mostly by dwellers in the hill country, by owners of mill sites and dam sites, by inhabitants of the waterpower manufacturing towns, is that rivers were, to a large extent, beneficently designed to carry off sewage, filth of all kinds and factory refuse. The other which is in vogue among the greater number, those living near tide-level, is that rivers might, with greater fairness, be looked upon, not as sewers, but as gigantic aqueducts, designed by nature to supply, at small cost, dense centres of population with an adequate supply of pure drinking water.[92]

Reason no less than legislative authority rested on the side of urban water supply. The association between dug wells, the traditional urban source, which too commonly existed in close proximity to privies and other flagrant sources of pollution and contagious diseases such as typhoid fever, dysentery, scarlet fever, and diptheria, was increasingly a matter of concern to local and state health authorities.[93]

[91] George I. Rockwood, "On the Value of a Horse-Power," *TASME* 21 (1899–1900):601. Rockwood's article (pp. 590–626) discusses a famous case involving diversion of streamflow from mills on the Blackstone River by the City of Worcester, Massachusetts. See also Fall River, Mass., Reservoir Commission, *Report upon Improvement of the Quequechan River* (Fall River, 1910), p. 6. An important early instance of expropriation was the buying up of the water privileges at the outlet to Lake Cochuit extending to the Concord River, together with the manufacturing establishments thereon, in order to enlarge the water supply of Boston. The purchase, at a cost of some $150,000, was part of a project to carry water by aqueduct 14.5 miles to a reservoir in Brookline, 4.5 miles from the center of Boston (Hayward, *Gazetteer of Massachusetts,* pp. 446–47).

[92] Albert R. Leeds, "Report by the Committee of Water Supply, New Jersey State Sanitary Commission: Water Supply of the State of New Jersey," *JFI* 105 (1878): 194.

[93] See Massachusetts, State Board of Health, *Seventh Annual Report* (Boston, 1876)

Protest and litigation rarely blocked seizure by higher authority; one engineer remarked in 1894 that water had many other and greater values than driving mills or serving navigation as in olden times.[94] The philosophy of the colonial mill acts favoring the millmen was now turned against them; top dog was now on the bottom. Many indeed seemed less concerned with the fact of seizure than the amount of damages awarded. An engineer active in this field of litigation declared that "the only real value existing in [many of] these water powers is the chance of unloading them upon the towns which have arrived at the necessity of building water works." If the men who originally put their money into the waterpowers at such places as Lowell, Nashua, and Lewiston had that money today, "they would build their mills convenient to the market and to the sea, and run them by steam power."[95]

The fate of the waterpower at Fall River, Massachusetts, points up the low estate to which waterpower had fallen in the closing years of the century. Fall River was a town whose industrial beginnings rested primarily upon the unusual and felicitous combination of seaport facilities and an excellent waterpower of some 1,300-horsepower capacity. Rising in the Watuppa Ponds, which were converted into a reservoir for the regulation of streamflow, the Quequechan River fell some 130 feet in the last half-mile of its short course in a succession of readily developed waterpowers. Some dozen mills established at these powers chiefly during the 1820s and 1830s marked the beginnings of a manufacturing center that by 1880 had replaced Lowell as the nation's leading producer of textiles, early outstripping the very

for stream pollution from industrial and sewage sources. See also Edwin B. Goodell, *Review of Laws Forbidding Pollution of Inland Waters,* U.S. Geological Survey, Water-Supply Paper no. 152, 2d ed. (Washington, D.C., 1905). The appalling situation on the Merrimack is described in Frances G. Jewett, *Town and City* (Boston, 1906), chaps. 15, 16.

[94] Kimball, "Water Power," p. 112; A. F. Nagle, "The 'Commercial' Value of Water Power per Horse-Power per Annum," *TASME* 24 (1902–3):286 ff.; Main, "Computation of the Values of Water Powers," pp. 68 ff. These articles contain numerous examples of awards by urban authorities to owners of water privileges.

[95] Tidd, in discussion of Kimball, "Water Power," pp. 112–13. British policy and practice were different. Waterpower was too scarce a resource to be liquidated, even to meet water needs of a densely populated, rapidly urbanizing island. A basic condition of diversion of surface waters to public water supply was the provision of "compensation water," typically a first charge upon reservoirs built to capture rainfall and runoff. See *Encyclopaedia Britannica,* 11th ed., s.v. "Water-Supply." All concerned were spared the costs and delays of litigation, though engineers lost welcome fees and a place in the public view.

limited power capacity of the river. By 1900 the once clear stream had sadly fallen from its onetime preeminence, its power role now almost negligible. Hemmed in by massive steam-powered mills, filled with noisome sewage and refuse, rank with weedy growths and all but denied access to the few mills they continued to serve, the Watuppa Ponds had in effect become little more than a stagnant cooling pond for the condensing apparatus serving some 40,000 horsepower of mill engines. In seasons of summer drought they failed at times even in this pedestrian role when pond temperatures over 100°F compelled the halting of some engines.[96] For many years, too, the Watuppa Ponds' waters had been diverted, not without protest and litigation, to serve public needs. The continuing use of the diminished supply for power was viewed by some as a public misfortune. The city and its great manufactories, declared a report upon the improvement of the river, "cannot afford to have the water drawn for power when its value is so great in other directions. The loss of water power merely means burning so much more coal."[97]

Fall River supplied both the symbol and the substance of changing times, serving through successive generations of the ruling Borden family to mark the climax of New England's textile advance and the triumph of steam power over waterpower. In the very years when the great Niagara project was preparing the obsequies of direct-drive power, water and steam alike, the Bordens' Fall River Iron Works brought to completion Cotton Mill Number 4, opened in 1895 to a fanfare of publicity, banqueting, and oratory. Mill Number 4

[96] A classic account of the degradation of a basic resource, documented and illustrated with photographs, is the Watuppa Ponds and Quequechan River Commission, *Report . . . to the City Council, City of Fall River.* This should be supplemented by the earlier report by Arthur R. Safford in *Report upon Improvement of the Quequechan River,* giving a fuller historical background. That Fall River's experience was far from an isolated instance there is little doubt. An almost identical change was reported of the North Branch of the Nashua River, which provided the power base for the early industrial development of Fitchburg, Massachusetts. "Today," ran a later account, "steam and electrical energy operate the factories of the city, the river being used more as a medium through which the industrial wastage can be absorbed than as a source of power" (Orra L. Stone, *History of Massachusetts Industries,* 4 vols. (Boston, 1930), 2:1820. See also reports of the sanitary boards of Massachusetts and New Jersey cited in nn. 92 and 93 above.

[97] *Report upon Improvement of the Quequechan River,* p. 272. What was described as "a classic reference on the general subject" was *Watuppa Reservoir Company* v. *Fall River,* 147 Mass. (1888), p. 548. See also Main, "Computation of the Values of Water Powers," p. 73.

increased the Iron Works' textile capacity by nearly one-third and raised the overall capacity of Fall River to a point approaching that of Lowell, Lawrence, and Manchester combined. The new mill's twin, triple-expansion Corliss-built 3,000-horsepower engine may well have established a record capacity for mill engines. It was more than twice the size and power of the Corliss Centennial engine and placed in motion, through almost one hundred tons of millwork, four floors and nearly six acres of machinery.[98] Approaching if not establishing the ne plus ultra of the direct-drive period now nearing its end, engine and millwork alike were virtually obsolete when installed.

[98] The massive conventional millwork began with a 45-ton main shaft bearing a 30-foot flywheel of nearly equal weight and having the 15-foot width of face necessary to carry the great belts conveying power to the four floors (John G. Speed, *A Fall River Incident; or a Little Visit to a Big Mill* [New York, 1895], esp. pp. 22–33). See also Smith, *The Cotton Textile Industry of Fall River,* chaps. 1, 2; J. D. Van Slyck, "The Bordens of Fall River," *Representatives of New England Manufacturers,* 2 vols. (Boston, 1879), 1:112 ff. Richard and Joseph, two sons of John Borden, believed progenitor of the family in America, settled in Fall River. Two of the sons of one and one son of the other, according to Van Slyck, "inherited the land and water power" respectively on the south and north sides of the river, a small stream "two miles long and a rod wide." "When Fall River became a town, in 1803, it contained eighteen families; and nine of these were Bordens" (ibid., p. 112).

Epilogue

The Age of Mechanical Power.—Man is supposed to have passed through several epochs in the march of progress from savagery to civilization. . . . The present epoch is called the "Age of Mechanical Power," and is considered by many as more important than the epochs which followed the introduction of the use of fire, the domestication of animals, the cultivation of fruits and grains, the discovery of iron or even the invention of the printing press. . . . Mechanical power in the short interval of a little more than a century, by transferring reliance from animate to inanimate energy, has revolutionized the whole environment of human life by enabling man to utilize the energy and materials of his environment more effectively.[1]

For two and one-half centuries waterpower was the chief source of mechanical energy in colonial and independent America. Falling water was the main reliance for stationary power at all levels of capacity, in most branches of industry, and throughout the greater part of the country brought under settlement. Down to the 1860s waterpowers in their vast numbers and wide range of capacities ranked among the basic natural resources of the nation: agricultural land, timber upon the land, navigable rivers and coastal waterways, and mineable deposits of coal and metal ores. In 1860 the ratio of developed waterpower capacity to steam-power capacity in manufacturing in the United States stood at 56 to 44. Ten years later the balance had shifted to place steam power ahead in the ratio of 52 to 48. The succeeding decades witnessed a continued and progressive decline in the importance of waterpower in the manufacturing industries, although varying from industry to industry.[2]

For the greater part of the nineteenth century, American entrepreneurs and engineers led the industrial world in the development and effective use of waterpowers of the largest capacity and on a scale unapproached elsewhere. Those powers provided the energy base for most of the larger

[1] U.S. National Resources Committee, Subcommittee on Technology, *Technological Trends and National Policy* (Washington, D.C., 1937), p. 261.

[2] Carroll R. Daugherty, "The Development of Horsepower Equipment in the United States," in *Power Capacity and Production in the United States,* U.S. Geological Survey Water-Supply Paper no. 579 (Washington, D.C., 1928), Table V, p. 49; U.S., Census Office, Ninth Census, 1870, vol. 3, *Statistics of Wealth and Industry* (Washington, D.C., 1872), Table VIII-B, pp. 394 ff.

industrial cities in the United States before the 1850s. At Lowell and its successor textile centers of the Northeast, American hydraulic engineers pioneered not only in devising the basic hydraulic facilities of dams, canals, and supporting structures, but they successfully developed and employed means for the measurement, management, and commercial disposal of waterpower in quantities numbered in thousands of horsepower.

While contributing relatively little to hydraulic theory, American wheel designers and builders made important innovations in waterwheels of the reaction type, and from the 1850s adapted with great success to American needs the Fourneyron and Jonval turbines, the leading nineteenth-century innovation in waterwheel design before hydroelectricity. In a largely pragmatic fashion those designers and builders developed and perfected a new wheel type, the American mixed-flow turbine. During the post–Civil War years the mixed-flow wheels progressively displaced waterwheels of the traditional type save in water mills of the smallest capacity. In the more successful forms the mixed-flow wheel by its small size and weight relative to capacity, its high efficiency at both full and part gate, and its low unit cost served admirably the industrial needs of the day. Similarly, the tangential waterwheel as developed and widely used in the mining and coastal regions of the Far West was an innovation which under the prevailing conditions of high heads and limited volumes of streamflow made another important contribution to hydraulic technology.

In short, taking advantage of the great natural wealth of waterpower, which stood in striking contrast to the limited waterpower of Great Britain, Americans demonstrated the economy and effectiveness with which waterpower could be applied in the service of industrial and economic development. This was accomplished, moreover, within the framework of a westward-expanding settlement, a thinly distributed population, and a predominantly agricultural economy, conditions which added to the complexity and difficulties of the achievement. Although in several instances the industrially more advanced European nations matched the United States in their reliance on waterpower rather than on steam power during the early stages of industrialization, in no instance was the diversity and the scale of the American experience approached.[3]

[3] See Louis C. Hunter, "Waterpower in the Century of the Steam Engine," in Brooke W. Hindle, ed., *America's Wooden Age* (Tarrytown, N.Y., 1975), pp. 186-89. For a more extended discussion of this subject, see David S. Landes, *The Unbound Prometheus* (Cambridge, 1969).

Despite the abundant witness to its widespread availability, utility, and employment, the much lower cost and greater ease of development, the fundamental role of waterpower in American economic development has, with occasional exceptions, been overlooked, misapprehended, or simply ignored. This is most apparent in general surveys of American economic, industrial, or technological development, including especially the single-volume textbooks for college and university use, as a check of indexes and a scanning of relevant chapters will demonstrate.[4] This inattention may reflect a failure to grasp, especially to visualize, the role of stationary power in industrial production as conducted during the nineteenth century. The neglect of stationary power stands in striking contrast with the extended attention accorded the mobile applications of steam power in steam navigation and railway locomotion. Indeed, among laymen, today as in the past, who is not more familiar with transportation and travel than with industrial production? It may be that, historically, waterpower is viewed chiefly in terms of the simple water mills of the medieval and preindustrial periods, as enumerated by the thousands in the often-cited Domesday account or as illustrated by the splendid wood engravings of water mills in the *theatra machinarum* of the sixteenth and seventeenth centuries, of which Agricola's *De Re Metallica* is a familiar if special example. More probably the failure to deal more adequately with waterpower in nineteenth-century America reflects the common identification of steam power with the Industrial Revolution in Britain and the industrialization of the Western world that it set in motion.

Britain success in the application of steam power to the mining and manufacturing industries, 1750–1850, was a matter of universal admiration and acclaim by men prominent in public affairs, literature, and philosophy hardly less than by leaders in business and industry. The

[4] Hunter, "Waterpower in the Century of the Steam Engine," pp. 161-63. Two examples from comparatively recent years illustrate the point, "It is well known how the Watt-Boulton steam engine freed England and then all nations from the geographic and climactic vagaries of water power and how it permitted men for the first time to concentrate great quantities of efficient motive power in one location" (F. M. Scherer, "Invention and Innovation in the Watt-Boulton Steam-engine Venture," *Technology and Culture* 6 [1965] :165). "In the early days of America's industry, water wheels were used to activate machinery, but these contraptions were subject to the vagaries of the weather, and their usefulness was restricted to the site of waterfalls. It was the development of the electrical industry that brought water power into its own" (Shepard B. Clough and Theodore F. Marburg, *The Economic Basis of American Civilization* [New York, 1968], p. 30). For a more detailed consideration of the matter see "Seasonal Aspects of Industry and Commerce before the Age of Big Business," in Louis C. Hunter, *Studies in the Economic History of the Ohio Valley* (Northampton, Mass., 1933-34), pp. 5-49.

American success in steam navigation coming in the first decade of the new century was almost as widely admired in Great Britain and on the Continent as here. Robert Fulton was to be assigned a place among the heroes of invention not greatly inferior to that of James Watt. The adulation of steam power in this country reached a new peak with the coming of the railroad. "By 1844," declares one historian, "the machine had captured the imagination. The invention of the steamboat had been exciting, but it was nothing compared to the railroad. In the 1830s the locomotive, an iron horse, or Titan, is becoming a kind of national obsession."[5] Soberer minds appraised steam power in more general terms; they were wont to calculate the blessings of steam power by the millions of mechanical slaves placed at the disposal of Britain or America.[6] Yet the big news that makes the headlines does not always find its way into written history. The waterpower that led in industrial production in the United States until the 1860s passed unnoticed by most historians.

The central fact that emerges from this study is that industrialization, which in Great Britain was at virtually every stage inextricably bound up with and dependent upon steam power, alike in mining, manufacturing, and transportation, was in the United States no less decisively based upon waterpower. Industrialization did not await the general introduction of steam power in the manufacturing industries. "There can be no doubt," Douglass North declared in his 1961 study, "that American industrialization was well under way before the Civil War. Since 1810 . . . the value of manufacturing output had increased approximately tenfold, while population had increased only four and one-half times." The upsurge of industrial expansion that began in the early 1840s "was an era in which the Northeast had ceased being a marginal manufacturing area and could successfully expand into a vast array of industrial goods."[7]

It is clear that the traditional identification of steam power with industrialization has no general validity outside the British Isles. So far as figures presently available indicate, the British experience was exceptional, and for obvious reasons: scarcity and inadequacy of waterpower and great abundance of coal. The primacy of waterpower during the early stages of American industrialization is an important fact in the history of Western technology, refuting the widespread assumption

[5] Leo Marx, *Machine in the Garden* (New York, 1964), p. 191.

[6] Examples are found in *Hunt's Merchants Magazine* 18 (1847):328, *Manufacturer and Builder* 11 (1874):288; and Robert H. Thurston, "An Era of Mechanical Triumphs," *Engineering Magazine* 6 (1893–94):456 ff.

[7] Douglass C. North, *The Economic Growth of the United States, 1790–1860* (Englewood Cliffs, N.J., 1961), p. 176.

that steam power reduced waterpower to obsolescence. The fact is all the more interesting because in this country, the largest class of waterpowers excepted, the reliance upon waterpower before the 1860s rested upon the traditional technology of waterwheels and supporting facilities going back hundreds of years. The reasons were simple and clear: waterpower was abundantly available and effective; it was familiar; and it was cheap.

The tens of thousands of common water mills at one end of the industrial spectrum and the score of small industrial cities based on waterpowers measured in thousands of horsepower at the other are the more obvious features of the American industrial economy of the 1840s and 1850s. The community water mills afforded relief to farm families of the pioneering years and often stimulated the transition to a market economy by providing an outlet for surplus production. Lowell and its kind at the other extreme were in many ways impressive as examples of large-scale mechanized production. Yet while acknowledging their important contribution to hydraulic engineering, the technology of the textile machinery, and the immense output of cheap textiles, such concentration of industrial production was hardly indispensable. Within fifteen years of the first Lowell mills there were over a hundred steam cotton mills in the United States, the largest of which were comparable to those clustered about the great waterpowers of the Merrimack River. Whether, as asserted, the steam mills made better quality goods at equal or lower cost remained a bone of contention. The progressive adoption at Lowell and sister centers of steam to meet expanding power requirements made the issue academic.

Of much greater importance for the country at large was the role of waterpower in providing a practicable energy base for the innumerable small centers of manufacturing and trade that in the prerailway age brought manufacturing within the reach of a large part of the predominantly rural population of the country that lacked access to main-line channels of communication and trade. The waterpowers on which such communities were based may well have comprised as much as one-third to one-half of the 650,000 total water horsepower in use as estimated for 1849.[8] The scale of operations and services at these small industrial centers was narrowly limited to the highly localized markets into which the country was divided, owing to the difficulties and expense of common road transport in the prerailway years. While this condition was aggravated by the wide dispersion of settlement and population, it was at the

[8] Daugherty, "Development of Horsepower Equipment," p. 49; and author's observations.

same time alleviated by the presence in virtually every township of waterpowers on the mill creeks and streams that blanketed the countryside. As improvements in local transportation facilities made wider markets accessible, the opportunities for specialization and for growth occasioned further development about waterpowers of larger capacities. Everywhere waterpower provided the indispensable base for the mechanization upon which lower costs of production and wider use so greatly depended. Only a small proportion of the literally innumerable small waterpowers were ever brought under development. Yet manufacturing above the craft or workshop level brought on the one hand opportunities for ambition, enterprise, and labor, and on the other outlets for local produce and materials in exchange for manufactured articles not otherwise readily available.

As shown in the concluding chapter of this volume, the advance of industrialization in the United States after the Civil War depended increasingly upon steam power. This was due in some part to a growing dissatisfaction with the irregularity and unreliability of waterpower under the changed conditions of the railway age. When competitors were subject to the same limiting conditions, the situation was tolerable. Under the pressures of competition the limitations of waterpowers in the long run, if not always the short term, became a serious handicap. More importantly, the shift to steam power was a consequence of the progressive concentration of manufacturing in cities where, save in infrequent instances, waterpower was not available. The primary attraction was the railway service over an ever-expanding network, especially at the nodal points offering more varied and wider outlets. To the advantages of access to markets and raw materials were added the increasingly recognized benefits of commercial, banking, and other services of an urban center and a more flexible supply of labor.

The other side of the coin of industrialization was less attractive. The urbanization of manufacturing during the railway age in effect condemned to gradual obsolescence and decline the great majority of innumerable small manufacturing centers in the older, settled parts of the United States. In the face of competition from the ever larger and more rationalized industrial enterprises, they were in most instances beyond the redemptive influence of steam power alone. Deprived of the protective barrier of distance by the spreading network of railroads, the position of such small communities became increasingly precarious. For much of what has here been termed "grass roots industrialization," there was no future. Yet during the half century before the Civil War these small communities had made possible for the limited areas served a meaningful rise in the condition of the work and living above the bare survival existence of drab subsistence agriculture.

Appendixes
Index

Appendix 1
Grain-Milling Practices among the American Indians

The native peoples of North America were nearly as dependent upon bread grains as the European migrants who settled here. The Indian's chief reliance by far was upon maize, or Indian corn, a plant indigenous to this continent, with Mexico believed to be the principal region of origin. Well-documented accounts of its antiquity, cultivation, and use are found in the literature of history, archaeology, and ethnology. Linton has distinguished three "maize complexes" in North America, identified with Mexico, the American Southwest, and the extensive area of Iroquois influence in the Northeast.[1] To prepare the grain, the Indians usually first reduced it to meal. In Mexico and the Southwest, as in many parts of Africa and Asia, this grinding was done in what in America is known as the metate, a slightly hollow, sometimes flat stone of various forms and sizes. The grain was ground with a matching handstone, or mano, in a back-and-forth rubbing, shearing action, with the kneeling worker's body moving from the hips. One form of this grinding as observed in Africa by an American traveler was described as follows: "In shallow depressions worn into the boulder's surface, each girl, down on her knees, ground the [broom corn] seeds to a grayish powder with a stone held in both hands and worked back and forth, as if she were washing clothes on a rubbing board."[2]

[1] Ralph Linton, "The Significance of Certain Traits in American Maize Culture," *American Anthropologist*, n.s. 26 (1924):345 ff. See also Paul Magelsdorf et al., "Domestication of Corn," *Science* 143 (1964):538-45, on archaeological evidence of the origins of maize. For the wide range of early mesoamerican grain-grinding devices, see Richard S. MacNeish et al., "Food-Preparation Artifact," *The Prehistory of the Tehuacan Valley,* vol. 2 (Austin, Tex., 1967), pp. 101 ff.; *Handbook of Middle American Indians*, ed. Robert Wauchope, vol. 1 (Austin, Tex., 1964), pp. 449 ff. Among well-illustrated studies dealing with the Iroquois country of the Northeast are Lucien Carr, "The Food of Certain American Indians," *The Antiquarian* 1 (1897):1 ff., 34 ff., 69 ff.; O. F. Cook, "Food Plants of Ancient America," *Annual Report of the Smithsonian Institution, 1903* (Washington, D.C., 1904), pp. 491 ff.; Arthur C. Parker, *Iroquois Uses of Maize and Other Food Plants*, New York State Museum Bulletin no. 144 (Albany, 1910); M. R. Harrington, "Some Seneca Corn-Foods and Their Preparation," *American Anthropologist,* n.s. 10 (1908):575 ff.

[2] Seth K. Humphrey, *Loafing through Africa* (Philadelphia, 1929), p. 240. See the account of Alfred Friendly, Jr., on the first of a series of power mills built to replace mortar and pestle grinding in Nigeria in *New York Times*, 4 Feb. 1968.

In the northeastern part of what is now the United States, Indian tribes from the Iroquois in upper New York State and Canada to the Osage and Kansas of the upper Mississippi Valley used the ancient mortar and pestle, with its simpler pounding, crushing action. In some areas made of stone, both parts were more commonly made of wood. The mortar was conveniently shaped from a section of tree trunk by alternately applying fire and then scraping with some stone or metal tool to obtain a deep, bowllike cavity. The pestle, or pounder, was usually of hardwood, three to four feet in length, several inches thick, with rounded ends. It was usually held in both hands at a somewhat reduced middle section. For larger mortars two persons often worked together, using their pounders in alternating strokes. While metates and mortars often belonged to individual households, multiple or community metates were found among the Indians of the Southwest. Similar practices occurred among the Indian tribes of the Northeast. In his 1824 account of life in captivity among the Kansas and Osage tribes, James Dunn Hunter reported mortars among household equipment: "In addition . . . each village has one or two large stone mortars for pounding corn; they are placed in a central situation, are public property, and are used in rotation by the different families." This practice was found elsewhere among American Indian tribes.[3]

Almost from the beginning European settlements along the Atlantic seaboard made the maize culture their own, frequently with the aid of nearby Indians. As oft repeated accounts have told us, the settlers adopted the native methods of planting and cultivation, some at least of the methods of preparation, and, under circumstances delaying the introduction of gristmills, the native method of grinding by mortar and pestle. With possibly less time and labor at their disposal than the Indians, they introduced improvements, which evidently failed to interest the natives. The principal such device was variously termed a "sweep and mortar," a "stump and sapling," or a "springpole mortar" mill. A stump for the mortar was found, or placed, close to a sturdy sapling, from the bent top of which a wooden pestle of appropriate size and weight was suspended. The merit of such "mills" was not, as sometimes explained, that they were laborsaving, "where half [the] work was done" by the upward thrust of the bent sapling raising the pestle for

[3] *Memoirs of Captivity among the Indians of North America,* ed. Richard Drinnin (1824; reprint ed., New York, 1973), pp. 128-29. "In many parts of the United States . . . remain many large, hollowed stones on which the grain of a community or tribe was pounded . . . some tribes had hollowed sections of tree trunks" (*Knight's American Mechanical Dictionary,* s.v. "Mortar").

the next stroke.[4] The work was easier only in the sense of being more convenient. To the downstroke of the pestle the workers brought body weight as well as muscular exertion, combining the pounding action of the pestle with energy storage in the bent sapling for the lifting action of the return stroke. To the greater diffusion of effort less arduously felt in the downstroke was added a certain momentary relief attending the intermittent action. As remarked in chapter one, hand grinding by such methods was tiring and tedious at best; the frequently long distances covered in trips to mill, often on foot, reflect the eagerness to abandon the stump mortar, however contrived, and its coarse, unpalatable product.

[4] Marion Nichols Rawson, *Little Old Mills* (New York, 1935), p. 83. This useful volume, in most respects carefully done, has numerous drawings of original artifacts.

Appendix 2
Survivals: The Tub Mill

In the fall of 1867 John Muir abandoned his successfully launched career as inventive mechanic and mill manager in Indiana and with backpack and plant press took off from Louisville, Kentucky, on his "thousand mile walk to the gulf," the opening episode of his new life as a naturalist. With map and dead reckoning he made his way on a southeastward course that was to bring him to the seacoast at Savannah, Georgia. On September 17 in the highlands of east Tennessee he made this entry in his journal:

> Spent the day in botanizing, blacksmithing, and examining a grist mill. Grist mills, in the less settled parts of Tennessee and North Carolina, are remarkably simple affairs. A small stone, that a man might carry under his arm, is fastened to a vertical shaft of a little home-made, boyish looking, back-action water-wheel, which, with a hopper and box to receive the meal, is the whole affair. The walls of the mill are of undressed poles cut from seedling trees and there is no floor, as lumber is dear. No dam is built. The water is conveyed along some hillside until sufficient fall is obtained, a thing easily done in the mountains. On Sundays you may see wild, unshorn, uncombed men coming out of the woods, each with a bag of corn on his back. From a peck to a bushel is a common grist. They go to the mill along verdant footpaths. . . . The first arrived throws his corn into the hopper, turns on the water, and goes to the house. After chatting and smoking he returns to see if his grist is done. Should the stones run empty for an hour or two, it does no harm. This is a fair average in equipment and capacity of a score of mills that I saw in Tennessee. . . . All the machines of Kentucky and Tennessee are far behind the age.

This was, Muir continued, "the most primitive country I have seen. . . . The remotest hidden parts of Wisconsin are far in advance of the mountain regions of Tennessee and North Carolina" where there was "scarce a trace of that restless spirit of speculation and invention so characteristic of the North."[1]

Some thirty years later northern literary folk, scholars, and magazine editors were to discover Appalachia. A spate of articles described the findings of brief forays into the surviving past—enlarging considerably upon Muir's brief references to the region—under such titles

[1] John Muir, *A Thousand Mile Walk to the Gulf* (Boston, 1916), pp. 35–37.

as "A Peculiar People," "Hobnobbing with Hillbillies," and "A Retarded Frontier." What was to become the classic account of this surviving fragment of the frontier past was that by Ellen Churchill Semple, geographer, "The Anglo-Saxons of the Kentucky Mountains: A Study in Anthopogeography." Semple wrote from an extended journey and sojourn in the rugged terrain of the Cumberland plateau, a land "devoted by nature to isolation." Here was a region, she said, "where people are still living the frontier life of the backwoods, where the civilization is that of the eighteenth century, where the people speak the English of Shakespeare's time, where a large majority of the inhabitants have never seen a steamboat or railroad, where money is as scarce as in colonial days, and all trade is barter." Exponents of a retarded civilization, the people "show the degenerate symptoms of an arrested development."[2]

The technology of this people was coeval with its institutional equipment. In a trip of 350 miles, the trails were so wretched, typically following the course of dry or half-dry streambeds, as frequently to require double teaming. The Semple party met only a single wheeled vehicle, apart from a few trucks hauling railroad ties, a principal export. "All passenger travel is on horseback. . . . Almost every cabin has its blacksmith forge under an open shed or in a low outhouse. Every mountaineer is his own blacksmith." The typical dwelling was the windowless one-room cabin of logs, roughhewn or in the round. Nearly all the furnishings were hand-made of wood. The exceptions noted in one cabin were the iron-bound well bucket, an iron stove with its few utensils, and some table knives. "Clay lamps of classic design, in which grease is burned with a floating wick, are still to be met with." Every home had its spinning wheel and usually a hand loom. "Some cabins are still provided with hand—mills for grinding their corn, when the water-mills cease to run in a dry summer . . . or when the water-mill is too remote."[3]

In their explorations few of these northern outlanders failed to remark upon, and often photographed or drew, the water mills and the more primitive hand mills, or querns. "To see some of these toylike mills working away, with a bushel of corn in the hopper and no one near," wrote William Perry Brown in 1888, ". . . impresses one strangely as though nature had gone to work by herself in sheer disgust at the incapacity of her children." "At intervals of a few miles

[2] Ellen Churchill Semple, "The Anglo-Saxons of the Kentucky Mountains: A Study in Anthropogeography," *Geographical Journal* 17 (1901): 588–623, esp. pp. 588–93.

[3] Ibid., pp. 598–607.

along the streams there are log dams and small grist-mills which look more like rustic summer houses than places of manufacture," reported George E. Vincent in the *American Journal of Sociology*. "Do you have to grind all your meal that way?" inquired one visitor, impressed by the slow stream of meal from the hand mill. No, there was a little tub mill down the creek a mile or so. "When a man lives on a branch or a prong of the creek whar the water's lasty and thar's a right smart trickle all the time," came the answer, "he puts him in a tub mill and lets the water grind fer him."[4]

When you first travel in the southern mountains, declared Horace Kephart some years later, "one of the first things that will strike you is that about every fourth or fifth farmer has a tiny tub-mill of his own," the capacities ranging from a half to two bushels a day. The Hazel Creek settlement of forty-two households was served by seven such mills in the vicinity. The appurtenances of these mills, "even to the buhr-stones themselves," he continued, "are fashioned on the spot. . . . A few nails and a country-made iron rynd and spindle were the only things that [the neighbor] had not made for himself, from the raw materials." The completed mill was actually offered to Kephart for six dollars.[5] Sometimes these small mills were not only roofed but fully enclosed; more often than not they were open at the sides, and on occasion, they were without roofs and completely in the open. One such mill in western North Carolina was described as "forming with roof, hopper and all, a structure no larger than a hackney coach." Some tub mills in eastern

[4] "A Peculiar People," *Overland Monthly* 12 (1888):507; "A Retarded Frontier," *American Journal of Sociology* 4 (1898–99):6–7; James Watt Raine, *The Land of Saddle-Bags* (New York, 1924), pp. 80–81.

[5] Horace M. Kephart, *Our Southern Highlanders* (New York, 1913), pp. 132–33; see also pp. 30–31 and opposite p. 152. See also John Fox, Jr., "The Southern Mountaineer," *Scribner's Magazine* 29 (1901):387 ff., 556 ff. "Among the strange anomalies of our day," declared a prominent New England industrialist in 1880, "reference may be made to the singular fact that a section of country may be reached within forty-eight hours' ride from Boston which is now only just being opened by the railroad, and which neither the factory system nor the fabrics of the factory itself have yet penetrated. In the very heart of the country, a sparse population inhabiting the fertile valleys of the Alleghanies and the hill-sides of the lateral mountain ranges of Virginia, Kentucky, Tennessee, and the Carolinas still clothe themselves in the product of the hand-loom, and the whirr of the spinning-wheel may be heard now, as it was of old in New England in every household. . . . In a seven days' journey the citizen of Boston may still study the industrial facts of two hundred and fifty years' progress, and witness with his own eyes most of the arduous conditions of life to which his ancestors were subjected" (Edward Atkinson, "Boston as a Centre of Manufacturing Capital," in vol. 4 of Justin Winsor, ed., *The Memorial History of Boston, 1630-1880* [Boston, 1881], pp. 107-8).

Pennsylvania a century earlier were arranged so that during low water they could be turned by hand and produce as much as three bushels of corn daily. By one account "a simple tub-mill answers the purpose best as the meal *least perfectly ground* is always preferred." In the contrary view "tub mills and mortar and pestle had to suffice until mills were available."[6]

It is probable that tub mills survived in other parts of the hill country of the South, especially in the somewhat similar terrain of the Ozarks.[7] Of particular interest are the diagrammatic sketches and photographs in Eliot Wigginton's *Fox Fire 2* that show the construction and method of operation of a surviving tub mill in Georgia in 1973. In contrast to the many curved, hand-forged iron "buckets," or floats, on a tub wheel obtained by the Bucks County Historical Society from North Carolina in 1917, the Georgia tub wheel has eight slightly curved inclined buckets hewn from the solid wood in a manner resembling the "home-made four bladed turbine" described by Mary Verhoeff as in use in the Kentucky River Valley in 1917. This type of wheel was also found in the Pine Mountain region of eastern Kentucky.[8]

The isolation and remoteness of Appalachia were to be broken forever in the years before World War I when discoveries of the wealth of timber upon the land and of coal beneath its surface brought northern capital to the region, with railroads and resource development in its wake.[9] Under the impact of the new industrial forces the traditional culture with its meager self-sufficiency crumbled and a new mode of obloquy settled upon the land. In 1916–17 a northern folk and farmer's museum sent scouts down to western North Carolina to find and ship home specimens of the mountaineers' mills before they disappeared. Some

[6] W. G. Zeigler and B. S. Grosscup, *The Heart of the Alleghanies, or Western North Carolina* (Raleigh, N.C., 1883), pp. 104 ff., 129–30, 304–5; *Proceedings and Collections of the Wyoming [Pennsylvania] Historical and Geological Society* 5 (1900):126; J. G. M. Ramsay, *The Annals of Tennessee to the End of the Eighteenth Century* (Charleston, S.C., 1853), pp. 718–19; Inez E. Burns, *History of Blount County, Tennessee, 1795–1955* (Nashville, 1957), pp. 217–20.

[7] C. M. Wilson, *Backwoods America* (Chapel Hill, N.C., 1934), pp. 186–87.

[8] Eliot Wigginton, *Fox Fire 2* (New York, 1973), plates 151–62; Mary Verhoeff, *The Kentucky River Navigation* (Louisville, 1917), p. 138 n.

[9] Semple remarked in 1901, "Large tracts of Kentucky mountain lands are owned by persons outside the state, by purchase or inheritance or original pioneer patents, and these are waiting for the railroads to come into the country, when they hope to realize on the timber and the mines" ("The Anglo-Saxons of the Kentucky Mountains," p. 600). See also the comprehensive scholarly survey edited by Thomas R. Ford, *The Southern Appalachian Region: A Survey* (Lexington, 1962), and a classic of a different kind, Harry M. Caudill, *Night Comes to the Cumberlands: A Biography of a Depressed Area* (Boston, 1962).

difficulty was met in locating in working condition the hand mills so frequently pictured in action in magazine articles hardly a decade earlier. With tub mills a choice was readily had, although some were dismantled and all evidently were in disuse.[10] Thus was finis written to what was perhaps the most ancient of water mills, which had served so long, so humbly, and so well in the land to which it was an early immigrant.

[10] See the account of Horace M. Mann, "Gristmills of an Ancient Type Known as Norse Mills," *A Collection of Papers Read before the Bucks County Historical Society* 5 (1926): 68–75, and above, Fig. 23. Cf. description of a tub mill antedating the Revolution in the Juniata Valley in eastern Pennsylvania, ibid. 7 (1937):118–19. It was reported of Norway, land of the "Norse" wheel's presumed origin, that "where the nature of the terrain was favorable, each farm had its own tub wheel or at least a share in one, and in many ways water power has served to save labor on farms . . . particularly as power for flour mills, sawmills and grindstones" (Norwegian National Committee, "Utilization of Small Water Powers," *Transactions of the Third World Power Conference, 1936,* 10 vols. (Washington, D.C., 1938), 7:450. On the search for a typical hand-operated corn mill in North Carolina for the Mercer Museum of Bucks County, Doylestown, Pennsylvania see William A. Labs, "Survival of Corn Querns of an Ancient Pattern in the Southern United States," *A Collection of Papers Read before the Bucks County Historical Society* 4 (1917):740 ff. and Frank R. Swain, "Hand Corn Mill at Georgetown, South Carolina," ibid., pp. 735–39.

Appendix 3

List of Ironwork for Gristmill

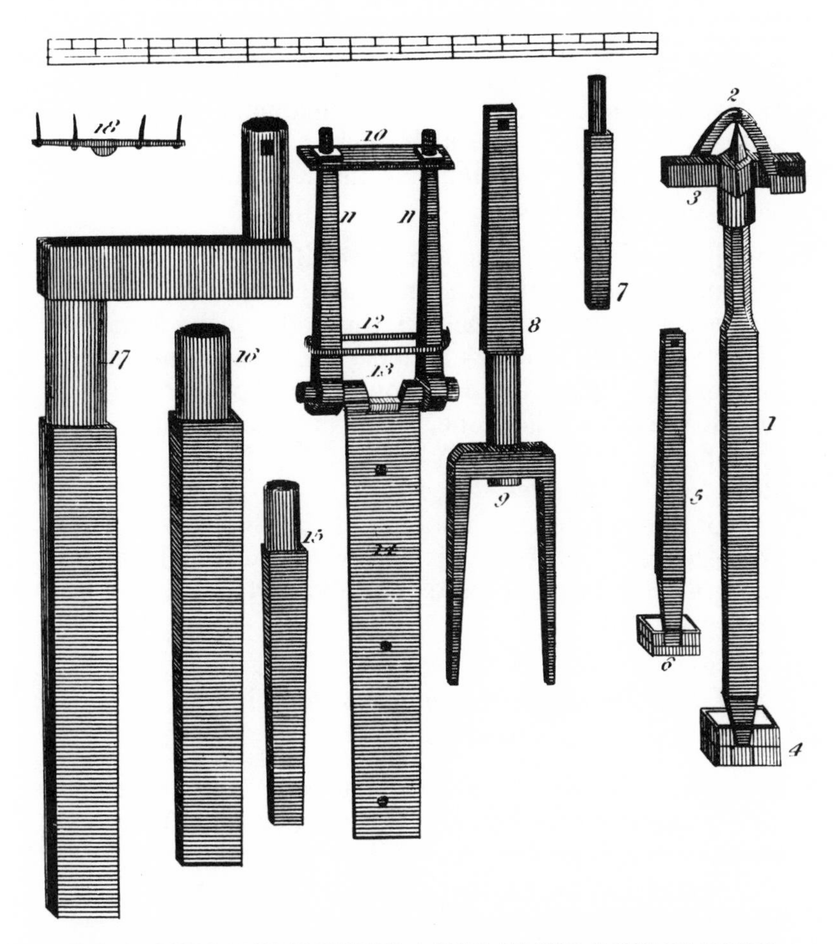

BILL OF THE LARGE IRONS FOR A MILL OF TWO PAIRS OF STONES.

2 gudgeons, 2 feet 2 inches long in the shaft; neck $4\frac{1}{4}$ inches long, 3 inches diameter, well steeled and turned. (See fig. 16, Plate XXIV.)

2 bands, 19 inches diameter inside, $\frac{3}{4}$ thick; and 3 inches wide, for the ends of the shaft.

2 do. $20\frac{1}{2}$ inches inside, $\frac{1}{2}$ an inch thick, and $2\frac{1}{2}$ inches wide, for do.

From Oliver Evans, *The Young Mill-wright and Miller's Guide* (1850 ed.), Pages 334–36

2 do. 23 inches do. ½ an inch thick, and 2½ inches wide, for do.
4 gudgeons, 16 inches in the shaft, 3½ inches long, and 2½ inches diameter in the neck, for wallower shafts; (See fig. 15, Plate XXIV.)
4 bands, 12 inches diameter inside, ½ an inch thick, and 2 wide, for do.
4 do. 12 inches do. ½ an inch thick, and 2 wide, for do.
4 wallower bands, 3 feet 2 inches diameter inside, 3 inches wide, and ¼ of an inch thick.
4 trundle bands, 2 feet diameter inside, 3 inches wide, and ¼ of an inch thick.

2 spindles and rynes; spindles 5 feet 3 inches long from the foot to the top of the necks; cock-heads 7 or 8 inches long above the necks; the body of the spindles 3¾ by 2 inches; the neck 3 inches long, and 3 inches diameter; the balance rynes proportional to the spindles, to suit the eye of the stone, which is 9 inches diameter. (See fig. 1, 2, 3, Plate XXIV.)
2 steps for the spindles, fig. 4.
2 sets of damsel-irons, 6 knockers to each set.
2 bray-irons, 3 feet long, 1¾ inch wide, ½ an inch thick: being a plain bar, one hole at the lower, and 5 or 6 at the upper end.

Bill of Iron for the Bolting and Hoisting Works, in the common way.

2 spur-wheel bands, 20 inches diameter from outsides for the bolting spur-wheel, ¾ths of an inch wide, and ¼th thick.
2 spur-wheel bands, 12 inches diameter from outsides, for the hoisting spur-wheel.
2 step-gudgeons and steps, 10 inches long, 1⅛ inch thick in the tang or square part; neck 3 inches long, for the upright shafts. (See fig. 5 and 6, Plate XXIV.)
2 bands for do. 5 inches diameter inside, 1¼ wide, and 1¼ thick.
2 gudgeons, 9 inches tang; neck 3 inches long, 1⅛ square for the top of the uprights.
8 bands, 4½ inches diameter inside.
1 socket gudgeon, 1⅛ of an inch thick; tang 12 inches long; neck 4 inches; tenon to go into the socket 1½

inch, with a key-hole at the end. (See fig. 8 and 9.)

14 gudgeons, neck $2\frac{1}{2}$ inches, tangs 8 inches long, and I inch square, for small shafts at one end of the bolting-reels.

10 bands for do. 4 inches diameter inside, and 1 inch wide.

4 socket-gudgeons, for the 4 bolting reels $1\frac{1}{4}$ square; tangs 8 inches; necks 3 inches, and tenons $1\frac{1}{2}$ inch with holes in the ends of the tangs for rivets, to keep them from turning; the sockets one inch thick at the mortise, and 3 inches between the prongs. (See fig. 8 and 9.) Prongs 8 inches long and 1 wide.

8 bands, $3\frac{1}{4}$ inches, and 8 do. 4 inches, diameter, for the bolting-reel shafts.

For the Hoisting Wheels.

2 gudgeons, for the jack-wheel, neck $3\frac{1}{2}$ inches, and tang 9 inches long, $1\frac{1}{8}$ square.

2 bands for do. $4\frac{1}{2}$ inches diameter.

2 gudgeons, for the hoisting wheel, neck $3\frac{1}{2}$ inches, tang 9 inches long, and $1\frac{1}{4}$ inch square.

2 bands for do. 7 inches diameter.

6 bands for bolting-heads, 16 inches diameter inside, $2\frac{1}{4}$ wide, and $\frac{1}{6}$th of an inch thick.

6 do. for do. 15 inches do. do.

N. B. All the gudgeons should taper a little, and the sides given are the largest part. The bands for shafts should be widest at the foremost side, to make them drive well; but those for heads should be both sides equal. Six picks for the stones, 8 inches long, and $1\frac{1}{4}$ wide, will be wanted.

Appendix 4
The Tenth Census Reports on Waterpower

The Tenth Census of 1880 was planned and organized by Francis Amasa Walker. The most comprehensive and thorough of the decennial surveys up to its time and possibly for the entire nineteenth century, it was distinguished not only by its wide coverage but also by the high level of competence of the technical, industrial, and scientific specialists on its staff. One commentator has remarked that the Tenth Census in its day was "a statistical and economic achievement of the highest order."[1] Two of the twenty-two volumes published as the Tenth Census, volumes 16 and 17, comprise the *Reports on the Water-Power of the United States* (Washington, 1885, 1887).

The director of the waterpower survey—better termed a "reconnaissance" in view of the limited personnel and time available for the project—was Professor William P. Trowbridge of the School of Mines, Columbia College, New York City. Sharing the field work and preparation of reports were George F. Swain, assistant professor of civil engineering, Massachusetts Institute of Technology; Dwight Porter, instructor of civil engineering, School of Mines, Columbia College. The survey covered the eastern seaboard states, the entire Mississippi Valley, the eastern Gulf slope, and the principal drainage basins tributary to the Great Lakes. It did not extend to the western mountain and Pacific Coast regions. Swain, as ranking special agent, prepared the general introduction and the reports covering New England, except the region tributary to Long Island Sound, and the Middle and South Atlantic watersheds.

The detailed descriptions of the waterpowers, developed and undeveloped, of each of the drainage basins covered by the reports are preceded by introductory sections describing characteristic geological, topographical, and hydrological features, with some account of the problems encountered in surveying and measuring the region. The usual procedure followed was to begin at the mouth of the trunkline stream of each drainage basin and advance to the tributaries at first, second, and further remove, noting the distinctive characteristics of

[1] Meyer Fishbein, "The Industrial Reports of the Tenth Census," National Archives.

the successive waterpowers and the manner of their development, ownership, and use. The broad purposes of the survey were stated by Swain in the general introduction:

> The main object of this report is to give an idea of the available power of the country, describing privileges actually in use, and calling attention to locations where power could be advantageously developed. . . . it is not technical availability alone which determines the value of a privilege . . . many sites where considerable power could easily be developed are commercially valueless, perhaps on account of inaccessibility, remoteness from markets or other reasons; and it has been an important part of the object of this report to call attention to conditions affecting the commercial value and availability of the locations described.[2]

"The aim," he concluded in a letter of transmittal, "has been to give a reliable account of the great powers in the region considered, as well as a general discussion of its water power in general."[3]

Unfortunately, the limited resources of the survey in personnel and in time, combined with a lack of streamflow records over long periods for all but a handful of streams, prevented calculation of power potential save of the most general kind. There were extant in public or private hands only a half-dozen records of streamflow covering periods of a decade or more. The more precise and reliable of these records were gathered for municipal water-supply systems and were limited to small streams in the environs of the cities concerned. The few relatively extensive records in private hands, chiefly the waterpower corporations such as the Locks and Canals at Lowell, were typically unavailable for consultation. In the absence of such records, hydraulic engineers well after 1900 were obliged to calculate streamflow and power potential using as sources the more widely available records of rainfall maintained by the Smithsonian Institution and eventually other government agencies, calculations of the square mileage of drainage basins, and estimates of runoff. No attempt was made in the 1880 census survey to undertake even sample gaugings of streamflow. The civil engineers in charge of the project combed the professional and technical literature, including reports of the Corps of Engineers and papers of the recently established United States Geological Survey. They consulted and corresponded with railroad engineers, state geologists, and members of the engineering profession generally; the result was in many ways an impressive body of information for the regions covered by the survey.[4] Emphasis was

[2] U.S., Census Office, Tenth Census, 1880, vols. 16–17, *Reports on the Water-Power of the United States* (Washington, D.C., 1885, 1887), pt. 1, p. xi.

[3] Ibid., p. 47.

[4] See ibid., pp. xxvii–xxix.

on stream slopes, area of drainage basins, annual and seasonal distribution of rainfall, and the description, river basin by river basin, of power sites, developed and undeveloped, and the manner and extent of utilization. Estimates of volume of streamflow were attempted for the major river systems, but with warnings as to the dubious data on which it was necessary to rely in such estimates.

Most important of all, perhaps, was the belated recognition and overall view of a major national resource that had long been taken for granted, even while it was exploited to great advantage. The report provided a comprehensive survey of the hydrology of the eastern half of the continent, conducted by professional engineers of competence and concern. In the second decennial census to include systematic coverage of waterpower and steam power employed nationally in manufacturing, there was now added a comprehensive view of the geographical base and of the power installations that underlay the statistical measure of what till but recently had provided the greater part of the mechanical energy of American industry. How widely the two large and heavy volumes comprising *Reports on the Water-Power of the United States* were read and consulted and the extent of their practical usefulness for millowners and manufacturers, millwrights, and engineers are largely a matter for conjecture. Some forty years after their completion, John R. Freeman, a prominent hydraulic engineer and past president of both the American Society of Mechanical Engineers and the American Society of Civil Engineers, referred to the *Reports* as "a thoroughly admirable review of the state of the art and extent of development [of waterpower] in the United States" at the time and urged that a similar power survey be undertaken at the next census.[5]

[5] John R. Freeman, "The Fundamental Problems of Hydroelectric Development," *TASME* 45 (1923):51-532.

Appendix 5

Tidal Power along the Atlantic Seaboard

The harnessing of tidal flow at seacoast points for driving water mills was very practicable where the conditions for their adoption were present, as chiefly from Long Island northward, especially along the coast of Maine. Their number was quite small; in the careful survey of Maine, 1868–69, fewer than ten are reported (Maine, Hydrographic Survey, *The Water-Power of Maine, by Walter Wells* [Augusta, 1869]). For Wells's discussion of tidal power, see the reprint below. For information on the laying out and building of tide mills in Europe, see John Nicholson, *The Operative Mechanic, and British Machinist*, 2d Am. from 3d London ed., 2 vols. (Philadelphia, 1831), 1:94–104. The development of the milldam scheme at Boston was followed by an even larger tidal-power project, never realized, described in George L. Vose, *A Sketch of the Life and Works of Loammi Baldwin, Civil Engineer* (Boston, 1885), p. 14. For further references to tidal power, see U.S., Census Office, Tenth Census, 1880, vols. 16–17, *Reports on the Water-Power of the United States* (Washington, D.C., 1885, 1887), pt. 1, pp. 54, 529. References to a number of tidal gristmills are found in Horatio G. Spafford, *A Gazetteer of the State of New York* (Albany, 1824), pp. 62–63, 70.

From *The Water-Power of Maine, by Walter Wells,* Pages 33–34

The value of tidal water-power.

The tides upon the coast of Maine are so remarkable for volume as to require special notice in the discussion of its water-power. Tidal force is not considered available for ordinary manufacturing purposes in which large numbers of hands are employed, because of the unseasonable and constantly changing hours in which it is necessarily used: this at least without considerable expense in the way of supplementary basins for the reception of the mill discharge during the time of high tide, by the use of which the season of work can be regulated to convenience. But on the other hand for certain forms of manufacturing, as sawing lumber, grinding grain, grinding slate for paint or plaster of Paris, sawing marble, etc., in which large amounts of power are used and

only a few attendants are required, it is of decided value. The flow of the tide is so great on our coast, that with suitable wheels it can be operated to advantage sixteen hours out of the twenty-four. The supply of water of course never fails, and the extreme fluctuations of volume are very small as compared with those experienced upon even the most constant of fresh water privileges. There is less trouble from ice, and the privileges are obviously accessible to market by the cheapest form of transportation.

The coast of Maine is in nearly all parts iron-bound, so that dams and mill structures can be planted upon firm foundations. Upon the vast reaches of sandy and swampy coast south of us, the tide, however great might be its rise and fall, would be unavailable for power to any considerable extent, except at great cost, on account of the impossibility of securing solid basis for superstructures.

The following table exhibits the *mean* range from low to high water at various points upon the coast:

MEAN HEIGHT OF THE TIDE ON THE COAST OF MAINE.

Localities.	Feet.	Localities.	Feet.
Eastport,	18.1	Castine,	12.0
Machias,	13.0	Camden,	9.8
Machiasport,	16.0	Thomaston,	9.4
Jonesport,	15.0	Damariscotta,	8.1
Columbia Falls,	14.0	Wiscasset,	9.4
Steuben,	13.0	Harpswell, (Basin)	9.0
Winter Harbor,	10.5	Portland,	8.9
Cold Harbor, Swan's Island,	12.0	Saco,	8.5

Mean, 11.6 feet.

MEAN TIDE BEYOND STATE ON THE ATLANTIC COAST.

Localities	Feet.	Localities.	Feet.
Portsmouth, N. H.,	8 6	Brunswick, Ga.,	6.8
Boston, Mass.,	10.0	Fernandina, Fla.,	5.9
Nantucket,	3.2	Charlotte, Fla.,	1.1
Providence, R. I.,	5.1	Mouth Mississippi,	1.3
New York, N. Y.,	4.8		

Mean, 5.2 feet.

Effect of the broken shore line.

The value of this form of power in its sum for the State is indefinitely increased by the great length and broken outline of our shore, which, as before remarked, is extended to the estimated length of 2,000 to 3,000 miles in a direct horizontal distance of 226 miles. The contour of the coast is such, that coves, inlets, and reëntering arms of the sea in great numbers and of large size, are susceptible of inexpensive improvement as storage basins and reservoirs of power.

Prospective employment of tide power.

For many years, doubtless, and until our fresh-water power is measurably put to use, salt-water power will be employed only in small amount, and only where insignificant outlay will be required. But it is in nowise improbable that in time the large sites will be utilized, works constructed on a great scale, and enormous amounts of labor accomplished by this great natural motor. If anywhere in the world it can be done to advantage, it is here, by reason of favoring physical conditions, and to all the higher advantage in connection with the great manufacturing system to be established on our rivers.

Appendix 6

Data on Streamflow

From George A. Kimball, "Water Power: Its Measurement and Value," *JAES* 13 (1894):96.

The following concrete example is based on the records of the pioneer Boston Manufacturing Company, Waltham, Massachusetts, on the Charles River, with a drainage basin of 185 square miles.

Showing Horse Power of Charles River at the Dam of the Boston Manufacturing Company in Waltham.

	I.	II.	III.	IV.	V.	VI.	VII.	VIII.
	Rainfall collectible.	Amount collected and flowing off per month. (Area 184.78 square miles.	Amount flowing off per minute.	Amount of flow available for power in river.	Amount of flow not available for power.	Amount of pondage. (To be added to amount in Col. IV.)	Gross amount available for power.	Power with effective head of 11 feet. 75 per cent. efficiency at the wheels.
	Ins.	Cubic feet.	Cubic feet per minute.					Horse Power.
January	2.005	860,700,000	19,281	18,000	1,281	22,400	40,400	630.24
February	3.206	1,376,260,000	33,831	26,000	7,831	22,400	48,000	755.04
March	4.997	2,145,100,000	48,053	26,000	22,053	22,400	48,400	755.04
April	3.609	1,549,260,000	35,862	26,000	9,862	22,400	48,400	755.04
May	1.987	852,970,000	19,107	18,000	1,107	22,400	40,400	630.24
June	0.864	370,894,500	8,585	8,000	585	14,154	22,154	345.64
July	0.335	143,807,000	3,221	3,000	221	5,266	8,266	128.90
August	0.549	235,670,000	5,279	5,000	279	8,777	13,777	214.90
September	0.457	196,180,000	4,541	4,500	41	7,961	12,461	194.40
October	1.053	452,030,000	10,126	9,500	626	16,677	26,177	408.36
November	1.617	694,130,000	16,068	15,500	568	22,400	37,900	591.24
December	1.940	832,790,000	18,656	18,000	656	22,400	40,400	630.24

12)6,039.28

Average daily horse-power for the year = 503.27

From Joseph P. Frizell, *Water–Power* (New York, 1901), Page 6

TABLE 1.—DATA REGARDING THE FLOW OF STREAMS.

Stream.	Place of Measurement.	Drainage and Square Miles.	Mean Rainfall, Inches. Spring.	Summer.	Autumn.	Winter.	Year.	Extremes of Flow, Cu. Ft. per Sec. Max.	Min.	Minimum Flow per Square Mile.	Ordinary Low-water Flow of Stream.	Ordinary Low-water Flow per Sq. Mi.
Merrimac	Lawrence, Mass.	4 599	10	11	13	9	43	96 000	1 400	0.30	2 800	0.60
"	Lowell, Mass.	4 085	10	11	13	9	43	81 000	1 275	0.31		
Concord	" "	361	11	11	12	10	44	4 449	60	0.17	126	0.35
Sudbury	Framingham, Mass.	78	11	11	12	10	44	3 228	2.8	0.036	12.5	0.16
Connecticut	Hanover, N. H.	3 316	10	12	12	10	44		1 006	0.303	1 210	0.365
"	Hartford, Conn	10 154	10	12	12	10	44	205 464	5 208	0.503		
Croton	Croton Dam, N. Y.	339	12	13	13	10	48	25 380	60	0.178		
Passaic	Paterson, N. J.	813	12	14	12	10	48	17 913	195	0.24		
Delaware	Lambertville, N. J.	6 820	11	13	11	9	44	350 000	2 000	0.29		
Potomac	Cumberland, Md.	920	10	12	9	8	39	17 900	25	0.022		
"	Great Falls, Md.	11 476	12	13	9	8	42	175 000	1 063	0.093		
James	Richmond, Va.	6 800	12	12	9	10	43		1 300	0.191		
Allegheny	Roberts Run, Pa.	6 020	10	11	9	10	40		2 070	0.34		
Ohio	Near Pittsburgh, Pa	18 732	10	12	9	10	41		2 271	0.12		
Kanawha	Charleston, W. Va	8 900	12	13	9	10	44	118 291	1 100	0.123		
Red River of the North	National Boundary	39 577	5	8	3	2	18				2 800	0.07
Mississippi	Grand Rapids, Minn	3 636	7	12	6	3	28				969	0.27
"	Rock Island, Ill.	87 842	7	12	7.5	3.5	30				19 000	0.216
Illinois	Mouth	29 013	11.5	11	9	8	39.5		1 600	0.055	1 750	0.06
Kansas	Topeka, Kan.	56 354	8	7	6	3	24		2 000	0.035		0.043
Arkansas	Mouth	160 000									3 000	0.019
White	Mouth	27 925	13.5	11	10	9	43.5				3 000	0.1074
Potomac, 1895-6	Cumberland, Md.	891						17 600	250			
James, 1896	Buchanan, Va.	2 058						31 950	300			
Cape Fear, 1896	Fayetteville, N. C.	4 493						52 340	489			
Yadkin, 1896	Salisbury, N. C.	3 400						64 200	1 000			
Catawba, 1896	Rock Hill, S. C.	2 987						62 550	1 330			
Ocmulgee, 1894-5-6	Macon, Ga.	2 425						36 200	380			
Tennessee, 1890-96	Chattanooga, Tenn	21 832						445 120	16 360			
Platte, 1895	Columbus, Neb.							27 200	96			
" Aug. 17 to Nov. 1, 1896	" "							0	0			
Solomon, 1895	Beloit, Kan.	5 539						24 000	7			
Kansas, 1895-6	Lawrence, Kan.	59 841						53 308	700			
Arkansas, 1895-6	Hutchinson, Kan.	34 000						19 600	16			
Rio Grande, 1895-6	San Marcial, N. Mex.	28 067						11 300	0			
Gila, 1889-96	Buttes, Ariz.	13 750						12 000	1			
Humboldt, 1896	Oreana, Nev	13 800						1 508	20			

Appendix 7

James B. Francis on Estimating the Power of a Stream Based on Watershed Area, Rainfall Records, and Estimates of Runoff

From George A. Kimball, "Water Power: Its Measurement and Value," *JAES* 13 (1894):72–73

The area of the water-shed at the mill dam was twenty-one and a half square miles.

Mr. James B. Francis, the eminent hydraulic engineer, was called to testify in behalf of the mill owner, and furnished the commission with the data given in table No. 1, showing the quantity and commercial value of the water power taken, based on the data furnished him by the mill owner. The explanation of the table, as given by Mr. Francis and taken from the stenographer's report, is substantially as follows:

"There are various ways of estimating the flow of streams. The flow *at any particular moment* can be determined by measuring through weirs, also by measuring the velocity of a given section.

"To get at the *average* flow, which is the chief point, is a very different matter. To take into account the variations of the seasons would require a long series of measurements, which is seldom convenient; or, proceeding upon another method, to ascertain the area of the water-shed supplying the stream, and the amount of the rainfall upon it. In some regions this would not do, but it does very well in this part of the country, and I think it is the usual method.

"For a number of years the city of Boston has gathered valuable information, which is applicable, I think, to this case. The Sudbury River and the Stony Brook water-sheds are not far apart, and are alike in general character of country. I should think it would be fair to apply to this place the information gained from the Sudbury, that is, I think it is better for our purpose than any information I know of. I have made a table or statement in regard to this power, but it is based on data taken from other parties. I have made no observations myself. I take twenty-two square miles as the area of water-shed.

"The rainfall given in column 3 is the average of eleven years, and is taken from the published observations of the city of Boston. Column 4 gives the percentage of this rainfall, averaging about 45 per cent., which runs off in the brooks and streams. The percentage is usually taken by engineers at about 50 per cent., but I should have taken 45 per cent. without reference to these observations. It will be noticed that it varies widely from month to month.

"From the area of twenty-two square miles, the amount of rainfall, and the percentage gathered, I get the average quantity flowing into the brook per day. This is given in column 5. The average flow in the brook for the year is about 2,881,700 cubic feet per twenty-four hours. Now there are in the year three or four months, which I have taken as February, March and April, when there is an excess of water over what can be advantageously used, and when there is consequently a waste. I have taken May as being a month when, as a rule, the entire flow may be utilized. The average wastage during February, March and April is 3,563,030 cubic feet per day.

TABLE No. 1. By JAMES B. FRANCIS.

Estimate of the Commercial Value of the Water Power furnished by Stony Brook at the Sibley Privileges, of 26 feet and 10 feet respectively, in Waltham and Weston, Mass. Water derived from a Water-shed of 22 square miles. Data obtained from observations made on the Boston Water Works during the eleven years from 1875 to 1885 inclusive, and recorded in Table 8 of the Report by Desmond Fitz Gerald, dated May 30, 1887, on the capacity of the Sudbury River and Lake Cochituate Water Sheds in time of drought. The water was used during twelve hours per day on water wheels giving a useful effect of 75 per cent. of the total power of the water expended.

Month.	Average Number of Working Days in Each Month.	Rainfall. Average for 11 years, 1875 to 1885, inclusive. Inches.	Flow in the Brook. Percentage of Rainfall.	Average Flow in Brook per Day. Cubic Feet.
January	26.57	4.135	38.3	2,611,100
February	24.21	4.280	74.7	5,836,000
March	26.57	4.352	115.2	8,265,880
April	24.71	3.365	106.4	6,099,790
May	26.57	3.092	62.2	3,170,860
June	25.71	3.146	27.2	1,457,860
July	25.57	3.666	8.3	501,670
August	26.57	4.007	10.3	680,460
September . . .	25.71	2.600	9.4	416,380
October	26.57	4.001	12.7	837,760
November . . .	24.71	4.027	31.1	2,133,680
December . .	26.43	3.470	44.9	2,568,760
	309.90	44.141	45.06	
	Total.	Total.	Average.	

The following estimates are based upon the assumption that 75 per cent. of the gross power is effective. The falls are taken at 1 foot less than their actual height. The power is supposed to be used 12 hours per day, and the night flow stored.

	H. P.
Power utilized during the eight dry months, 207.84 working days.	
On the fall of 26 feet	68.91
" " " " 10 "	24.81
Total	93 72
During the four wet months, 102.06 working days, available flow, 3,170,860 cubic feet per day	212.11
Average for the entire year	132.71

The coal required to give 132.71 (steam) horse-power for 309.90 working days, 12 hours per day, 3 pounds of coal per hour, per horse-power, would be 660.97 gross tons.

Cost of 660.97 tons of coal at $5.00 per ton, delivered at mill and $1.00 per ton for expenses at mill	$ 3,965 82
Capitalized value of the coal, at 6 per cent.	66,097 00
" " " " " 5 "	79,316 40
" " " " " 4 "	97,145 50

Appendix 8
The Franklin Institute's Experiments on Waterwheels

In 1829 the Franklin Institute of Pennsylvania announced its "Proposals for a series of experiments" on waterwheels, followed in due course by the more detailed program of "Experiments on waterpower," with names of the committee in charge.[1] The intention of these experiments was to bring up to date, correct, and supersede John Smeaton's celebrated experiments with waterwheels carried out in 1752–53 and published in 1759, which demonstrated the marked superiority of bucket wheels over impact wheels.[2] The announcements of the Franklin Institute reflected certain doubts about the adequacy of experiments conducted on small-scale models of wheels, as by Smeaton, and proposed to relate the new experiments to actual working practice in this country. They were intended to provide the millwright "with a sure and safe guide in his practice," thereby promoting "one of the most important of the mechanic arts." The institute's experiments employed full-scale working waterwheels with diameters ranging from twenty feet to six feet. The wheels were equipped with different forms of buckets, floats, and gates under different distances of free fall from forebay to wheel. An illustrated description of the methods and apparatus employed and a detailed tabulation of the results of the experiments were published serially in volumes 11 through 14 of the *Journal* in 1831 and 1832.

Then followed a long period of unexplained inactivity. In 1841 the experiments and reports were resumed under the heading "Part Second." Of the four planned divisions of this part, dealing with overshot, undershot, and breast wheels and "general inferences," only the first two were completed. The committee had already been reported to have been prevented by "circumstances" from carrying out experiments on reaction wheels.[3] The "Third Part" of the report, to contain "rules founded on the general conclusions, and applicable directly to the questions occurring in the use of water as a moving power," never appeared. The *Journal*'s review of the American translation of Jean F. d'Aubuisson's

[1] *JFI* 7 (1829):217 ff., 364 ff.

[2] Ibid.; Paul N. Wilson, "The Water-wheels of John Smeaton," *TNS* 30 (1955–57):25 ff.

[3] *JFI* 14 (1832):297.

A Treatise on Hydraulics for the Use of Engineers in 1853 referred to the report as "a work yet incomplete."[4]

The institute's experiments gave materially different results from those of Smeaton in respect to the efficiency of overshot and undershot wheels of twenty-foot diameter: 84 percent efficiency for the overshot and 28.5 percent for the undershot wheels, as compared to 69 percent (cited by the committee) or 63 percent (Smeaton's figure) for Smeaton's overshot wheel and 22 percent for his undershot wheel. Among other findings of the institute the following were of particular practical significance: (1) there was a difference in efficiency of only a few percentage points between the largest and the smallest of the overshot wheels tested; (2) variations in gate openings between eight and twenty-four square inches changed overshot wheel efficiency less than 1 percent; and (3) changes in velocity of large overshot wheels, within certain limits, had little influence on their efficiency.[5]

The Franklin Institute experiments became a subject of some controversy. Although George Rennie, the distinguished British engineer, spoke highly of them, Robert Mallett declared that they had "added nothing important to our knowledge beyond the experiments of our great Smeaton." James Whitelaw, another British engineer, contended that Smeaton performed experiments that "the Franklin Institute could do no more than repeat."[6] The unnamed author of the article "Water-Wheels" in *Appleton's Dictionary of Machines, Mechanics, Engine Work and Engineering* (1852), ignored the Franklin Institute experiments and reprinted tables, not clearly explained, from the Smeaton report of 1759, which were described as the only trustworthy figures available.

[4] Ibid. 55 (1853):215–16.

[5] Ibid. 31 (1841):145 ff., 217 ff., 289 ff., 361 ff.

[6] See ibid. 36 (1843):62 ff., 37 (1844), 38 (1844):73 ff., 78 ff., 34 (1842):390.

Appendix 9

Comparison of an Early Boyden Turbine and the Burden Overshot Wheel at Troy, New York

The most obvious difference between turbines and old-style waterwheels of comparable capacity was the small size and general compactness of the former. One of Benoît Fourneyron's early turbine wheels, employed on a well-above-average head, weighed only 180 pounds, less than the weight of a single arm of the overshot wheel it replaced, and it developed one-third more power.[1] With an outside diameter under

Fig. 120. Burden wheel, Troy, N.Y., as it looked c. 1900. (Courtesy of the Smithsonian Institution.)

[1] "Practical Experiment Made by M. Fourneyron . . ." *JFI* 34 (1842):378–79.

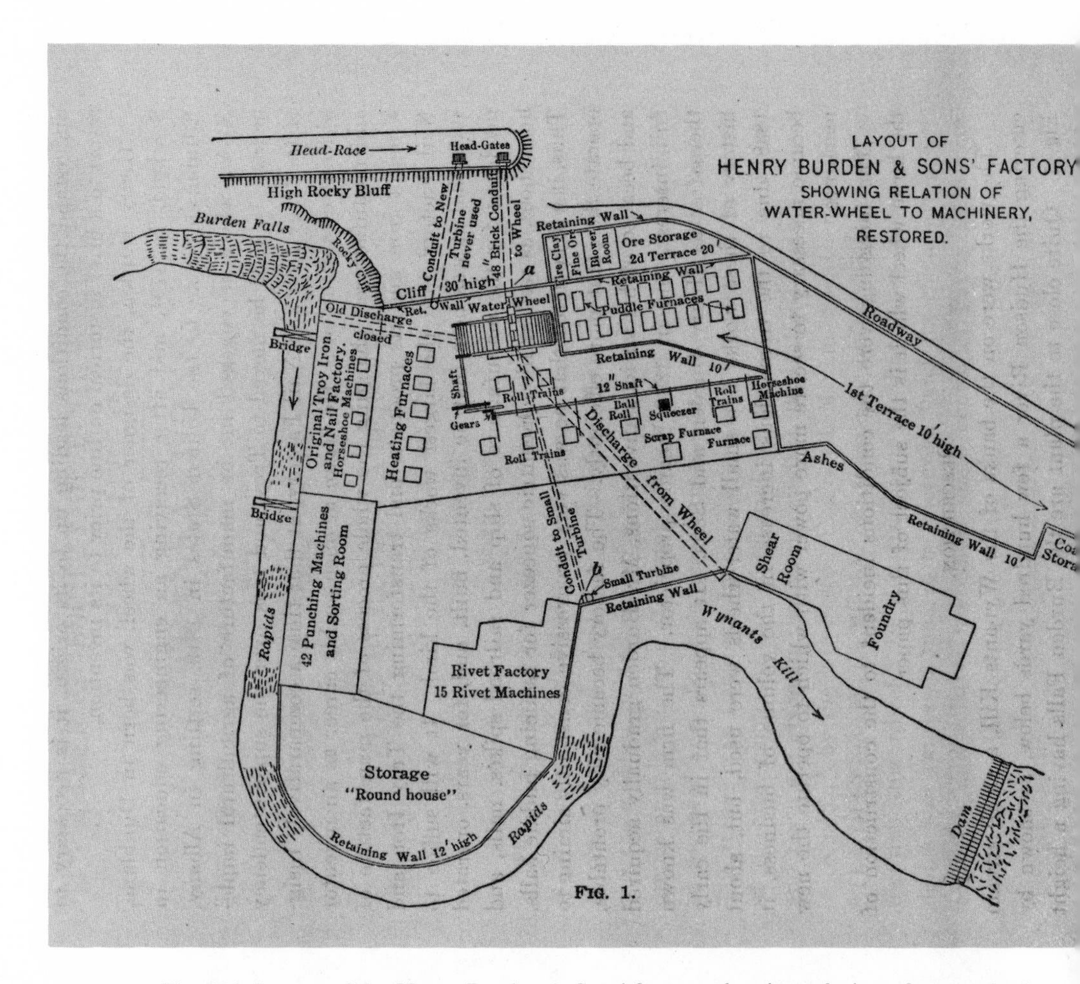

Fig. 121. Layout of the Henry Burden & Sons' factory, showing relation of waterwheel to machinery. (F. R. I. Sweeney, "The Burden Water-Wheel," *TASCE,* 79 [1915].)

five feet and a vertical dimension under two and a half feet, one of the early turbines installed by Ellwood Morris in Pennsylvania took the place of two breast wheels, ten and sixteen feet in diameter and fourteen and eight feet, respectively, in breadth of face. Smaller dimensions meant not only lower cost for a given capacity but lesser requirements in space, foundations, settings, and cost of installation.[2] Since overshot wheels, to be used to greatest advantage, had to be given a diameter approximating the distance of fall, the higher the fall the larger the wheel and the lower the rotating speed. For this reason bucket wheels forty to fifty feet in diameter were rare and not only because of difficulties of construction—falls of this height were uncommon not only in the eastern United States but in Europe outside of the mountain or piedmont regions.

The most impressive example of an American overshot wheel was that built by Henry Burden in 1851 to drive his celebrated automatic horseshoe and spike manufactory at Troy, New York. This was not the largest wheel of its type so far as diameter was concerned, being exceeded in this respect—though probably not in power—by the celebrated wheels at Laxey on the Isle of Man and at Greenock, Scotland, the latter supplied by Shaw's Waterworks with water from an elevated reservoir.[3] The Burden wheel, sixty-two feet in diameter and twenty-two feet in breadth, was supplied by a small stream, the Wynantskill, whose natural fall of some fifty feet was increased substantially—together with provision of storage capacity for year-round operation—by a dam and related structures of conduit and penstock of ingenious design. The wheel itself was of what came to be termed the "suspension" type, familiar to us in the bicycle wheel, with iron rods in tension replacing the usual arms. It was made almost entirely of iron, save for floor, or soling, of the wheel and its buckets. The appearance of the

[2] Morris, "Experiments on the Useful Effect of Turbines in the United States," ibid. 36 (1843):377–79. See also Emerson, *Treatise Relative to the Testing of Water-Wheels and Machinery*, 3d ed. (Springfield, Mass., 1881), pp. 26–27.

[3] The best account of the Burden wheel is F. R. I. Sweeney, "The Burden Water-Wheel," *TASCE* 28 (1893):237 ff. A reprint of this article with additional information, photographs, and an introduction by Robert M. Vogel appears in *Society for Industrial Archeology Occasional Publications* no. 2, Apr. 1973. The Burden wheel was made the subject for graduating theses of students at Rensselaer Polytechnic Institute, 1855–67: see theses of Howard Crosby, Frederick Grinnell, A. B. Cox, and Charles McMillan. Sweeney's figure of 278 horsepower for the wheel as normally operated does not reflect the heavy losses that must have occurred in the gearing through which the power was taken off and transmitted to the mill. Data for the Tremont turbine are from Francis, *Lowell Hydraulic Experiments* (Boston, 1855), pp. 7 ff.

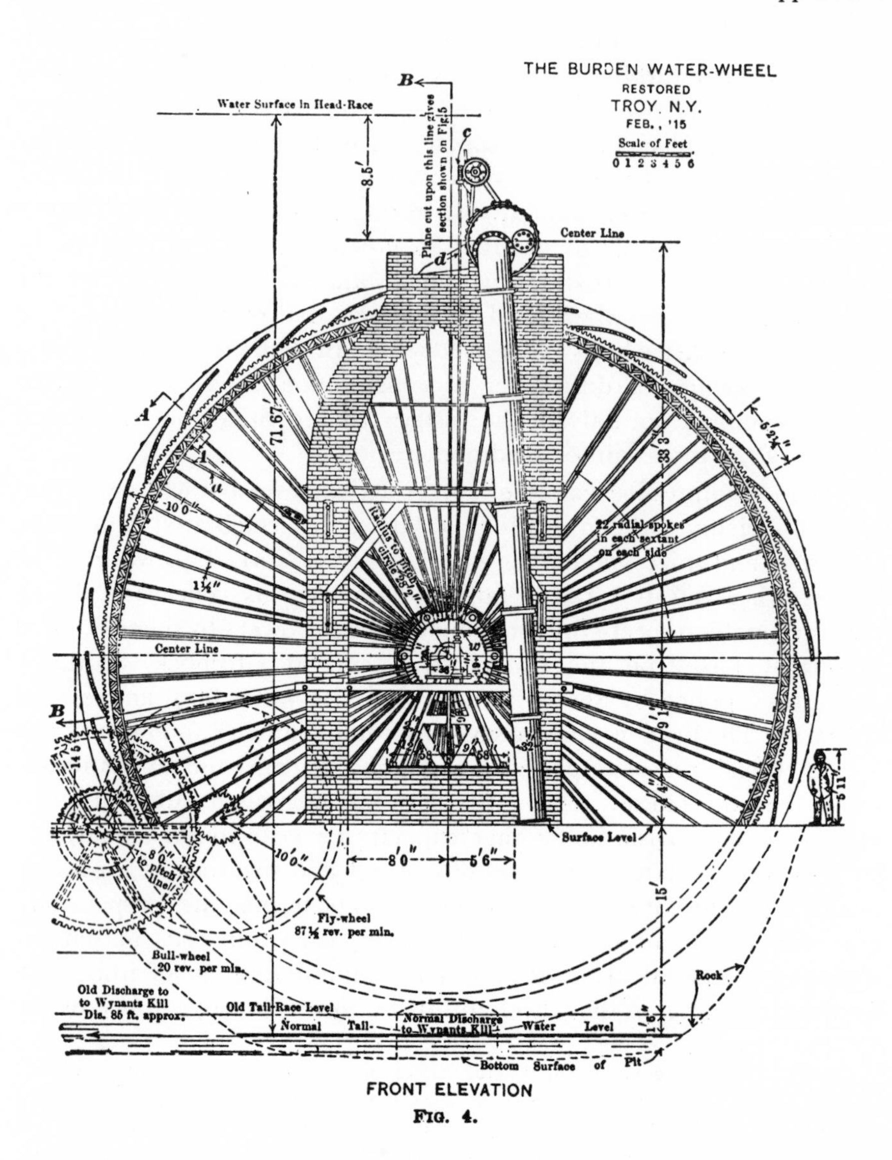

Fig. 122. Front and side elevations of the Burden Waterwheel. (F. R. I. Sweeney, "The Burden Water-Wheel," *TASCE,* 79 [1915].)

wheel and details of its construction and of the elaborate gearing by means of which the power was taken off and conveyed to the mill are shown in fig. 122. This gigantic prime mover was

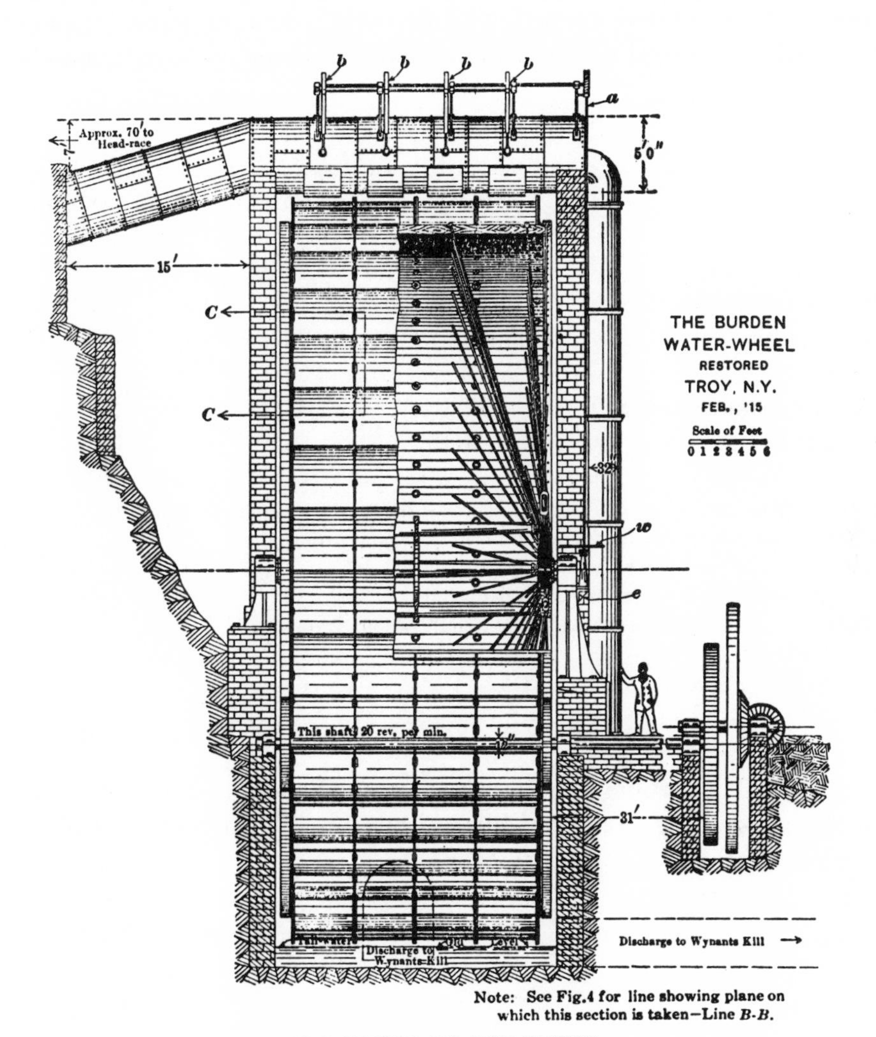

SIDE ELEVATION AND PART SECTION.

continuously in service night and day for nearly a half century. Following its abandonment in the 1890s, it lay idle for another twenty years before its final collapse.

In the accompanying tabulation the differences between the old-style gravity wheel and the new-style turbine—with which the greater the fall the smaller the wheel required to develop a given power—are strikingly

	1. Burden overshot wheel	2. Tremont turbine	Ratio of col. 1 to 2
Diameter (feet)	62.0	5.0	12
Breadth or depth (feet)	22.0	3.0	7
Bulk (cubic feet)	56,000.0	60.0	900
RPM	2.5	150.0	60
Capacity (horsepower)	300.0[a]	500.0	3/5
Weight (tons)	250.0[b]	4.0[c]	60

[a] Estimated variously at 200–500 horsepower, depending upon the amount of water available and supplied; Sweeney reported the power developed in normal working practice as 278 horsepower at 2.5 rpm.
[b] Weight with wooden members water-soaked.
[c] Does not include weight of the 26-foot shaft, about 2 tons.

illustrated. The turbine cited here is one of the Boyden–Francis type that was installed at the Tremont Mills at Lowell in 1851, the same year the Burden wheel was installed at Troy.

Index